# *Practical Antenna Design*
## *-Revised*

PRACTICAL ANTENNA DESIGN



This book is printed on acid free paper

A Shoemaker Labs Book
Indian Harbour Beach, Florida

www.Shoemakerlabs.com



ISBN 978-9815092-8-2
ISBN 0-9815092-8-2

This book is registered with the Library of Congress
Front and back cover art courtesy of NASA

Grateful acknowledgement is made to those who have given permission for the use of previously copyrighted material in this book. Every reasonable care has been taken to correctly acknowledge copyright ownership. The author and publisher would welcome information that will enable them to rectify any errors or omissions in succeeding printings.

First Edition February, 2011
Second Edition September, 2016

Printed in the United States of America

*To Peg, Leah and Stephen for their inspiration and support*

*And to my late great father, Donald, who reviewed a draft of the first edition of this book near the end of his life and pronounced it worthy. An avid reader and professor of political science and economics, his vast knowledge of history and the history of science enabled him to inspire many generations of students, family, and friends. A low estimate of the number of books he read is around 10,000. As a result, he had perspectives on the evolution of human thought and accomplishment few others have shared.*

**Acknowledgment**

I would like to sincerely thank the Electromagnetic Engineers I have had the pleasure to work with in general and Tom Milligan, Jason Coder and Roger Strange in particular for their comments. Also, I would like to thank my friends Bob Johnk, Russ Mellon and Eliot Young for their guidance and comments. Finally, I would like to thank my family for their support and encouragement.

## Author's Note

This book is intended to allow Engineers to quickly understand fundamental antenna theory, antenna types and design methodologies. The design equations are easy to understand and many examples of typical designs are given. Selection matrixes, nomographs, design curves and diagrams are also provided to help the designer decide on the best solution for their application. A significant number of sample antenna designs are presented, with typical gain, bandwidth, patterns and additional design notes.

In addition, the history of antenna development is given, a comprehensive list of terms used in the industry and a large reference section are presented. Lastly a bibliography, design software and source reference section are included for those who want to pursue the subject further.

*Kevin O. Shoemaker*
*Indian Harbour Beach, Florida 2016*

# Table of Contents

Fighter Jet in Anechoic Chamber

NASA Deep Space Network Site in Australia

# Introduction to the Revised Practical Antenna Design Textbook

Although its been only a few years since the publication of the first edition of this book, much has changed in the field of electromagnetics. With the advent of very good field simulators and the beginning of the antenna magus, the field of antenna design is now capable of a very wide range of solutions for antenna requirements. Stealth, active systems, plasma, exotic materials and direct digitization (both transmit and receive) of antenna currents are but some of the new developments in this field. There are many more. Direct digitization of antenna current for instance is now allowing the designer to work closely with software developers to create the most capable antenna and antenna systems every conceived. Adaptability and reconfiguration are a few of the new options, where phase and amplitude parameters can be changed very quickly allowing for single and multi beam operation across wide bandwidths and within challenging environments.

Stealth antennas include those not only for use in military applications but also those in commercial and consumer venues where for aesthetic or other reasons, the antenna has to blend into the environment so as not be seen. Antennas have been developed that are made from flexible, conductive materials that can be worn, silk screened or even painted on non-conductive surfaces. The advent of many cell towers in an urban are for instance has led to the development of towers that look like trees or flagpoles. Houses near busy highways are sometimes made of plastic to hide a farm of antennas within. The increase in the number of antenna applications has caused so many of them to be used that it is sometimes considered to be a distraction. Even now it is almost impossible to drive down a busy highway and not see antenna after antenna, all for various communications needs.

Plasma antennas were conceived long ago but have not received the attention they deserve until recently. These antennas are able to be stacked onto each other without any interaction as the unused elements of such a stack are non conductive when not in use. Plasmas can be used as reflective shields as well as being able to change dimensions very quickly for broad band applications. For instance a conductive wall can rapidly change into several different antennas with microseconds. Plasmas can be blown into the air creating instant conductive constructs that can be used as antennas. Rocket plumes and other spaceborne plasmas can also be used for communication purposes.

Active antennas have now come down in price significantly allowing commercial and consumer product developers new options in beam control and interference mitigation. As the electromagnetic spectrum continues to fill, these capabilities are becoming more important.

Historically, antenna and their associated electronics have been developed starting at lower frequencies and progressing upwards. This trend has not stopped. For instance, radio astronomy was begun at 20.5 Mhz in 1931, today we routinely map the universe at 300 GHz and above. Communications first implemented below 1 MHz in the very early 20th century are now accomplished at optical frequencies in the TeraHertz region. This trend will continue as well as the use of antenna techniques to produce higher bandwidth data transfer and exploration.

The Atacama Radio Telescope Antenna Elements
capable of operations up to at least 300 GHz

# Chapter 1 - Introduction

## Need for Antennas

It has become fairly obvious to all people in general and technical people in particular that antennas are becoming more and more a necessity in our technological world. Increases in communications and information transfer are now requiring more and more bandwidth. Wire lines and cables can no longer handle the ever increasing demands for the technical public. Fiber Optics can handle the bandwidth but has the limitation of significant infrastructure buildup. As a consequence, antennas are required to transfer the high speed data at rates faster than can be handled by those traditional means.

If one compares the losses of coax, waveguides and an antenna link, for the short distances, the coax and waveguides have lower loss, at greater distances however the antenna to antenna link is far superior and significantly less expensive.

## Future for Antennas

There is no end in sight for the need of antennas. These components are capable of information transfer up into the Terahertz regions of the electromagnetic spectrum. In fact, operation beyond those frequencies can be considered by pointing out that quasi-optical principles are close relatives to the standard theories and principles employed by antenna designers. In the future, electrically steered phased arrays in the optical and inf;rared parts of the electromagnetic spectrum will be commonplace. In the future, modeling programs will be sophisticated enough to create artwork and drawings for first iteration prototypes after input of required specifications and these prototypes will have the expected performance. In the future, antenna testing will be significantly faster on spherical or hemispherical ranges, pre calibrated, and using high speed processing to characterize antenna performance in real time. In the future, antenna production will include more stealthy designs and adaptive techniques giving the user little or no clue as to the placement of an antenna in hand held device or vehicle.

It is conceivable that all houses, vehicles and places of work will be internally and externally connected to a network consisting of wireless connections for the purpose of information transfer. It is also conceivable that antennas will become "smarter" allowing phasing and other adaptive techniques to be directly applied to their operation.

Clearly, the future for antennas is "bright", and opportunities will be plentiful for the antenna designer.

## Applications

In the past, antennas were used for communications then radar and telemetry. As the need to transfer information evolved, antennas were employed to transfer the data in ever increasing rates. Presently, the above mentioned areas have expanded to telephony and entertainment. As a result, there has been a dramatic increase in the need to transfer time sensitive information. There has also been an increase in the need to transfer digital signals requiring careful control of the spectrum. Digital signals need to be filtered to avoid interference to other services, much of this requirement can be met by good antenna design. Analog signal need filtering as well, but square pulses imbedded in digital signals can have significant harmonic content. This facet is so important that the antenna engineer needs to be aware of the fact that most every house, car, and office has an antenna in it of some sort, and must design accordingly. Cell phones, computers, and many other household items can and do have multiple antennas imbedded in them in many cases. Single antennas sometimes need to operate with multiple channels simultaneously.

## Philosophy of this book

This book attempts to allow a technically adept person to design an antenna based on reasonable expectations. Although there are formulas throughout these pages, the intention is not to present a completely theoretical outline of antenna design. The true intent is to present a basis for the quick determination of antenna requirements, then proceed to a first approximation of the applicable antenna. Construction and testing techniques are discussed to give the designer the ability to determine if the design will meet the goals set out by the requirements.

This book is also intended to be useful, as a source of reference material as well as a manual on fundamental antenna design principles. Several categories of antennas are presented but be aware of the fact that there are thousands of antenna designs. Most of these individual designs incorporate one or more attributes presented in this book.

Finally, the history of antenna development is presented to allow the reader to get a sense of why certain antenna types have lasted for over a hundred years. The contributions of many brilliant scientists have brought what was once in the domain of the few to the appreciation of all.

VOR/DME/TACAN Antenna used for Aviation

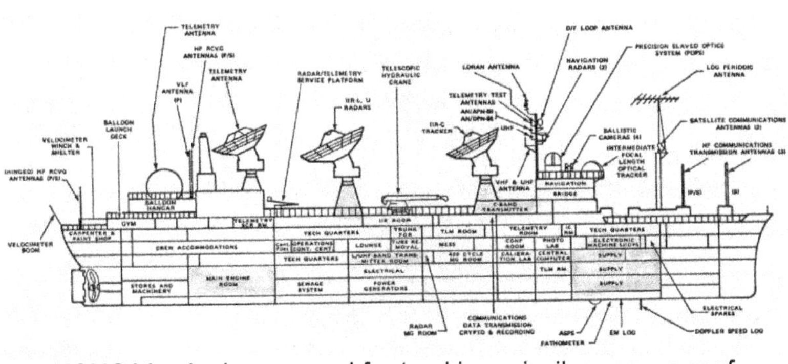

USNS Vandenberg, used for tracking missiles, now a reef

# Antenna Development

## 1880 - 1930

### Dipoles, Monopoles and Reflectors

Heinrich Hertz was arguably the first person to create an antenna. He understood the equations of Maxwell and Heaviside and was able to reduce the theory to practice. He created in essence the first dipole antenna and was able to transfer energy to another resonant structure through the air. In addition, Hertz created the first cylindrical reflector as well as demonstrated the principles of reflection and refraction in the radio regime.

Tesla and many that followed furthered Hertz's work by creating monopole antennas, loop antennas and transferred energy over greater and greater distances. By the time Tesla passed away, it was commonplace to have worldwide communications via shortwave. Local radio was available in all major cities and television was soon to follow.

## 1930 – 1950

### Arrays

Beginning with the need to broadcast radio station signals, much work was done in the area of array design. It is here where engineers understood the effects of phasing the many elements within an array of individual antennas. Typically, dipole arrays were designed with phasing networks setting the same phase to each element. In this way, the additive effects of the elements could combined to increase the gain and decrease the beamwidth. During WWII the English protected their shores with such arrays. Many ships at sea had similar pedestal mounted arrays used for radar. When the effects of varying the phase to the individual elements became understood, aircraft were equipped with crude *electrically* steerable arrays attached to their airframes. The radars contained inside the aircraft were engineered to steer the beams to scan ahead for enemy planes. Later, after the war, radio astronomer experimentation with large phased arrays was performed in an effort to further increase gain as well as measure the energy coming from small areas of the sky. By 1950, some of these arrays were getting very large with the ability to scan several tens of degrees. Shipboard arrays became electrically servoed in elevation and azimuth to compensate for wave motion. Lastly, operational frequencies became higher and higher allowing smaller physical structures to be made, or more elements to be used. Most of the early work on arrays was based on theories given by Schelkunoff and others a decade before the war. Another important innovation was the waveguide slot. This antenna allowed for carefully controlled beam shapes and the simplicity of machining the small slots allowed for large arrays to be built.

WWII Phased Array

**Parabolas**

The development of reflector antennas can be characterized as having a fast start, a waning period and finally, significant development. Heinrich Hertz, in 1888 designed a beam forming antenna known as a parabolic reflector, initially cylindrical in shape. He used these antennas to research the "optics" of microwave transmission and reception. A parabolic cylinder is a shape that allows focusing of radio waves to occur in a line along the longitudinal axis of the reflector. As a consequence, a reasonable amount of gain can be realized due to size of the reflector relative to the wavelength of the frequency used. Additionally, front to back ratios can be optimized as well as sidelobe performance. Hertz used this parabolic cylinder for fundamental measurements and it became a staple of antenna designers thereafter. By 1896, Marconi was using the reflector for radiotelegraphy at 25 cm wavelengths. By 1900, Marconi had moved to longer wavelengths and the cylinder fell out of favor. As designers opted for lower frequency usage (primarily for longer propagation distances) parabolic designs languished until about 1930. Again, it is interesting to note that microwave frequencies were developed very early followed by a roughly long period of disinterest. Ideas like remote control, point to point communications and collision avoidance for ships could have been significantly more developed had this not occurred. Shortwave band equipment and antennas had been developed by 1897 by Righi and others. By this time frequencies below 20 MHz (15 meter wavelength) became operational and designs of parabolic curtains became popular. It was only after 1930 that parabolas were developed into the fine antennas they are today. In 1930 Marconi developed a paraboloid at 50 cm for his experiments in beyond-the-horizon tropospheric communications in the Mediterranean. Hemispherical parabolas were developed around 1934 (Clavier and Gallant) for use in communications at 17 cm. Circular parabolas were developed for use in radio astronomy by Reber in 1937. This last development was the beginning of the development of the "big dishes". Reber made a dish 30 feet in diameter for use at several wavelengths to map the radio universe. Many significant discoveries came out of his research and as a consequence, parabolic dishes were and have been used for a variety of wide band, high gain applications. Many geometries have been created using the basic dish shape. Dishes have been used in tandem to create interferometers and other exotic antenna arrays. Interferometers are electrically connected antennas which are many wavelengths apart, causing a multi-lobed pattern, where each lobe is as small as a main lobe created by an antenna the size of the distance between elements. The use of interferometers allowed the realization of very high resolutions and immense increases in gain. More details will come in the radio astronomy antenna portion of this work. By 1950, very large parabolas built for research, and a proliferation of smaller versions had been made for airborne radar as well as other communication related applications.

Mars Reconnaissance Orbiter showing main antenna

Grote Reber's Parabolic dish, 1937

## Microwave

As mentioned earlier, microwave frequencies again became popular after 1930. During the second world war, the development of airborne radar required much higher frequencies to be employed, due the lack of space on aircraft. Magnetrons had been developed in the S and X bands (2 and 10 GHz) for use as high power transmitters. These transmitters in conjunction with small arrays and reflector antennas helped the allied forces achieve air superiority over the Germans. The further development of sensitive receivers allowed for the deployment of radars mounted in light British bombers. These planes flew primarily at night and were used to locate and destroy enemy aircraft using this new system. The primary laboratories for the development of these high frequency devices and antennas were Cambridge Labs in England and the Radlab in Boston. After the war, these significant developments in the microwave region were de-classified and optimized by these two great laboratories. As a result, microwave theory and techniques moved forward to allow a multitude of applications like space borne communications and weather radar.

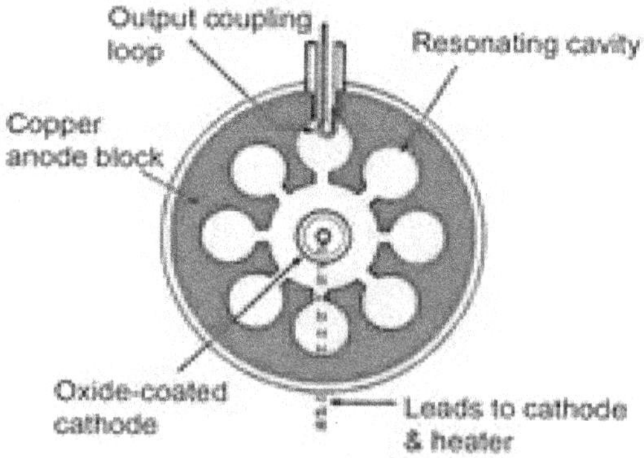


Resonant cavity magnetron high-power high-frequency oscillator

## Mechanically Steered Antennas

During this period, if an antenna of any sort needed to be scanned it was mechanically moved. Multi-axis positioners were developed that could be servoed or manually pointed. During the second world war, the Germans used a parabola to track the V-2 rockets. Several of these "Wurtzberg" antennas are in existence today and show the basic structure of a fully steerable altitude over azimuth mounted parabola. Later, shipboard steadied positioners were placed in feedback arrangements connected to the ship's gyroscope to allow smooth scanning of the ocean. In the air, the same basic techniques were employed to allow the forward scanning of radars for weather avoidance, target location, ground mapping, and beacon location. The latter being a navigation device where a transponding device was located on the ground to create a "bright" target for the airborne radar to find. This technique along with the use of retroreflectors (a device designed to reflect a return signal in the same angle as the incident signal) allowed for new capabilities such as poor weather landings and much more precise bombing operation.

Wurtzberg Radar from WWII

**High Power**

During this period, Magnetrons, Klystrons and other high power devices were invented to allow greater range for radars as well as longer, more reliable communication. These devices were primarily microwave in operation and used wave guide feeding structures to parabolic dishes. The combination of better antennas and higher power allowed multi-target characterization and positioning as well as interference rejection.

WWII Search Radar

**1950 – 1970**

This period of antenna development is best characterized as fast paced. The Cold War as well as the Vietnam and Korean conflicts required significant strides in technology development. Antennas became smaller, smarter and much more agile for military use. In radio astronomy antennas became huge in extent and a multitude of array concepts emerged to further into the universe and explain physical phenomena. Satellites became ever present and the our exploration of space and required the most innovative of antenna designs. Some of these developments are explained below.

Large Direction Finder Antenna, Wullenweber

**Phased Arrays**

During this period, significant advancements were made in the understanding of phased arrays. These are antenna systems that are made up of a plurality of individual element, typically in a square, circle or line. Using Schelkunoff's equations, arrays were made from waveguides and dipoles, with increased performance and flexibility. Later, non-uniform element distributions were made to examine properties of minimal element antenna beam forming. Amplitude tapering then followed allowing control of sidelobes and grating lobes. Taylor and others devised formulas to predict the sidelobe performance. Taylor, raised cosine and Chebycheff tapers were used to optimum performance of arrays on land, ships and aircraft. Arrays also made it into space for use in Apollo-Soyuz docking, landing Surveyors and LEMs on the Moon, and examining geologic/hydrologic phenomena from orbiting satellites. Conformal arrays around rockets bodies were developed. Some of these were steered, meaning that they had phase shifters attached to each element to affect the movement of the main beam. In radio astronomy innovations were created using individual phased elements. In the 1950's a linear array of many elements was crossed with another creating a "Mills" cross (after the Australian who invented it). Two intersecting fan beams were thus created. These arrays were sequentially fed in and out of phase creating a "blinking" main beam, this blinking was synchronously detected, with only the gain of the area of the intersection being recorded. A linear array necessarily has a fan beam shape, not useful for pinpoint measuring of celestial sources in the sky, when the synchronous detection takes place, the beamwidth of a full aperture antenna that has the extent of the cross is realized. Obviously building a cross was significantly cheaper than a full aperture dish, so they became very popular in the scientific community. Other forms of "aperture synthesis" became prevalent, notably American, Russian, Australian, English, and French astronomers developed a vast assortment of similar antenna designs that discovered such phenomena as quasars, pulsars, supernova remnants and many others. More discussion will follow in the Radio Astronomy Appendix.

HF Direction Finder Array using Log Periodics

Mill's Cross Array

**ECM**

ECM or Electronic Counter Measures were developed starting approximately in the second World War. When someone transmitted on the same frequency as another trying to receive information, the signals could be "jammed," causing loss of positional data in the case of radar or loss of the ability to communicate in the case of transceivers. During the second World War, jamming could be achieved with the use of " chaff". Chaff is made up of many pieces of reflective foils, nominally ½ wavelength long and can incapacitate or confuse a radar. Normally, thousands of these foils were dispensed from aircraft which would create very large areas of reflections that confused the radars. During the D-day invasion on June 6$^{th}$ 1944, this technique was used to make the Germans on the French coast believe that there was invading aircraft coming from a significantly different area other than Omaha beach. Other ECM techniques include quickly locating and jamming tracking radar signals by use of transmitters. During this period, wide band antennas were developed that could receive signals over many octaves, thus being able to detect many types of radar systems. These antenna typically were used in aircraft and ships for the purpose of evading attack. Arrays of wide band elements or parabolas with wide band feeds were used to locate the enemy radars or communications. An example of a wide band array is a Rottman Lens, named after the engineer who designed it. These lens have the unique property of surveying large angular areas over large bandwidths. The are used primarily to locate threats anywhere in the sky on an instantaneous basis.

ECM Antenna using Rottman Lens

**SAR**

SAR or Synthetic Aperture Radar antennas were also developed during this period with the unique ability to make high resolution images of the ground as the array was moved from one position to another, typically on the side of an aircraft. In this way, a larger antenna could be synthesized (by the distance traveled) and therefore finer beam dimensions could be realized. In the case of the radar versions, phase and amplitude of the reflected signals was detected, Fourier transformed and recorded. In the simplest case, the transformed images were recorded on moving photographic film. At the end of the "run" a swath of high resolution images could be seen of the area of interest. These antennas were place on aircraft in the form of "side looking radars". The antennas were conformal in nature and usually were attached to a large portion of the fuselage. Much later, in the 1980s, a synthetic aperture radar was developed for the very first flight of the Space Shuttle. The radar operated at L band and had the ability to penetrate the ground with radio waves for the purpose of imaging and as a result discovered a significant number of archeological sites including roads and buildings, during its fly by of the Near East.

SIR (Shuttle Imaging Radar) – Synthetic Aperture Radar Antenna

Today, SAR systems are in operation 24 hours a day in satellite operations as well as military and commercial aircraft applications.

**Interferometers**

Interferometers are made up of multiple antennas set at least several wavelengths apart. The advantage of such an arrangement is that it allows the measurement of signals with resolution of an antenna the size of the greatest distance between elements. Instead of main beam however, many "fringes" or antenna lobes appear, each with proportions of that of a single large antenna. These fringes are tapered by the pattern of the individual elements. Interferometers started appearing in the 1950s as tools for radio astronomers. In this way astronomers were able to closely examine astrophysical phenomena without the huge cost of a full size antenna miles in diameter. Later, during the 1960's, much progress was made using variable baselines and widely spaced arrays to further astrophysical research. More details on these antennas will be discussed later in this work.

The Keck Optical Interferometer

The VLTI Interferometer in Chile

Aerial Picture of the Very Large Array Interferometer

**Diversity**

When multiple antennas are used with a single receiver and they are switched back and forth so that the signal can be sampled for gain, then this system is known as a diversity antenna system. Orthogonally polarized antennas can also be used in this fashion. Recent commercial systems employ two antennas set ½ wavelength apart. Typically, a designer is trying to eliminate the effects of multi-path and variations in signals due to movement, sometime referred to as "picket fencing". Most diversity systems that were developed during the period of 1950-1970 were used for military purposes. This development allowed for a more reliable signal to be had in challenging environments. Later this technology was developed for use in cars with FM receivers to improve reception. Today, most interned base stations employ this technique.

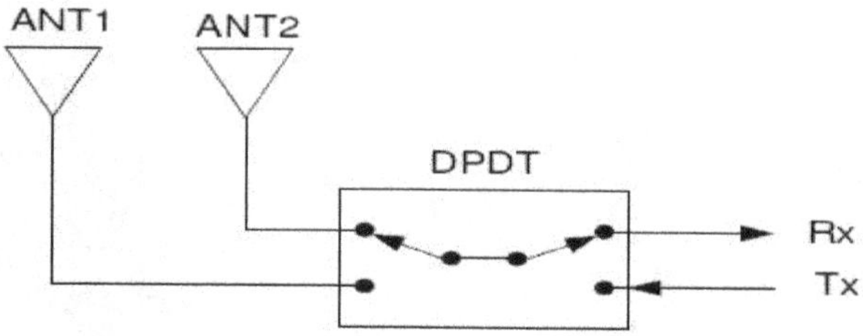


Simple Diversity Scheme

**Monopulse**

Monopulse antennas are a class of antennas with the purpose of signal location. They are designed in such a way as to present a sum and difference amplitude to a receiver. Based on the measurements of these two quantities, circuits following can determine a precise vector to a target. Fundamentally, the sum signal allows the reception of signal over a wide area, so the presence of a signal can be detected, the difference produced a very fine null in the center of the array. This allows for a high resolution determination of the position of the source. The magnitude of the sum and difference signals is measured and used to determine the pointing errors of the antenna array. The advantages of these designs were significantly quicker location of signal positions. This is particularly important when the emitter might only be on the air for a very small time, not allowing mechanical antennas to slew over and determine a fix. These antennas were developed initially for the military, however other, more commercial applications had been create by the end of this period. Later in this book there will be a more technical description of such antennas.

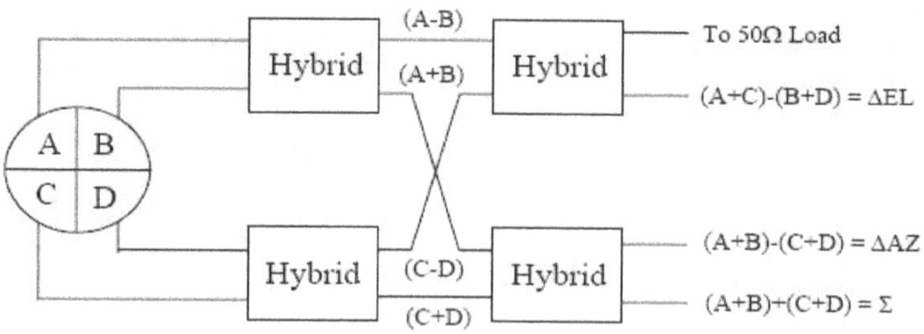


Typical Monopulse Setup. A,B,C and D are antennas

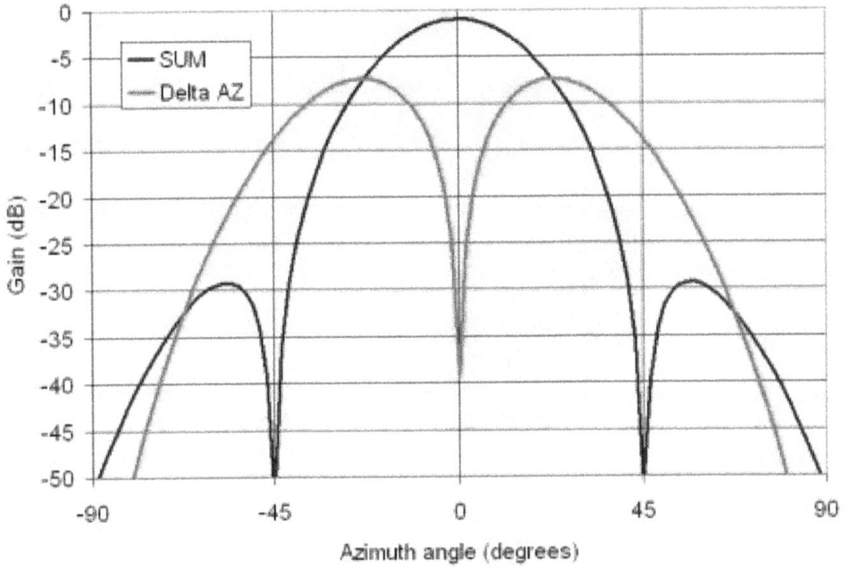


Antenna Patterns from a Monopulse system showing sum and difference, notice the fine resolution in the difference (Delta AZ) pattern, allowing for precise tracking or location

**Airborne**

During the second World War, there were such significant developments in airborne communication and radar that they became a distinct style in themselves. Along with the requirements of bandwidth and impedance match, other considerations set these antenna apart. Another important considerations was the need for conformal designs, radomes, hostile environment designs, aerodynamic designs, and lightning suppression. During the period of 1950-1979, designs were perfected for high mach airspeed aircraft, rockets, scanning radars, and other challenges. So unique were the requirements of these antennas that companies dedicated to this particular type of antenna were created and exist today.

Typical Aircraft Weather Radar

Aircraft Weather Radar antennas were initially dish type during this period but soon advanced to an array type approach as shown. Today this type is predominant.

**Spaceborne**

An extension of airborne design, space borne antennas require extra care considering the hostile environments they were expected to operated in. The temperature extremes pose certain threats that must be addressed. In addition, non outgassing, higher radiation immunity and lightweight designs must be employed for successful space borne applications. During the period 1950-1970, all of these challenges came into focus. First with the launch of the first satellites, then with Mercury, Gemini, Apollo, Shuttle and finally deep space probes. It was initially discovered that signals were "modulated" by what is now known as Faraday Rotation created by the upper atmosphere. This phenomena cases the rotation of the polarization vector and will cause cross pole conditions to exist, attenuating the signal greatly. Solutions had to be found to mitigate the loss of communications and telemetry. Higher frequencies had to be employed to save weight on the spacecraft. Also, sophisticated ground stations were designed and built that would allow the best chance of communicating with the spacecraft. During this period, major steps were taken in design that ultimately allowed the launching of deep space probes and sophisticated "man" qualified spacecraft that landed on the moon and other planets. The rate of progress was necessarily swift and testing facilities were very busy qualifying antennas for flight. These testing facilities included indoor and outdoor antenna ranges augmented by vacuum chambers,vibration simulators, radiation chambers and temperature chambers. The growth of the intercontinental ballistic missile defense system required similar testing. Only a few of the designs employed previously to this era could be used, most had to be designed on a custom basis depending on the application. The more interesting designs were those that could only deploy properly in space making earth bound testing problematic. Large reflectors were sent into space to experiment with relay communications like the Echo satellite. Inevitably, commercial enterprises understood the opportunities to be found in space, especially around transmitting and receiving television programs from around the world. Telestar was launched which allowed the United States population to see transcontinental television programs for the first time, including the Olympics. Obviously, this formative era blossomed into a multitude of commercial ventures we see today including Direct TV and Intelsat.

Direct TV Geostationary Satellite

One of the first Communications Satellites, Telstar, 1962

**1970 – 1999**

**Advanced Airborne**

Today, airborne antennas are used prolifically to allow passengers to use phone computers while en route. In addition, military aircraft have very sophisticated phased arrays that allow the mapping and location of targets as well as the establishment of secure communications at high altitudes. Millimeter frequencies are now used between aircraft for secure communications. Satellite communications allow advanced $C^3I$ (Communications, Command, Control and Intelligence) to be used by aircraft. Sentry aircraft have advanced SAR phased arrays as well as other types of arrays used for satellite communications. Powerful radars are now onboard aircraft in the forms of large rotating structures as well as in the form of phased arrays. Aircraft themselves have had fuselage separations using porcelain plugs that allow individual structures to be used as parts of a dipole. Lastly, most aircraft now have antennas for communications, navigation (including GPS or Loran) and emergencies. Additional applications include weather radar, ground proximity and beacon antennas. Today, student pilots must know the great variety of antennas on the small trainer aircraft used. Having 10 or more radiating elements is not uncommon in a standard aircraft.

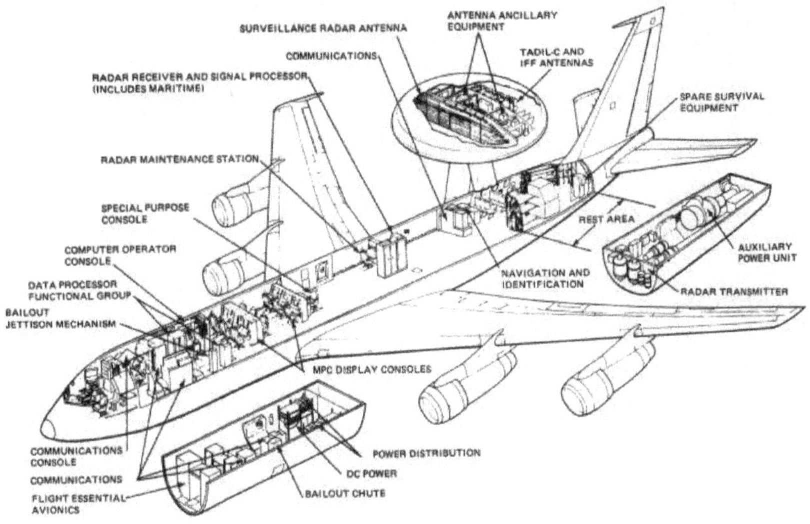


AWACS Aircraft

**Advanced Space Borne**

Going into space before the 1970's usually meant having a nose cone or other shroud protect the spacecraft antennas as they were launched. Today, the Space Shuttle must have a multitude of communications and telemetry antennas "deployed" while in launch phase. These antennas are typically conformal and capable of working in harsh environmental conditions without projecting into the airstream. In space there are hundreds of satellites that have an immense variety of antenna structures. Some of the more interesting ones are shaped like the areas or country they are designed to broadcast to. In this way very unique beam shapes are created to best use the power of the transmitters of these spacecraft. Very large arrays had been deployed in space for use as SAR system: intelligence gathering systems and $C^3I$ systems. Space Stations have been built with significant number of antenna structures.

Earth observation satellites

**Optical Phased Arrays**

One of the more advanced types of phased arrays that has been developed recently uses a combination of RF and Optical techniques to make phased arrays. For instance, the outputs of individual elements can be fed into optical transducers. Using these individual "points of light" allow for lenses and filters to be used to create such things as optical Fourier Transform and spacial filtering. Insofar as these conversions are instantaneous, a significant amount of processing can be accomplished without the use of sophisticated computers. This typically lowers the weigh and cost of these systems.

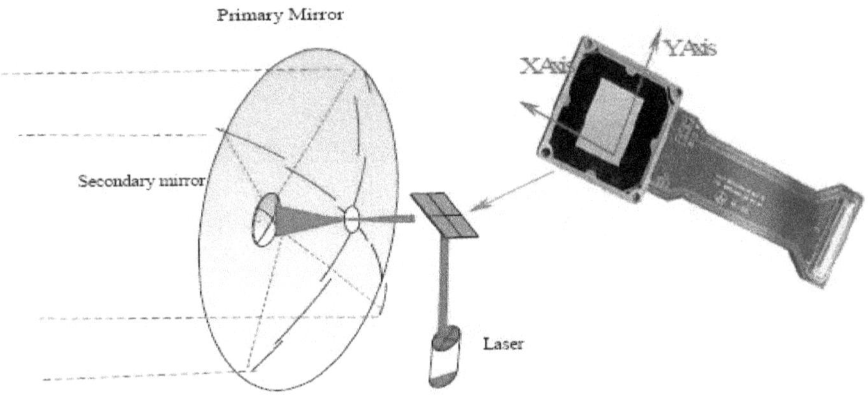


Optical Phased Array

Additionally, optical signals can be mixed, or hetrodyned to lower usable frequencies, by focusing two beams on the surface of a PIN diode, creating the addition and subtraction of the two optical frequencies. In this way the difference can be used to detect modulation or measure spectra.

Finally, integrated circuit fabrication techniques have been used to create resonant halt wave dipoles on silicon that receive or transmit at optical frequencies. In some cases these dipoles have been used in conjunction with diode rectifiers to create very efficient solar electric panels

**Adaptive Arrays**

Another advancement in phased arrays can be seen in the area of adaptive array technology. Using sophisticated algorithms, the amplitude and phase of individual elements be adjusted to allow multiple beams and nulls to be formed. The advantages of such agility allow for applications such as interference rejection and multiple target tracking. The military is a typical user of this, however because of the crowding of the airwaves due to cell phone and other uses, commercial ventures are now developing similar systems that can track individual users to allow better use of the spectrum.

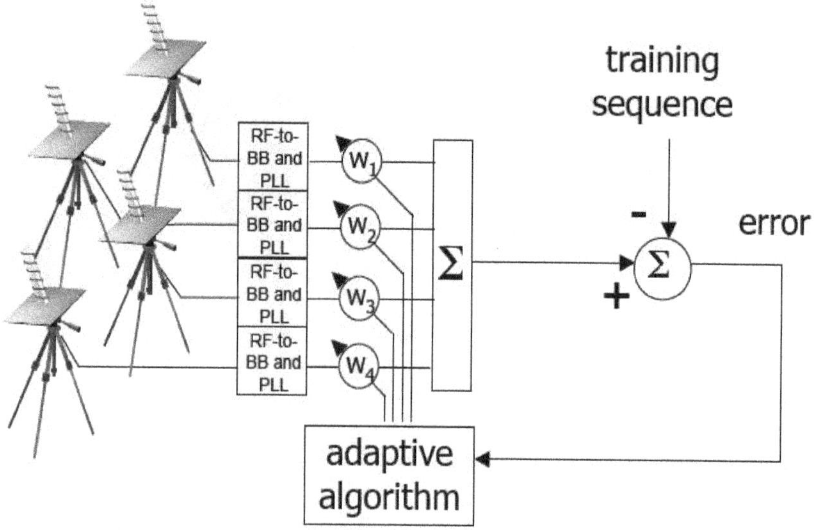


Adaptive Array

The adaptive array shown above shows some of the components used in creating this type of array. The "RF to BB and PLL" refers to the Radio Frequency to Base Band and Phased Lock Loop conversions. Base Band is where you mix the input frequency with an oscillator with the same frequency, the resultant subtraction allow only the resultant frequency of interest to move on down the signal path. This is different than a super hetrodyne approach whereby the initial signal is translated down to a lower frequency (and some times translated again) to a detection stage where the real frequency of interest, for instance audio, is delivered.

The W1-W4 blocks are weighting controls like phase and amplitude. The signal are summed then compared to the expected performance. The adaptive algorithm decides what weights are implemented.

Adaptive arrays are now coming on line for commercial and consumer uses.

**2000 - Future**

Although volumes could be written about recent advancements in antenna technology, even more could be written about the future of antenna development. I will attempt only to speculate about certain facets of this development. We can assume that there will be significantly more applications and enhancements in the future.

**Millimeter**

Millimeter Radio Telescope

As the requirements for bandwidth increase, there has been a steady increase in operating frequency. This has necessitated improvements in transmitting, receiving and antenna equipment operating above 30 GHz. Although the size of these system decreases for the same gain as lower frequency designs, the path loss increases to offset the benefits. In addition, absorption by the atmosphere increases and depending of the frequency of operation, can severely attenuate a signal. This being said however, massive amount of bandwidth can be had by going into the millimeter frequencies (characterized as between 30 and 300 GHz). These systems are becoming more and more prevalent for use in Internet and broadcast related applications. Their advantages in small size and high bandwidth make them ideal in these areas. In the future, these systems will proliferate and be found in many neighborhoods and businesses.

The ALMA array at Chajnantor, Chile

**Submillimeter**

The next step after millimeter technology will be the sub-millimeter. These frequencies are known as quasi-optical and are characterized as 300 GHz to 1 THz. Parabolics and phased arrays have been designed in this region. Again because of the increased need for bandwidth, expect to see increased development efforts in this range. Also, because optical techniques can be used here, much simplification in designs can be realized. For the most part, optical infrared as well as RF design techniques are used here. This has a wealth of advantages considering the amount of development in these areas. However, the disciplines have in the past not taken advantage of their complimentary mathematical approaches. This will change in the future.

Submillimeter Telescope Array

**Smart Antennas**

Smart antennas are distinguished by their use of integrated decision making electronics in close proximity to the antenna elements. Smart antennas will be used on a more frequent basis for applications such as spacial diversity systems, adaptive systems, and network systems. When frequencies of use increase, the application of integrated circuit technology lends itself well for both decision making circuitry and antenna systems. Included in these antenna systems are amplifiers, phasing components and distribution networks.

Typically, smart antennas have phase shifters integrated into them, allowing for the movement of the beam realized from an array of antenna elements. All possible beams are sampled, and the ones that need to be active represent themselves by the highest signal strength. A small computer keeps track of the positions of the active signals, and can with beam steering, ignore and attenuate an interfering sources. In essence this is adaptive beam forming, leading to more capability or bandwidth from a single cell tower for instance.

Another definition of smart antennas will be ones with integrated transmitters and receivers. These are used today in the military and are referred to as "TR Modules" or transmit/receive modules. They include a small transmitter, at least the front end of a receiver, phase shifters, an antenna and sometime a small processor to communicate with the antenna control system. A multitude of TR modules are assembled in close proximity to create a phased array. The operation of such an array is then determined by the antenna control system, either in an orthodox design or an adaptive design. In a radar the sum of the small transmitters determine the total radiated energy. If a TR module fails, the entire systems does not fail. This module is detected by the onboard electronics and is replaced during the next maintenance cycle.

Presently a significant amount of work has been done to create the simplest form of smart antenna. For reception, the antenna is connected to an amplifier and then to a analog to digital converter and finally a field programmable gate array (FPGA). For transmission, the FPGA is connected to a digital to analog converter and thence to a power amplifier. Software is able to define phase and amplitude for the TR elements, optimizing performance.

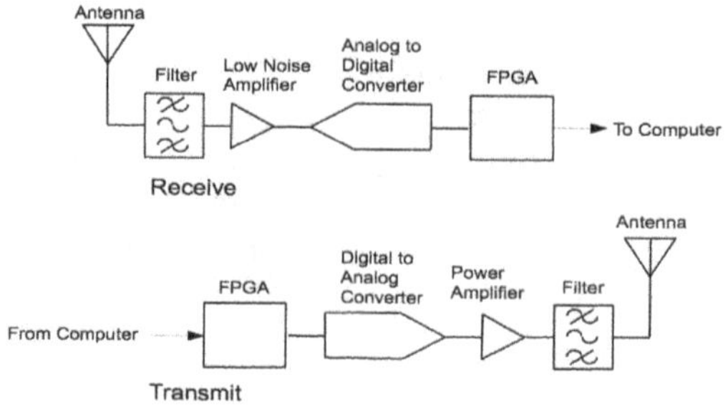


Direct Digital Conversion

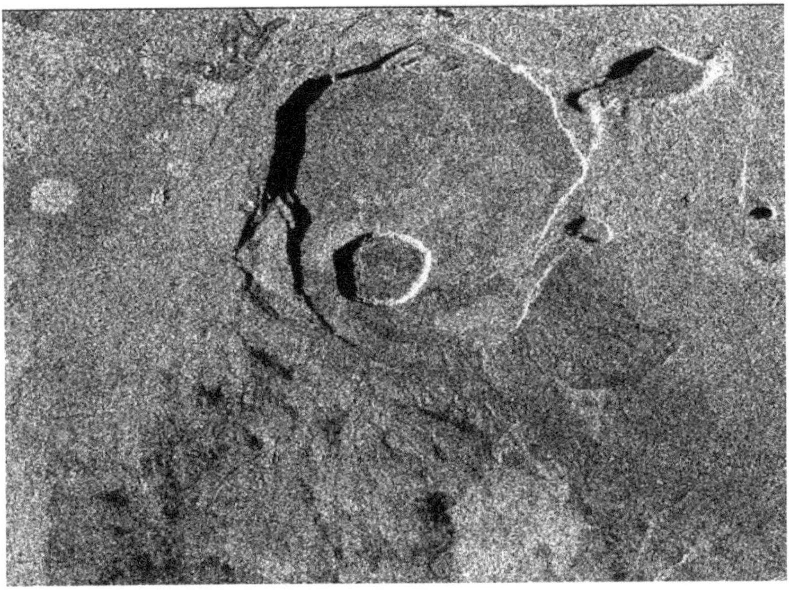

Radar image from an synthetic aperture radar of the Kilauea Volcano in Hawaii

SAR image of building and cars

# Chapter 2 – Fundamentals of Antenna Design

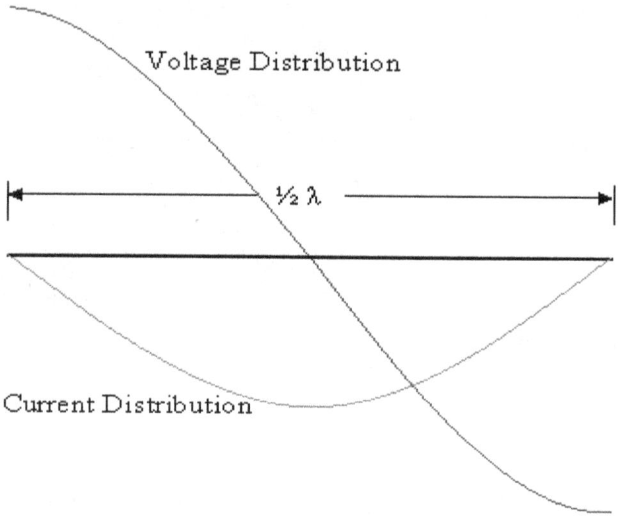


Voltage and Current on Dipole

**Definition**

A radio antenna may be defined as the structure associated with the region of transition between a guided wave and a free-space wave, or vice vera (Kraus). A guided wave is one that is contained in a structure for the purposes of moving a charge or voltage from one point to another. Wires, coax and waveguide are examples of structures that perform this operation. Typically, a transmitter or receiver is attached to one end and a resonator is attached to the other. This resonator allows the voltages present in the guided wave structure to be translated to or from the air. Once in the air, another resonator of similar frequency response can send or receive the voltage and again interface with a guided wave structure. These resonators typically have important relationships to wavelength (or frequency) and gain. Another definition of antenna is a structure that converts photons to currents and vice versa (Kraus).

**Basic Formulas**

The following descriptions should give the designer the basic tools required to build and analyze an antenna design. These formulae represent the procedures that all designers, experienced or not, must evaluate to begin the process of analyzing feasibility, cost, and development time with regards to building antennas.

**Link Analysis, the Real Starting Point**

To start any antenna design, the engineer has to understand the basics of the communications link. Fundamentally, this includes a transmitter, receiver, associated antennas and associated feed cables. Also, there is a distance between the components that dictates the amount of loss to be expected.

Knowing the characteristics of all but one of the components will give the designer the ability to calculate the missing value. For instance, if we know everything but the required receiver antenna gain, following the next set of formulas will show what is required to complete a successful link.

Keep in mind that the path loss is actually an ideal number, in real life designs it is important to provide a link *margin* to compensate for variations in weather and obstructions.

An Idealized Transmitter to Receiver Link

To calculate how much gain is needed for an antenna, one should look at both ends of the link, assuming that there is a transmit site and a receive site. In radar systems, one antenna typically does both duties, but due to some other system details a more refined radar equation is needed; refer to Appendix 2 for more details.

To understand the link analysis for a two antenna system, several factors need to be quantified:

Distance between antennas to calculate path loss

Power of Transmitter

Sensitivity of Receiver

Losses in feed cables

To calculate the space loss between antennas with existing antennas use the following formula:

$$Path\,Loss(dB) = C + (20 * \log(F * R) - G1 - G2) \qquad [1]$$

where,

F = Frequency in MHz
R = Range in units defined by C
G1 = Gain in dB of antenna 1
G2 = Gain in dB of antenna 2

C = 32.45 for Kilometers
37.80 for Nautical Miles
36.58 for Statute Miles
-27.55 for Meters
-37.87 for Feet

Calculating just Path Loss can also be done by using the following equations:

$$Path\,Loss(dB) = C + 20 * \log(F * R) \qquad [2]$$

where,

F = Frequency in MHz
R = Range in units defined by C

C = 32.45 for Kilometers
37.80 for Nautical Miles
36.58 for Statute Miles
-27.55 for Meters
-37.87 for Feet

Or,

$$Path\,Loss(dB) = ((4*pi*D*F)/C)^2 \qquad [3]$$

where,

F = Frequency in MHz
D = Distance in meters
C = Speed of Light in meters/second

Or,

$$Path\,Loss(dB) = ((4*pi*R)/\lambda)^2 \qquad [4]$$

where,

R = Range in meters
Lambda = Wavelength in meters

Or,

$$Path\,Loss(dB) = 92.4 + 20\log(F) + 20\log(R) \qquad [5]$$

where,

F = Frequency in GHz
R = Range in Kilometers

Adding in the effects of Transmitter Gain, Cable Loss, Receiver

Sensitivity and any conversion gain (added gain due to efficient modulation schemes) yields:

$$LM(dB) = PL(dB) - Pt - G1 + Tl - |(RTH)| - RG + RLO - CG \quad [6]$$

where,

LM = Link Margin
PL = Path Loss
$P_T$ = Transmitter Power in dB
G1 = Transmitter Antenna Gain (dB)
Tl = Losses in Transmitter Cable (dB)
RTH = Receiver Threshold in dB
RG = Receive Antenna Gain in dB
RLO = Losses in Receiver Cable (dB)
CG = Conversion Gain in dB

Given the examples above, lets calculate the required gain for a receiver antenna.

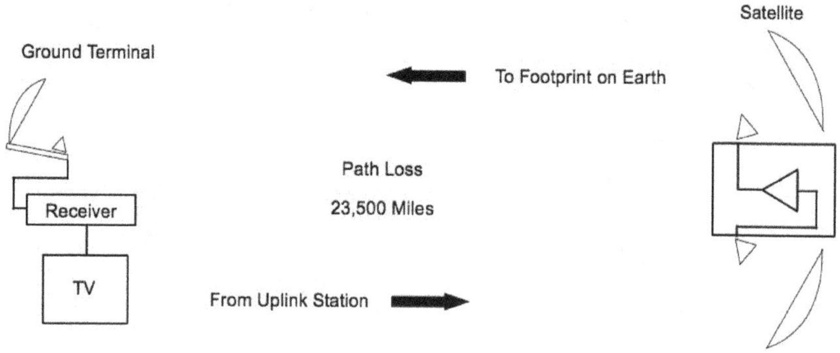


Satellite TV Link

<u>For a typical satellite link:</u>

| | |
|---|---|
| Transmitter Power | 100 Watts or 50 dB |
| Transmitter Antenna Gain | 30 dBi |
| Losses in Transmitter "cable" | 1 dB |
| Receiver Threshold | -100 dB |
| Receive Antenna Gain | Set to 0 to calculate Gain Required |
| Losses in Receiver cable | 1 dB |
| Conversion Gain | 10 dB |
| At 23,500 Miles the Path Loss is | 205.6 dB |

Therefore:

Link Margin = 205.6 - 50 – 30 + 1 – 100 – 0 + 1 – 10 from Eq. 3

or a minimum of 17.6 dBi Required, a typical Ku Band (12 GHz) backyard dish has approximately 36.3 dBi which is 18.7 dB higher in gain to compensate for antenna inefficiency (~ 1dB) and deep fading caused by clouds, rain and snow.

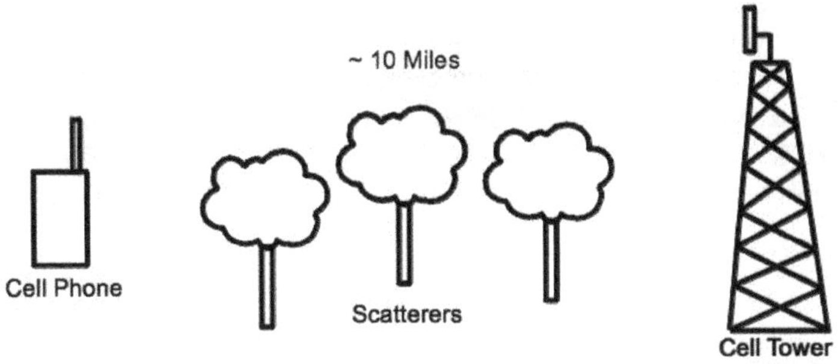


Cell Phone Link

For a typical cell phone link:

| | |
|---|---|
| Transmitter Power | 30 Watts or 44.8 dB |
| Transmitter Antenna Gain | 10 dBi |
| Losses in Transmitter "cable" | 5 dB |
| Receiver Threshold | -100 dB |
| Receive Antenna Gain | Set to 0 to calculate Gain Required |
| Losses in Receiver cable | 0.1 dB |
| Conversion Gain | 10 dB |
| At 10 Miles the Path Loss is | 115.2 dB |

Therefore:

Link Margin = 115.2 - 44.8 – 10 + 5 – 100 – 0 + 0.1 – 1 from Eq. 3

or -44.5 dBi required, a typical Cell phone (850 MHz) antenna has approximately 0 dBi which is in indication of how much fade margin is required to compensate for multi-path and deep fading caused by foliage, buildings and non line of sight operation. In addition, cell towers are placed as often as practical to minimize problems due to terrain and the urban "jungle." For a cell phone system based on the 1800 MHz band, the margin is -38.0, giving less margin for propagation conditions. Significant progress has been made in the area of processing gain which improves margins.

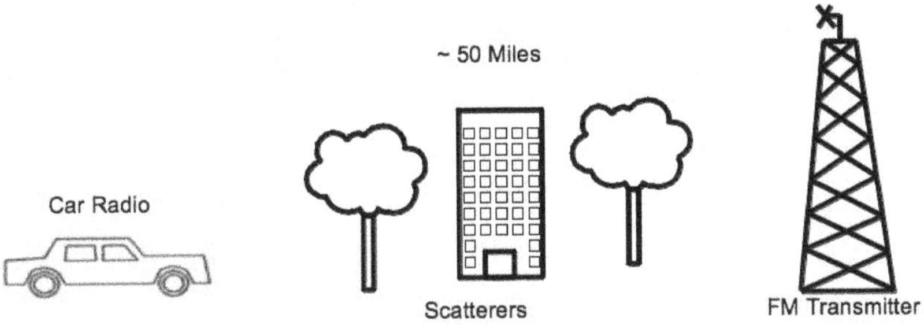


FM Car Radio Link

<u>For a typical FM Radio link:</u>

| | |
|---|---|
| Transmitter Power | 30,000 Watts (74.8 dB) |
| Transmitter Antenna Gain | 10 dBi |
| Losses in Transmitter "cable" | 3 dB |
| Receiver Threshold | -100 dB |
| Receive Antenna Gain | Set to 0 to calculate Gain Required |
| Losses in Receiver cable | 0.1 dB |
| Conversion Gain | 2 dB |
| At 50 Miles the Path Loss is | 110.6 dB |

Therefore:

Link Margin = 110.6 - 74.8 – 10 + 3 – 100 – 0 + 0.1 – 2 from Eq. 3

or -73.1 dBi Required, a typical FM Radio (100 MHz) antenna has approximately 1 dBi which is in indication of how much fade margin is required to compensate for multi-path and deep fading caused by foliage, buildings and non line of sight operation.

One way TV and FM stations mitigate fading is to broadcast in circular polarization, whereby both horizontal and vertical orientations emit power. This helps in complex environments where one polarization might work better than another.

And, just for fun:

Voyager-2 telemetry link from Neptune

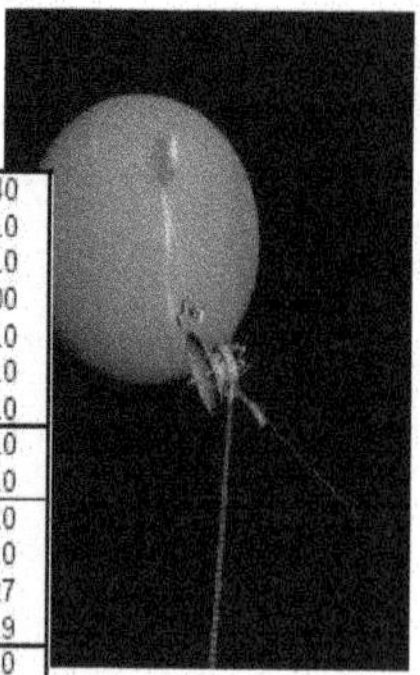

| | | |
|---|---|---|
| *Voyger-2 Spacecraft Transmitter* | Frequency (GHz) | 8.40 |
| | Transmitter Power (W) | 20.0 |
| | Transmitter Power (dBm) | 43.0 |
| | Antenna Diameter (m) | 3.700 |
| | Antenna Efficiency (%) | 60.0 |
| | Antenna Peak Gain (dBi) | 48.0 |
| | EIRP (dBm) | 91.0 |
| *Path* | Range (km) | 4500000000.0 |
| | Path Loss (dB) | -304.0 |
| *Ground Station Antenna* | Antenna Diameter (m) | 25.0 |
| | Antenna Efficiency (%) | 60.0 |
| | Number of Antennas | 27 |
| | Antenna Gain (dBi) | 78.9 |
| *Ground Station Receiver* | Total system Noise Temp (K) | 35.0 |
| | Boltzmann's Constant (dBW/Hz) | -228.6 |
| | Noise Spectral Density (dBm/Hz) | -183.2 |
| *Received Signal* | Received Power (dBm) | -134.0 |
| | Received Pr/No (dB-Hz) | 49.2 |
| | Data Rate (bps) | 21600.0 |
| | Eb/No (dB) | 5.8 |
| | Total System Losses (dB) | -1.3 |
| | Effective Eb/No (dB) | 4.5 |
| | Req'd Eb/No for BER=5x10e-3 w/Conv. Coding + RS | 2.3 |
| ***Link Margin*** | | **2.2** |

## Wavelength

The most fundamental task involved in analyzing an antenna design is understanding the wavelength of operation. Understanding this dimension allows the designer to then predict the size and shape of an antenna. Because the speed of radio wave propagation is the same as light, the following formula applies to both:

Wavelength (lambda) = Speed of Light ( c ) / Frequency (Hz) [7]

or simplified versions of this formula are as follows:

$$\lambda = 300{,}000{,}000 / Hz \qquad [8]$$

or,

$$\lambda = 300 / MHz \qquad [9]$$

and in inches,

$$\lambda = 11.803 / GHz \qquad [10]$$

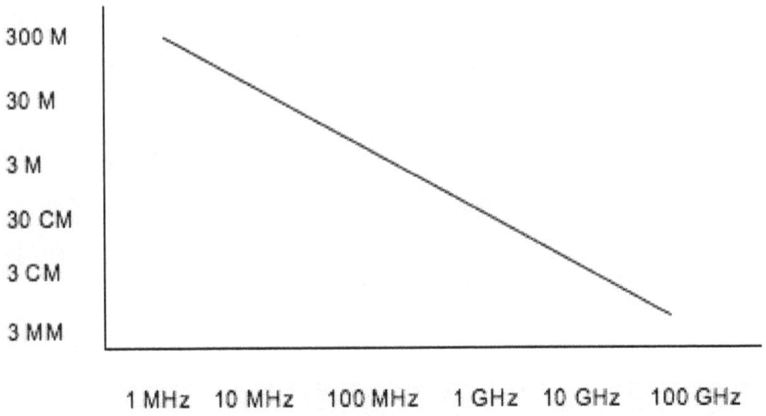


Relationship of Frequency to Wavelength

**Gain**

Gain is a number that is related to the effective aperture of an antenna in terms of number of wavelengths. There is also a relation ship between the beamwidth and the gain of an antenna. If the beamwidth is great, the gain is low and, when the beamwidth is small, the gain is high. To fully understand gain, we must first understand directivity. Directivity is defined as the ratio of the maximum radiation intensity (power per unit solid angle) to the average radiation intensity averaged over a sphere. The simplified formula for directivity is:

$$D=\frac{4\pi}{HP(\theta)*HP(\phi)} \quad \text{in radians} \qquad [11]$$

where $HP(\theta)$ and $HP(\phi)$ represent the beamwidths in two orthogonal directions.

Another simplification of formula 8 can be written as:

$$D=\frac{41{,}253}{HP(\theta)*HP(\phi)} \quad \text{in degrees} \qquad [12]$$

Here we have converted the 4*Pi number to the number of square degrees in a sphere.

By way of example, is we have an antenna with a 10 by 10 degree beam we get:

$$D=\frac{41{,}253}{10*10} \qquad [13]$$

or:

$$D=412.53$$

To convert to a more useful log measurement:

$$Gain(dB) = 10 * \log(D) \quad [14]$$

In this case the log is base 10.

Now, taking the log of 412.53, we get 2.615 and multiplying by 10 we get 26.15 which represents how much power is radiated by the main lobe maximum relative to the rest of the sphere. The log measurement is relative to an isotropic radiator and is thus referred to as dBi.

We now have part of the answer to how much gain an antenna has. Insofar as antennas are not perfectly efficient, we must multiply the directivity by a dimensionless number (between 0 and 1) that represents a more accurate approximation of the the true gain of the antenna. To go back to our example, if the antenna is 75% efficient we would multiply 0.75 times the directivity 412.53 to obtain gain or 24.9 dBi. This can be represented by:

$$Gain(dB) = k * D \quad [15]$$

Where k represents the dimensionless efficiency factor.

Another way of approximating gain is by analyzing an antenna's size and frequency of use. In this case, the higher the number of wavelengths across the antenna's surface, the higher the gain. This can be represented by:

$$Gain(dBi) = 10 * \log\left(\frac{4 * \pi * Ae}{\lambda^2}\right) \quad [16]$$

Where *Ae* is the aperture area, e.g. $Ae = \pi R^2$ for circular dish.

This last formula is perhaps the most used by designers to approximate gain, it is particularly useful when analyzing parabolas or phased arrays. Some difficulty will be had when analyzing the gain of a Yagi or other similar structure. Later in this work, we will discuss how to approximate these types of gains.

**Patterns**

All antennas have patterns, some are spherical in nature (or omnidirectional) and some are more "concentrated" (or directional). These patterns are measured by rotating an antenna in a particular plane an measuring it's response, this is then graphed and analyzed. Typical design requirements for antennas include pattern definitions. These can be in the form of beam widths, side lobe constraints, front to back ratios and polarization. Given good specifications regarding patterns allows the designer to back calculate required gain or directivity. From these quantities, we can calculate the required size of an antenna.

In reality, there are no typical patterns for size and gain requirements, there are only typical patterns for antenna types. Because of this, the designer must think about the preferred style of an antenna given certain pattern requirements. For instance, if the pattern requirements are for very low sidelobes, the designer should think about using a parabola or a phased array with amplitude weighting on the individual elements. Single elements or yogis could not give the performance needed.

Patterns are composed of several parts, generally speaking the following list comprises the important ones:

Beamwidth: Typically defined as the edges or 3 dB points of the main beam. The 3 dB points are the places were the power in the beam down a factor of 50%. Sometimes we see the notation as HPBW or Half Power

Sidelobes: Defined as the high points after the first null from the main beam. Normally, these will be relative to the main beam as in "20 dB down from the main beam". Depending on the complexity of the antenna, sidelobes can number from 2 or more. With squinted main beams, a single sidelobe can sometimes be seen, however for centered main beams one should expect an even number of sidelobes. The higher the number of wavelengths across an aperture, the more sidelobes.

Front-to-back: Defined as the ratio of power from the main beam to the back. Normally, this quantity, like sidelobes will be addressed as "30 dB down from the main lobe" for instance. Be careful in how this quantity if defined, some front-to-back ratios are defined as directly behind the main lobe, and some are defined as that area from 90 to 270 degrees behind the main lobe. This is case dependent, however one should use the IEEE definition as a standard.

Cross-pol ratio: This is a measurement of how much loss is realized when an antenna is set to a polarization orthogonal to its design. In other words, how well a vertically polarized antenna works with a horizontally polarized signal. Another example is how well a left hand circularly polarized antenna works with a right hand circularly polarized signal. Cross pol measurements are done both on axis (when the source antenna and the antenna under test are pointed directly at each other) and off axis, (all other areas).

Axial ratio: This is the measurement between the Horizontal and Vertical polarizations of a circularly polarized antenna. The lowest value is desirable as it indicates a proper balance of energy between the two polarizations. Measurements can be done by measuring the antenna under test with a spinning linear source or a modern approach is to calibrate a dual port source antenna and measure response from both polarizations simultaneously.

Realized Gain: No antenna is perfectly efficient, therefore many numerical simulation programs can calculate what gain will be realized given the specific design parameters. This could include the effects of nearby structures and radomes.

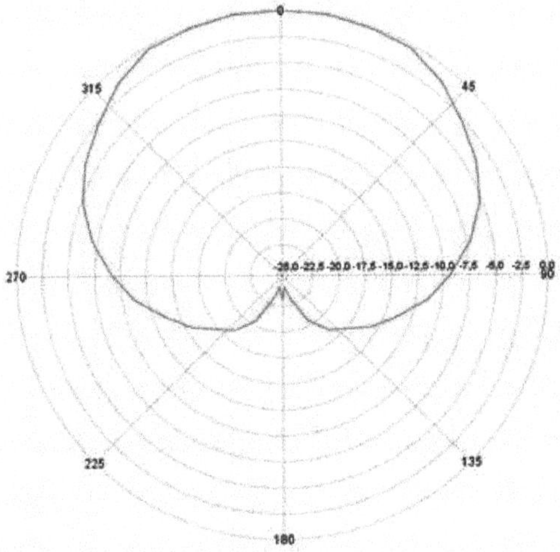


Antenna Pattern for Microstrip Patch on small ground plane

**Beamwidth**

As mentioned before, beamwidth, gain, antenna size and frequency of operation are all interrelated. Given a required beamwidth, a designer can calculate all of the other parameters necessary to complete the task. Given the size and frequency of operation of an antenna, one can calculate the beamwidth. A simple approximation of this is as follows:

$$Beamwidth(Degrees)=\frac{\lambda}{D}*57.3 \qquad [17]$$

Where D is the size of the aperture in x dimension and Lambda is the wavelength in x dimension. This formula works for parabolas or phased arrays best. In the case of a rectangular aperture, one can calculate the beamwidths in both the long and short side.

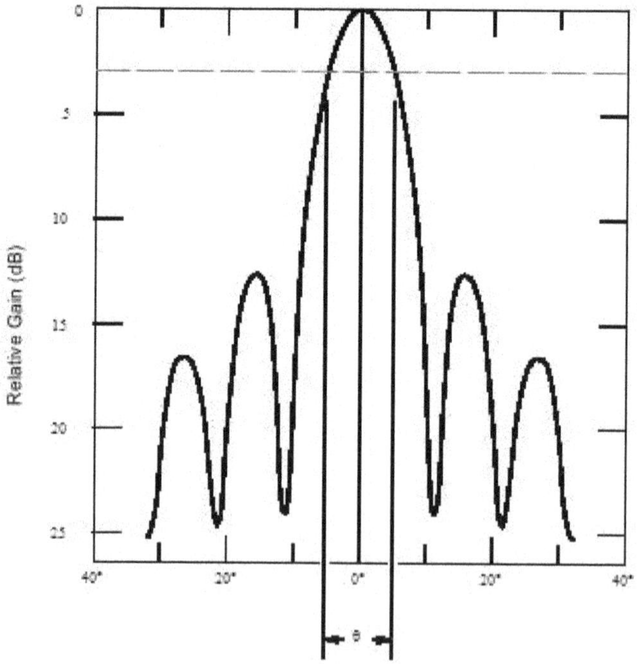


3 dB Beamwidth Sample Antenna Plot

**Impedance**

Impedance is a measurement of the match between two complex entities. In most cases, we talk about the impedance of an antenna and how well that matches the feed cable the antenna is attached to. For instance, it is very common to have coaxial cable designed for 50 ohms. Knowing this, the antenna designer will make an antenna that can match the feed cable to the best possible extent. In other word, the antenna impedance is also 50 ohms. By adjusting the real and imaginary part of the impedance nomenclature, the designer can transform an antenna best matched for gain or bandwidth etc, and allow it to be connected to the feed coax with minimal loss. Impedance is to be understood best as a complex number that combines a real part and an imaginary part. The real part contains a value of real ohms and the imaginary part characterizes the inductive or capacitive reactance component. This is written in the following form:

$$Z = R +/- j \qquad [18]$$

For instance, an antenna that has a 50 ohm real part and a capacitive reactance of 30 ohms the formula is written thus:

$$Z = 50 - j30 \qquad [19]$$

Similarly, and antenna that has a 50 ohm real part and an inductive reactance of 30 ohms is written thus:

$$Z = 50 + j30 \qquad [20]$$

By analyzing and understanding the impedance nomenclature, the designer can use a "Smith" chart to better understand the process of matching. This chart, shown on the next page, is a very useful tool for the designer. With this chart, one can add or subtract the correct reactance and real ohms to best optimize an antenna's match to the feed line. The center of the chart is set to the characteristic impedance of a system, typically 50 + jO. Any other number can be inserted with the same results. In any event the upper portion of the chart is the inductive reactance side, and the lower portion is the capacitive reactance side. By measuring the characteristics of an antenna, the designer can place a point on this graph that represents the action of the antenna a one frequency. The "solution" to optimizing the antenna than be seen by examining how much opposite reactance can be added to balance out the performance of the antenna and then the designer just has to work with the real portion of the impedance. If the designer has to add or subtract shunt capacitance or inductance (as apposed to series), one can mirror image the standard Smith chart to view the effect. The Smith chart also illustrates many other facets of an antenna (or other RF system).

If the designer draws concentric circles about the center point on the Smith chart, one can see the VSWR or reflective portion of the antenna. A perfect 1: 1 match is a point in the center. As the antenna degrades in performance, the measuring point moves outward radially illustrating a poorer and poorer match.

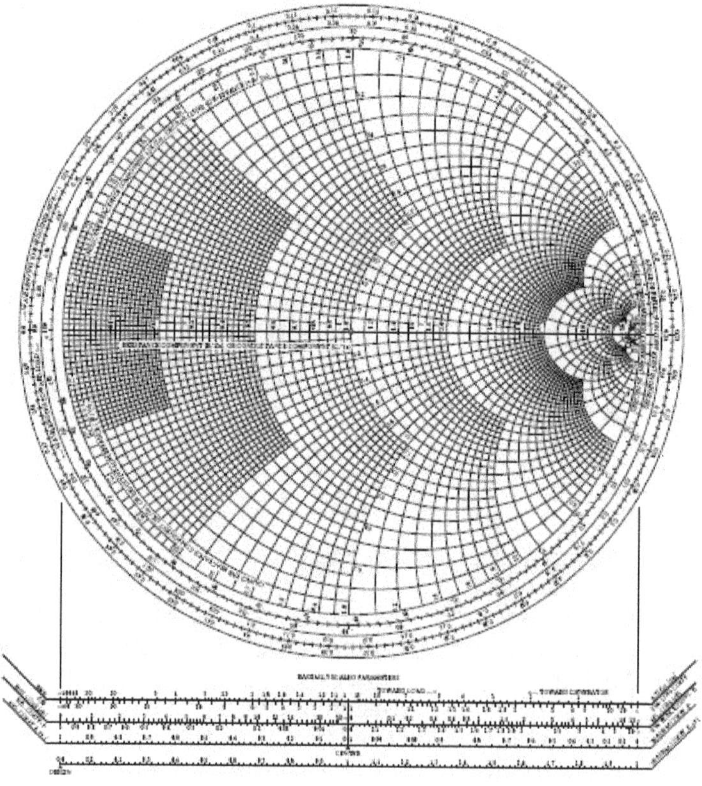

The Smith Chart

If the designer wants to know the phase of a cable or how an antenna changes phase relative to frequency, the outer perimeter of the Smith chart is calibrated in degrees to and from the antenna. For instance, using a Network analyzer that displays the Smith chart, the designer can attach an unterminated cable to the test port and set the test frequency to a single number of interest. The Smith chart will show a point on the outer perimeter that represents the number and portion of wavelengths to the analyzer. By cutting off portions of the cable, the point will move around the chart showing the effect in degrees. So, if the designer wants to make phase matched cables for a system, all one has to do is use a single frequency point on a Network Analyzer.

**Return Loss and VSWR**

Using the Smith chart or Network Analyzer, one can determine the quality of a match between an antenna and a feed coax. This is done by determining Return Loss, which is the measurement of the amount of energy going to the antenna compared to the amount of energy reflecting from the antenna.

Return Loss can be understood by realizing that a non perfect match as measured on the Smith Chart will move the measurement point out radially from the center. Concentric circles of equal return loss can be drawn on the chart to illustrate this. A non 50 ohm real resistance, a capacitive reactance, or an inductive reactance will "mis-match" the antenna.

One simple way of measuring this parameter is to us a directional coupler, a device commonly found in a Network Analyzer. This device has either three or four ports, with at least an input and output port and one "coupled" port. This coupled port enables one to measure the amount of energy going either to (forward) or from (reverse) the antenna under test. Measuring the reverse port magnitude allows the designer to optimize the antenna match.

Another way to measure match is to use a Wheatstone bridge with a diode across the middle to rectify the reflected energy. All the legs should be 50 ohm carbon resistors and on leg should be connected to the antenna under test. See Appendix 6 for more details.

VSWR (Voltage Standing Wave Ratio) is somewhat more of a common measurement than Return Loss, although is essence they are the same thing. They are related by the following formula:

$$Return\,Loss = 20 * \log \frac{(VSWR-1)}{(VSWR+1)} \qquad [20]$$

and

$$VSWR = \frac{1+10^{-RL/20}}{1-10^{-RL/20}} \qquad [21]$$

VSWR is usually stated as a ratio thus:

1.5:1 or 2:1

This number relates gives the designer a sense of quality of match and also give a measurement of how much energy is lost by a non perfect match (1:1). This Mismatch loss (ML) can be calculated by the following formulas:

$$ML = -10 * \log\left[1-\left(\frac{(VSWR-1)}{(VSWR+1)}\right)^2\right] \qquad [22]$$

and

$$ML = -10 * \log[1-(10^{(-RL/20)})^2] \qquad [23]$$

**Antenna Resistance**

Another method of determining antenna quality is by analyzing Antenna Resistance, which is defined as the power radiated by the antenna divided by the square of the antenna current (RMS). This calculation includes the radiated power, as well as the effects of power lost by dielectrics, ground resistance, insulator loss, etc.

**Field Strength**

Sometimes it is necessary to calculate the field strength at a point where an antenna is pointed at a certain distance. This is the case for determining harmful amounts of RF radiation, where for instance a radar system with power in the many kilowatts or even megawatts needs to have a safety zone and/or safety switches. This calculation is also important for determining safe limits for cell phones, which are usually in close proximity to the human head. The FCC, OSHA, ANSI and many foreign countries have maximum allowable field strength values for human safety. The calculation that follows assumes the radio waves propagate through the air from the transmitter to the area where the fields need to be known.

$$E = 1 * 10^6 * \sqrt{\frac{(377 * P)}{(4 * \pi * (D^2))}} \qquad [24]$$

where:

E = microvolts per meter

P = Power in watts, in this case it is the power of the transmitter times the gain of the antenna (ratio)

D = distance in meters

To calculate the power of the transmitter/antenna combination from the measured field strength use:

$$P = \left( \frac{4 * \pi * (D^2)}{377} \right) * \left( \frac{E}{(1 * 10^6)} \right)^2 \qquad [25]$$

To convert the power to decibels relative to milliwatt (dBm), take the log base 10 of the above equation and add 30.

simplified this becomes:

$$P(dBm) = 20 * \log(D * E) - 104.7713 \qquad [26]$$

If the distance is 3 meters as in the case of many US and foreign standards, the above equation becomes:

$$P(dBm)=20*\log(E)-95.2289 \quad [27]$$

For some commonly encountered field strengths, all set at 3 meter distance, use the following values:

| Field Strength | Power |
|---|---|
| 200µV/m | -49.20 dBm |
| 500µV/m | -41.25 dBm |
| 1250µV/m | -33.29 dBm |
| 12500µV/m | -13.29 dBm |
| 50 mV/m | -1.25 dBm |

**Polarization**

One attribute of antennas is the polarization of the wavefront emanating (assuming a transmit antenna) from the aperture. This polarization has useful properties and should be understood during the design of an antenna. For instance, a dipole sitting horizontal relative to the Earth's surface is considered a linearly polarized antenna where radiation is emitted or received best in the horizontal plane. Conversely a car antenna, made up of a wire placed vertically relative to the horizon is considered vertically polarized.

In a communications link, the antennas should always be polarized the same way for best performance. There are nuances to this rule which come into play if one or both antennas are moving relative to the other in terms of angle relative to the ground. In this case polarizations such as circular can minimize the effects of crossing linear polarization against each other. The effect of crossing polarizations is known as cross pole isolation and when the antennas are linearly polarized and are are 90 degrees relative to each other, or are circularly polarized and are left vs. right hand the attenuation of propagating signals can be very high. Fundamentally, the degree of isolation follows the following formula:

$$Cross\ Pole\ Isolation=\cos(Beam\ Angle) \quad [28]$$

where,

Beam angle is relative to boresite (in this case 0 Degrees) .

In the real world however, the cross pole isolation never goes to 0, or infinite amount of dB. Due to physical structures and imperfect matching, cross pole isolations in excess of 30 dB are considered good.

For Circularly polarized antennas, there is energy emitted by both the vertical and horizontal elements of the antenna, which in turn are delayed by 90 degrees relative to each other. A simple way to visualize this is to think of a standard bolt, sitting on a table, which is considered Right Hand Circular in form, whereby the polarization vector emanating away from an antenna will rotate to the right. The opposite is true for a Left Hand Circular antenna.

If the phase difference is not 90 degrees, then the polarization is considered elliptical.

If the phase difference is 0 degrees with both Horizontal and Vertical elements active, then the polarization is considered Slant Linear.

Just as in the linear systems above, a circular polarization isolation of > 30 dB between Left Hand and Right Hand is considered good.

In many designs, much care is used to isolated the orthogonal (opposite) polarizations to allow separate signals to be used simultaneously. In Radars, the degree of polarization isolation in the reflected beam is sometime measured to reveal important features of the targets of interest.

**Axial Ratio**

In Circularly Polarized Antenna systems the degree of polarization purity is sometimes measured by rotating a transmitting linear antenna in the far field (defined in Appendix 7) while receiving the signal on a circularly polarized antenna. Ideally, if the vertical and horizontal components of the circularly polarized antennas are matched well, there will be very little variation of the amplitude of the received signal.

Axial ratios of one dB or less are considered good, in satellite antenna designs, ratios of a few tenths of a dB are common.

**Integrated Cross Pol Ratio (ICPR)**

This value represents the complete 3D response of a cross pol measurement. Normally this measurement is taken over approximately 30 degrees from the main bore sight antenna beam at all angles radially from the center. For parabolic antennas, this value shows any anomalous cross pol sidelobes that could adversely affect performance. Again, for a parabolic dish, the ICPR will look like four distinct but small lobes at 45, 135, 225 and 315 degrees relative to the principal plane cuts of the antenna.

**Reciprocity**

Antenna Reciprocity is a concept that states "If an antenna transmits efficiently, it will receive efficiently" and vice versa. Generally true, the only caveat is trying to transmit too much power on an antenna designed for reception.

**Gain Bandwidth**

Gain Bandwidth is the measurement of the response of the antenna over frequency; which generally defines wide band vs. narrow band antennas. As the antennas are tested on a range, at boresite (facing each other) the frequency of the transmitter or signal generator is varied over a specific range centered at the design frequency of the antenna under test. The (calibrated) response is examined and the gain bandwidth is determined by finding the place where the signal is 3 dB down above and below the peak gain of the antenna. The difference in the upper and lower frequencies is the gain bandwidth.

Keep in mind that the match of the antenna influences this measurement. Make sure that impedance is optimal over the expected bandwidth of the antenna.

**Velocity of Propagation**

It is sometimes important to know the velocity of propagation for radio waves, in space and in conductors. In space of course it is the speed of light, but in any other intervening medium the velocity slows to a predictable value by evaluating the following formula:

$$Velocity\ of\ Propagation = \frac{1}{(\sqrt{(L*C)})} \qquad [29]$$

where:

L = Inductance/meter

C = Capacitance/meter

Coaxial cables are a prime example of when this velocity needs to be understood. In typical teflon or polyethylene insulated cables the velocity is on the order of 66% of the speed of light. For use in phasing antennas, this quantity must be used to adjust the length of cables for use in (for instance) a low frequency phase shifter design.

In the case of higher frequency designs, a phase shifter might be etched onto a substrate, where care must be taken to design the line lengths properly. In waveguides, the velocity of propagation increases above the cutoff frequency and slows significantly below it.

**Why 50 Ohms?**

Although there are many anecdotal stories about the origin of the 50 ohm standard impedance, there are actual electromagnetic reasons for this choice. Fundamentally its a compromise between low loss in a conductor and power handling capability.

The following graph illustrates this relationship. Today, 50 ohms is the standard impedance for most test equipment and most antennas. 75 ohms is a standard for the delivery of TV programming via cable and is second in popularity. 300 ohms and 600 ohms are older standards that are rarely used today.

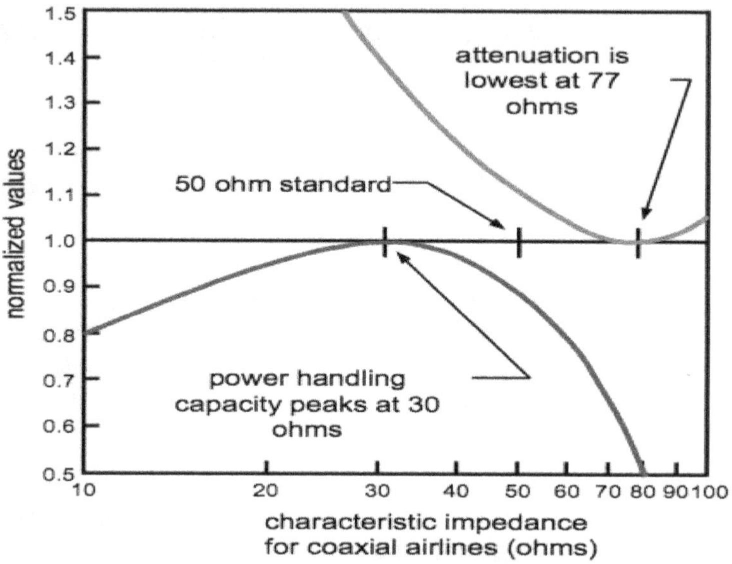


Relationship of power to attenuation in impedance

**Nomographs**

Located in Appendix 8 are a series of graphs and nomographs that can help the designer make the first approximations to certain antenna attributes. Included are:

1. Frequency vs. Wavelength
2. Gain vs. Frequency
3. Gain vs. Aperture size
4. Impedance
    a. Transmission losses due to mismatch
    b. VSWR vs. Return Loss
5. Power vs. DBm

6. Noise Figure vs. Noise Temperature

7. The Smith Chart

    a. Admittance
    b. Impedance

8. Waveguide dimensions

9. Coax vs. Attenuation

10. Distance vs. Attenuation vs. Frequency in:

    a. Km
    b. Miles
    c. Feet

11. Far Field vs. Near Field by:

    a. Frequency
    b. Aperture

12. Rotating Linear source vs. CP

**Visualizing Radio Waves**

Its important to have a fundamental understanding of how the radio waves are created and propagate from the radiating apertures. Modern simulation software can show the progression of currents and charges across an antenna in a movie like fashion and from this the designer can observe any discontinuities or unforseen issues with the antenna. It also helps the designer to see how the charges will form into an antenna beam in the far field.

Below is an example of the E and H fields (Electric and Magnetic) and how they are created by a simple dipole.

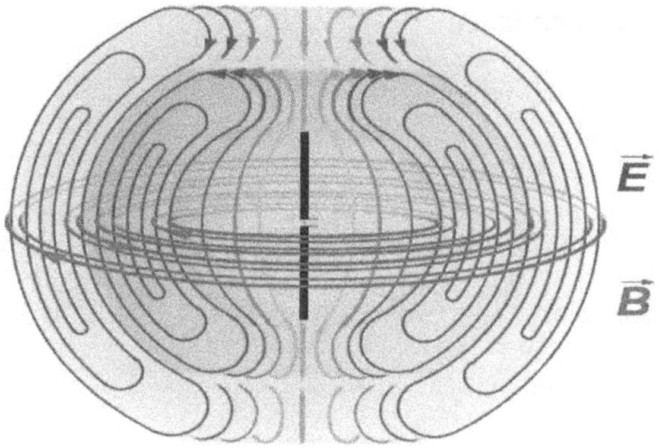


And,

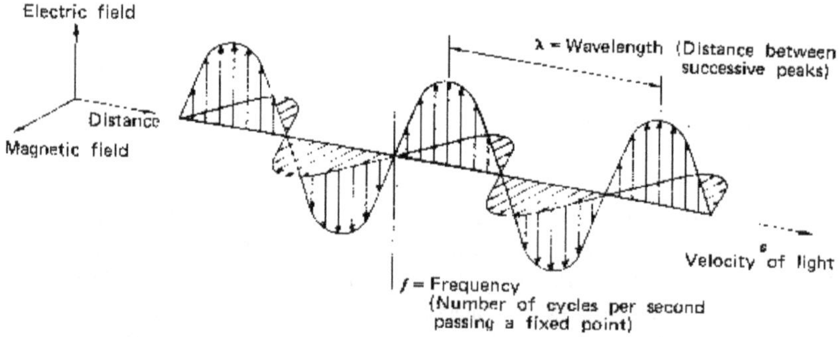


The next illustration show the progression of polarities as the oscillating wave departs the dipole.

**Next Level**

Understanding the fundamental design equations is important to get a feel for the expected operation of an antenna. This, followed by a judicious consideration of what type of antenna will best fit the requirements is also important. Finally, it is prudent to use antenna simulation software to best predict the dimensions and performance of the design. There are several very good software tools available today, refer to Appendix 10 for more details. Many of these tools are free and using them will save many, many hours of optimization, once the antenna prototype is build.

The very last step in antenna design is optimization, which takes the measured response of the design and compares it to the expected results. Slight variations in even the best designs occur, if the designer wants to remove the variations, then a "tweak" might be in order. For instance, if the optimum response of a microstrip patch is say, 1% low in frequency, then consider a design that is 1% higher to compensate for effects that the software is not capable of evaluating. These effects are typically due to parasitic inductances and capacitances in the environment that the antenna is design for. In other words, if a prototype antenna is made and measured by itself, its response will be different in the application if it is near other metallic objects or close the ground etc.

Optimization can sometimes be iterative, depending on the complexity of the environment. Take for instance an antenna on a satellite, where the proximity of solar arrays, the satellite bus and protruding cables, can cause a detuning of the expected response of the antenna. Several optimization cycles might need to be performed to realize expected results.

**Simulation Techniques**

During the course of antenna development, and after the designer has established the necessary gain, beamwidth etc., the next step should be to model the antenna. There are many methods available and selecting the most suitable algorithm depends on what antenna the designer has chosen to pursue. Typically, if one needs to examine the operation of wires, microstrips, monopoles, yagis, or log periodics, the moment method is used. If however, one needs to examine dish type reflectors, horns, or lenses the method used is typically GTD (Geometric Theory of Diffraction) or optical equivalents.

To begin this journey, it is important to understand the way these programs break down the complex problem of analyzing antenna structures. Unlike the previous chapter's technique of deriving answers based on the size of the total structure, these methods will examine and (typically) integrate very small portions of an antenna. In this way, the influence of each anomaly, corner or non-antenna structure can be assessed. Additionally, this technique allows for the exploration of arbitrary structures as radiating elements. This is (obviously) very helpful when there are physical constraints that cause complex radiating problems, e.g., space craft structures.

**Antenna Equivalent Circuit**

It has been shown that an antenna can be represented by its equivalent impedance Za where:

$$Za = (Rr + Rl) + jXa \qquad [30]$$

where:

Rr = Radiation Resistance of the antenna

Rl = Loss resistance of the antenna

jXa = Antenna Reactance

It turns out that the sum Rr + Rl is equal to the antenna resistance so this formula can be simplifies somewhat to:

$$Za = Ra + jXa \qquad [29]$$

As you remember in the previous chapter, this value can easily be displayed on a Smith chart. This fact allows the designer to analyze and change if necessary the constituent components of an antenna to optimize operation. In the transmitting as well as the receiving modes, other equivalent formulae have been developed, principally by Thevenin and Norton that allow a further understanding of the operation of a particular antenna.

Another way of looking at an antenna impedance (Za) is to define it as the ratio of the electric to the magnetic fields at the feed point. Yet another way is to define impedance as the ratio of voltage to current across the antenna terminals. Insofar as the designer is interested in knowing the impedance of the antenna, there are many methods to choose from to calculate this value. These methods can be broken into basically three types: (1) boundary-value method, (2) transmission-line method, and (3) Poynting vector method.

The boundary-value method is the most basic, and it treats the antenna as a boundary value problem. The solution to this is obtained by enforcing the boundary conditions (usually that the tangential electric field components vanish at the conducting surface). In turn the current distribution and finally the impedance (ratio of applied emf to current) are determined, with no assumptions as to their distribution, as solutions to problem. The principal disadvantage of this method is that it has limited applications. It can only be applied and solved exactly on simplified geometrical shapes where the scalar wave equation is separable.

The transmission-line method, which has been used extensively by Schelkunoff, treats the antenna as a transmission line, and it is most convenient for the biconical antenna. Since it utilizes tangential electric field boundary conditions for its solution, this technique may also be classified as a boundary-value method.

The basic approach to the Poynting vector method is to integrate the Poynting vector (power density) over a closed surface. The closed surface chosen is usually either a sphere of a very large radius r (r >= 2 D/\ 2 / lambda where D is the largest dimension of the antenna) or a surface that coincides with the surface of the antenna. The large sphere closed surface method lends itself to calculations only of the real part of the antenna impedance (radiation resistance). (77)

**Simulating Antenna Patterns**

The antenna pattern from an extended aperture can be simulated by using a method referred to as line source analysis. Consider an linear array of n elements, this can simulate the single plane pattern of a parabolic dish or a flat phase array. The line source is simulated by a series of radiating elements set at a specific distance apart, typically $\lambda/2$. For a phased array this is optimal although rarely practical. For the purpose of evaluating optimal array layouts, the distance between radiating elements can be changed and the pattern evaluated.

Many electromagnetic simulation programs start with the analysis of the near field including element interaction (or mutual coupling) and extrapolating the far field performance from these calculated values. Keep in mind that directivity, a value often calculated by these methods, is not as accurate as realized gain, where the effects of various inefficiencies is taken into account.

Using the following formula for broadside patterns:

$$E=\left(\frac{1}{n}\right)*\left[\frac{\left(\sin\frac{(n*\psi)}{2}\right)}{\left(\sin\left(\frac{\psi}{2}\right)\right)}\right] \qquad [30]$$

where:

E = Field Strength normalized to 1

n = number of elements

$\psi$ = scan angle

For endfire patterns use:

$$E=\sin\left(\frac{\pi}{2n}\right)*\left[\frac{\left(\sin\left(\frac{(n\psi)}{2}\right)\right)}{\left(\sin\left(\frac{\psi}{2}\right)\right)}\right] \qquad [31]$$

This routine is used in the web site programs to evaluate various apertures, it does not compensate for mutual coupling and other incidental effects, but is useful in first order evaluation of antenna characteristics.

Beamwidth, Sidelobe position and Sidelobe Magnitude can be estimated by evaluating the above formulas.

**Simulating Gain**

For aperture type antennas like parabolas, phased arrays and other distributed types, the use of the standard gain formula is very useful. It is also useful to evaluate the effects of efficiency for these types of antennas as well. For gain estimates use:

$$Gain(dBi) = 10 * \log\left(\frac{4 * \pi * Ae}{\lambda^2}\right) \qquad [32]$$

Where Ae is the aperture area, e.g. $Ae = \pi R^2$ for circular dish.

**Simulating Antenna Match**

The best method to evaluate the performance of RF circuits including antennas is to take advantage of the many software packages available today from the Internet. The ARRL (Amateur Radio Relay League), Ansoft, Eagle, Nittany Sciences and many others offer evaluation software that work very accurately. There are in fact numerous software packages that are freeware, using NEC (Numerical Electromagnetic Code) and other approaches that work nicely as well.

**Moment Method**

The impedance of an antenna can also be found using a numerical technique which is widely referred to as the Method of Moments. This method, which in the late 1960s was extended to include electromagnetic problems, is analytically simple, it is versatile, but it requires large amounts of computation. (77)

Put simply, an integral equation can be transformed into a set of simultaneous linear algebraic equations (or matrix equations) which may then be solved by numerical techniques. Roger Harrington has unified the various procedures into a general moment method now widely used with computers solving electromagnetic field problems.

By segmenting an antenna into small parts, then applying formulae to calculate the electric field of charges and currents we can derive the first coefficients of interest. Now, by using Green's function to calculate the vector potential of the charge from a certain observation point we can derive the second coefficient. By arranging these coefficients in matrix form and solving by using ohms law, we can calculate the impedance, current, and voltage of the antenna (at the point of interest). This method is best accomplished with a computer, the results of which are to be able to visualize the charges on the surface of the antenna. As a result, one can then derive the near field antenna patterns and thus by using the Fourier Transform, can then calculate the far field of the antenna and thus the gain and beam characteristics. This technique is used in a variety of popular antenna modeling programs, like NEC, MiniNec, Ensemble and others. It tends to run slowly on 486 or lower machines due to the significant amount of computations required.

**Geometric Theory of Diffraction**

The Geometric Theory of Diffraction (GTD) is in reality an extension of classical Geometric Optics (GO) which includes direct, reflected, and refracted rays. In addition, it includes formulas on the diffraction mechanism.

The diffracted field, which is determined by a generalization of Fermat's principle is initiated at points on the surface of the object where there is a discontinuity in the incident GO field (incident and reflected shadow boundaries). The phase of the field on a diffracted ray is assumed to be equal to the product of the optical length of the ray (from some reference point) and the phase constant of the medium. Appropriate phase jumps must be added as a ray passes through a caustic (a point or a line through which all the rays of the wave pass. Examples of it at the point of a paraboloid and the focal line of a parabolic cylinder. The field at the caustic is infinite because, in principle, an infinite number of rays pass through it). The amplitude is assumed to vary in accordance with the principle of conservation of energy in a narrow tube of rays. The initial value of the field on a diffracted ray is determined from the incident field with the aid of an appropriate diffraction coefficient (which, in general, is a dyadic for electromagnetic fields). (77)

Some of the advantages of GTD are:

1. It is simple to use.

2. It can be used to solve complicated problems that do not have exact solutions.

3. It provides physical insight into the radiation and scattering mechanisms from the various parts of the structure.

4. It yields accurate results which compare extremely well with experiments and other methods.

Although a comprehensive discussion of this approach can be found elsewhere, is should be obvious that this method is very useful for large structures as well as monopoles on ground planes, etc. It also should be noted that the computational time is less and therefore more applicable.

Several methods of analysis have briefly been presented here as a way to show the reader that methods exist to analyzer antenna performance before hardware is actually built. Certainly with the moment method, using modem PCs the computational time is significant. In addition, if you include the time to enter the geometries, calculate the fields, and display the results, relatively simple antenna structures can take a lot of time to analyze. Using GTD methods, the calculation part is somewhat faster however it is important to understand that with either method a designer can spend a substantial amount of time analyzing proposed structures. We need to balance this fact with many designers can approximate the structures with simpler formulas, make a quick sample, and adjust the design as necessary to meet the required specifications. The best designers can make good decisions on how much analysis time should be done before "cutting metal". The ultimate goal, obviously is to proceed from specifications to working antenna in the shortest amount of time.

Using the moment method for instance, the designer can quickly analyze the performance of a single microstrip patch. If however, the designer starts adding more patches to form an array, the time to compute the performance can rise significantly. The designer should start with the patch or small array characteristics and then revert to simpler formulae that analyzes array performance, doing this will allow design time to be cut down considerably. This assumes of course that the element or sub array is repeated many times and does not change.

**Other approaches**

BOR - Body of Rotation. This technique takes advantage of the symmetry around the antennas beam to make computations more efficient. This is used on dish antenna analysis primarily.

FDTD – Finite Difference Time Domain. This technique is known as a full wave analysis approach and can be used to predict performance in microstrip patches and other complex shaped antenna apertures.

NEC – Numerical Electromagnetic Code. This code is the basis of several analysis programs and approaches including Mininec. It is based on the Moment Method and uses an iterative approach to analyzing the charge and interaction of small descrite components of an antenna. Much like Finite Element Analysis approaches, it divides a radiating aperture into small components, analyses their performance adds the effects of their neighbors and computes the total near field power distribution which can then be extrapolated into the far field.

**Examples of Antenna Design Software**

The following is a list of numerical codes currently being used for antenna design and analysis. Codes like these enjoy constant updating and support. They are the basis for designs used in aerospace, nuclear physics and other state of the art applications.

HFSS – High Frequency Structure Simulator from Ansoft
CST – Computer Simulation Technology from CST
Sonnet – from Agilent (includes ADS and EMPro)
EM Works – from Microwave Office
Microworks – from Spectra
EagleWorks – from Gensys
Mspice
IE3D
Pulse EM
Polar

These codes, when properly used can create near perfect simulation of antennas and antenna systems. This includes the effects of nearby conductive or absorptive objects. In a satellite environment for instance, this is very useful.

### Microwave Design is about Vector Addition

Specifically, the design of waveguide components and microwave antenna components must be seen in the light of vector mathematics. It is imperative to consider both amplitude and phase when designing such devices. The placement of junctions, irises, tuning stubs and other waveguide components are not arbitrary but considered in terms of phase position along a propagating pathway.

Keeping this rule in mind allows the designer to best match different components in a complex systems which optimized its performance. An example is a combination of two devices where the first has a relative phase relationship of 0 degrees and the second device has a relative phase relation ship of 180 degrees, in this case a poor impedance match will be the result, even if the individual components appeared good in their separate impedance measurements.

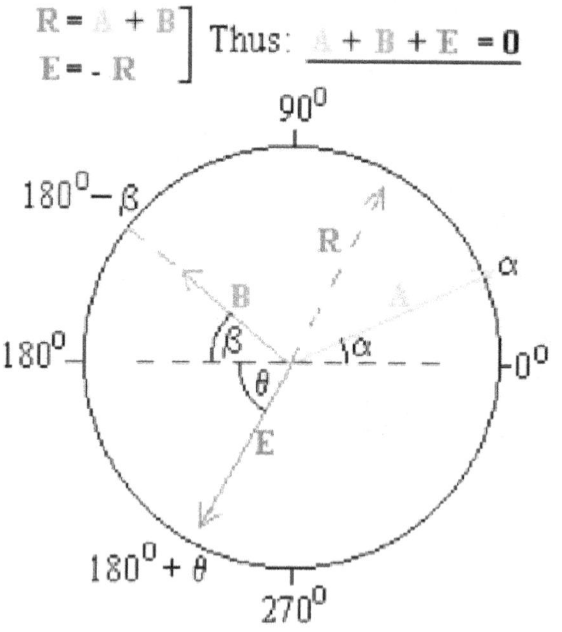


Vector Addition visualized, showing resultant [R]

**Questions:**

1. Calculate the amount of antenna gain needed for a cell phone assuming:
   a. Transmitter Power of 40 watts
   b. Transmitter Antenna Gain of 15 dBi
   c. Processing Gain of 12 dB
   d. Range of 4 Miles
   e. Cell Phone Receiver Sensitivity of -105 dBm
   f. Cell Phone Transmitter Gain of 0.7 Watts
   g. Cell Tower Receiver Sensitivity of -110 dBm

2. Calculate the Loss of power with a 2:1 Mismatch

3. Calculate the gain and beamwidth of an antenna with the following characteristics:
   a. 100 Meters in Diameter
   b. Frequency of 1 GHz

4. Calculate the gain and beamwidth of an antenna with the following characteristics:
   a. 0.1 Meter in Diameter
   b. Frequency of 100 GHz

5. Calculate the beamwidth of an antenna with a diameter of 2,500 Miles. Assume a frequency of 1,420 MHz.

6. Locate a free antenna design software package on the Internet and design a dipole antenna with the following characteristics:
   a. Frequency of 30 MHz
   b. 10 Meters above ground
   c. 75 ohms impedance

   Provide Plots of Near Field Patterns, Far Field Pattern and VSWR.

7. Assuming a safety limit of 1 μV/m, calculate the field strength of a 0.7 watt transmitter at 2 cm.

8. What is the path loss to the star Vega?

# Chapter 3 – Antenna Selection

## Selection Criteria

One of the most interesting and challenging aspects of antenna design is choosing what style of antenna is required for a particular application. Generally, it is helpful to answer a few fundamental questions as the process is started:

1. Is this a high gain (say > 10 dBi) or low gain antenna?
2. Does it have to be flat or conformal?
3. Will it be used for transmitting?
4. Is weight an issue?

5. Is cost an issue?

6. What kind of environment will it be operating in?

7. What is the frequency of usage?

8. What kind of bandwidth is required?

9. What kind of polarization is required?

10. What kind of size constraints are there?

11. Any other unusual considerations (stealth, placement, sidelobes, patterns, etc.)?

By answering the above questions the engineer has a good start on the design process. For instance if the answer to the first questions is "greater than 10 dBi," this narrows the antenna selection process down to an array, large dish, or large yagi, etc. In other words, high gain would preclude the use of most single element antennas. In addition, answering questions like those above serve several purposes. For instance, due to the interrelationship of gain and beamwidth, answering a question on gain necessarily effects selection of beamwidth and sidelobe performance. So a matrix can be assembled to give the designer some ideas as to what kind of antenna can be selected for a specific purpose.

**Design Matrix**

By using the variables presented above, a selection graph can be made showing the relative merits of several popular antennas. By analyzing the deciding factors for placement of antenna in the tables, other antennas not presented here can be considered.

Keep in mind when using these graphs and tables, that they are based on generalized performance criteria of the antennas presented. Special conditions always exist that will allow some of these antenna to exceed the expectations presented.

The following tables will also help the designer to decide what kind of antenna type will fit the required specifications:

**Gain:**

| Antenna Type < 10 dBi | Approximate Gain (dBi or dBc) |
|---|---|
| Monopole | 0 to 3 |
| Wire | 0 to 3 |
| Dipole | 2.14 |
| Slot | 2 |
| "F" | 4 |
| Patch | 0 to 9 |
| Small Yagi (< 4 Elements) | 3 to 10 |
| Small Dish (< 10 Wavelengths across) | 5 to 10 |
| Small Phased Array (< 4 Elements) | 3 to 10 |
| Spiral | 0 to 9 |
| Small Lens (< 10 Wavelengths across) | 3 to 10 |

| Antenna Type > 10 dBi | Approximate Gain (dBi) |
|---|---|
| Backfire | 18 |
| Yagi (> 4 Elements) | 11 to 20 |
| Yagi Array | 11 to 30 |
| Microstrip Array | 10 to 33 |
| Dipole Array | 10 to 33 |
| Parabolic Dish (> 10 Wavelengths across) | 10 to 50 |

**Conformal vs. Non-Conformal**

| Conformal | Non-Conformal |
| --- | --- |
| Wire | Parabolic Dish |
| Microstrip | Lens |
| Dipole Array | Horn |
| Slot Array | Corner Reflector |
| Single Slot | Backfire |
| Cavity | Yagi |
| Spiral | Helix |

**Power Considerations:**

**Low Power:**

| Antenna Type: | Approximate Maximum Power |
| --- | --- |
| Dipole | 2 KW |
| Slot | 1 KW |
| Yagi | 2 KW |
| Patch | 1 KW |
| Monopole (Wire) | 1 KW |
| Wire | 2 KW |
| Spiral | 1 KW |
| Backfire | 1 KW |
| Log Periodic | 1 KW |

**High Power:**

| Antenna Type | Approximate Maximum Power |
|---|---|
| Waveguide Array | 1 MW |
| Dish | 2 MW |
| Monopole (Tower) | 500 KW |
| Dipole Array | 1 MW |

**Weight Issues:**

| Low | Medium | High |
|---|---|---|
| Wire | Yagi | Dish |
| Monopole (Wire) | Log Periodic | Lens |
| Microstrip | Backfire | Phased Array |
| Spiral | Helix | |
| Slot | | |

**Cost Issues:**

| Inexpensive | Mid-Priced | Expensive |
|---|---|---|
| Wire | Patch | Large Dish |
| Slot | Yagi | Phased Arrays |
| Loop | Log Periodic | Lens |
| "F" | Backfire | |
| Monopole | Spiral | |
| | Helix | |

**Environmental Concerns:**

| Harsh (e.g., Aerospace) | Non-Harsh |
|---|---|
| Monopole | Wire |
| Slot | Loop |
| Microstrip Phased Array | Lens |
| Patch | "F", without radome |
| Dish | Yagi |
| Horn | Log-Periodic |

**Frequency of Usage:**

| 1 Khz – 1 MHz | 1 MHz – 100 MHz | 100 MHz – 1 GHz | 1 GHz – 100 GHz |
|---|---|---|---|
| Wire | Wire | Yagi | Dish |
| Loaded Tower | Yagi | Log Periodic | Lens |
| Loaded Monopole | Log Periodic | Monopole | Patch |
| Ferrite Loop | Loaded Monopole | Patch | Slot |
| | | Slot | |
| | | Backfire | |
| | | Helix | |
| | | Dish | |

**Bandwidth Issues:**

| Narrow Band | Wide Band |
|---|---|
| Lens | Log Periodic |
| Yagi | Dish |
| Patch | Vivaldi |
| Slot | Horn |
| Loop | |
| Wire | |

**Polarization Issues:**

| Linear Polarization | Circular Polarization |
|---|---|
| Wire | Crossed Yagis |
| Monopole | Crossed Log Periodic |
| Patch (Asymmetric) | Helix |
| Yagi | Backfire (dual feed) |
| Log Periodic | Patch (Symmetric) |
| Vivaldi | Spiral |
| Horn (rectangular waveguide) | Horn (square waveguide) |
| Backfire (single feed) | Dual Port Patch |

**Size Constraint Issues:**

| Large Antennas | Small Antennas |
|---|---|
| High Gain Dishes | Patch |
| Lens | Slot |
| Log Periodic | Spiral |
| Yagi | |
| Wire | |
| Helix | |

**Unusual Considerations:**

| Stealth | Low Sidelobe | Custom Pattern |
|---|---|---|
| Patch | Backfire | Adaptive Array |
| Slot | Amplitude Tapered Array | AESA |
| Microstrip Phased Array | Dish (> 10 lambda) | Interferometer |
| Plasma | Lens | |

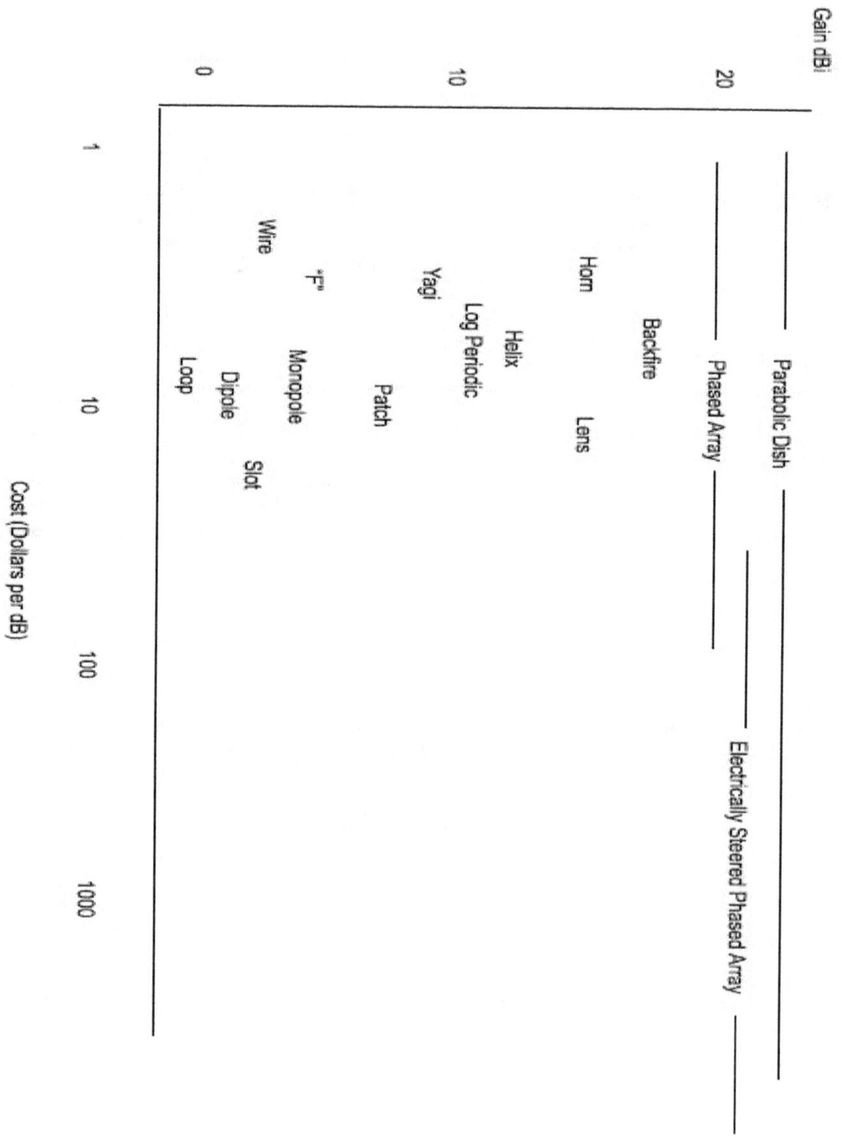


Cost vs. Gain Selection Chart for Common Antennas

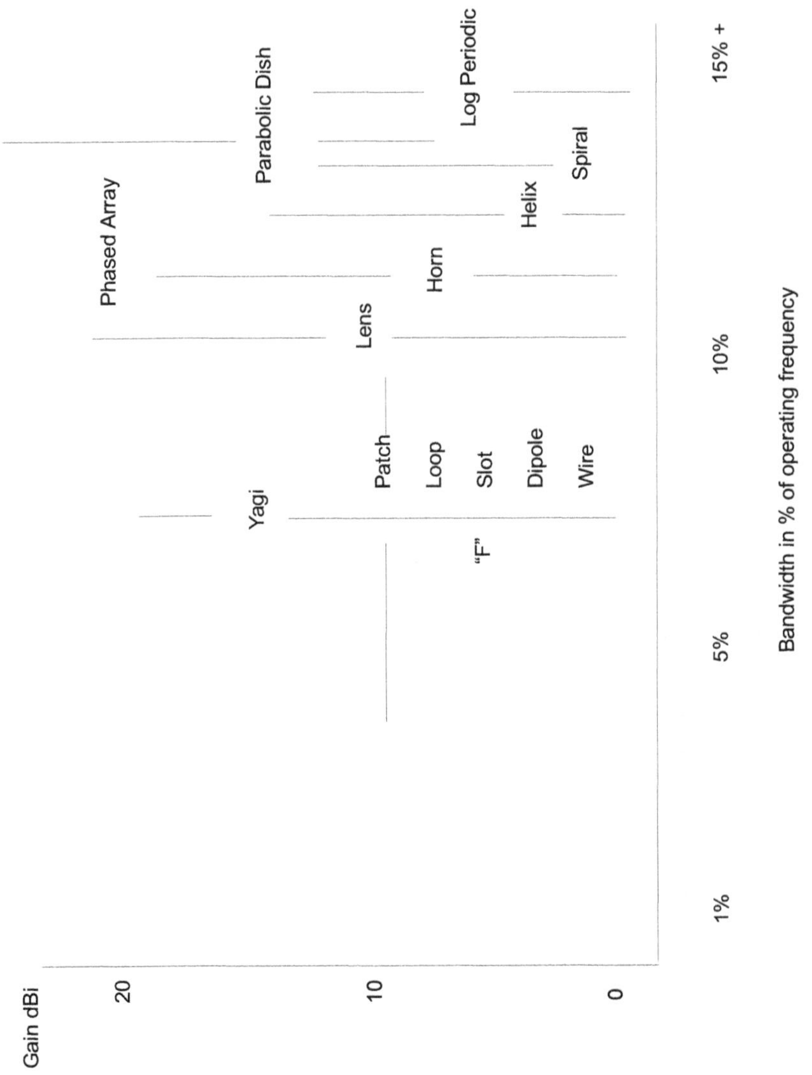


Bandwidth vs. Gain Selection Chart for Common Antennas

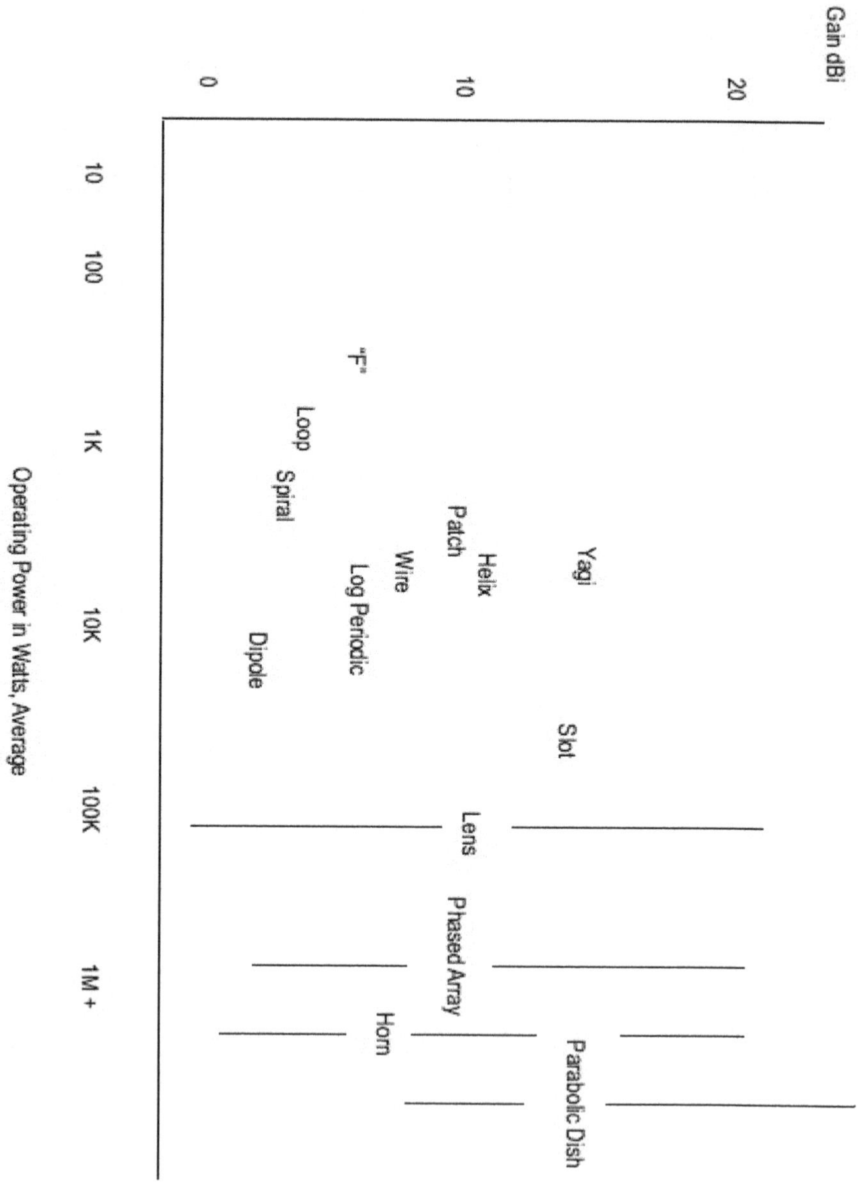


Average Power vs. Gain for Common Antennas

**Questions:**

1. Suggest an antenna type that has the following characteristics:

   a. Frequency of 200 to 400 MHz
   b. Gain of 15 dBi
   c. Linear Polarization

2. Your boss tells you price is a huge factor in the design of antenna with the following characteristics:

   a. Frequency of 20 MHz
   b. Gain of 20 dBi
   c. Linear Polarization

3. Suggest an antenna design for the next generation rocket booster that is capable of going to the moon, with the following properties:

   a. Frequency of 2200 MHz
   b. Gain of 3 dBi
   c. Circular Polarization

4. For a stealth application, is there a way to create an antenna that appears and disappears from a fields point of view (but not by mechanical means) ?

5. What options could you consider if you need to have sub arc-second beam resolution at 100 MHz?

6. If you have an antenna that is tuned very finely, in other words, very high Q, what types of environment would effect it's performance ? How ?

7. What type of antenna would you suggest that could withstand hurricane like winds ? For instance, made for use on an aircraft.

# Chapter 4 – Wire Antennas

Titanic, 1912, note long wire antenna between masts, used to sent Morse code distress signal

This style of antenna has been used successfully for long wave and shortwave applications for well over 100 years.

Specific uses include:

1. Ham Radio HF (High Frequency) Operations
2. Ship to shore and ship to ship

3. International weather and news transmissions
4. Local AM broadcasts
5. Transoceanic aircraft operations
6. Transoceanic telephone
7. Ionospheric propagation determinations
8. Thunderstorm activity studies
9. Low frequency radio and radar astronomy
10. Over the horizon radar
11. Sea State studies

There are several types of wire antennas, as depicted below. One of these types can be seen on most any automobile, aircraft or boat.

Specific types and uses include:

| | |
|---|---|
| 1. Long Wire | Short Wave, Ham Radio, AM Radio |
| 2. Dipole | Short Wave, Ham Radio, WWV, VHF |
| 3. V | Broadcast, Military |
| 4. Rhombic | Broadcast, Military |
| 5. Turnstile | Military, VHF |
| 6. Beverage | Low Frequency, Short Wave, Ham Radio |
| 7. Inverted V | Short Wave, Ham Radio |
| 8. Loaded Dipole | Short Wave, Ham Radio |
| 9. Folded Dipole | Short Wave, Ham Radio |
| 10. Collinear Array | Radar, Broadcast, Cell Tower, VHF |
| 11. Double Zepp | Short Wave, Ham Radio |
| 12. Quad Loop | Short Wave, Ham Radio |
| 13. Delta Loop | Short Wave, Ham Radio |
| 14. Doublet | Short Wave, Ham Radio |
| 15. Bowtie | Short Wave, Ham Radio, Television, VHF, UHF |
| 16. Loop | Ham Radio, DF, Military, VHF, UHF, Aircraft |

| | | |
|---|---|---|
| 17. | Log Periodic | Short Wave, Military, Television, VHF, UHF |
| 18. | Curtain | Broadcast, Military |
| 19. | Yagi | Ham Radio, Radio Astronomy, VHF, UHF |
| 20. | Inverted V | Short Wave, Ham Radio, Military |
| 21. | Discone | Surveillance, Military, Aircraft |
| 22. | Bi-Conical | Surveillance, Military, Aircraft |
| 23. | Wire Reflectors | Broadcast, Radio Astronomy, VHF, UHF, Microwave |
| 24. | Monopole with Ground Counterpoise | VHF, UHF |
| 25. | Monopole over Ground Plane | VHF, UHF |
| 26. | 5/8 (Cellular) | Cell, VHF, UHF |

**Theory of Operation**

Wire antennas work by transitioning electromagnetic energy from a cable or other suitable feed system to a radiator element. The wire element then transfers (or transduces) this energy into free space. Simple analysis of the performance of wire antennas can be realized by using formulas that show dependance on physical dimensions. A more rigorous understanding of the operation of a wire antenna can be had by the use of moment methods and NEC (Numerical Electromagnetic Code). Several popular software packages use the latter method and can be found easily on the Internet. Refer to the Bibliography and References section for further details.

In general, multiples of ½ wavelengths suffice, with appropriate matching hardware, good performance can be expected. Lengths over ½ wavelength tend to create complex radiation patterns, sometimes with multiple lobes, which must be taken into consideration. The vast experience radio amateurs have should be taken advantage of by the antenna designer considering using this style of antenna.
Good reference material can be found with the ARRL (Amateur Radio Relay League).

Matching networks optimize the performance of these antennas although for a simple dipole the typical impedance is 72 ohms, a close match to existing 75 ohm cable TV coax and a reasonable match to common 50 ohm cable. Taking the ratio of the impedance of the wire vs. the cable will give the VSWR of the system. Any mismatch will cause surface currents to occur so it is recommended that matching techniques be used to optimize performance.

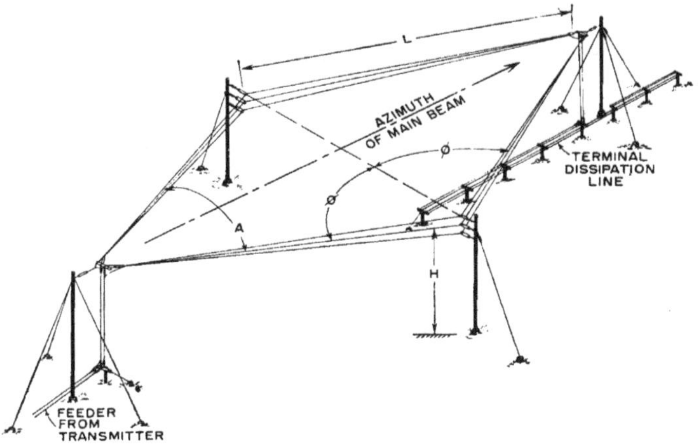


Wire Rhombic Antenna

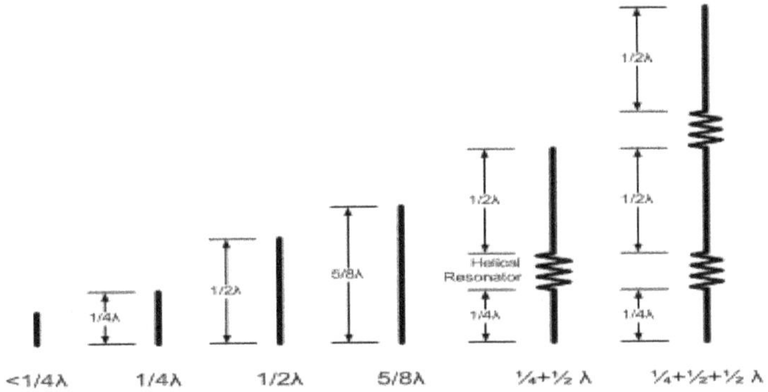


Evolution of the wire whip antenna

**Design Examples**

The following pages contain examples of antennas discussed in this chapter. These examples represent a first order design approach. For a more complete and optimized antenna solution, use the example closest to the requirement needs and optimize by using good simulation software, creating prototypes, testing and finally modifying prototypes to optimize performance.

Type: ***Long Wire***

Frequency of Operation: *1 Khz to 100 MHz*
Gain Range: *2 to 9.5 dBi*
Bandwidth: *~ 5%*
Approximate Power Rating: *10 Kw*
Design Software: *Nec-Win Pro, Antennamax, Moment*
*Method*

*NEC*

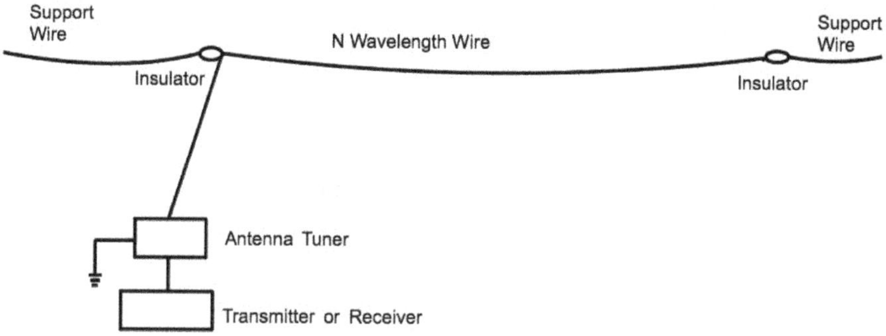


Configuration:

Design Approach:

*To Calculate Length:*

$$L(Meters) = \frac{(300*(N-.025))}{(F(MHz))}$$

*where N = antenna length in wavelengths*

Construction:

*A wire of sufficient strength is supported on insulated stands and insulated ends and fed from one end. The length of the antenna element should be calculated using the above formula. The feed end should be brought into the "radio room" with a good earth ground and be connected to an impedance matching device before connection to the "radio". Lightning protection is recommended.*

*Azimuth Pattern* *Elevation Pattern*

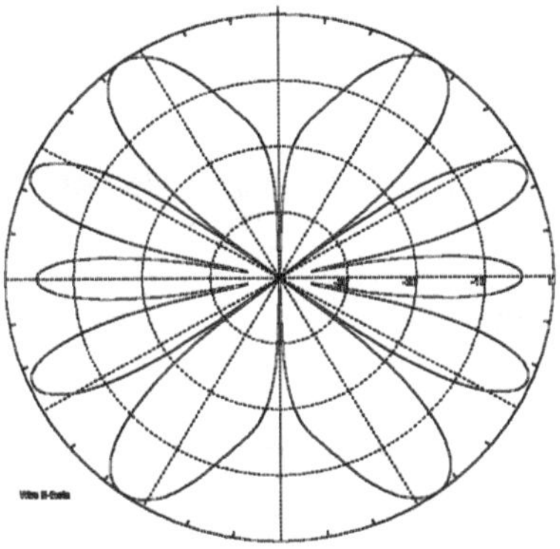

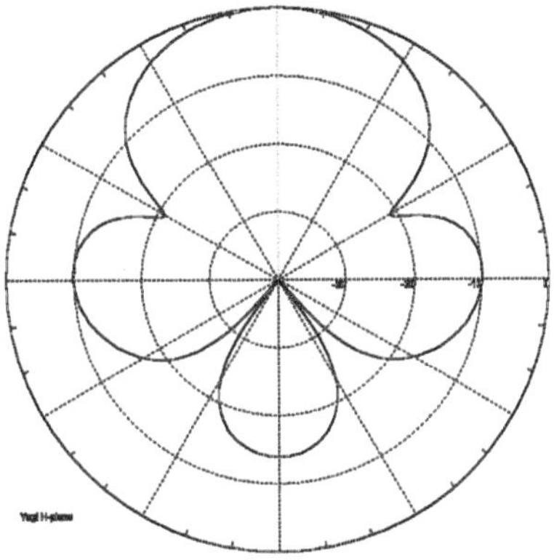

Special Considerations:

*Use Antenna match circuity for best performance.*

*Use lightning protection*

*Height effects, the higher off of the ground, the lower the elevation pattern main lobes, chart shown above is for minimal ground effect.*

Testing:

*Walk around with signal strength meter, plot values on polar graph paper*

*Scaling of antenna to higher frequency can allow more convenient testing in anechoic chamber for instance.*

*Plot signal reports from Ham operators*

*2nd Edition*

| | |
|---|---|
| Type: | *"L"* |
| Frequency of Operation: | *1 Khz to 100 MHz* |
| Gain Range: | *2 to 9.5 dBi* |
| Bandwidth: | *~ 5%* |
| Approximate Power Rating: | *10 Kw* |
| Design Software: | *Nec-Win Pro, Antennamax, Moment Method,* NEC |

Configuration:

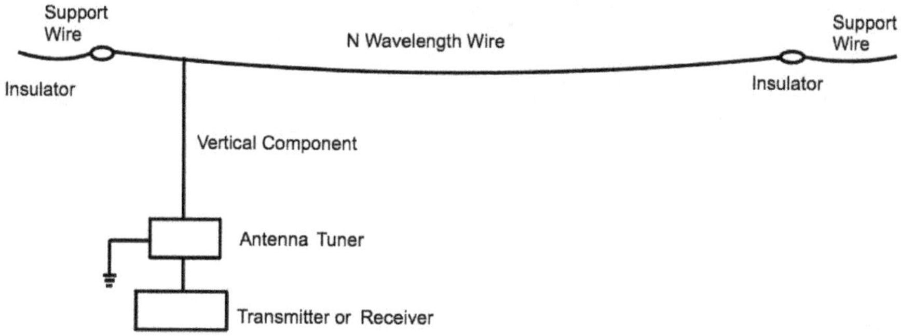


Design Approach:

*To Calculate Length:*

$$L(Meters) = \frac{(300*(N-.025))}{(F(MHz))}$$

*where N = antenna length in wavelengths*

Construction:

*A wire of sufficient strength is supported on insulated stands and insulated ends and fed from one end. The length of the antenna element should be calculated using the above formula. The feed end should be brought into the "radio room" with a good earth ground and be connected to an impedance matching device before connection to the "radio". Lightning protection is recommended.*

*Azimuth Pattern*

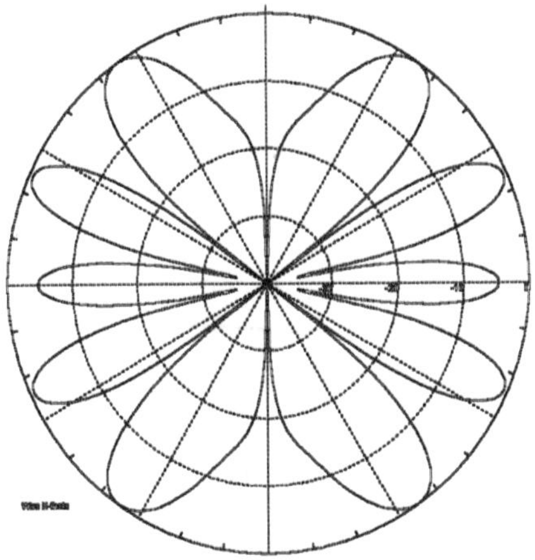

*Elevation Pattern*

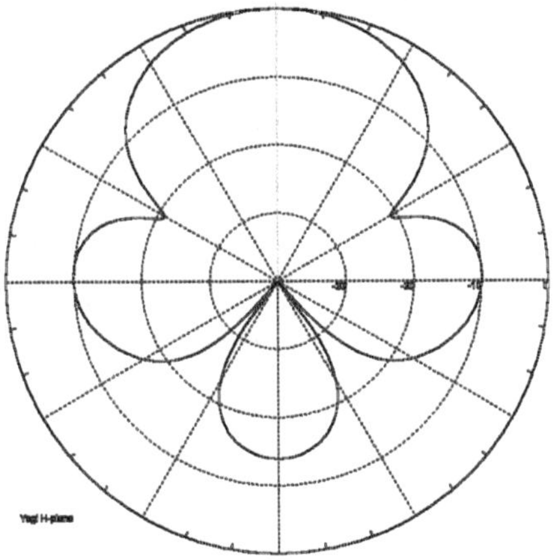

Special Considerations:

*Use Antenna match circuity for best performance.*

*Use lightning protection*

*Height effects, the higher off of the ground, the lower the elevation pattern lobes*

Testing:

*Walk around with signal strength meter, plot values on polar graph paper*

*Scaling of antenna to higher frequency can allow more convenient testing in anechoic chamber for instance.*

*Plot signal reports from Ham operators*

Type: ***Beverage***

Frequency of Operational: *5 Khz to 5 MHz*
Gain Range: *0 to 6 dBi*
Bandwidth: *~ 5%*
Approximate Power Rating: *10 Kw*
Design Software: *Nec-Win Pro, Antennamax, Moment Method*

Configuration:

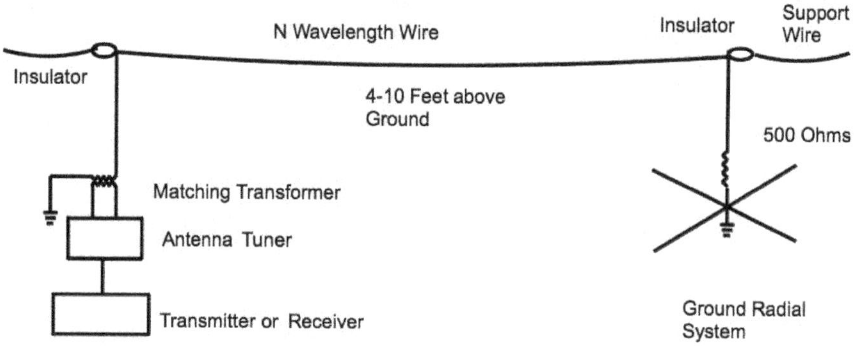


Design Approach:

*To calculate length:*

$$L(Meters\ per\ Side) = \left(\frac{300}{(F(MHz))}\right) * N$$

*where: N = number of wavelengths per side (integer value)*

Construction:

*This antenna, developed by Beverage, Rice, and Kellogg, is unique as a directive antenna for reception of low and very low frequencies. It was the first such antenna to use the traveling-wave principle. In its simplest form, it consists of a single wire transmission line running in the direction of expected wave arrival and terminated in its characteristic impedance. The electric vector of the equiphase front, tilted forward, produces a component of electric force parallel to the wire, inducing a current in the wire. Wave tilt increases with frequency and with ground resistivity.*

*Azimuth Pattern* *Elevation Pattern*

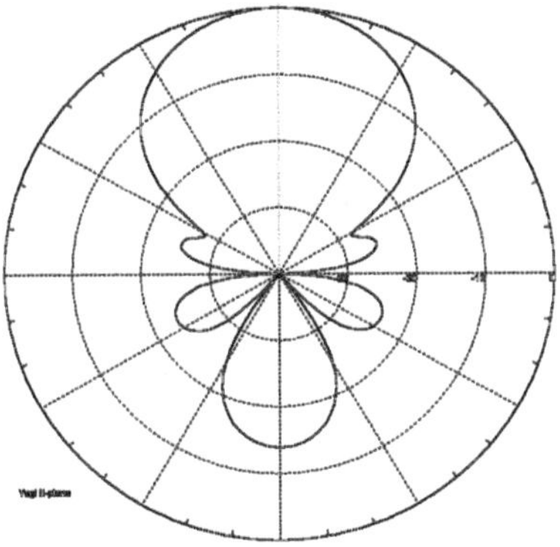

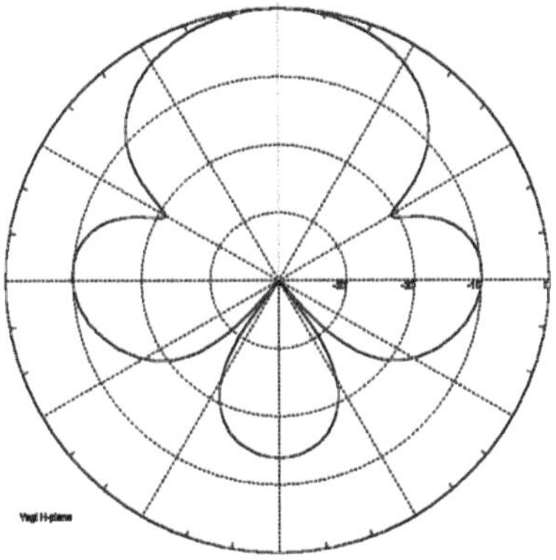

Special Considerations:

*Beverage antennas are directional but inefficient and best used for reception.*

*Length must be at least 1 lambda at lowest frequency.*

*Use Antenna match circuitry for best performance.*

*Use lightning protection*

*Height effects, the closer the antenna is to the ground, the higher the elevation pattern*

Testing:

*Walk around with signal strength meter, plot values on polar graph paper*
*Plot signal reports from Ham operators*

*Scaling of antenna to higher frequency can allow more convenient testing in anechoic chamber for instance.*

Type: ***Dipole***

Frequency of Operational: *5 Mhz to 1 GHz*
Gain Range: *2.12 or 0 dBd*
Bandwidth: *~ 5%*
Approximate Power Rating: *10 Kw*
Design Software: *Nec-Win Pro, Antennamax, Moment Method*

Configuration:

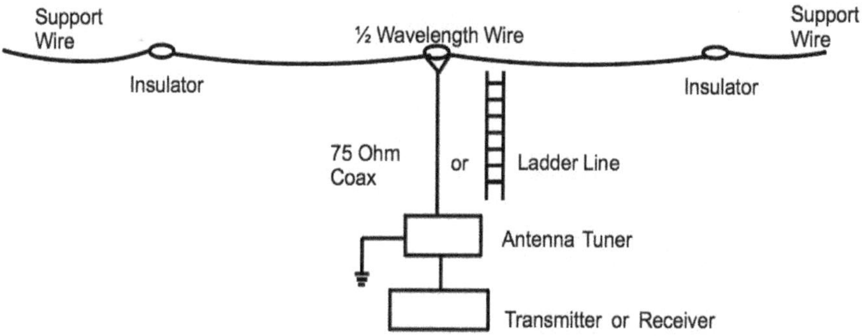


Design Approach:

*To Calculate Length:*

$$L(Meters) = \frac{150}{(F(MHz))}$$

Construction:

*A wire of sufficient strength and L feet long is suspended approximately one half wavelength above ground between two insulated supports. The center of the wire is separated and an insulated feed assembly is inserted. The feed assembly splits the coax or twin line so that the elements of the dipole are each fed by one side of the feed line. The feed impedance match can be achieved by inserting a match system (such as a balun) at the antenna feed assembly. In the case of a coax feed, the match system will have to be designed to accommodate a balanced to unbalanced configuration. In the case of a twin lead feed, one end of the twin lead can be attached to the dipole and the other end to an impedance matching device in the "radio room".*

*Azimuth Pattern*

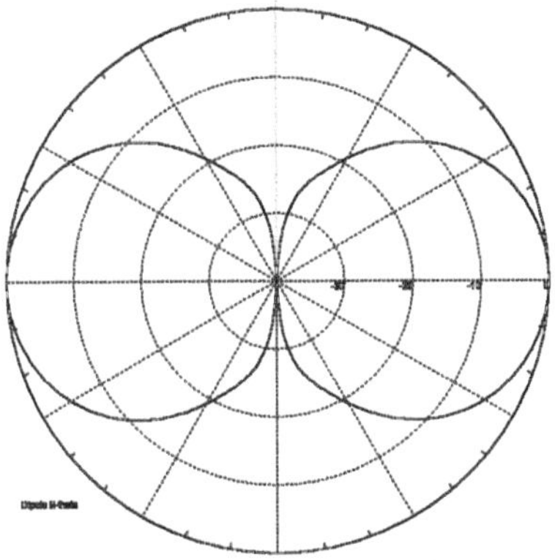

*Elevation Pattern*

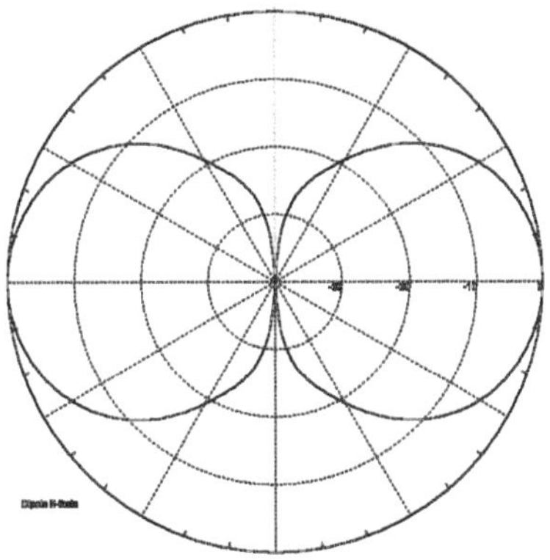

Special Considerations:

*Standard Impedance of Dipole is 72 ohms, close match to 75 ohm cable, use matching transformer, balun or antenna tuner to optimize.*

*Use lightning protection*

*Height effects, the higher off of the ground, the lower the elevation pattern*

*Useful for 2.12 dBi Gain Standard*

Testing:

*Walk around with signal strength meter, plot values on polar graph paper*

*Scaling of antenna to higher frequency can allow more convenient testing in anechoic chamber for instance.*

*Far Field Range with rotator at higher frequencies*

Type: **Bowtie**

Frequency of Operational: *5 MHz to 500 MHz*
Gain Range: *0 to 3 dBi*
Bandwidth: *~ 15%*
Approximate Power Rating: *10 Kw*
Design Software: *Nec-Win Pro, Antennamax, Moment Method*

Configuration:

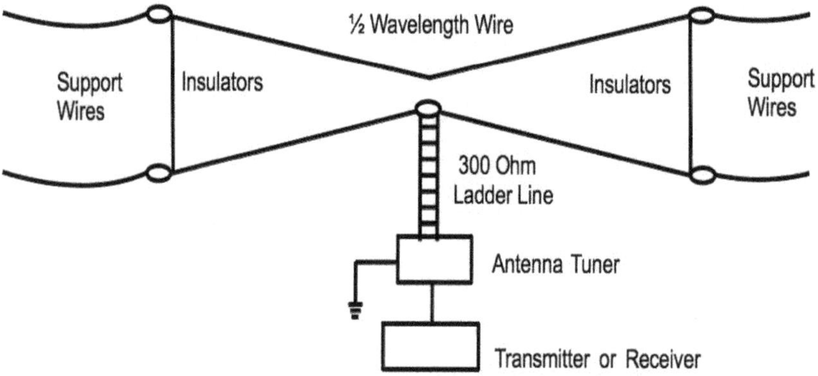


Design Approach:

*To Calculate Length:*

$$L(Meters) = \frac{150}{(F(MHz))}$$

Construction:

*This antenna is essentially a dipole in length, however it has the additional property of having better bandwidth capabilities. It is recommended that a matching transformer be added to the antenna system to optimize performance. Typical impedance for this type is 300 ohms.*

Azimuth Pattern

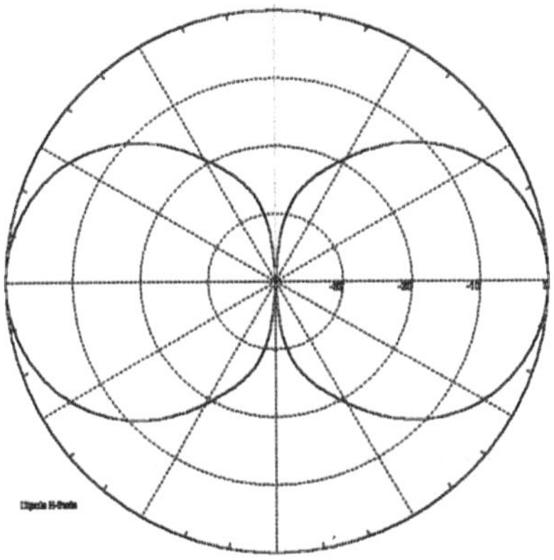

Elevation Pattern

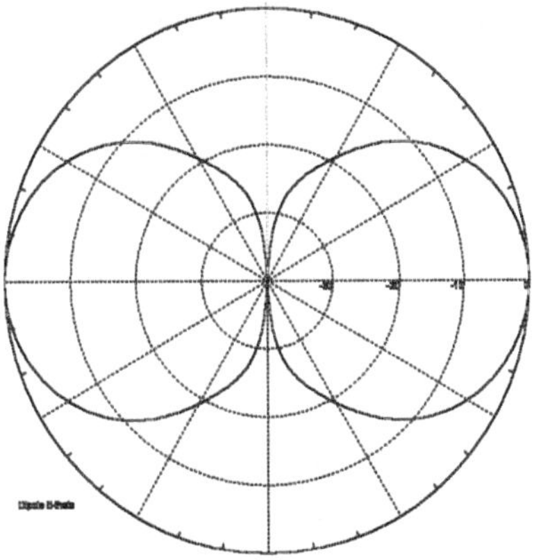

Special Considerations:

*Good wideband antenna with 300 ohm input impedance.*

*Use lightning protection*

*Height effects, the higher off of the ground, the lower the elevation pattern, towards the horizontal.*

Testing:

*Walk around with signal strength meter, plot values on polar graph paper*

*Plot signal reports from Ham operators*

*Scaling of antenna to higher frequency can allow more convenient testing in anechoic chamber for instance.*

*Far Field Range with rotator at higher frequencies*

Type: ***Folded Dipole***

Frequency of Operational: *5 Mhz to 500 MHz*
Gain Range: *2.12 or 0 dBd*
Bandwidth: *~ 10%*
Approximate Power Rating: *10 Kw*
Design Software: *Nec-Win Pro, Antennamax, Moment Method*

Configuration:

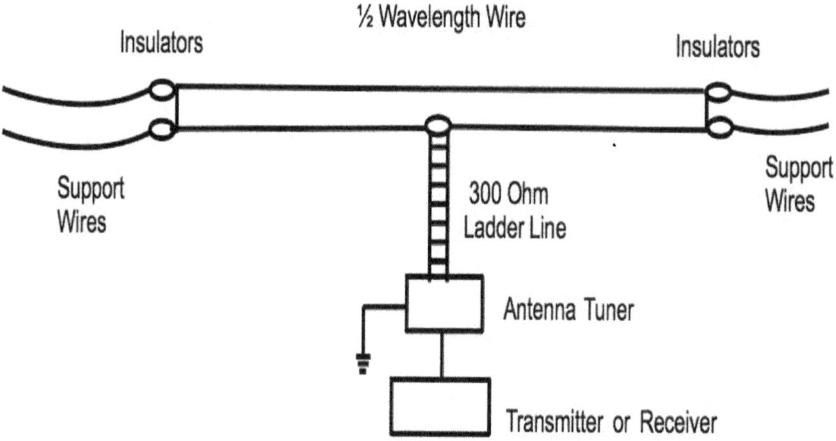


Design Approach:

*To Calculate Length:* $L(Meters) = \frac{150}{(F(MHz))}$

Construction and Theory of Operation:

*Essentially a dipole above another dipole connected at the ends. This antenna is held in place on both ends with insulators and and quy wires. Impedance is 300 ohms, polarization is horizontal with a figure 8 azimuth patterns. This antenna has broader bandwidth characteristics than the dipole.*

| *Azimuth Pattern* | *Elevation Pattern* |
|---|---|
| 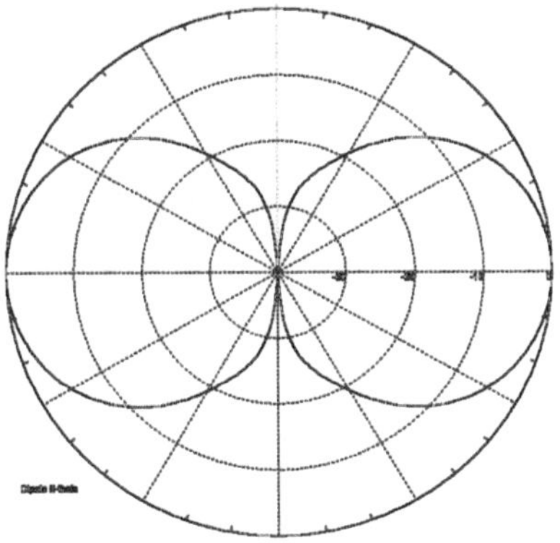 | 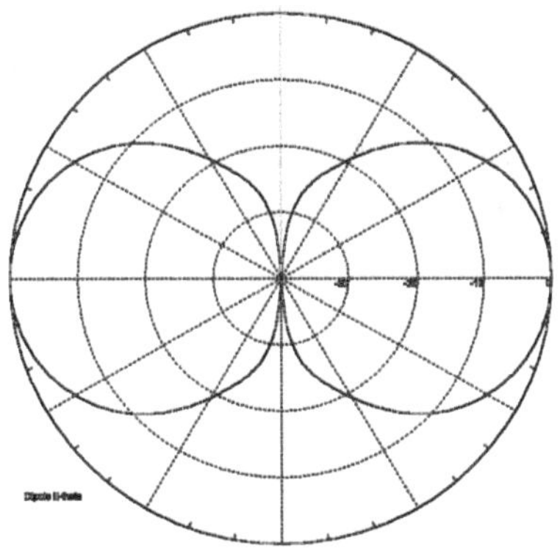 |

Special Considerations:

*Good wideband antenna with 300 ohm input impedance.*

*Use lightning protection*

*Height effects, the closer the antenna is to the ground, the higher the elevation pattern*

Testing:

*Walk around with signal strength meter, plot values on polar graph paper*

*Plot signal reports from Ham operators*

*Scaling of antenna to higher frequency can allow more convenient testing in anechoic chamber for instance.*

*Far Field Range with rotator at higher frequencies*

Type: ***Large Yagi***

Frequency of Operation: *1 Mhz to 100 MHz*
Gain Range: *5 to 15 dBi*
Bandwidth: *~ 10%*
Approximate Power Rating: *1 Kw*
Design Software: *Nec-Win Pro, Antennamax, Moment Method*

Configuration

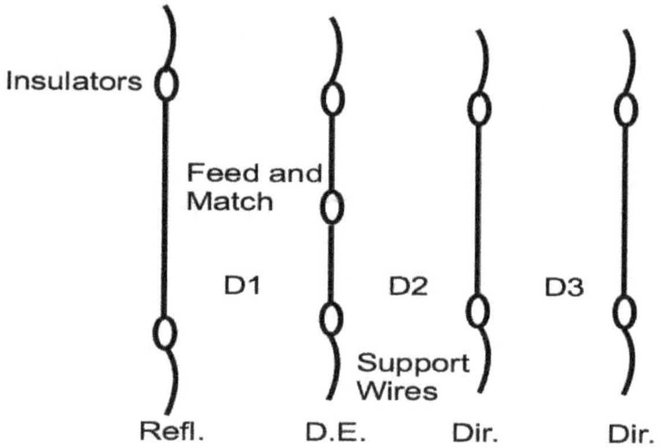


Design Approach:

*To calculate length:*

*Reflector Length = 0.6 lambda*
*Driven Element Length =0 .5 lambda*
*Director Length = 0.4 lambda*
*D1 Distance = 0.23 lambda*
*D2 Distance = 0.24 lambda*
*D3 Distance = 0.24 lambda*

Construction:

*Wires of sufficient strength are supported on insulated stands and insulated ends and fed into the Driven Element with an appropriate match. The length of the antenna elements should be calculated using the above formulas. The feed end should be brought into the "radio room" with a good earth ground and be connected to an impedance matching device before connection to the "radio".*

*Azimuth Pattern* *Elevation Pattern*

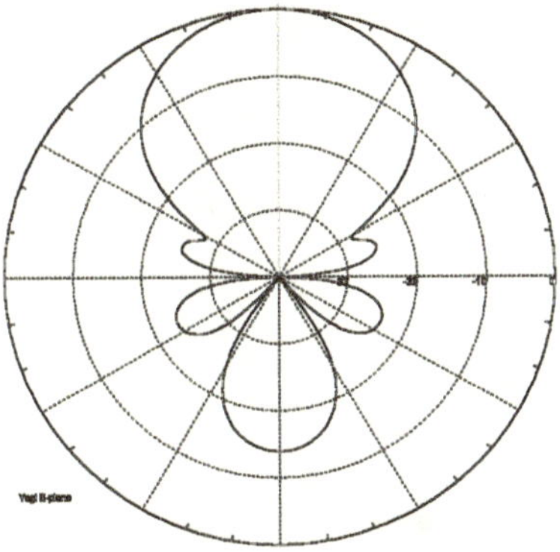

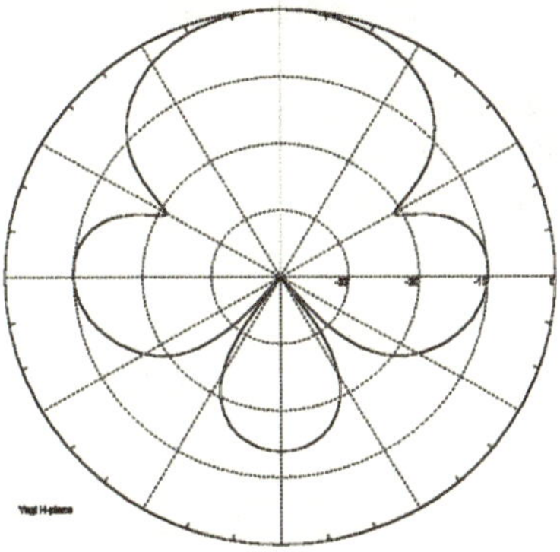

Special Considerations:

*Use Gamma, Delta or Capacitive Match (see Yagi section) to optimize performance.*

*Use Lightning protection.*

*Keep antenna at least ½ wavelength above ground for best results.*

Testing:

*Walk around field strength meter, plot on polar graph paper*

*Ham radio reports*

*Scaling of antenna to higher frequency can allow more convenient testing in anechoic chamber for instance.*

*Far Field Range in case of higher frequency designs*

Type: ***Rhombic***

Frequency of Operational: *5 Mhz to 50 MHz*
Gain Range: *0 to 6 dBi*
Bandwidth: *~ 5%*
Approximate Power Rating: *10 Kw*
Design Software: *Nec-Win Pro, Antennamax, Moment Method*

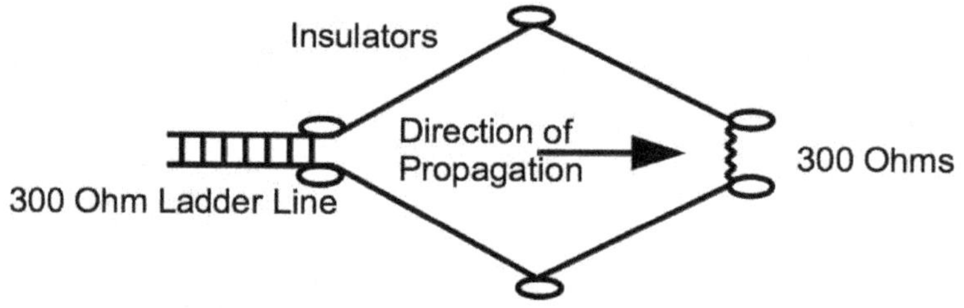


Configuration:

Design Approach:

*To calculate length:*

$$L(meters\ per\ side)=\left(\frac{300}{(F(MHz))}\right)*N$$

*where: N = number of wavelengths per side (integer value)*

Construction:

*This antenna is constructed as an elevated diamond with sides from two to many wavelengths long. It is used for transmission and reception of high-frequency waves propagated via the ionosphere. When terminated with a resistance equal to its characteristic impedance at its forward apex, it functions as a traveling-wave antenna; the termination suppresses reflections of transmitted power and absorbs signals from the contrary direction. To improve the uniformity of distributed line constants and to reduce the rhombic's characteristic impedance to convenient values such as 600 ohms, three wires expand from and to each apex on each side. They are at their greatest spread at the sides.*

*Azimuth Pattern*

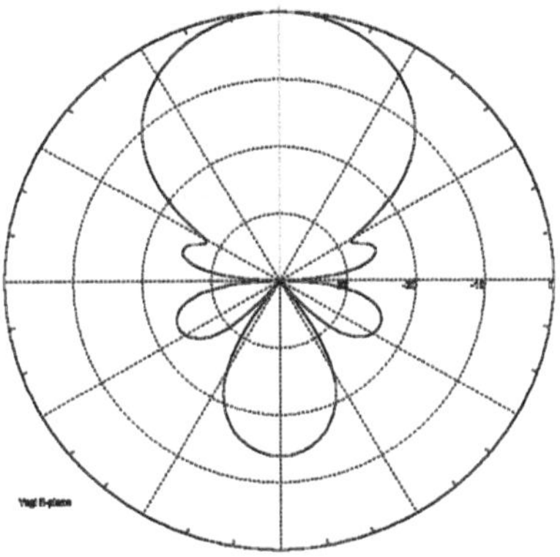

*Elevation Pattern*

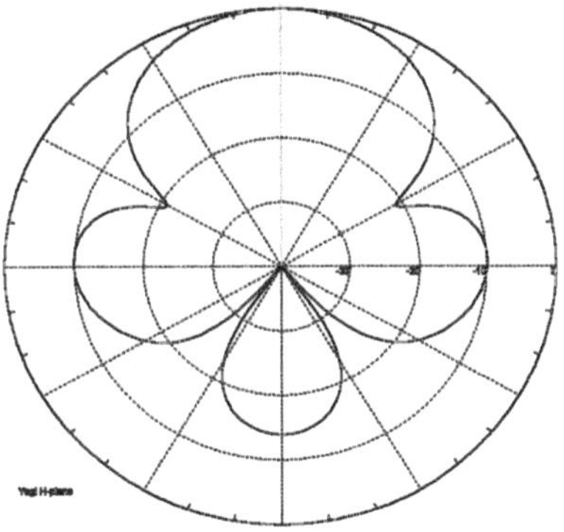

Special Considerations:

*Use lightning protection*

*Height effects, the closer the antenna is to the ground, the higher the elevation pattern*

Testing:

*Walk around signal strength meter*

*Scaling of antenna to higher frequency can allow more convenient testing in anechoic chamber for instance.*

*Plot signal reports from Ham operators*

| | |
|---|---|
| Type: | ***Quad Loop*** |
| Frequency of Operational: | 15 *Mhz to 500 MHz* |
| Gain Range: | *5 dbi* |
| Bandwidth: | *~ 10%* |
| Approximate Power Rating: | 2 *Kw* |
| Design Software: | *Nec-Win Pro, Antennamax, Moment Method* |

Configuration:

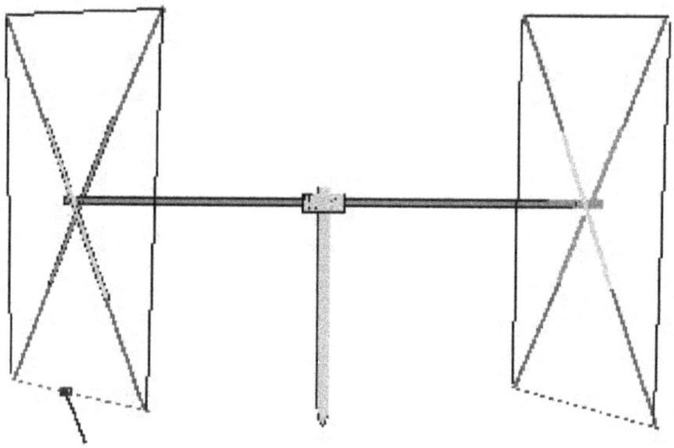

Design Approach:

*To Calculate Length of sides:*

$$L(Meters) = \frac{306}{(F(MHz))}$$

*Separation between squares is also the same as length of sides*

Construction and Theory of Operation:

*This antenna is like a yagi with four sides and exhibits good directional gain as a result. This antenna is held in place on both ends with insulators and crossed poles. Impedance is 72 ohms, polarization is horizontal with a directional azimuth patterns. This antenna can be used with a balun or antenna tuner to increase useful bandwidth.*

*Azimuth Pattern*

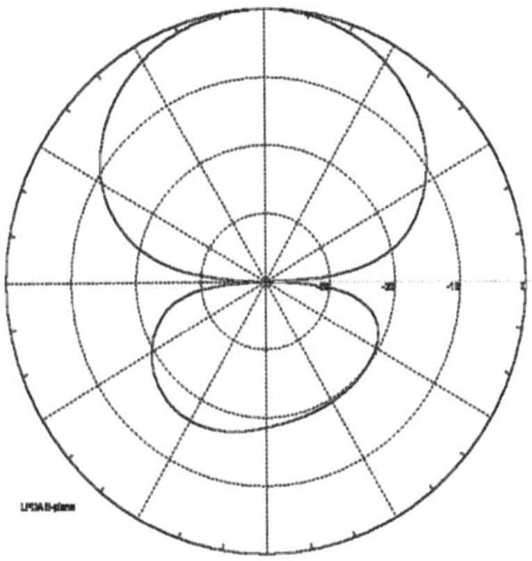

*Elevation Pattern*

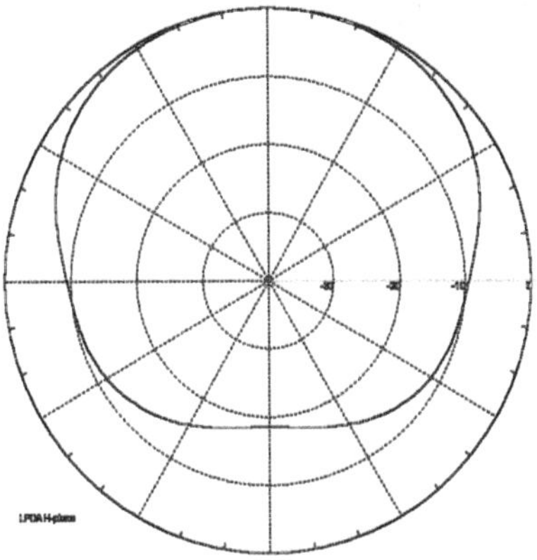

Special Considerations:

*Use lightning protection*

*Height effects, the higher off of the ground, the lower the elevation pattern (shown by the dashed lobe in the elevation plot)*

Testing:

*Walk around signal strength meter*

*Scaling of antenna to higher frequency can allow more convenient testing in anechoic chamber for instance.*

*Plot signal reports from Ham operators*
*Use rotator to measure WWV or other permanent stations*

2nd Edition

Type: **Delta Loop**

Frequency of Operational: *5 Mhz to 500 MHz*
Gain Range: *3.5 dBi*
Bandwidth: *~ 5%*
Approximate Power Rating: *10 Kw*
Design Software: *Nec-Win Pro, Antennamax, Moment Method*

Configuration:

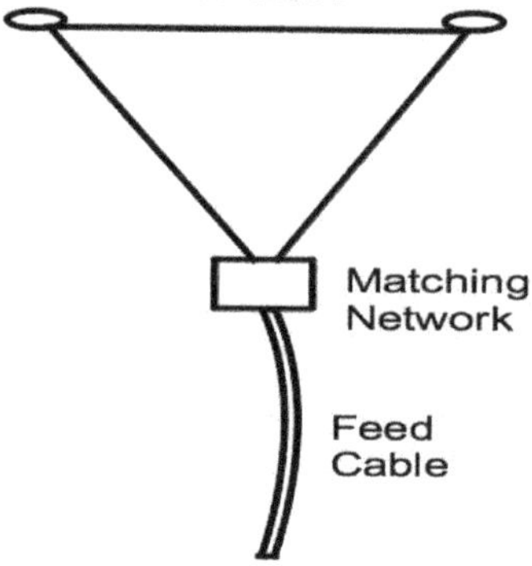


Design Approach:

*To Calculate Length of sides:*

$$L(Meters) = \frac{300}{(F(MHz))}$$

Construction and Theory of Operation:

*Essentially a loop 1 wavelength long fed in the center. This antenna is held in place on both ends with insulators and and guy wires. Impedance is 200 ohms, polarization is horizontal with a figure 8 azimuth patterns. This antenna has broader bandwidth characteristics than the dipole. Use a good balun to match the 200 to 50 ohms*

*Azimuth Pattern*

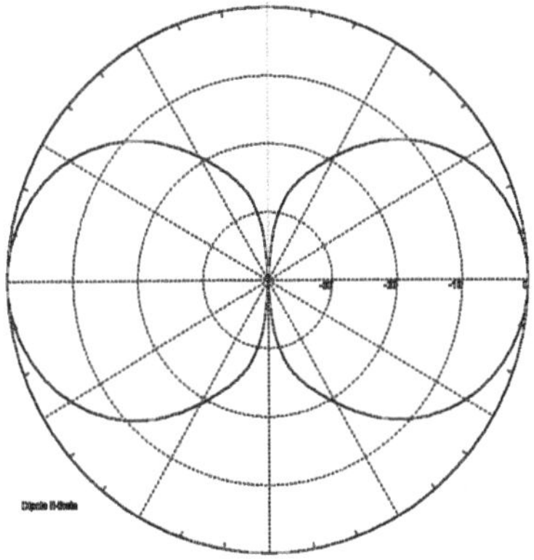

*Elevation Pattern*

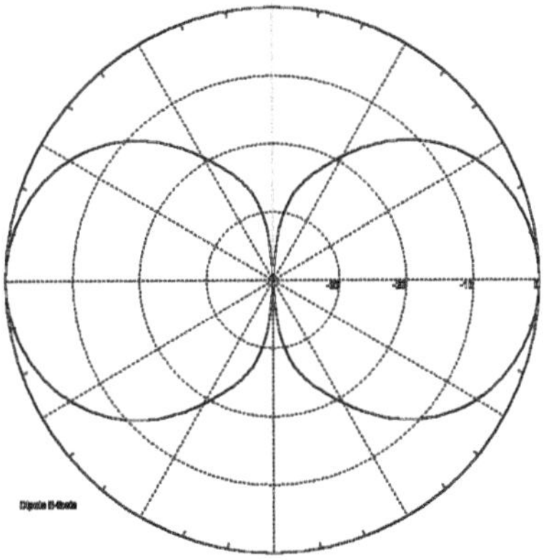

Special Considerations:

*Use lightning protection*

*Height effects, the closer the antenna is to the ground, the higher the elevation pattern*

Testing:

*Walk around signal strength meter, plot on polar graph paper*

*Scaling of antenna to higher frequency can allow more convenient testing in anechoic chamber for instance.*

*Plot signal reports from Ham operators*

| | |
|---|---|
| Type: | ***5/8 Monopole*** |
| Frequency of Operation: | 50 *Mhz to 2 GHz* |
| Gain Range: | 3 *to 9.5 dBi* |
| Bandwidth: | ~ *5%* |
| Approximate Power Rating: | *1 Kw* |
| Design Software: *Method* | *Nec-Win Pro, Antennamax, Moment* |

Configuration

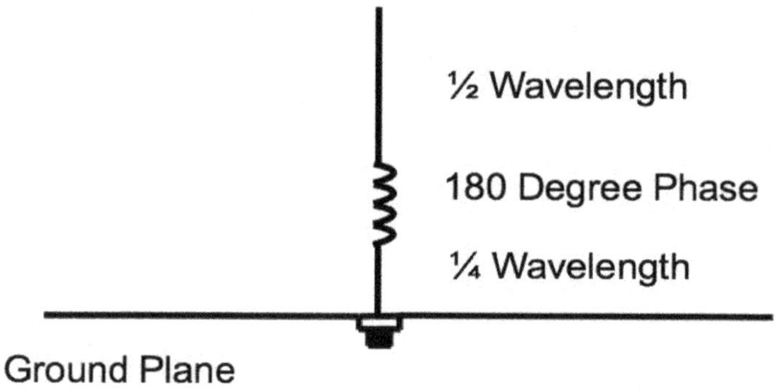


Design Approach:

*To calculate length:*

$$L(Meters) = \frac{150}{(F(MHz))}$$

*where L is ½ wavelength, divide in half for lower element.*

Construction:

*A wire of sufficient strength is supported on insulated stands and insulated ends and fed from one end. The length of the antenna element should be calculated using the above formula. The feed end should be brought into the "radio room" with a good earth ground and be connected to an impedance matching device before connection to the "radio". Lightning protection is recommended.*

*Azimuth Pattern*

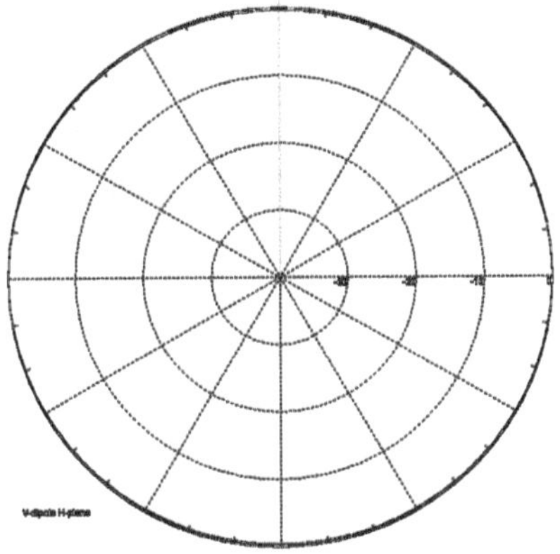

*Elevation Pattern*

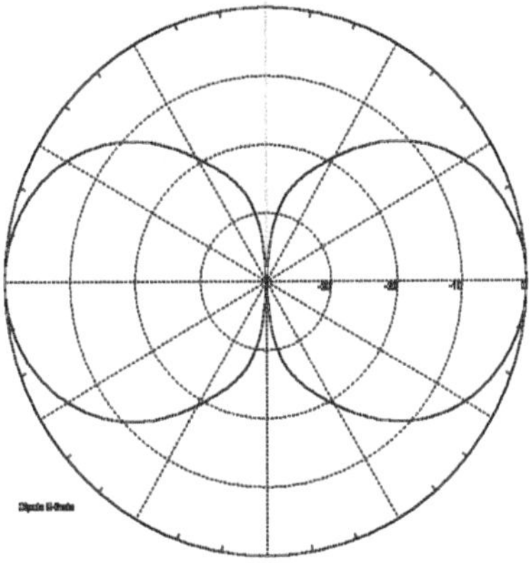

Special Considerations:

*Build over ground plane several wavelengths in size. Use 50 ohm feed. Design can be stacked to increase gain. Example depicted has 3 dBi gain*

*At frequencies above about 300 MHz, azimuth patterns become more and more complex due to interaction with surrounding materials, especially in a vehicle type application.*

*Stacking this type of monopole can be done, with phasing coils in between radiating elements. This approximates the design of a coaxial collinear (co-co) array, which increases gain. This approach also lowers beamwidth, which needs to be considered in a moving vehicle application.*

Testing:

*Far field range measurements with Azimuth and Elevation patterns*

Type: ***Wire Monopole***

Frequency of Operation: *100 Mhz to 5 GHz*
Gain Range: *2 to 3 dBi*
Bandwidth: *~ 5%*
Approximate Power Rating: *1 Kw*
Design Software: *Nec-Win Pro, Antennamax, Moment Method*

Configuration:

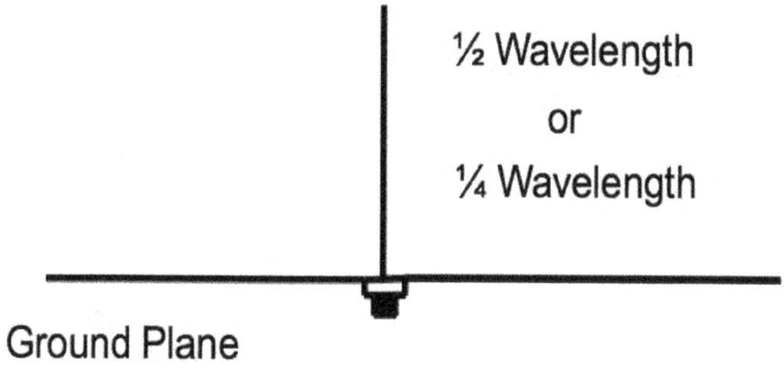


Design Approach:

*To calculate length:*

$$L(Meters) = \frac{150}{(F(MHz))}$$

*where L is ½ wavelength, divide by 2 for quarter wavelength.*

Construction:

*A wire of sufficient strength is supported on insulated stands and insulated ends and fed from one end. The length of the antenna element should be calculated using the above formula. ¼ Wavelength has 2 dBi of gain, ½ has 3 dBi. The ground plane should be at least 1 wavelength in size.*

*Azimuth Pattern*

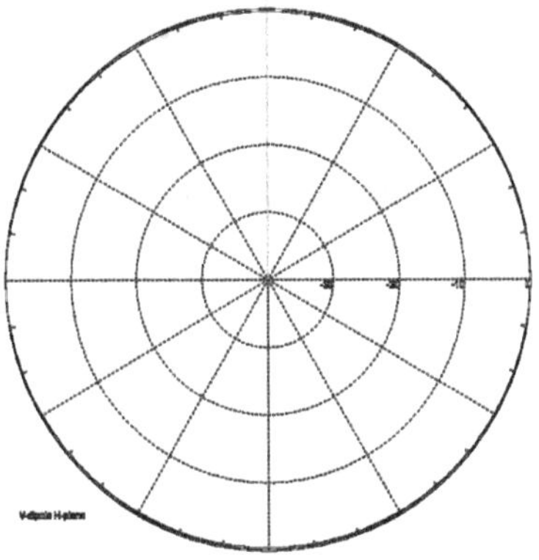

*Elevation Pattern*

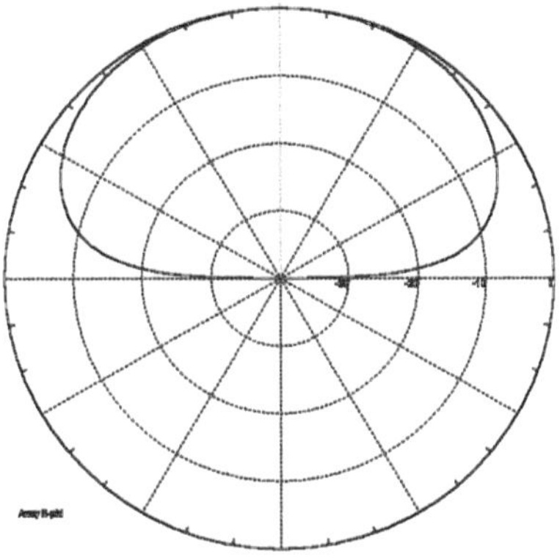

Special Considerations:

*Build over ground plane several wavelengths in size. Example depicted has 2 dBi gain at ¼ Wavelength and 3 dBi at ½ Wavelength.*

*At frequencies above about 300 MHz, azimuth patterns become more and more complex due to interaction with surrounding materials, especially in a vehicle type application.*

Testing:

*Far field range measurements with Azimuth and Elevation patterns*

*2nd Edition*

| | |
|---|---|
| Type: | *"V"* |
| Frequency of Operational: | *5 Khz to 50 MHz* |
| Gain Range: | *0 to 6 dBi* |
| Bandwidth: | *~ 5%* |
| Approximate Power Rating: | *10 Kw* |
| Design Software: | *Nec-Win Pro, Antennamax, Moment Method* |

Configuration:

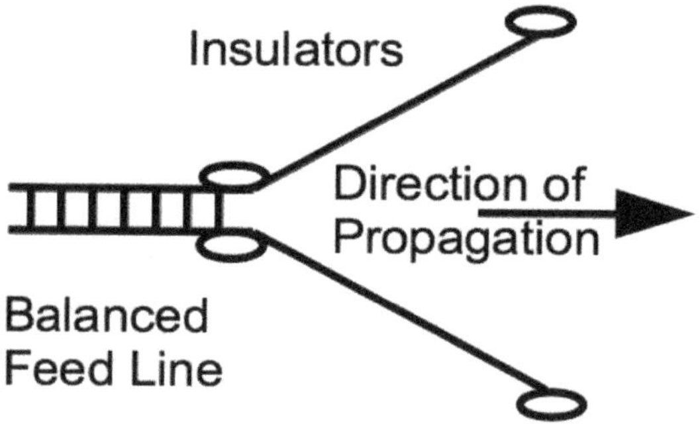


Design Approach:

*To calculate length:*

*L(meters per side) = (300 / f(MHz)) * N*
*where: N = number of wavelengths per side (integer value)*

Construction:

*This antenna is quite directional with good gain. Fed with an open wire (typically 300 ohm) balanced feed line it consists of sides of N wavelengths set at about 45 degrees. Place antenna wire supports as high as practical, using insulators. Use antenna tuner with balanced to unbalanced capability to optimize performance.*

*Azimuth Pattern* *Elevation Pattern*

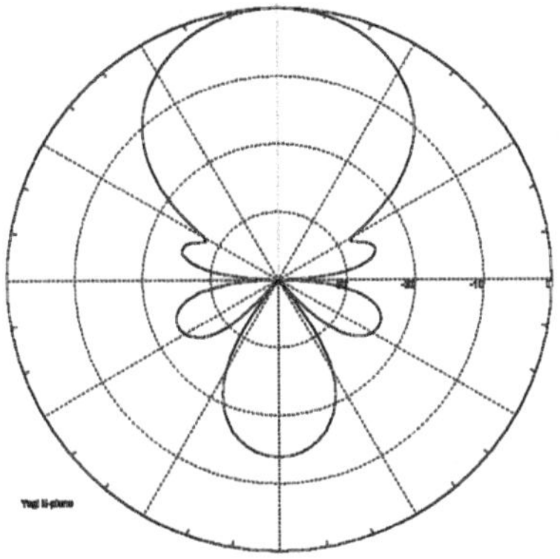

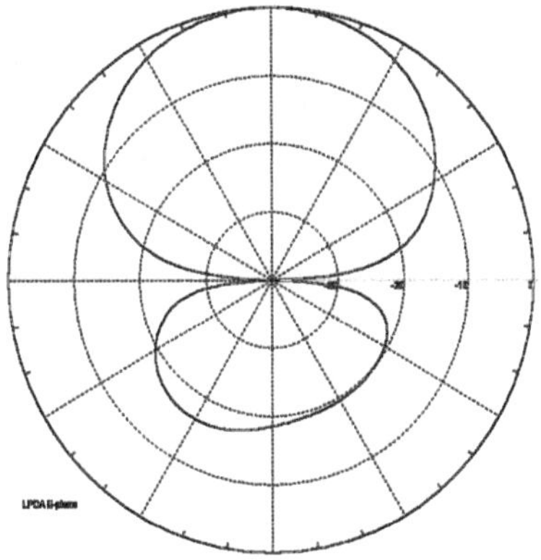

Special Considerations:

*Use lightning protection*

*Height effects, the closer the antenna is to the ground, the higher the elevation pattern*

Testing:

*Walk around signal strength meter*

*Plot signal reports from Ham operators*

Type: ***Turnstile***

Frequency of Operational: *50 Mhz to 500 MHz*
Gain Range: *0 to 6 dBi, can be stacked for more*
Bandwidth: *~ 5%*
Approximate Power Rating: *10 Kw*
Design Software: *Nec-Win Pro, Antennamax, Moment Method*

Configuration:

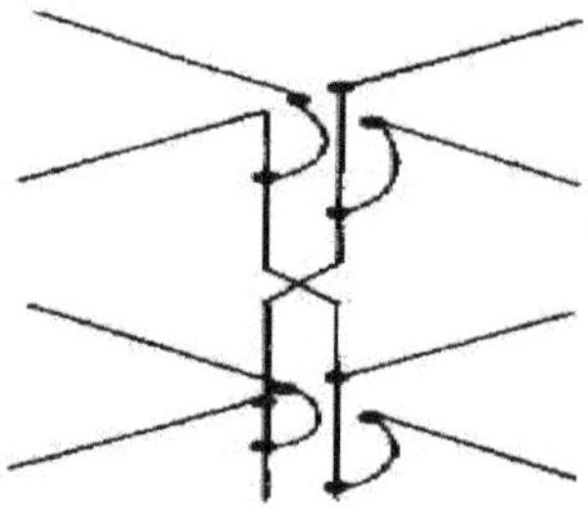

Design Approach:

*To calculate length:*

*L(meters per side) = (150 / f(MHz))*

Construction:

*The simplest turnstile consists of two horizontal half-wave antennas mounted at right angles to each other in the same horizontal plane. When these two antennas are excited with equal currents 90 degrees out of phase, the typical figure-eight patterns of the two antennas merge to produce the nearly circular pattern shown. Pairs of such antennas are frequently stacked. Each pair is called a BAY. In the figure above two bays are used and are spaced 1/2 wavelength apart, and the corresponding elements are excited in phase. These conditions cause a part of the vertical radiation from each bay to cancel that of the other bay. This results in a decrease in energy radiated at high vertical angles and increases the energy radiated in the horizontal plane. Stacking a number of bays can alter the vertical radiation pattern, causing a substantial gain in a horizontal direction without altering the overall horizontal directivity pattern.*

*Azimuth Pattern*

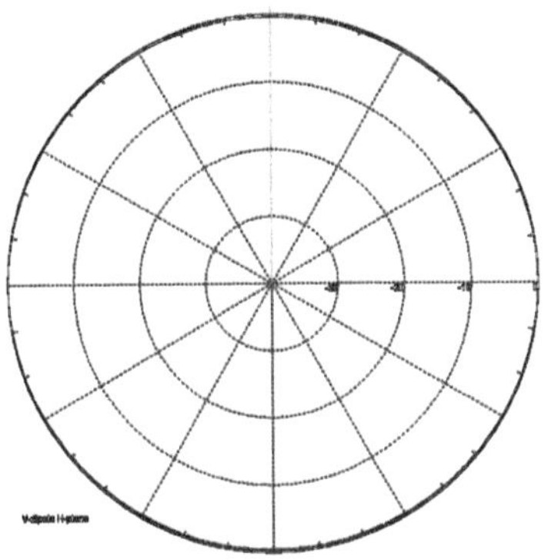

*Elevation Pattern*

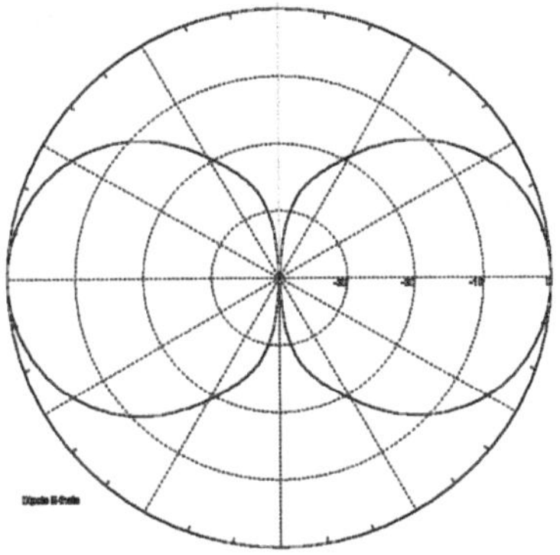

Special Considerations:

*Use similar material to assemble structure. Beware of PIM (Passive Intermod) products which can create signals out of band.*

*Use coax and impedance matched phasing sections. Phase will be effected by conductive mounting hardware mounted in close proximity to radiating elements.*

Testing:

*Far field range measurements, Azimuth and Elevation patterns*

*2nd Edition*

Type: ***Inverted V***

Frequency of Operational: *5 Mhz to 50 MHz*
Gain Range: *2.12 or 0 dBd*
Bandwidth: *~ 5%*
Approximate Power Rating: *10 Kw*
Design Software: *Nec-Win Pro, Antennamax, Moment Method*

Configuration:

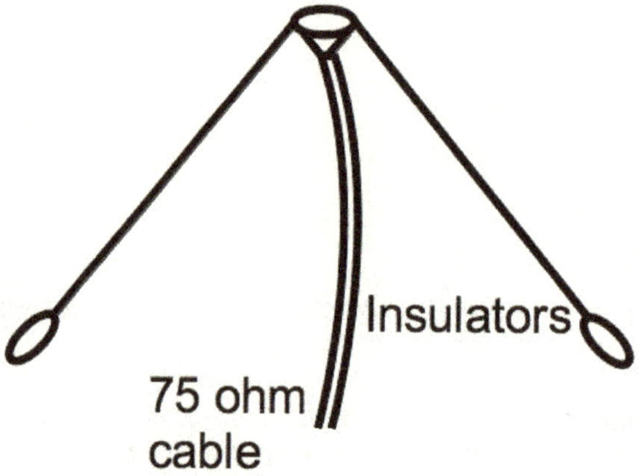


Design Approach:

*To calculate total length:* *L(meters) = 150 / f (MHz)*

Construction and Theory of Operation:

*Essentially a dipole at a 90 degree angle, this antenna is supported in the center and held down on both ends with insulators and and guy wires. Impedance is 72 ohms, polarization is both horizontal and vertical with uniform azimuth patterns.*

*Azimuth Pattern*

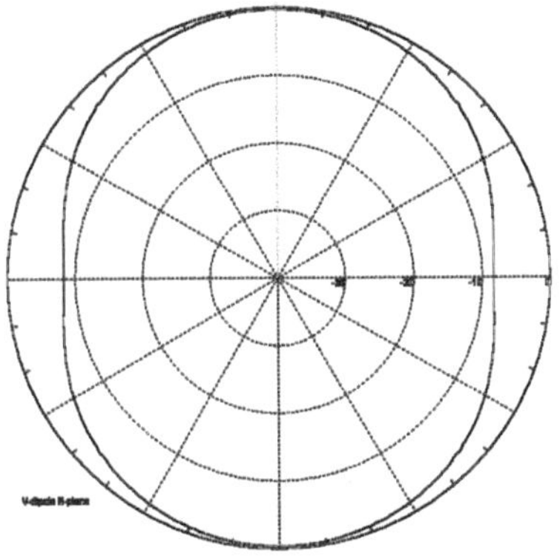

*Elevation Pattern*

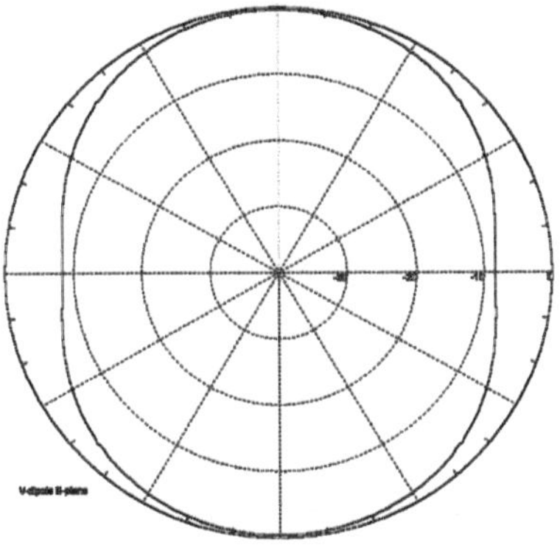

Special Considerations:

*Use lightning protection*

*Height effects,  the closer the antenna is to the ground, the higher the elevation pattern*

Testing:

*Walk around signal strength meter*

*Plot signal reports from Ham operators*

Type: ***Loaded Dipole***

Frequency of Operational: *5 Mhz to 50 MHz*
Gain Range: *0 to 2.12 dBi*
Bandwidth: *~ 5%*
Approximate Power Rating: *1 Kw*
Design Software: *Nec-Win Pro, Antennamax, Moment Method*

Configuration:

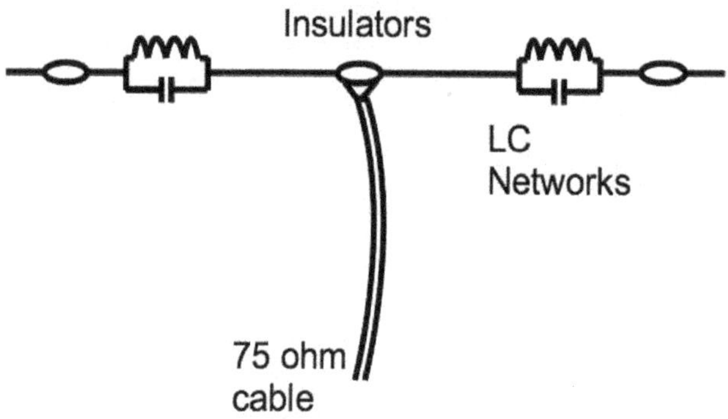


Design Approach:

*To calculate length:* *L(meters) = 150 / f (MHz) – j*

*where j = effective reduction in size due to capacitance and inductance*

Construction and Theory of Operation:

*Essentially a dipole with a pair of inductors in parallel with capacitors resonant at the frequency of interest and held down on both ends with insulators and and guy wires. Impedance is 72 ohms, polarization is horizontal. with a broad figure 8 azimuth pattern.*

*Azimuth Pattern*

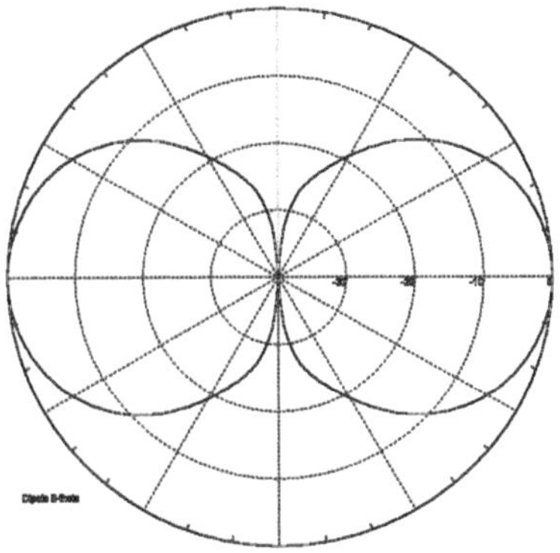

*Elevation Pattern*

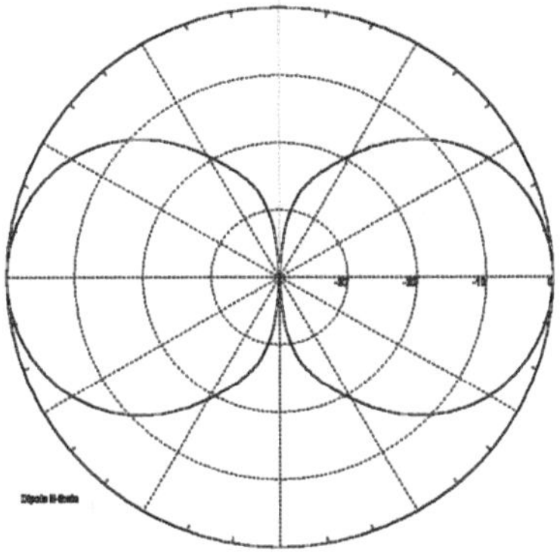

Special Considerations:

*Standard Impedance of Dipole is 72 ohms, close match to 75 ohm cable, use matching transformer, balun or tuning box to optimize.*

*Use lightning protection*

*Height effects, the closer the antenna is to the ground, the higher the elevation pattern*

Testing:

*Walk around signal strength meter, plot on polar graph paper*

*Plot signal reports from Ham operators*

*Far Field Range with an antenna rotator at higher frequencies*

*2nd Edition*

| | |
|---|---|
| Type: | ***Discone*** |
| Frequency of Operational: | *5 Mhz to 500 MHz* |
| Gain Range: | *0 to 3 dBi* |
| Bandwidth: | *~ 20%* |
| Approximate Power Rating: | *1 Kw* |
| Design Software: | *Nec-Win Pro, Antennamax, Moment Method* |

Configuration:

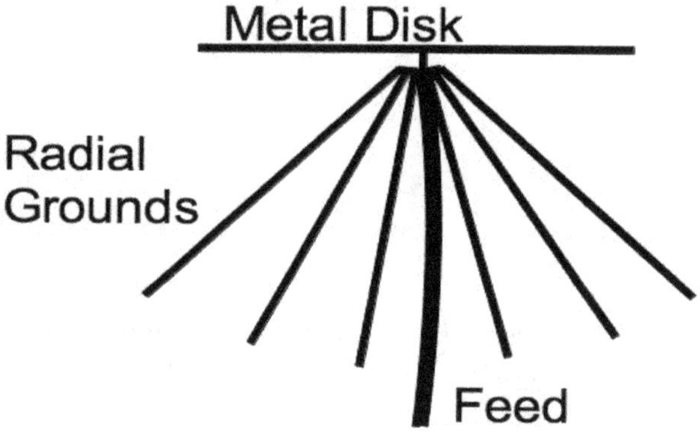


Design Approach:

$$Length\ of\ Radial = \lambda / 4\ at\ lowest\ frequency$$

$$Width\ of\ Disk = \lambda / .175\ at\ lowest\ frequency$$

Construction and Theory of Operation:

*The antenna is designed to be broadband and useful for frequency scanning. Set the divergence angle to about 90 degrees. Set initial width of disk to .175 wavelength at lowest frequency. Adjust disk to radial separation for best match.*

*Azimuth Pattern*

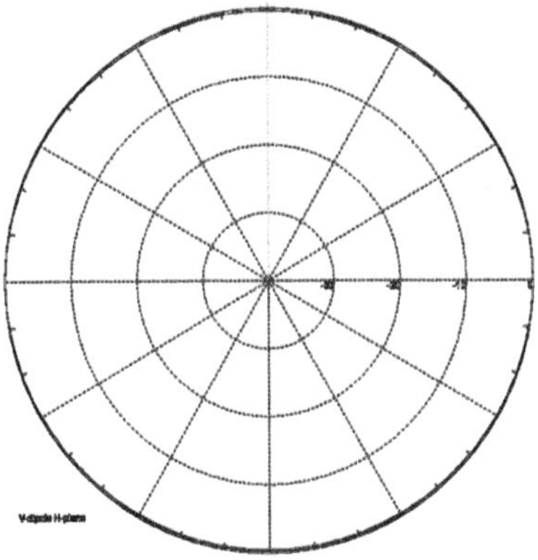

*Elevation Pattern*

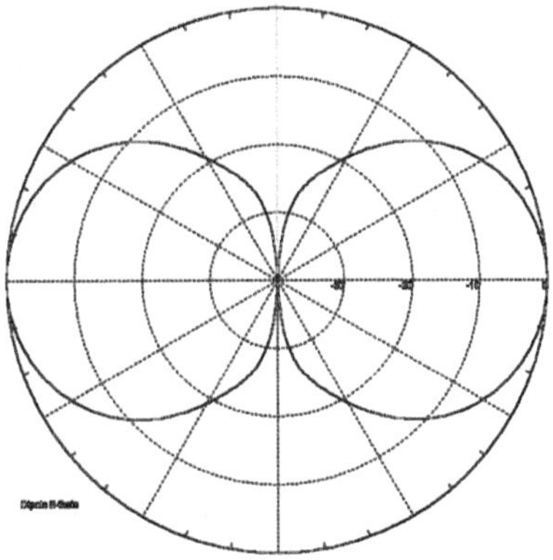

Special Considerations:

*Use disk size to match impedance to 50 ohms.*

*All radials are the same length*

Testing:

*Walk around signal strength meter*

*Plot signal reports from Ham operators*

*Far Field Range with an antenna rotator at higher frequencies*

Type: ***Bi-Conical***

Frequency of Operational: *5 Mhz to 50 MHz*
Gain Range: *0 to 3 dBi*
Bandwidth: *~ 25%*
Approximate Power Rating: *1 Kw*
Design Software: *Nec-Win Pro, Antennamax, Moment Method*

Configuration:

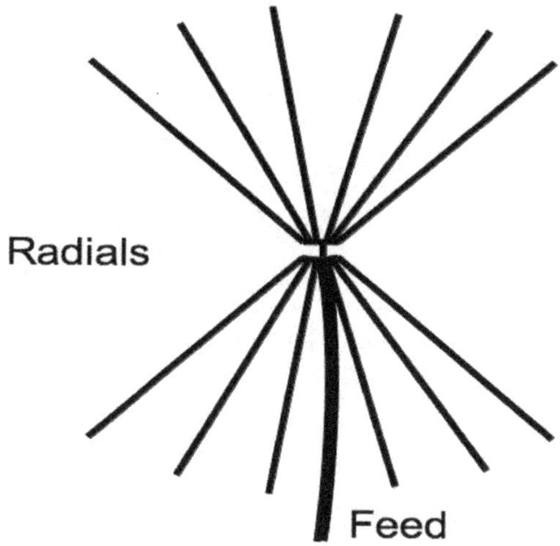


Design Approach:

$$Length\ of\ Radials = \lambda / 4\ at\ lowest\ frequency$$

Construction and Theory of Operation:

*The antenna is designed to be broadband and useful for frequency scanning. Set the divergence angle to about 90 degrees. Adjust upper to lower radial separation for best match.*

*Azimuth Pattern*

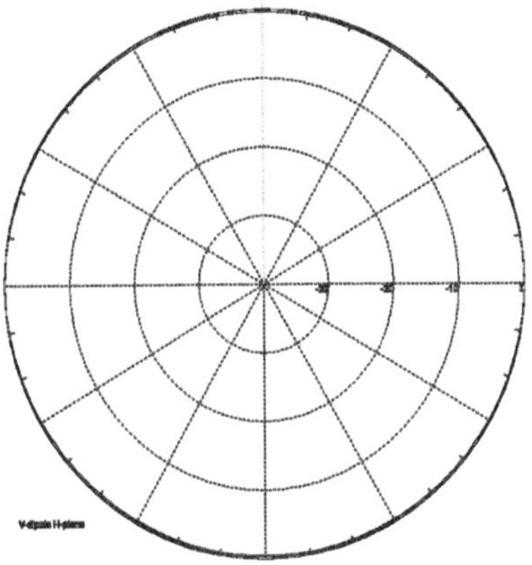

*Elevation Pattern*

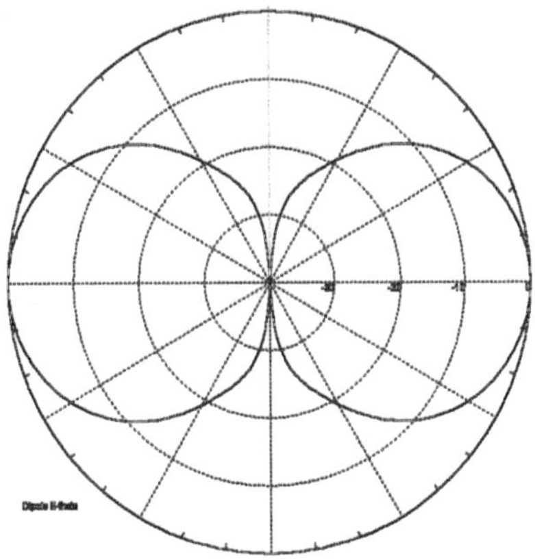

Special Considerations:

*Use separation of radials to match impedance to 50 ohms.*

*All radials are equal length.*

Testing:

*Walk around signal strength meter*

*Plot signal reports from Ham operators*

*Far Field Range with an antenna rotator at higher frequencies*

| | |
|---|---|
| Type: | ***Wire Array*** |
| Frequency of Operational: | *5 Mhz to 500 MHz* |
| Gain Range: | *9+ dBi* |
| Bandwidth: | *~ 5%* |
| Approximate Power Rating: | *10 Kw* |
| Design Software: | *Nec-Win Pro, Antennamax, Moment Method* |

Configuration:

Design Approach:

$$Length\ of\ Dipoles = \lambda/2$$
$$Dipole\ height\ above\ ground\ plane = \lambda/4$$

$$Separation\ between\ dipoles = .5\,\lambda - .75\,\lambda$$

Construction and Theory of Operation:

*This antenna is actually an array of dipoles equi-spaced over a ground plane. Do not exceed .75 lambda separation as over this value the array will endfire. This antenna is highly directional and should be fed by a corporate feed structure to optimize bandwidth.*

*Essentially a dipole array either with or without a ground plane. Used for many years as a short wave broadcast antenna. This antenna is held in place on both ends with insulators and and guy wires. Impedance is 50 ohms, polarization is horizontal or vertical depending on dipole orientation.*

*Azimuth Pattern*

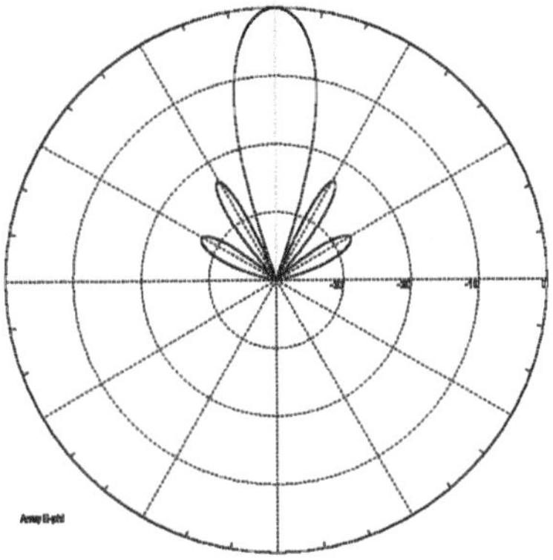

*Elevation Pattern*

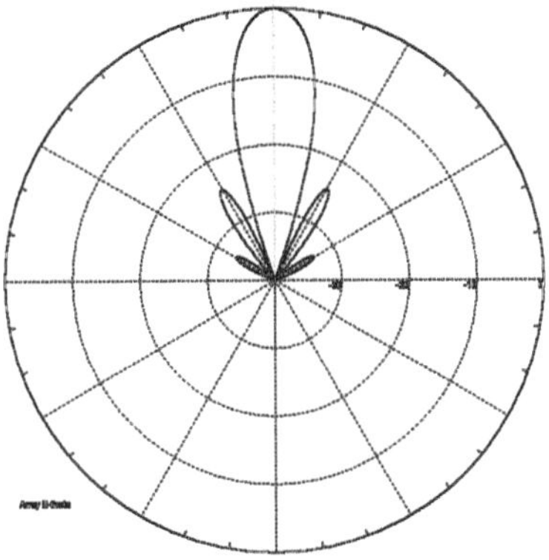

Special Considerations:

*Place antenna on rotator to optimize pointing. Antenna is scalable in terms of gain, doubling the number of elements increases gain by 3 dB.*

*Ground plane should be at least ½ wavelength from edge of outside dipole to edge of metal. Better front to back ratios can be obtained by enlarging ground plane to be up to several wavelengths beyond edge of outside dipole.*

Testing:

*Using rotator, calculate far field and use receiver or spectrum analyzer to measure antenna pattern.*

| | |
|---|---|
| Type: | ***Curtain*** |
| Frequency of Operational: | *5 Mhz to 500 MHz* |
| Gain Range: | *5+ dBi depending on size* |
| Bandwidth: | *~ 5%* |
| Approximate Power Rating: | *100 Kw* |
| Design Software: | *Nec-Win Pro, Antennamax, Moment Method* |

Configuration:

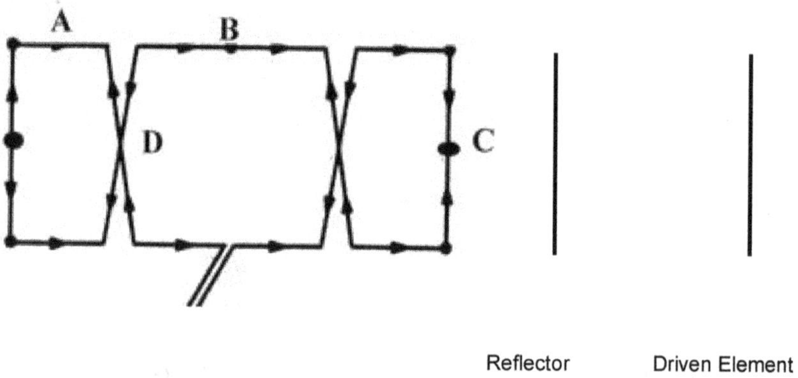


Design Approach:

*To calculate length:* *L(meters) = 150 / f (MHz)*

*where L is length of segments B,C, and D*
*L/2 works for segment A*

Construction and Theory of Operation:

*Essentially a dipole array either with or without a ground plane. Used for many years as a short wave broadcast antenna. This antenna is held in place on both ends with insulators and and guy wires. Impedance is 50 ohms, polarization is horizontal or vertical depending on dipole orientation.*

*Azimuth Pattern*

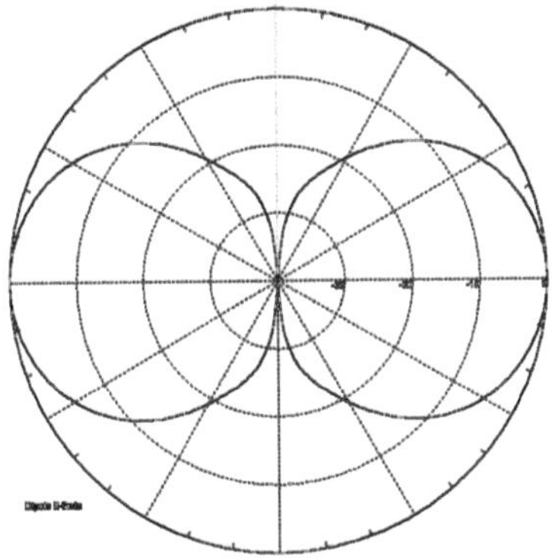

*Elevation Pattern*

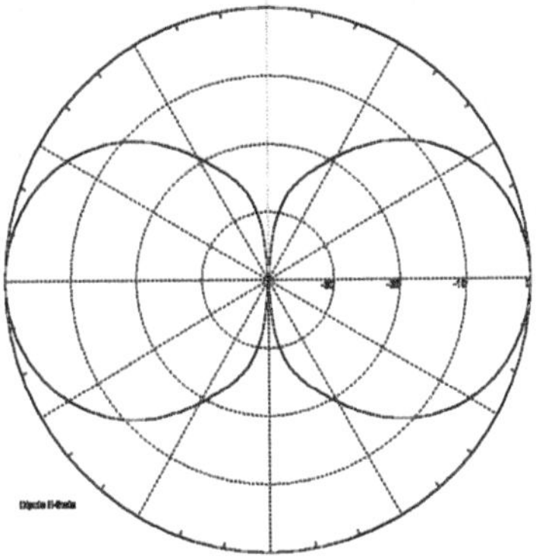

Special Considerations:

*Use lightning protection*

*Height effects, the closer the antenna is to the ground, the higher the elevation pattern*

*The addition of a reflector at approximately ¼ wavelength will add gain in the forward direction and increase the front to back ratio. Pattern will be like yagi.*

Testing:

*Walk around signal strength meter and plot values on polar graph paper*

*Plot signal reports from Ham operators*

*Far Field Range with rotator at higher frequencies*

| | |
|---|---|
| Type: | ***Reflector - Corner*** |
| Frequency of Operational: | *5 Mhz to 5000 MHz* |
| Gain Range: | *9 = dBi* |
| Bandwidth: | *~ 5%* |
| Approximate Power Rating: | *10 Kw* |
| Design Software: | *Nec-Win Pro, Antennamax, Moment Method* |

Configuration:

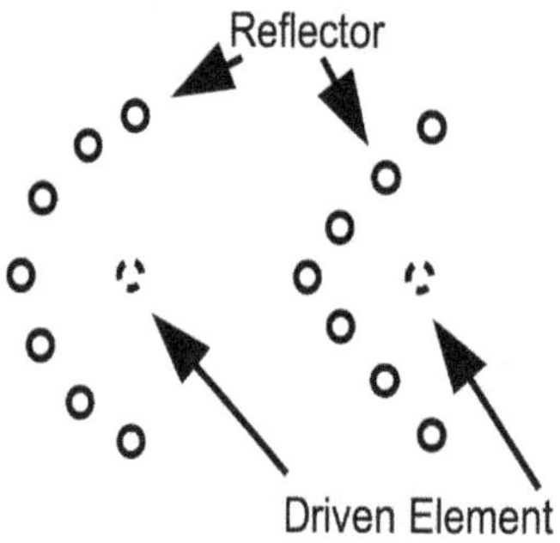


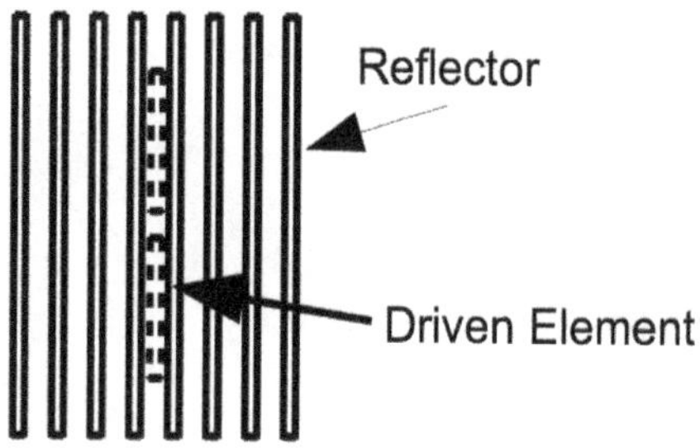


Design Approach:

*To calculate length of driven element:* *L(meters) = 150 / f (MHz)*

*Place feed ¼ wavelength from reflectors arc*

Construction and Theory of Operation:

*Essentially a dipole with either a 90 degree or parabolic reflector. Inexpensive and very low wind loading makes this antenna attractive for many outdoor applications. Orient dipole driven element in parallel with reflector components to obtain very good polarization characteristics.*

*Azimuth Pattern*

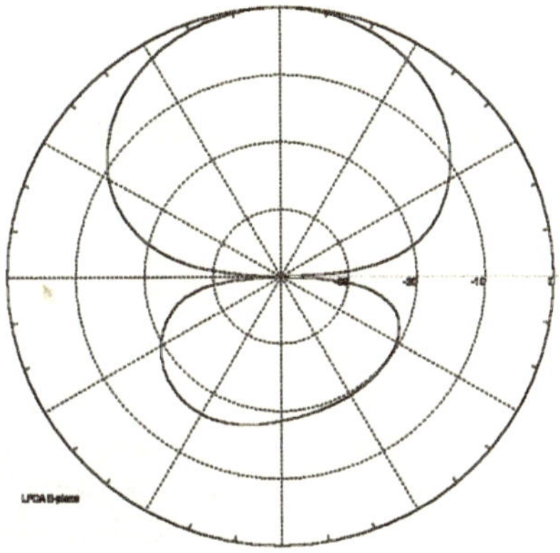

*Elevation Pattern*

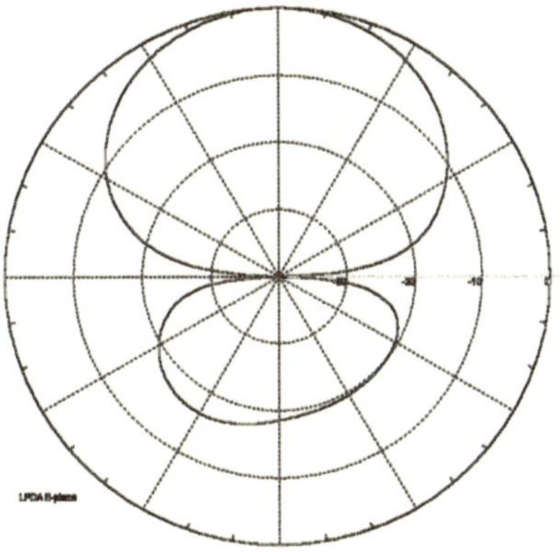

Special Considerations:

*Use Gamma, Delta or Capacitive Match (see Yagi section) to optimize performance.*

*Feed from center of reflector structure*

Testing:

*Walk around signal strength meter at lower frequencies*

*Plot signal reports from Ham operators*

*Far Field Range with rotator at higher frequencies*

Type: ***Reflector – Parabolic (Catenary)***

Frequency of Operational: *50 Mhz to 500 MHz*
Gain Range: *12+ dBi*
Bandwidth: *~ 25%*
Approximate Power Rating: *10 Kw*
Design Software: *Nec-Win Pro, Antennamax, Moment Method*

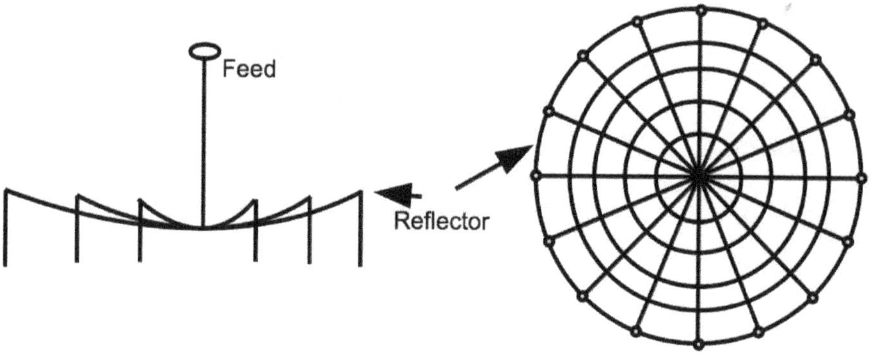


Configuration:

Design Approach:

*To calculate length of feed:* *L(meters) = 150 / f (MHz)*

Construction and Theory of Operation:

*This antenna is made from suspending wires from poles and forming a natural catenary curve. The reflector wires are held in place on both ends with insulators and and guy wires. To a small degree, the beam can be moved by displacing the feed from the vertical. Used for radio astronomy, this antenna was the first to detect radio emissions from the Andromeda galaxy.*

*Spacing between wires should be less than 1/5 wavelength for best performance.*

*Azimuth Pattern* *Elevation Pattern*

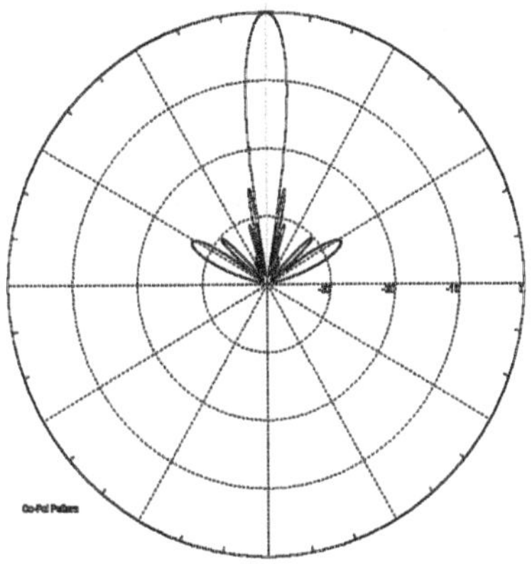

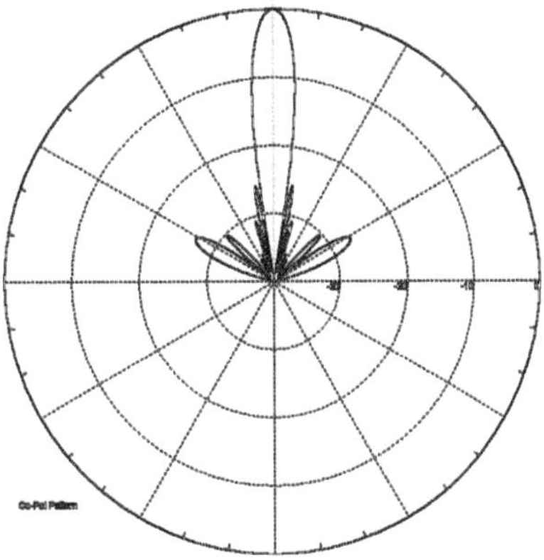

Special Considerations:

*Interlace wire to have grid squares no greater than 1/5 wavelength.*

*Focal point is elongated along axis of beam direction, requiring line feed type*
*feed system, refer to work done at Arecibo 300 m radio telescope.*

Testing:

*Use radio stars for calibration. Extended sources are usable as well as long as the source size is smaller than the predicted beam width*

| | |
|---|---|
| Type: | ***Collinear Array*** |
| Frequency of Operational: | *5 Mhz to 10 GHz* |
| Gain Range: | *2.12 +, depending on length* |
| Bandwidth: | *~ 10%* |
| Approximate Power Rating: | *1 Kw* |
| Design Software: | *Nec-Win Pro, Antennamax, Moment Method* |

Configuration:

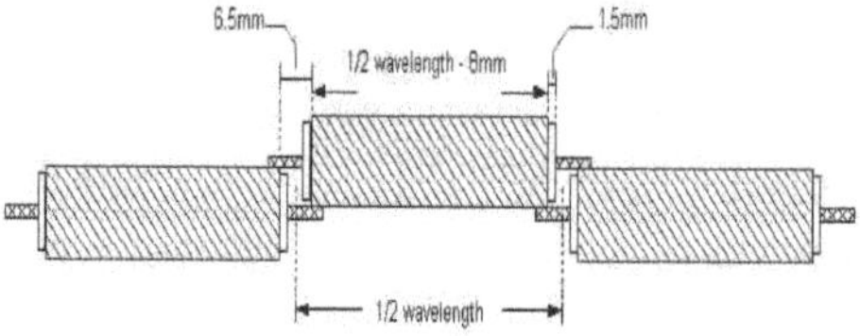


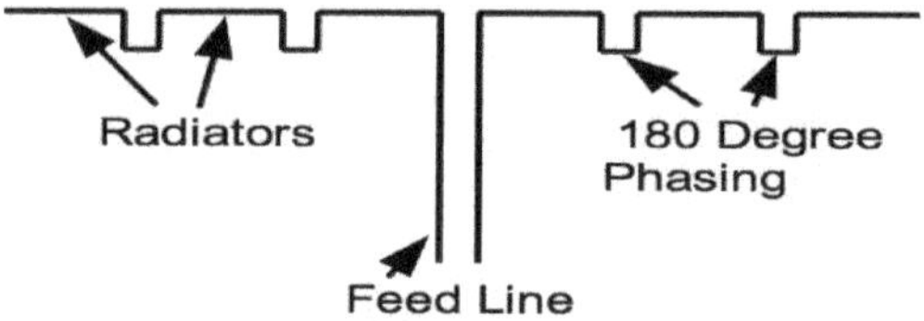


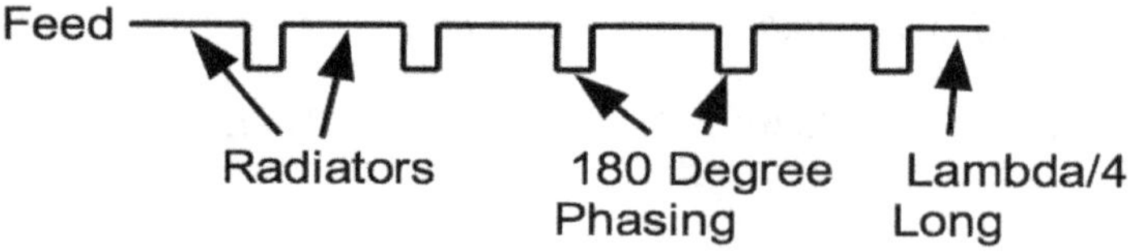


Design Approach:

*To calculate length of segments:* *L(meters) = 150 / f (MHz)*

Construction and Theory of Operation:

*Essentially a series of half wave elements connected together to form an array. This antenna is held in place on both ends with insulators and and guy wires or can be placed in a plastic tube for vertical polarization. Impedance is 50 ohms, polarization is horizontal or vertical with a figure 8 azimuth patterns.*

*Azimuth Pattern* *Elevation Pattern*

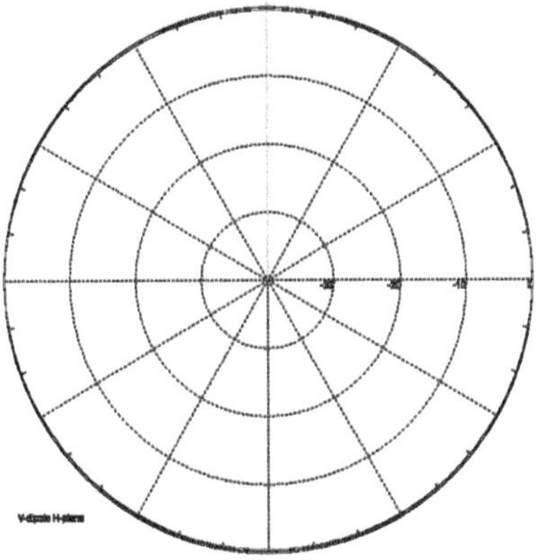

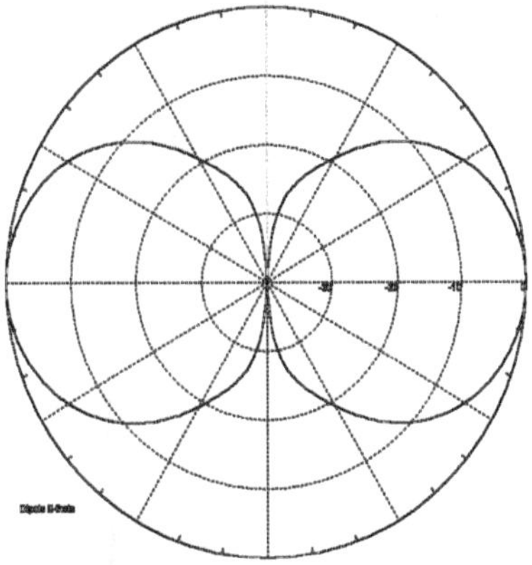

Special Considerations:

*On the coaxial collinear type, place a ¼ wavelength portion on end of antenna, opposite of feed. This will allow efficient match. Adjust length of this section for best match.*

*Be aware of frequency dependent beam steering. Useful for down-tilt applications.*

Testing:

*Far Field Range with rotator, test both planes*

Type: ***Log Periodic***

Frequency of Operational: *5 Mhz to 15 GHz*
Gain Range: *5+ dBi*
Bandwidth: *~ 30%*
Approximate Power Rating: *1 Kw*
Design Software: *Nec-Win Pro, Antennamax, Moment Method*

Configuration:

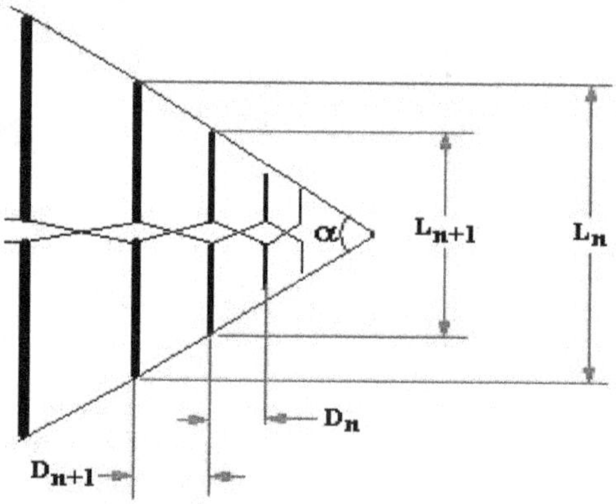


$$L_N = \frac{500}{f_{MIN}} \qquad L_1 = \frac{360}{f_{MAX}} \qquad \sigma = \frac{1-\tau}{4\tan(\alpha)} \qquad \tau = \frac{D_n}{D_{n+1}} = \frac{L_n}{L_{n+1}}$$

Design Approach:

*To calculate length:* *L(meters) = 150 / f (MHz)*

*where: L is length of each element at a specific frequency*

Construction and Theory of Operation:

*Essentially a series of dipoles at a specific angle with cross fed and termination at the larger end.. This antenna is held in place on both ends with insulators and and guy wires. Impedance is 300 ohms, polarization is horizontal or vertical depending on orientation and is very directional.. This antenna also has very broad bandwidth characteristics.*

*Azimuth Pattern*

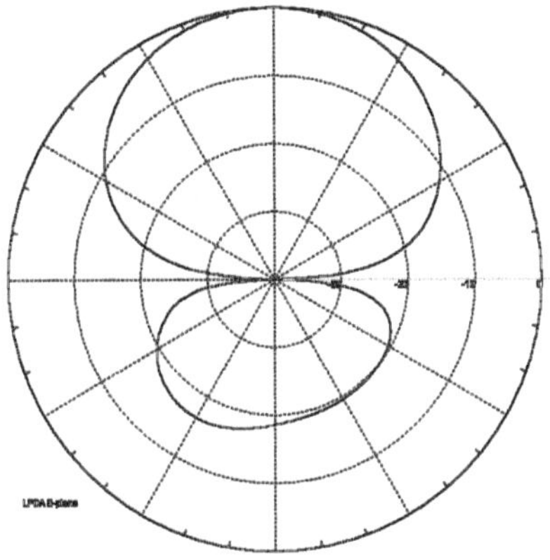

*Elevation Pattern*

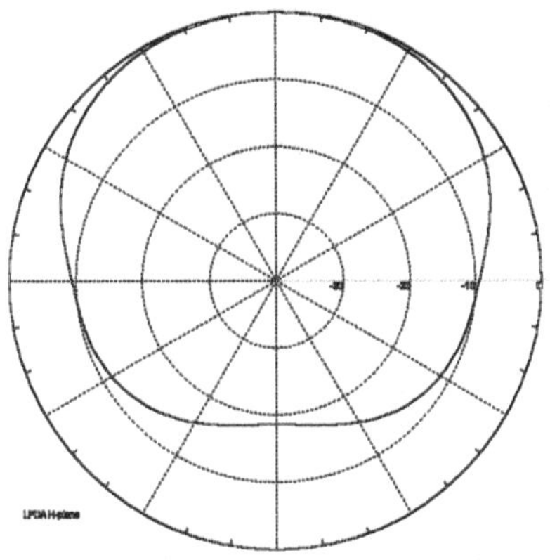

Special Considerations:

*Termination Resistor can be used to improve match. Several variations on this design are described in Chapter 5.*

*Feed line is cris-crossed down support structure. Another option is to place two feed rails, one above the other with every other element connect to top rail and vice versa on lower rail.*

Testing:

*Far Field Range with rotator, test both planes*

*Scaling of antenna to higher frequency can allow more convenient testing in anechoic chamber for instance.*

Type: ***Double Zepp***

Frequency of Operational: *5 Mhz to 500 MHz*
Gain Range: *2.12 or 0 dBd*
Bandwidth: *~ 10%*
Approximate Power Rating: *10 Kw*
Design Software: *Nec-Win Pro, Antennamax, Moment Method*

Configuration:

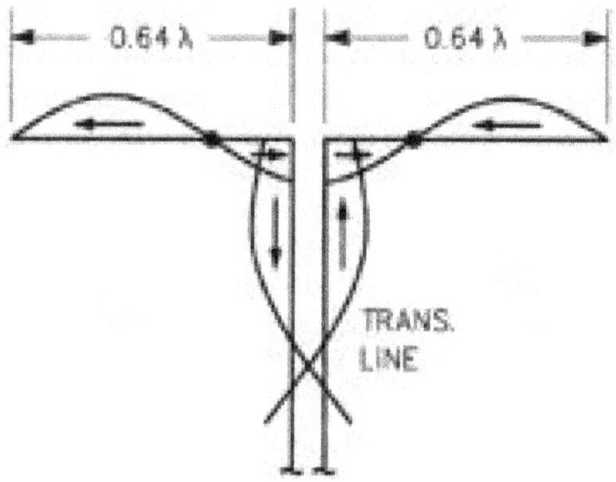


Design Approach:

*To calculate length:* *L(meters) = 192 / f (MHz)*

*where L is the length of each arm of dipole. Monopole section below dipole can be ½ wavelength at another frequency of interest.*

Construction and Theory of Operation:

*Essentially a dipole above a monopole. This antenna is held in place on both ends with insulators and and quy wires. Impedance is 300 ohms, polarization is horizontal and vertical with a figure 8 azimuth patterns. This antenna has broader bandwidth characteristics than the dipole. Typically this antenna is fed with a a balun to convert 300 ohms to 50 ohms*

*Azimuth Pattern*

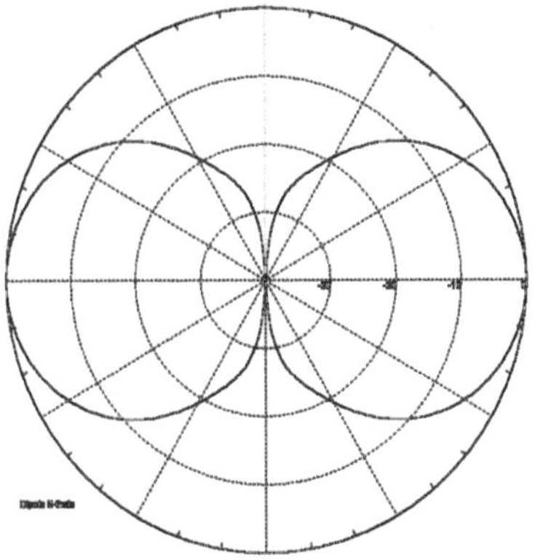

*Elevation Pattern*

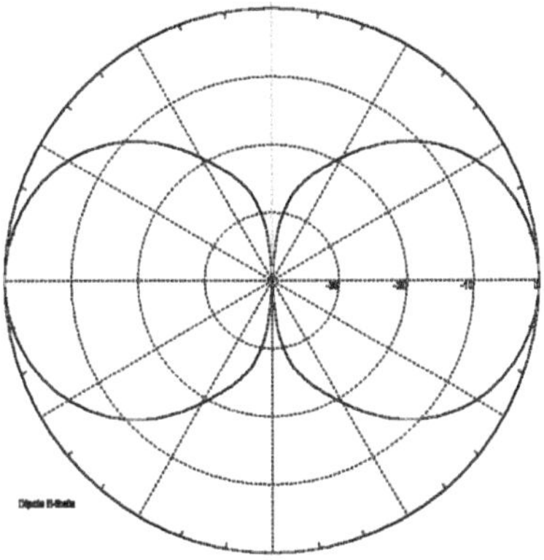

Special Considerations:

*Good wideband antenna with 300 ohm input impedance. Use appropriate matching electronics for best performance.*

*Use lightning protection*

*Height effects, the higher off of the ground, the lower the elevation pattern*

Testing:

*Walk around signal strength meter at lower frequencies*

*Plot signal reports from Ham operators*

*Far Field Range with rotator at higher frequencies*

*Scaling of antenna to higher frequency can allow more convenient testing in anechoic chamber for instance.*

Type: ***"F" or "J" pole***

Frequency of Operational: *5 Mhz to 5 GHz*
Gain Range: *0 to 4 dBi*
Bandwidth: *~ 10%*
Approximate Power Rating: *1 Kw*
Design Software: *Nec-Win Pro, Antennamax, Moment Method*

Configuration:

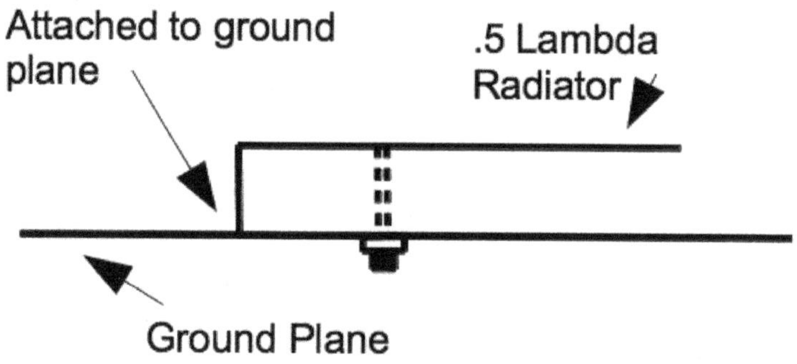


Design Approach:

*Length of radiating element = .5 lambda*
*Separation from ground plane = .125 lambda*
*Distance of shorting point to feed = .125 lambda*

Construction and Theory of Operation:

*This antenna is made with wire, tubing or can be etched into copper clad dielectric material. The shorting bar adjusts resonance and should be moved to tune to optimum performance.*

*This style of antenna is typically used for applications above 100 MHz.*

*Azimuth Pattern*

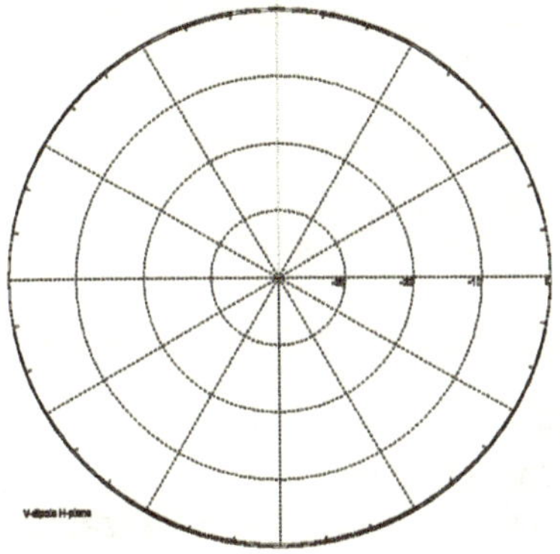

*Elevation Pattern*

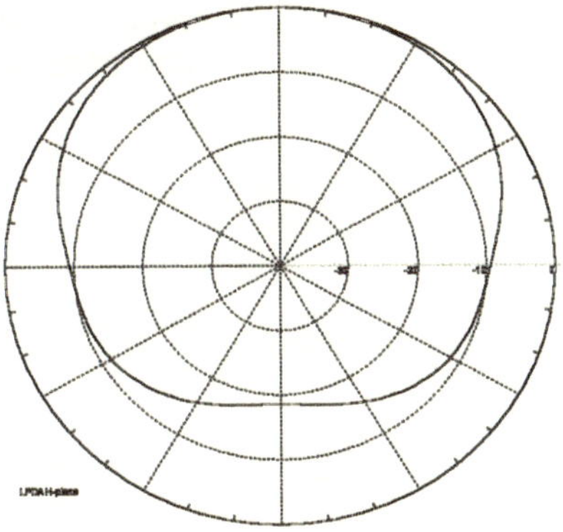

Special Considerations:

*Antenna has a direct connection to ground design, good for inherent lightning protection*

*Can be adjusted for a wide range of impedances.*

*Used commonly as Wi-Fi antenna in computers, marker beacon antenna on aircraft and 2 meter elements for Ham radio operators.*

Testing:

*Walk around signal strength meter at lower frequencies*
*Plot signal reports from Ham operators*

*Far Field Range with an antenna rotator at higher frequencies*
*For higher frequencies, anechoic chambers, see antenna test appendix.*

*Scaling of antenna to higher frequency can allow more convenient testing in anechoic chamber for instance.*

**Questions:**

1. Suggest a wire antenna design for the following characteristics:

   a. Frequency of 100 MHz
   b. Gain of 20 dBi
   c. Horizontal Polarization

2. How would you approach the following problem?

   a. Frequency of 40 Mhz
   b. Gain of 60 dBi
   c. Right Hand Circular Polarization

3. Suggest an antenna for the next generation rocket system,capable of going to the moon. Use the following details:

   a. Frequency of 28 MHz
   b. Conformal
   c. Gain 3 dBi

4. As wire can be bent into 3D shapes, how could this help the designer?

5. Suggest an antenna design for 60 Hz. Are they they already being made ?

6. Very Low Frequency Radio Astronomy is an exciting field in astronomy. Where would you place an antenna with requirements below 30 MHz.

7. Short Wave antennas often use wire antennas for long distance propagation. How could you use this type of antenna for medium (not line of sight) distances ?

8. Traditional car antennas have been replaced by integrated wires. Where are they placed ?

# Chapter 5 – Loop Antennas

Loop antenna used for Maritime Navigation

Loop antennas have been around since the time of Hertz. From about 1915 to 1920 they were the dominate antenna type for radio receivers. After that period, wire antennas took over in popularity. The loop reappeared about 1938 to allow smaller, self contained AM radios to be produced. From this time on, they have been the dominate choice for AM radio reception. They can be categorized into two major groups, large and small. The small loops are generally no more than .1 lambda in diameter. These loops have the advantage of being directional with good null performance. Hence these antennas are used extensively in direction finding applications. Another version of the small loop uses a ferrite rod to loop the active element around. This has the effect of concentrating the magnetic field and increasing the open circuit voltage. This antenna also has directional properties and is seen extensively in small low frequency radios (e.g. AM, 550 to 1600 Khz) Large loops consist of antennas greater than .1 lambda in diameter. These antennas are used in yagi like structures typically and can have multi lobed antenna patterns giving a degradation in directional performance.

Specific uses include:

1. Small loops are used extensively for direction finding receivers for aircraft and ships. Frequencies from 100 Khz to 500 MHz can employ such antennas.
2. Very small loops are used for RF tag applications.
3. Arrays of large loops using a ground plane have been designed to take advantage of the simple construction and relatively good performance of these antennas.
4. Crossed small loops can be joined with a goiniometer to create a non moving direction finder. This design has been used on ships for navigation.
5. Multi-turn loops on a ferrite rod has produced one of the most popular antennas in history, this being for AM (500 to 1700 Khz) receivers. This design gained popularity in the 1930s and is still in use today.
6. Large half loop antennas have been made a very low frequencies to take advantage of the relatively low propagation losses.

**Small Loops**

A loop is considered a small loop if it is less than ¼ of a wavelength in circumference. Most directional receiving loop are about 1/10 of a wavelength in diameter. The small loop is also called magnetic loop because it is more sensitive to the magnetic component of the electromagnetic wave. As such, it is less sensitive to near field electric noise when properly shielded.[25] A variation on this is a very small loop, or maybe in more accurate terms, an inductor with a lot of space between conductors. These are used for RF tag applications at frequencies around several hundred Khz. At these frequencies, antennas have been made with very small footprints, around a few centimeters in diameter. Used with a large capacitor, these resonant circuits resonate quite well. They are used typically with very low power and short operational range. RF tag technology is found at all major sales outlets to control theft as well as track goods through the manufacturing to sale process.

**Large Loops**

Large loop antennas can be thought of as a dipole. Except that the ends of the dipole are connected to form a circle, triangle or square. Typically a loop is a multiple of a half or full wavelengths in circumference. It has more gain (about 10%) than the other forms of large loop antenna, as directly proportional to the area enclosed by the loop. Strongest signal in the plane of the loop and nulls in the pattern exist perpendicular to the plane of the loop. This is the opposite orientation of the small loop. [25]

Typical Applications:

1. Small Loop — RF Tags, Direction Finding
2. Large Loop — Communications
3. Crossed loop — Direction Finding, Military, Ships
4. Delta loop yagi — Ham radio, Shortwave listening
5. Quad loop yagi — Ham radio, Shortwave listening
6. DDR — Limited space omni antenna
7. Ferrite loop — AM radio, Direction finding
8. Crossed Ferrite loop — Lightning detection, range and bearing.

**Design Examples**

The following pages contain examples of antennas discussed in this chapter. These examples represent a first order design approach. For a more complete and optimized antenna solution, use the example closest to the requirement needs and optimize by using good simulation software, creating prototypes, testing and finally modifying prototypes to optimize performance.

Type: ***Small Loop***

Frequency of Operation: *2 Khz to 30 MHz*
Gain Range: *-10 to 3 dBi*
Bandwidth: *~ 3%*
Approximate Power Rating: *10 W*
Design Software: *Nec-Win Pro, Antennamax, Moment Method*

Configuration:

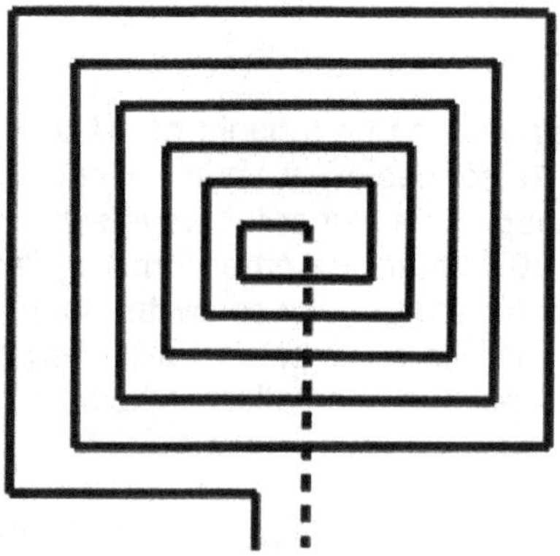

Design Approach:

*To calculate length:* *L(meters) = 150 / f (MHz) or L(meters) = 75 / f(MHz) Use loading capacitance if necessary*

Construction and Theory of Operation:

*This is a multiple turn loop less than .1 wavelength in diameter. If this type is very small, the Q will be very high leading to narrow bandwidth applications. This design has been used extensively in the rf tag business as they typically need to work on only one frequency. Use inductors and capacitors to tune in the frequency that is desired.*

*Azimuth Pattern*

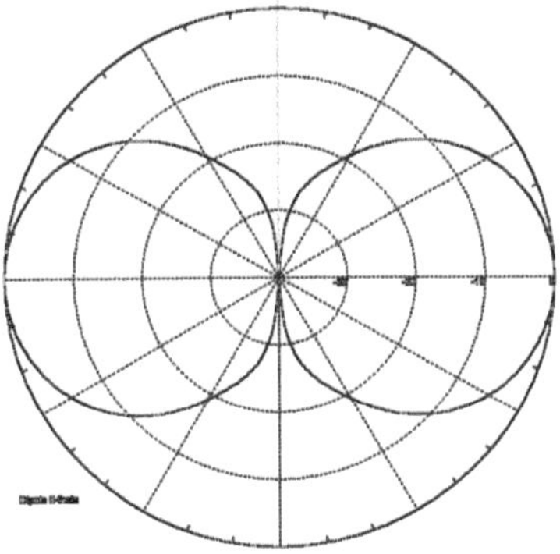

*Elevation Pattern*

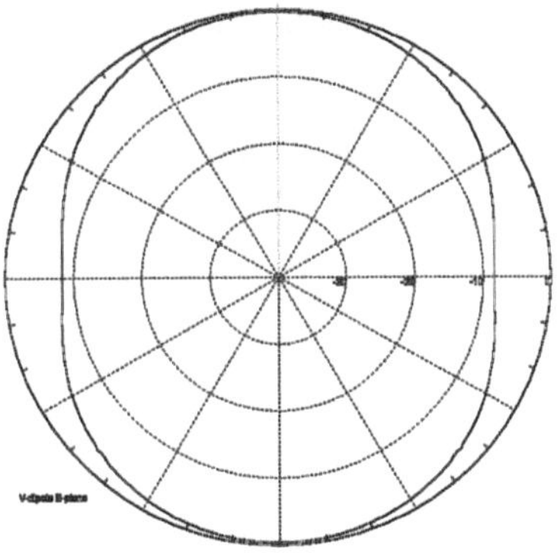

Special Considerations:

*The nulls on a small loop like this are oriented 90 degrees from that of a larger loop. In other words, the plane of this loop is also the plane of the null positions.*

*For very small loops, like RF tags, nulls are not produced, leading to omni patterns.*

Testing:

*Use far field range to measure pattern. Considering the small aperture relative to wavelength, the distance to the source antenna will not be great. Check Appendix 6 for more details.*

Type: ***Large Loop***

Frequency of Operational: 20 *Mhz to 300 MHz*
Gain Range: *-10 to 3 dBi*
Bandwidth: *~ 5%*
Approximate Power Rating: *100 W*
Design Software: *Nec-Win Pro, Antennamax, Moment Method*

Configuration:

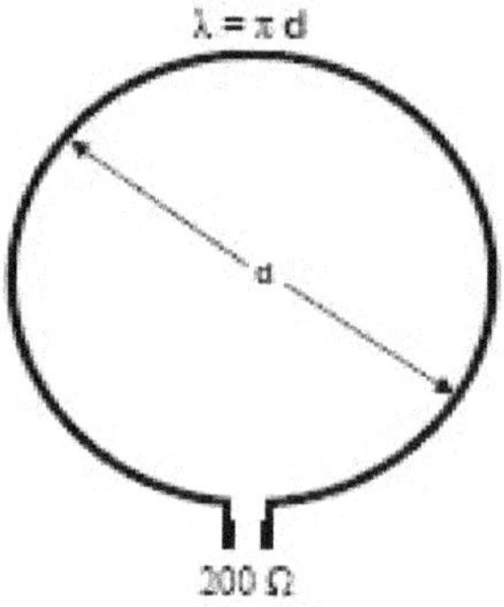


Design Approach:

*To calculate length:* *L(meters) = 150 / f (MHz) or*
*L(meters) = 75 / f(MHz)*

Construction and Theory of Operation:

*This loop is greater than .1 wavelength in diameter and as a result is more sensitive to the electric field of a radio wave. The nulls are perpendicular to the plane of the antenna. Bandwidth is greater than a small loop so this antenna can be used as a direction finding component over many MHz of frequency.*

*Azimuth Pattern*

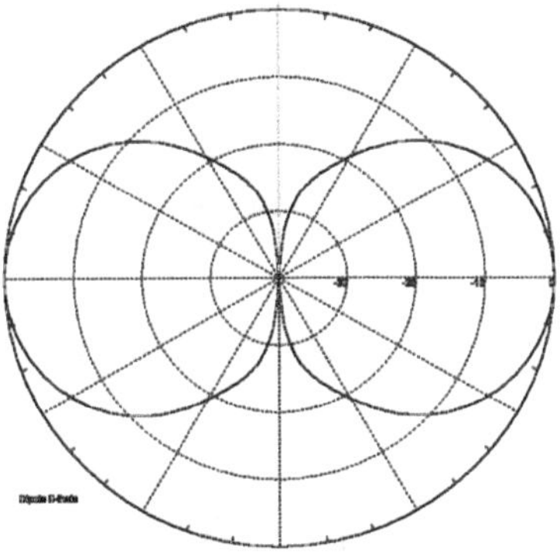

*Elevation Pattern*

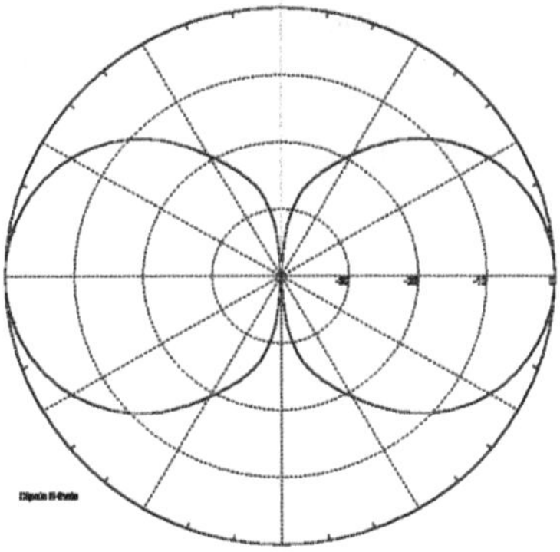

Special Considerations:

*Impedance match with a broadband device like a transformer or balun to optimize performance. Place on a rotator to use as a direction finding element, along with a "sense" antenna to resolve the 180 degree ambiguity from a two lobed pattern.*

Testing:

*Use far field range to measure pattern. Considering the small aperture relative to wavelength, the distance to the source antenna will not be great. Check Appendix 6 for more details. Examine the details of the null as this will allow direction finding optimization. The null should be deep and have a limited angle.*

*2nd Edition*

Type: ***Crossed Loops***

Frequency of Operational: *5 Mhz to 50 MHz*
Gain Range: *0 to 3 dBi*
Bandwidth: *~ 5%*
Approximate Power Rating: *1 Kw*
Design Software: *Nec-Win Pro, Antennamax, Moment Method*

Configuration:

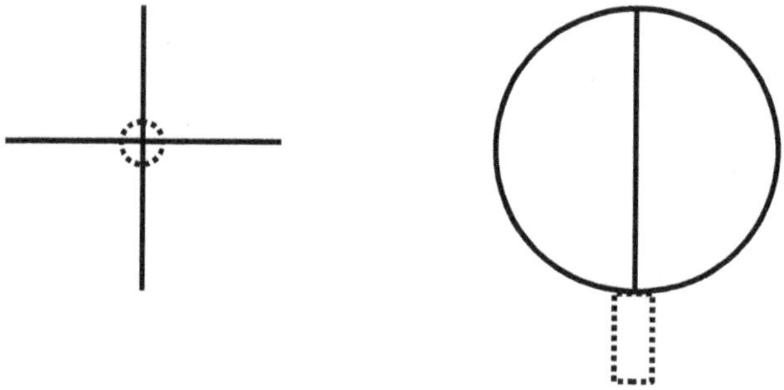

Design Approach:

*To calculate length:* *L(meters) = 150 / f (MHz) – j*

*where j = effective reduction in size due to capacitance and inductance*

Construction and Theory of Operation:

*These loops are greater than .1 wavelength in diameter and as a result are more sensitive to the electric field of a radio wave. The nulls are perpendicular to the plane of the antennas. Bandwidth is greater than a small loop so this antenna can be used as a direction finding component over many MHz of frequency. The advantage of using a crossed loop is that there is no need for mechanical scanning, azimuth to a signal of interest is obtained by connecting the loops to a "goniometer" or software defined radio with a fast analog to digital converter to measure phase and amplitude to calculate bearing.*

*Azimuth Pattern*

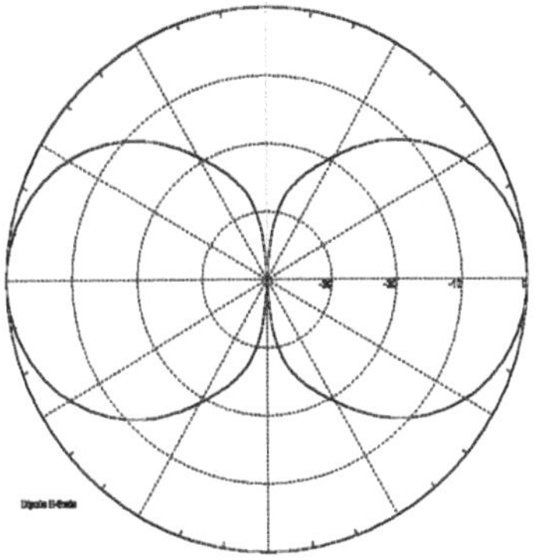

*Elevation Pattern*

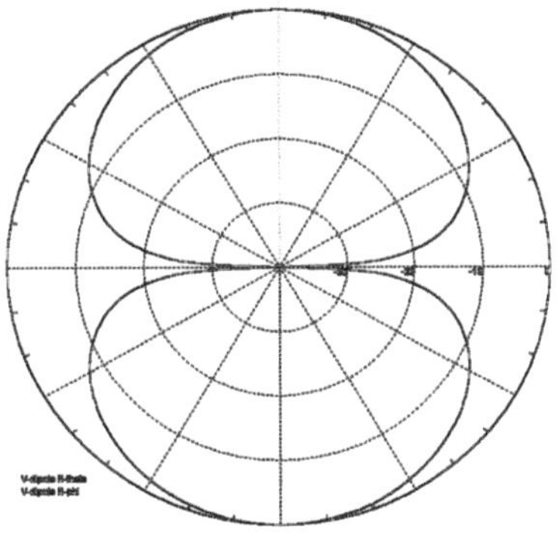

Special Considerations:

*Attach loops to electronics that discriminate phase and amplitude for high speed bearing determination.*

Testing:

*Use far field range to measure pattern. Considering the small aperture relative to wavelength, the distance to the source antenna will not be great. Check Appendix 6 for more details. Examine the details of the nulls as this will allow direction finding optimization. The nulls should be deep and have a limited angle.*

*$2^{nd}$ Edition*

| | |
|---|---|
| Type: | ***DDRR*** |
| Frequency of Operational: | *5 Mhz to 50 MHz* |
| Gain Range: | *0 to 3 dBi* |
| Bandwidth: | *~ 5%* |
| Approximate Power Rating: | *100 W* |
| Design Software: | *Nec-Win Pro, Antennamax, Moment Method* |

Configuration:

Design Approach:

*To calculate length:* *L(meters) = 150 / f (MHz) – j*

*where j = effective reduction in size due to capacitance*

Construction and Theory of Operation:

*Also known as a Direct Discontinuity Ring Radiator.*

*This is a large loop design which is located parallel and close to a ground. A capacitor is used at one end to tune the antenna to the proper frequency. Narrow band in nature, this antenna has been used to detect strong signals from Jupiter and other terrestrial sources.*

*Azimuth Pattern*

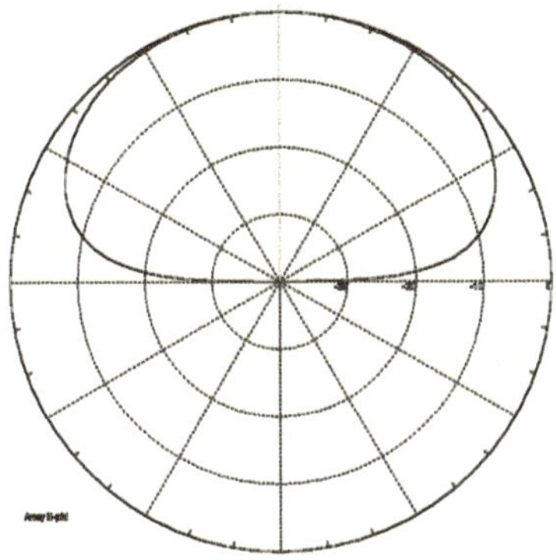

*Elevation Pattern*

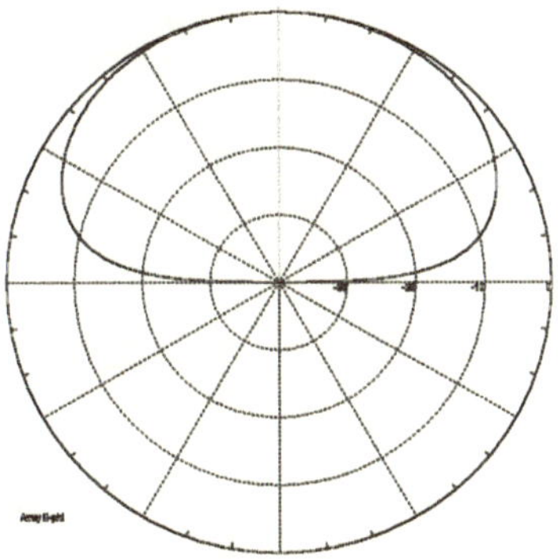

Special Considerations:

*Narrowband in nature, use a good variable capacitor to tune to desired frequency. Conductive vs. earth ground will vary capacitor setting. This antenna can be used for low earth orbiting satellites as it has good sky coverage.*

Testing:

*Use far field range to measure pattern if using a conductive ground plane; place antenna on side. Considering the small aperture relative to wavelength, the distance to the source antenna will not be great. Check Appendix 6 for more details.*

**Questions:**

1. Suggest an antenna type that has the following characteristics:

   a. Frequency of 20 to 40 MHz
   b. One degree azimuthal resolution
   c. RHC Polarization

2. Your boss tells you price is a huge factor in the design of antenna with the following characteristics:

   a. Frequency of 20 MHz
   b. Gain of 2 dBi
   c. LHC Pol

   Suggest a loop type antenna that will be inexpensive yet useful.

3. Suggest an antenna design for tracking the next generation rocket booster that is capable of going to the moon, with the following properties:

   a. Frequency of 220 MHz
   b. 1 degree azimuthal resolution
   c. Tracks emitter while spinning

4. What would be the effect of placing multiple loops about 1/8 lambda apart in a line ?

5. Design a loop array with 16 elements on a single ground plane. What are the advantages / disadvantages of using this approach ?

6. What happens to a loop antenna if you extend it in the z plane?

7. How does a half loop antenna work ?

# Chapter 6 – Patch Antennas

## Introduction

Patch antennas or Microstrip antennas consist of a radiating surface above a ground plane, separated by a dielectric. The radiating surface can be made by etching or cutting out a good conductor like aluminum or copper. The ground plane is typically larger than the radiating surface and is made of the same conductive materials. The dielectric layer can be composed of anything from air or vacuum to dense materials such as Alumina. Selection of dielectric layer is based on dielectric constant and loss tangent. The higher the dielectric constant on a material, the smaller the radiating surface can be for a given frequency. The use of minimal loss tangent will optimize the efficiency and thus the gain of the antenna. Both of these qualities are available from the manufacturer of the material or from a reference book that lists physical properties of materials.

The idea of this type of antenna was first proposed by Deschamps around 1953 but did not see practical usage until Howell and Munson developed it around the early 1970's. Since then there have been thousands of manifestations of this design including planar, phased array, wrap around (or conformal) and "smart" (integrated with other electronics).

Typically, the Microstrip antenna is about .5 wavelength long (with Er of 1.0) and .7 wavelength wide and is fed either from the bottom or at an edge. Normally the feed point is in the center however offset feeds have been used in the past for unusual applications.

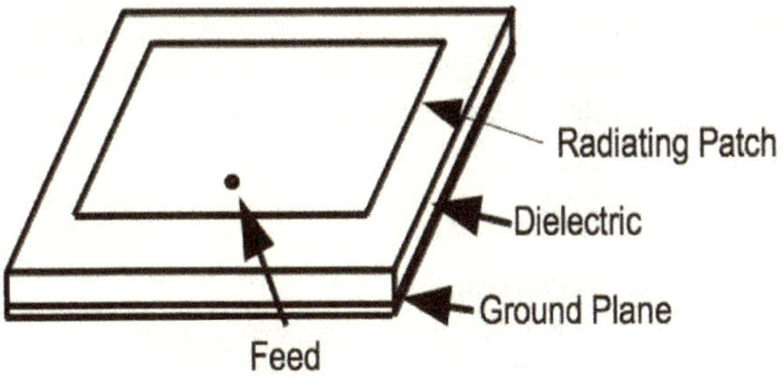


Typical half wave microstrip radiator

To best calculated the dimensions of a $\lambda/2$ radiating "Patch" use

$$Length = .49 * \left( \frac{\lambda}{\sqrt{(\epsilon r)}} \right) \quad [35]$$

The width of the patch must be less than one wavelength in the dielectric substrate material so that higher order modes will not be excited.

The feed is typically centered and about $.1\,\lambda$ from the edge of the patch. Positioning this feed along the centered line, moving closer to the edge will increase the impedance and conversely moving it away will lower the impedance. The very center of a patch in this configuration is around 0 ohms and the very edge is around 140 ohms. The polarization of the radiated fields will be vertical in the example cited. Cross-pole isolation for a single patch could exceed 30 dB and efficiency could be in excess of 90%, making this type of antenna very popular.

It is important to have at least one board thickness separation from any edge of the patch to the edge of the ground plane, two or more would be optimal.

Bandwidth of a microstrip patch can be estimated with:

$$BW = \left( \frac{height}{\sqrt{(\epsilon r)}} \right) \quad [36]$$

**Microstrip Lines**

Microstrip lines are important to understand while designing patch antennas and arrays. Several components will be discussed here to give the designer enough tools to complete a design.

They are:

Microstrip Lines
¼ wave matching transformers
Champers
Wide bandwidth transformers
Stub

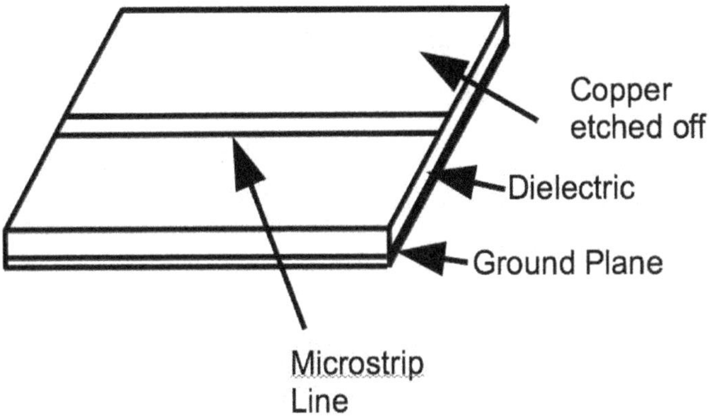


To calculate the required width of a microstrip line, use the following equation [36]:

$$Zm = \frac{Zo}{((2*\pi)*\sqrt{(2*(1+Er))})}*\ln\left(\left(1+\frac{(4*h)}{Weff}\right)*\left((\frac{(14+(\frac{8}{Er}))}{11})*(\frac{(4*h)}{Weff})+Zw1\right)\right)$$

and:

$$Zw1 = \sqrt{\left(\left((\frac{(14+(\frac{8}{Er}))}{11})*(\frac{(4*h)}{Weff})\right)^2+(\pi)^2*(\frac{(1+(\frac{1}{Er}))}{2})\right)} \qquad [37]$$

The effective width, Weff, can be calculated thus:

$$Weff = W + T * \left( \frac{(1+\frac{1}{Er})}{(2*\pi)} \right) * \ln \left[ \frac{(4*e)}{\sqrt{\left( (\frac{t}{h})^2 + \left( (\frac{1}{\pi}) * (\frac{1}{(\frac{w}{t}+\frac{11}{10})}) \right)^2 \right)}} \right] \qquad [38]$$

where:

Er = dielectric constant of substrate

w = width of strip

h = thickness of substrate

t = thickness of strip metallization

Zo = Impedance of free space (377 ohms)

## Microstrip Components

### Transformers

These are used to couple lines efficiently and thus transfer the most amount of energy from one place to another. Because impedance is always a consideration, coupling lines from for instance, two antennas to a single point, requires that a transformer structure be designed to best transfer the energy. Typically, most lines on a microstrip array are of the same impedance, usually around 100 ohms. This value is easily made on a substrate and is convenient for the final impedance of the feed connection which is usually 50 ohms.

When a 100 ohm microstrip line is bisected by a feed connection, the impedance will be 50 ohms, just like two 100 ohm resistors in parallel. If the energy needs to be transferred further, a transformer structure can be designed to interface between (in this case) the 50 ohms from a bisection, to another 100 ohm line, see diagram.

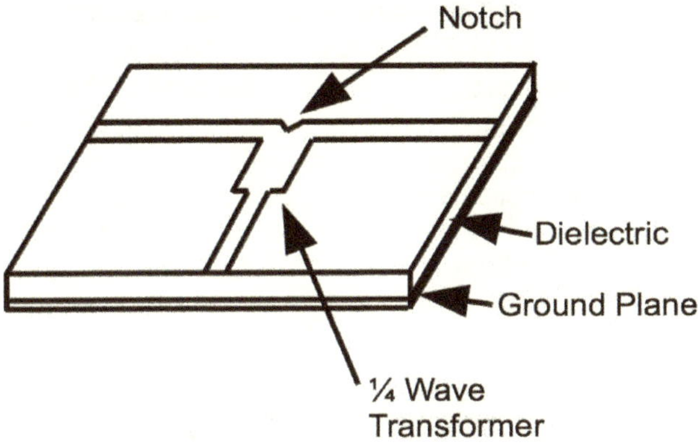


Transformer example, coupling two microstrip lines

In this case a quarter wave (in the material) microstrip line is built with an impedance that follows the formula:

$$Zt = \sqrt{(Z1 * Z2)} \qquad [39]$$

where:

Zt = the impedance of the transformer

Z1 = the input end of the transformer

Z2 = the output end of the transformer

To calculate the length of the transformer use:

$$Length = .24 * \lambda / \sqrt{(\epsilon r)} \qquad [40]$$

where:

$\lambda = \textit{free space wavelength}$
$\epsilon r = \textit{dielectric constant of substrate material}$

These types of transformers are useful for about 7% of bandwidth, for more wide band applications, tapered or curved transformers have been used for bandwidths of up to 30% of operating frequencies. Each transformer must be at least $\lambda/4$ long, in the case of the curved transformer, the usual method to design the angles is to use an exponential taper approach. Abrupt changes in line direction will cause reactive responses to occur, complicating design. Use smooth transitions and keep the distance between one microstrip line and any other structure to at least one board (dielectric) thickness.

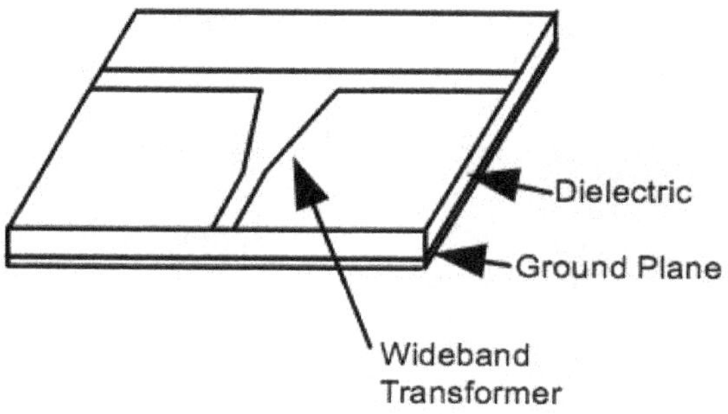


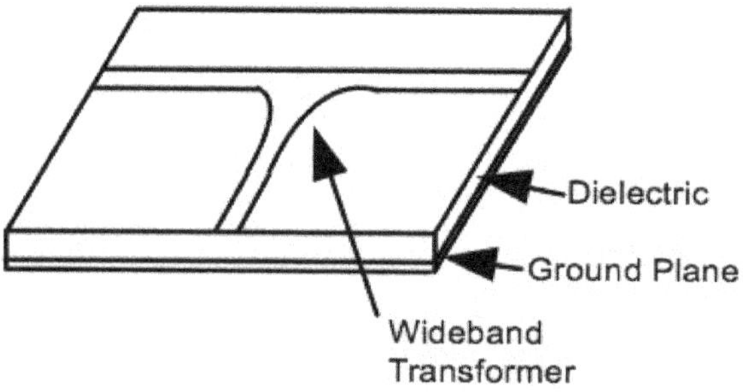


Wide Bandwidth Transformer examples

### Microstrip Line Bends and Turns

Because these structures need to be a consistent impedance, care must be taken to turn or bend the lines to avoid significant reactance and radiation. Lines can be bent using a bend radius of several line widths. In addition, if a 90 degree bend is required, care must be taken to mitre the corner in such a was as to allow the greatest transfer of energy. 1.5 time the line width for the mitre is a good starting point.

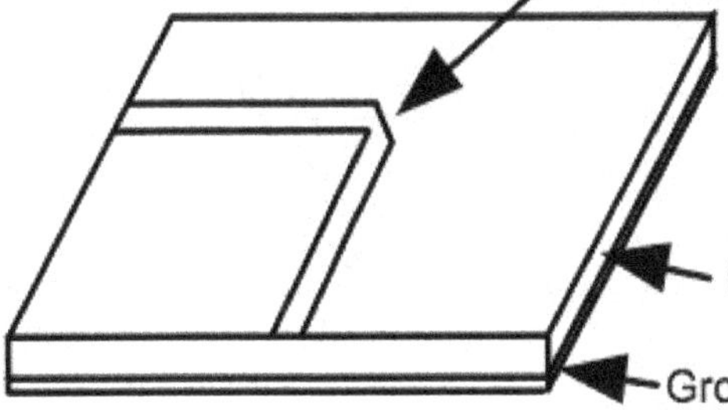


Optimal configuration of microstrip line turns

### Stubs

Stubs on microstrip lines are sometime employed to act as filtering elements when optimizing performance of microstrip arrays and other complex structures. The length of the first example stub is usually around $\lambda/4$ and the placement along the horizontal microstrip line causes changes in the phase and amplitude of the complex impedance. This gives a designer a valuable tools for optimizing performance. In the case of the second stub example the width is typically $\lambda/4$ and again placement along the primary microstrip line can affectively change the impedance. Be careful not to place too many stubs on a microstrip structure as the stubs themselves can radiate, lowering the gain and causing pattern distortion.

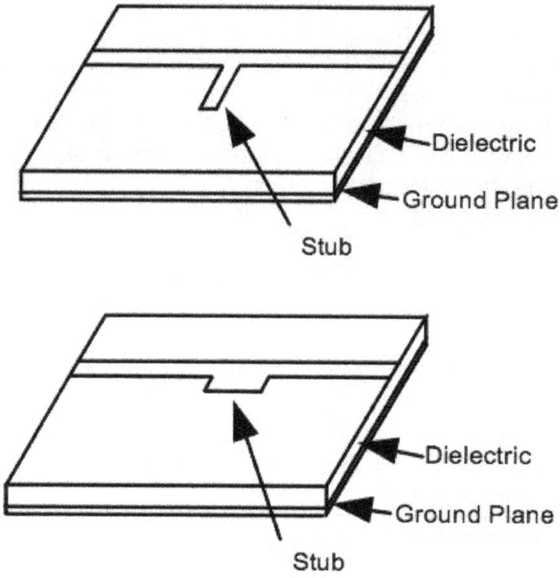


Stub Examples

Of the Antenna types, several will be highlighted in the following pages.

Specific types include:

½ Wave Linear
¼ Wave Linear
Circular Polarized
Stacked
Aperture Coupled
Wrap Around
Phased Array

**Arrays**

Phased arrays using microstrip patches are economical and efficient. The example below can have gain in excess of 12 dBi, with cross pole values above 25 dB. The patches are separated by approximately .7 lambda, the microstrip lines are set to 140 ohms for the lines connecting the patches and 100 ohms for the lines with the transformers. The center feed is therefore set to 50 ohms.

Another approach to making an array is called a "wrap around" where a single patch is extended in the horizontal plane and fed periodically using a corporate feed. This type of antenna is used extensively in missile and rocket applications.

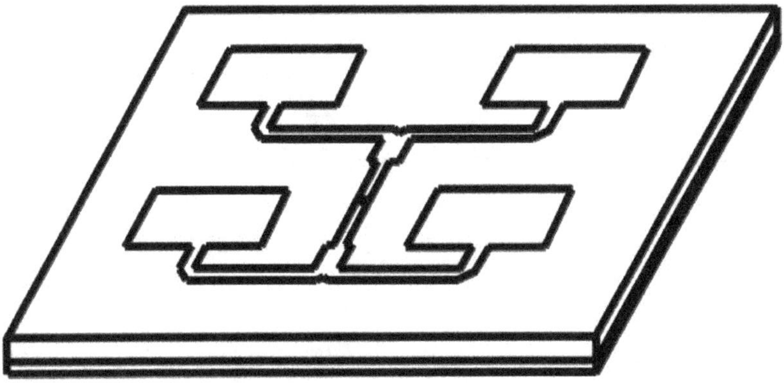

Microstrip Phased Array.

**Design Examples**

The following pages contain examples of antennas discussed in this chapter. These examples represent a first order design approach. For a more complete and optimized antenna solution, use the example closest to the requirement needs and optimize by using good simulation software, creating prototypes, testing and finally modifying prototypes to optimize performance.

*2nd Edition*

| | |
|---|---|
| Type: | ***½ Lambda Patch*** |
| Frequency of Operational: | *5 Mhz to 50 GHz* |
| Gain Range: | *3 to 10 dBi* |
| Bandwidth: | *~ 5%* |
| Approximate Power Rating: | *1 Kw* |
| Design Software: | *Ensemble, IE3D, Moment Method* |

Configuration:

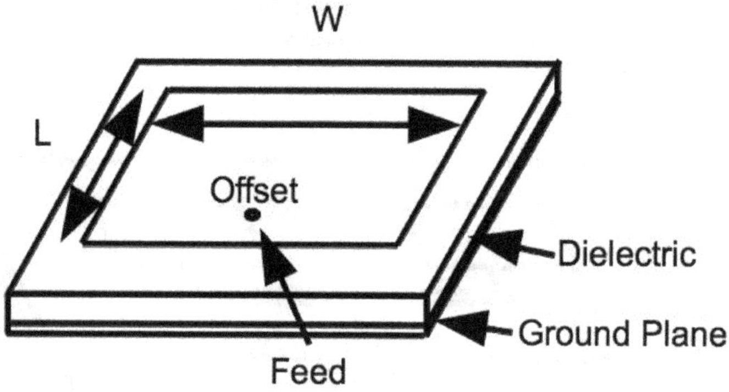


Design Approach:

*To calculate length:* *See calculations at beginning of chapter*

*For calculations for width, see same area*

Construction and Theory of Operation:

*This antenna can be thought of as 2 slots radiating in phase, one at each long edge. Together this makes a very efficient antenna, usable in a multitude of applications. Conformal and inexpensive to manufacture, this antenna can be used in an array for a very good, flat aperture. Dielectrics are typically air, teflon-fiberglass and foam. Pick a dielectric substrate that is low in loss tangent (details available from manufacturer), then use a permitivity (or Er) that is suitable for bandwidth and size. Higher Er will make the antenna smaller, less gain and higher Q (or less bandwidth).*

*Azimuth Pattern*

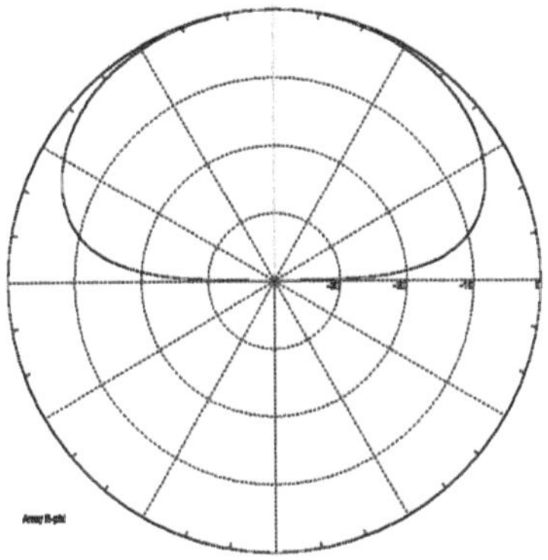

*Elevation Pattern*

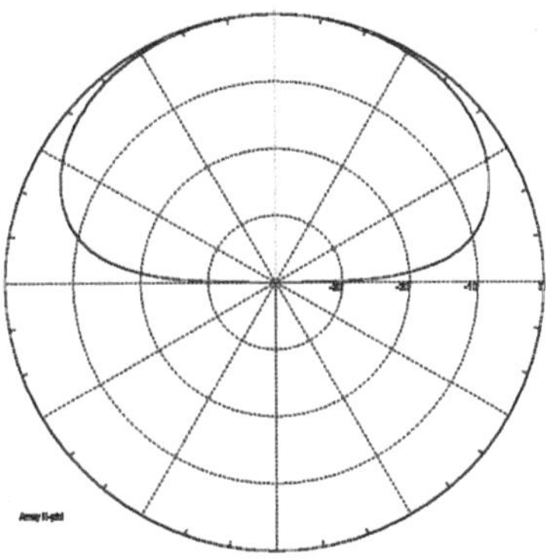

Special Considerations:

*Choose ground plane made of a good conductor, like aluminum or copper. Element can be etched, cut out, silk screened (using conductive ink) or die cut. Substrate needs to be low loss, use sheet adhesive with foams, consider over etching when using teflon fiberglass or similar materials. Ground plane should be at least 2 board thicknesses away from edge of patch.*

Testing:

*For mid sized and small apertures use far field range with transmitter/receiver or network analyzer. Near field and compact ranges work also. Use calibrated reference antenna. Use axial ratio measurements to confirm quality of required polarization. For large apertures, use radio star techniques, holography or satellites with known EIRP. Check Appendix 6 for details.*

*Use VSWR bridge, Network Analyzer or similar device to optimize match.*

Type: ***¼ Lambda patch***

Frequency of Operational: *5 Mhz to 50 GHz*
Gain Range: *2 to 6 dBi*
Bandwidth: *~ 5%*
Approximate Power Rating: *1 Kw*
Design Software: *Ensemble, IE3D, Moment Method*

Configuration:

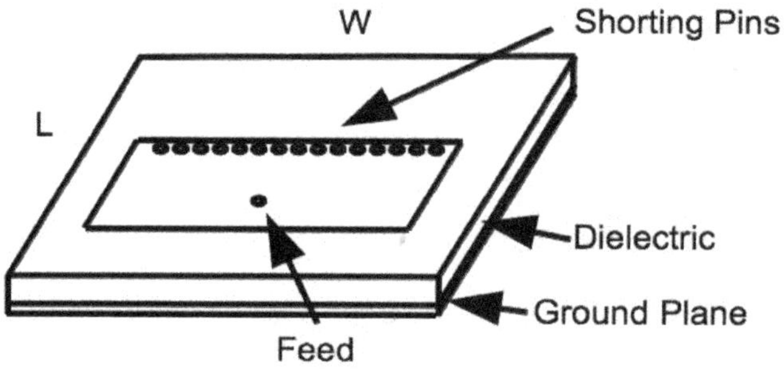


Design Approach:

*To calculate length:* *See calculations at beginning of chapter, cut this value in half. Add shorting pin length length of patch, for a complete length of ¼ lambda.*

*For calculations for width, see same area*

Construction and Theory of Operation:

*This antenna can be thought of as 1 slot radiating into space, at the long edge closest to the feed. This makes a very efficient antenna, usable in a multitude of applications. Conformal and inexpensive to manufacture, this antenna can be used in an array for a very good, flat aperture. Dielectrics are typically air, teflon-fiberglass and foam. Pick a dielectric substrate that is low in loss tangent (details available from manufacturer), then use a permitivity or Er that is suitable for bandwidth and size. Higher Er will make the antenna smaller, less gain and higher Q (or less bandwidth).*

Azimuth Pattern

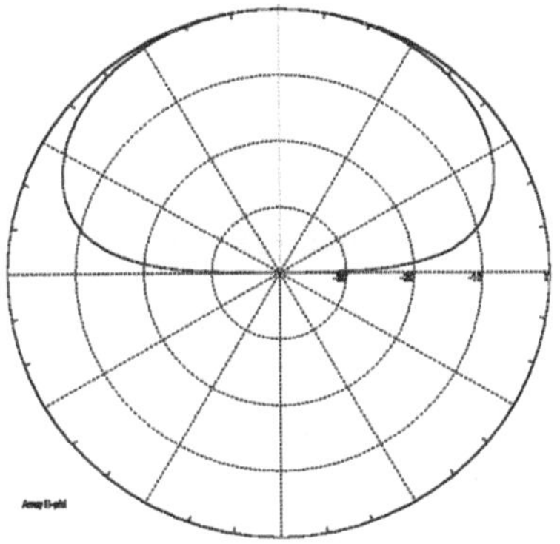

Elevation Pattern

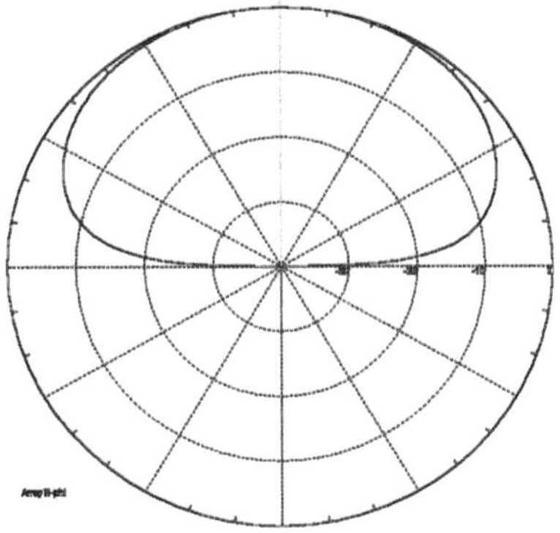

Special Considerations:

Choose ground plane made of a good conductor, like aluminum or copper. Element can be etched, cut out, silk screened (using conductive ink) or die cut. Substrate needs to be low loss, use sheet adhesive with foams, consider over etching when using teflon fiberglass or similar materials. Ground plane should be at least 2 board thicknesses away from edge of patch.

Testing:

For mid sized and small apertures use far field range wit transmitter/receiver or network analyzer. Near field and compact ranges work also. Use calibrated reference antenna. Use axial ratio measurements to confirm quality of required polarization. For large apertures, use radio star techniques, holography or satellites with known EIRP. Check Appendix 6 for details. Use VSWR bridge, Network Analyzer or similar device to optimize match.

*2nd Edition*

Type: ***CP Patch-1***

Frequency of Operational: *5 Mhz to 50 GHz*
Gain Range: *0 to 7 dBic*
Bandwidth: *~ 5%*
Approximate Power Rating: *1 Kw*
Design Software: *Ensemble, IE3D, Moment Method*

Configuration:

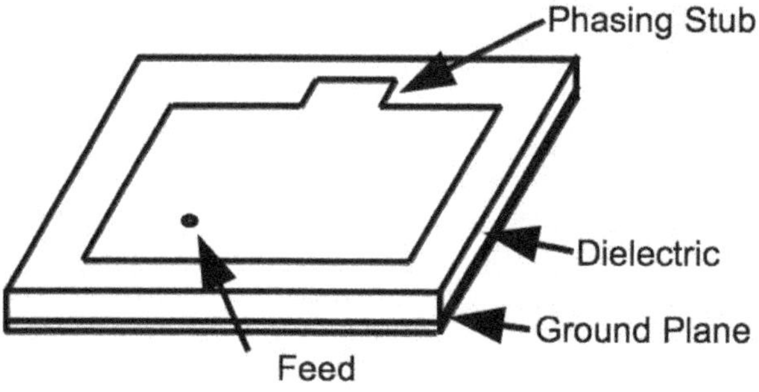


Design Approach:

*To calculate length:* *See calculations at beginning of chapter, use resonant lengths for both length and width. Use same distance from feed to edge on two sides. Adjust stub to optimize axial ratio.*

Construction and Theory of Operation:

*This antenna can be thought of as 4 slots radiating, two in phase and two 90 degrees out of phase to create the circular polarization one at edge. Together this makes a very efficient antenna, usable in a multitude of applications. Conformal and inexpensive to manufacture, this antenna can be used in an array for a very good, flat aperture. Dielectrics are typically air, teflon-fiberglass and foam. Pick a dielectric substrate that is low in loss tangent (details available from manufacturer), then use a permitivity or Er that is suitable for bandwidth and size. Higher Er will make the antenna smaller, less gain and higher Q (or less bandwidth).*

*Azimuth Pattern*

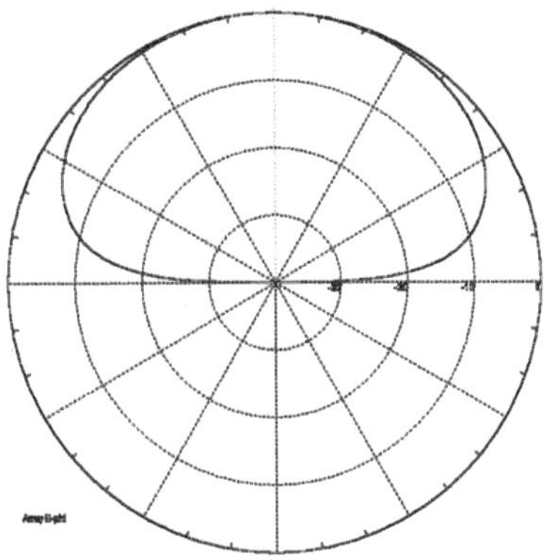

*Elevation Pattern*

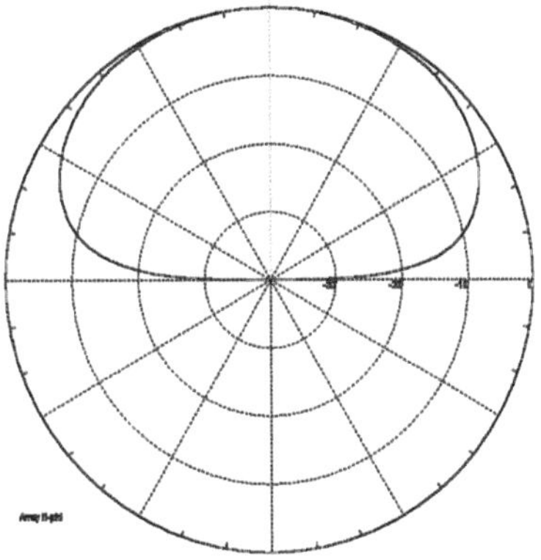

Special Considerations:

*Choose ground plane made of a good conductor, like aluminum or copper. Element can be etched, cut out, silk screened (using conductive ink) or die cut. Substrate needs to be low loss, use sheet adhesive with foams, consider over etching when using teflon fiberglass or similar materials. Ground plane should be at least 2 board thicknesses away from edge of patch.*

Testing:

*For mid sized and small apertures use far field range with transmitter/receiver or network analyzer. Near field and compact ranges work also. Use calibrated reference antenna. Use axial ratio measurements to confirm quality of required polarization. For large apertures, use radio star techniques, holography or satellites with known EIRP. Check Appendix 6 for details.*

*Use VSWR bridge, Network Analyzer or similar device to optimize match.*

Type: ***CP Patch-2***

Frequency of Operational: *5 Mhz to 50 GHz*
Gain Range: *0 to 7 dBic*
Bandwidth: *~ 5%*
Approximate Power Rating: *1 Kw*
Design Software: *Ensemble, IE3D, Moment Method*

Configuration:

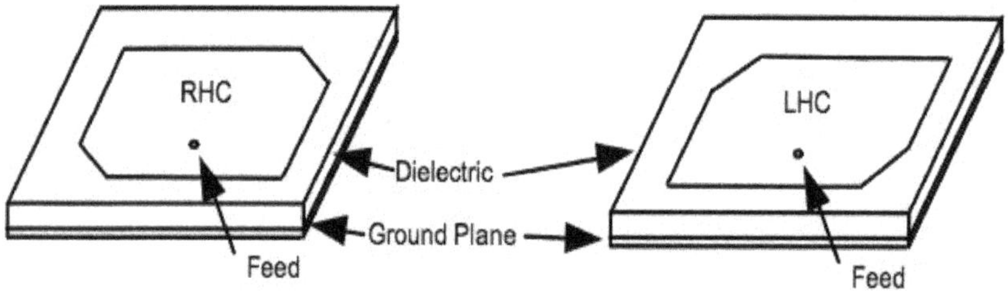


Design Approach:

*To calculate length:* *See calculations at beginning of chapter, use resonant lengths for both length and width. Use same distance from feed to edge on two sides. Adjust chamfers to optimize axial ratio.*

Construction and Theory of Operation:

*This antenna can be thought of as 4 slots radiating, two in phase and two 90 degrees out of phase to create the circular polarization one at edge. Together this makes a very efficient antenna, usable in a multitude of applications. Conformal and inexpensive to manufacture, this antenna can be used in an array for a very good, flat aperture. Dielectrics are typically air, teflon-fiberglass and foam. Pick a dielectric substrate that is low in loss tangent (details available from manufacturer), then use a permitivity or Er that is suitable for bandwidth and size. Higher Er will make the antenna smaller, less gain and higher Q (or less bandwidth).*

*Azimuth Pattern*

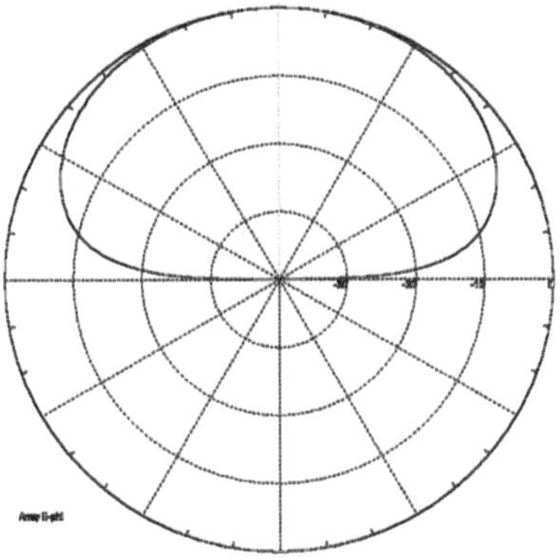

*Elevation Pattern*

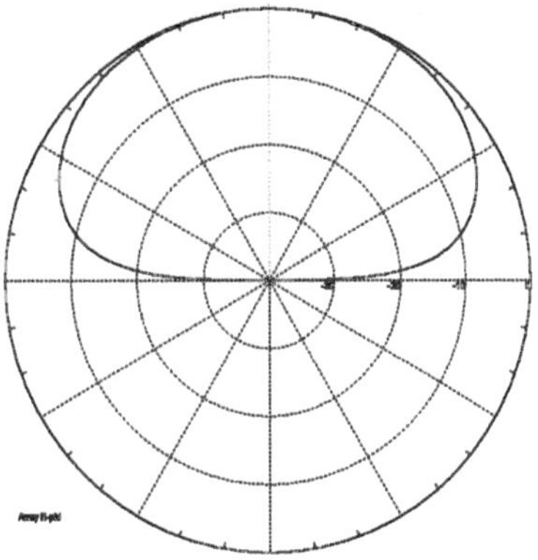

Special Considerations:

*Choose ground plane made of a good conductor, like aluminum or copper. Element can be etched, cut out, silk screened (using conductive ink) or die cut. Substrate needs to be low loss, use sheet adhesive with foams, consider over etching when using teflon fiberglass or similar materials. Ground plane should be at least 2 board thicknesses away from edge of patch.*

Testing:

*For mid sized and small apertures use far field range with transmitter/receiver or network analyzer. Near field and compact ranges work also. Use calibrated reference antenna. Use axial ratio measurements to confirm quality of required polarization. For large apertures, use radio star techniques, holography or satellites with known EIRP. Check Appendix 6 for details.*

*Use VSWR bridge, Network Analyzer or similar device to optimize match.*

Type: ***Stacked Patch***

Frequency of Operational: *5 Mhz to 50 GHz*
Gain Range: *0 to 12 dBi*
Bandwidth: *~ 5%-30%*
Approximate Power Rating: *1 Kw*
Design Software: *Ensemble, IE3D, Moment Method, HFSS*

Configuration:

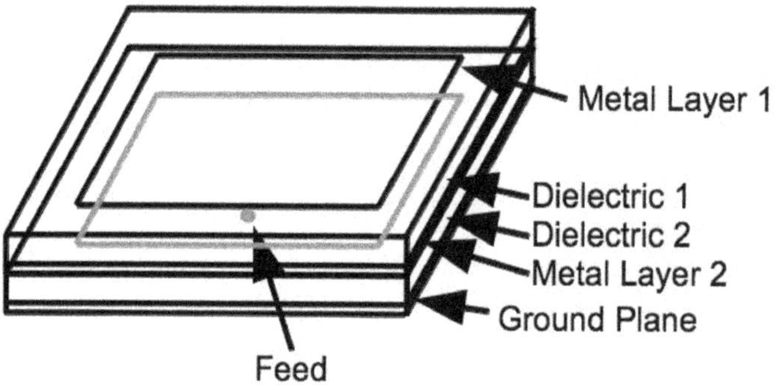


Design Approach:

*To calculate length:* *See calculations at beginning of chapter. For calculations for length and width, see same area*

Construction and Theory of Operation:

*This antenna can be thought of as 2 slots radiating in phase, one at each long edge. Together this makes a very efficient antenna, usable in a multitude of applications. Conformal and inexpensive to manufacture, this antenna can be used in an array for a very good, flat aperture. Dielectrics are typically air, teflon-fiberglass and foam. Pick a dielectric substrate that is low in loss tangent (details available from manufacturer), then use a permitivity or Er that is suitable for bandwidth and size. Higher Er will make the antenna smaller, less gain and higher Q (or less bandwidth).Stacking allows wider bandwidth operation.*

*Azimuth Pattern*

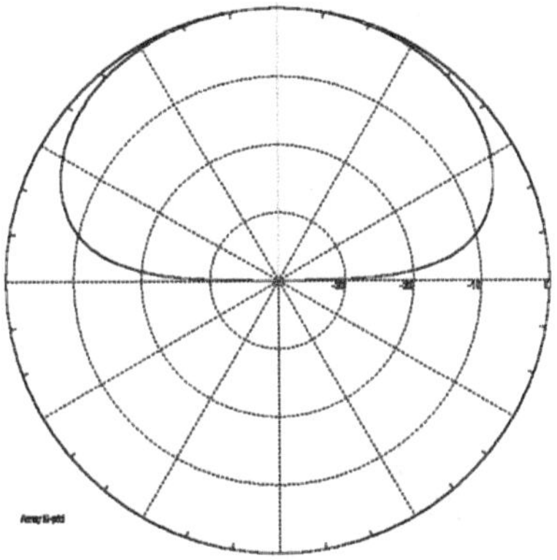

*Elevation Pattern*

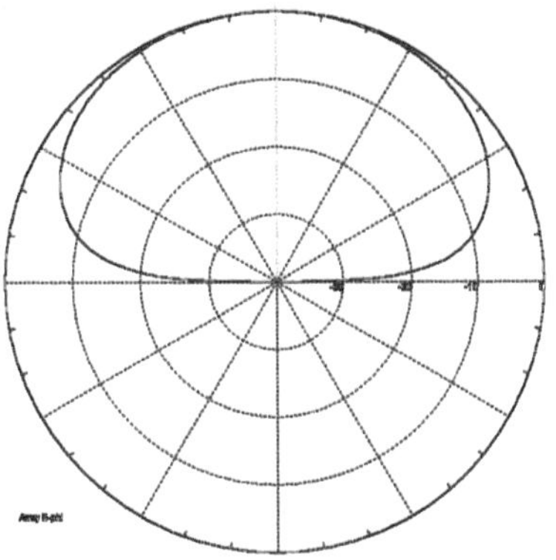

Special Considerations:

*Choose ground plane made of a good conductor, like aluminum or copper. Element can be etched, cut out, silk screened (using conductive ink) or die cut. Substrate needs to be low loss, use sheet adhesive with foams, consider over etching when using teflon fiberglass or similar materials. Ground plane should be at least 2 board thicknesses away from edge of patch.*

*Using lower dielectric constant for upper patch dielectric allows for higher gain. Model configuration first, make several small variation on dimensions as this design is very sensitive to position and size of upper radiator.*

Testing:

*For mid sized and small apertures use far field range with transmitter/receiver or network analyzer. Near field and compact ranges work also. Use calibrated reference antenna. Use axial ratio measurements to confirm quality of required polarization. For large apertures, use radio star techniques, holography or satellites with known EIRP. Check Appendix 6 for details.*

*Use VSWR bridge, Network Analyzer or similar device to optimize match.*

| | |
|---|---|
| Type: | ***Proximity fed Patch*** |
| Frequency of Operational: | *5 Mhz to 50 GHz* |
| Gain Range: | *0 to 2.12 dBi* |
| Bandwidth: | *~ 5%* |
| Approximate Power Rating: | *1 Kw* |
| Design Software: | *Ensemble, IE3D, Moment Method* |

Configuration:

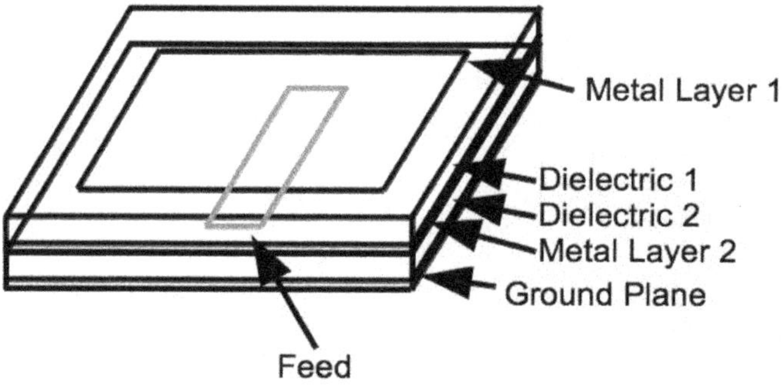


Design Approach:

*To calculate length:* *See calculations at beginning of chapter For calculations for length and width, see same area*

Construction and Theory of Operation:

*This antenna can be thought of as 2 slots radiating in phase, one at each long edge. Together this makes a very efficient antenna, usable in a multitude of applications. Conformal and inexpensive to manufacture, this antenna can be used in an array for a very good, flat aperture. Dielectrics are typically air, teflon-fiberglass and foam. Pick a dielectric substrate that is low in loss tangent (details available from manufacturer), then use a permitivity or Er that is suitable for bandwidth and size. Higher Er will make the antenna smaller, less gain and higher Q (or less bandwidth). Stacking allows wider bandwidth operation.*

*Azimuth Pattern*

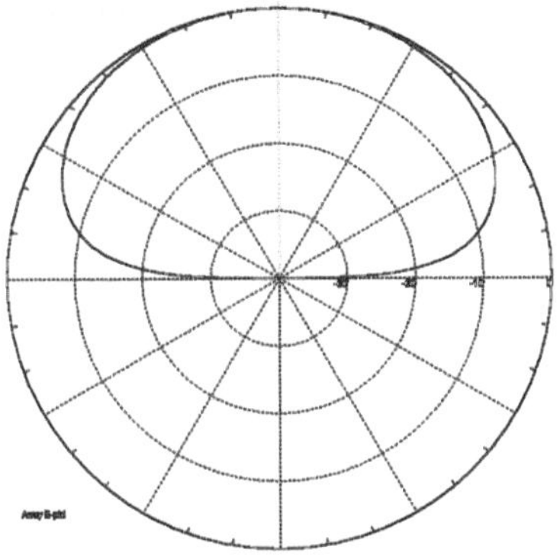

*Elevation Pattern*

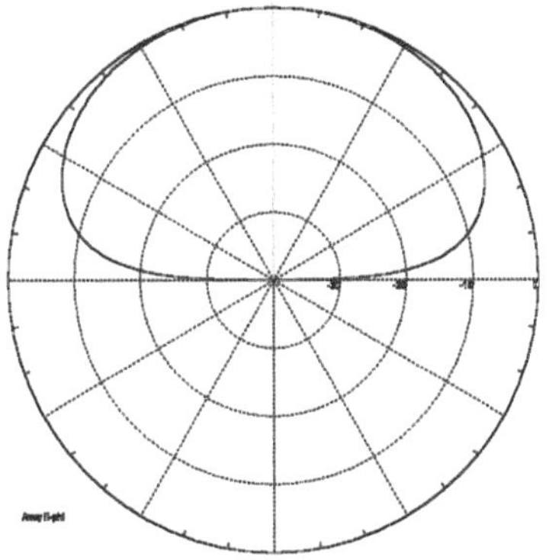

Special Considerations:

*Choose ground plane made of a good conductor, like aluminum or copper. Element can be etched, cut out, silk screened (using conductive ink) or die cut. Substrate needs to be low loss, use sheet adhesive with foams, consider over etching when using teflon fiberglass or similar materials. Ground plane should be at least 2 board thicknesses away from edge of patch.*

Testing:

*For mid sized and small apertures use far field range with transmitter/receiver or network analyzer. Near field and compact ranges work also. Use calibrated reference antenna. Use axial ratio measurements to confirm quality of required polarization. For large apertures, use radio star techniques, holography or satellites with known EIRP. Check Appendix 6 for details.*

*Use VSWR bridge, Network Analyzer or similar device to optimize match.*

**Questions:**

1. Suggest an antenna type that has the following characteristics:

   a. Frequency of 200 to 220 MHz
   b. Gain of 15 dBi
   c. Linear Polarization

2. Your boss tells you price is a huge factor in the design of antenna with the following characteristics:

   a. Frequency of 2000 MHz
   b. Gain of 20 dBi
   c. Linear Polarization

   Keep in mind that material choice is the best way to control costs.

3. Suggest an antenna design for the next generation rocket booster that is capable of going to the moon, with the following properties:

   a. Frequency of 2200 MHz
   b. Gain of 3 dBi
   c. Circular Polarization

4. What would the effect of multiple stacks do to bandwidth ?

5. If the frequency of operation of two bands are far apart, can you use the radiating surface of one microstrip patch as the ground plane of another ?

6. What happens when you feed the corner of a square patch and do not use a tuning stub, e.i., not a CP design ?

7. What is the best, most efficient dielectric ?

8. When designing a patch for use in space, what coating would you apply to the surface ?

# Chapter 7 – Slot Antennas

Slot Antenna Types

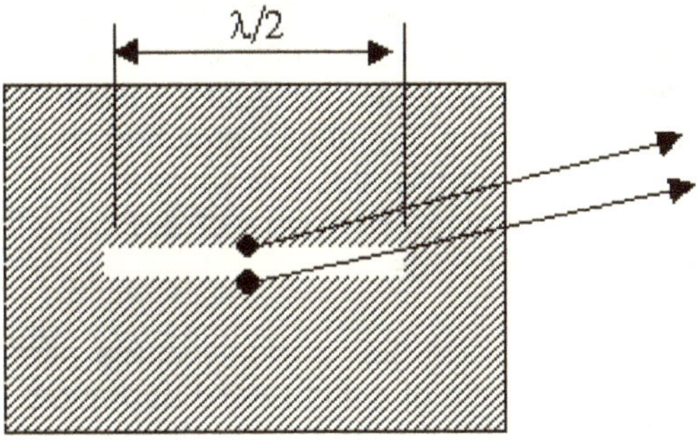


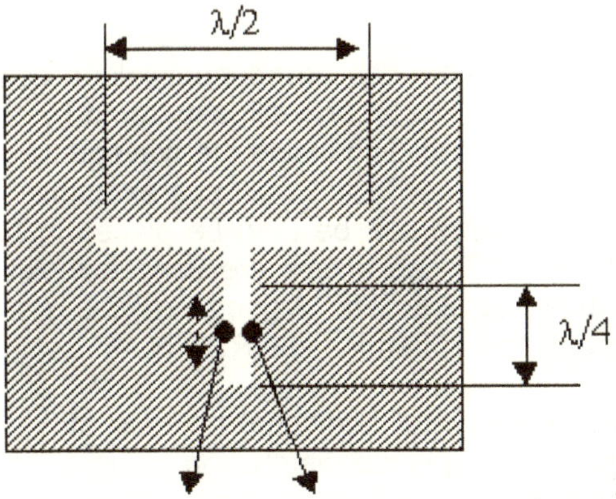

**Babinet's principal**

In optics, this is the principle that two diffraction screens, one being exactly the negative of the other, will form the same diffraction pattern.

Extended to electromagnetics, this principle can be used to find complementary electrical characteristics of antennas. This is where a large ground plane with a slot ½ wavelength long and 1/10 wavelength wide has the same pattern and gain as a dipole of similar dimensions. The dipole is fed in the center with the center conductor (assuming coax) going to one ¼ wavelength segment and the shield attached to the other ¼ wavelength segment. In the case of the slot, the same coax has its shield attached at the midpoint of the slot on one side of the gap, the center conductor is attached on the opposite side.

Adaptations to the slot design include alternate shapes both with equal width or in the case of a wide band approach, flaring out of the slot in complement to the bowtie design type of dipole.

For best match, adjust the width of the slot as necessary.

**Slot array**

An array of slots can be economically produced using waveguide. The slots are cut into the broad wall, separated by ½ wavelength. Due the phase of the currents residing inside the waveguide, the slots alternate from upper broad wall to lower broad wall, see examples in the following pages. A series of waveguides can be then be placed in parallel to create a two dimensional array. These arrays are used extensively in aircraft weather radars operating near 8.5 GHz. A good example is shown in the "Interesting and unusual antenna appendix; the RDOW.

There are several additional benefits of a slot array including the ability to taper the amplitude distribution using a Taylor taper for instance, thus lowering sidelobe amplitudes. An addition, the arrays can be scanned along the long dimension of the waveguide by simply changing frequency. As the wavelength changes so does the time delay of the radiation pattern from slot to slot, thereby changing the beam position. This technique is used in military and commercial radars as a beam height controller while the array is scanned mechanically in azimuth.

Several satellite designs incorporate slotted circularly polarized omni antennas for use as beacons. The slots are set a 45 degrees relative to each other in a circular fashion, as if a waveguide was bent into a circle. Ground planes are used to further shape the beam as necessary. Left Hand and Right Hand arrays can be attached in close proximity using a common ground plane to minimize interaction and optimize cross pole isolation.
One array is attached above and one below the ground plane, which extends at least one wavelength beyond the face of the arrays.

**Conformal Applications**

Another advantage of a slot antenna or slot array is its ability to be conformal, useful in aircraft design or other vehicles. Many of these antennas have been used in aerospace, from rockets to airliners. The vertical stabilizers on many intercontinental airliners have a slot built in for HF use over the oceans, this can be found on the leading edge of the structures.

**Polarization**

The polarization of a slot is orthogonal to the opening, opposite that of a dipole, whose polarization vector coincides with its physical structure. Crossed slots have been designed to accommodate both vertical and horizontal polarization. Circularly polarized slot arrangements have also been designed, useful for GPS and other space borne applications.

**Slotted Cylinders**

Taking a slot design and folding the ground plane into a cylinder allows for some good designs where a monopole layout is needed for restricted space. In this case a vertically positioned slot cylinder antenna has horizontal polarization.

Typical dimensions are:

| | |
|---|---|
| Cylinder Diameter: | .125 lambda |
| Slot Length: | .75 lambda |
| Slot width: | .02 lambda |

The small diameter makes the slotted cylinder antenna usable at frequencies too low for waveguide to be practical.

The patterns for this type of antenna are omni in the horizontal plane and dipole like in the vertical plane. Feeding the slot entails using a balun, normally 50 ohm coax to 180 ohm feed, that is placed across the center of the slot. Gain for this type of antenna is around 3 dBi. Instead of a solid, folded ground plane, a skeleton ground plane can be made with good performance. The "ribs" are placed about .25 lambda apart, with a slot made by a gap in the ribs. Wind resistance is low in this configuration, and many have been been designed for use in the television industry.

**Alford Slot Antenna**

If the slot is extended to around 2 lambda, the cylindrical antenna have have more gain, up to 6 dBi. The dimensions become critical however and adjustments must be made in small increments. Again several TV and FM broadcast antennas have been made in this way, where there is minimum bandwidth and performance can be optimized. The pattern is again omni in the horizontal plane, with horizontal polarization. The beamwidth in the vertical plane is restricted but dipole like in appearance. A balun like feed system is recommended for use in this type antenna. The slot impedance is around 200 ohms in this approach.

An example Alford Slot antenna at 1296 MHz has the following dimensions:

| | | |
|---|---|---|
| Slot Width: | 8 mm | (.035 lambda) |
| Slot Length: | 510 mm | (2.2 lambda) |
| Tube O.D. | 35 mm | (.15 lambda) |
| Tube thickness | 1.1 mm | (.002 lambda) |

Slotted Waveguide Radar Antenna fed with a slotted waveguide which enables frequency dependent beam steering

**Design Examples**

The following pages contain examples of antennas discussed in this chapter. These examples represent a first order design approach. For a more complete and optimized antenna solution, use the example closest to the requirement needs and optimize by using good simulation software, creating prototypes, testing and finally modifying prototypes to optimize performance.

| | |
|---|---|
| Type: | ***½ Wave Slot*** |
| Frequency of Operational: | 100 *Mhz to 5 GHz* |
| Gain Range: | *0 to 2 dBi* |
| Bandwidth: | *~ 5%* |
| Approximate Power Rating: | *1 Kw* |
| Design Software: | *Ensemble, IE3D, Moment Method* |

Configuration:

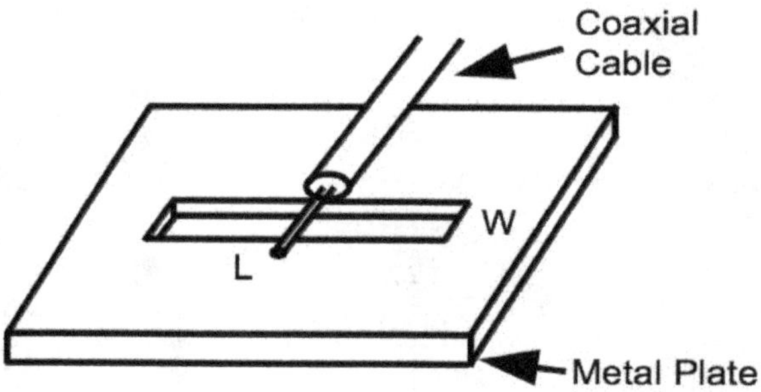


Design Approach:

*To calculate length:* *L = ~.5 lambda*
*W = ~.1 lambda*

Feed Point = Centered across gap with ground of coax on one side and center of coax on opposite side.

Construction and Theory of Operation:

*This antenna is one of the simplest to design and build. It can be manufactured by cutting a slot on a metal ground plane and feeding it across the center with coax. The characteristics of the slot are very similar to a dipole in both gain and pattern. Obviously this antenna is one of the least expensive to build. It can be made into round or other conformal structures. It can also be made in the form of arrays as well as covered with radome material. Babinet's principle applies to the operation of these antennas; including the interesting fact that the primary polarization vector is 90 degrees from the orientation of the slot axis.*

*Azimuth Pattern*

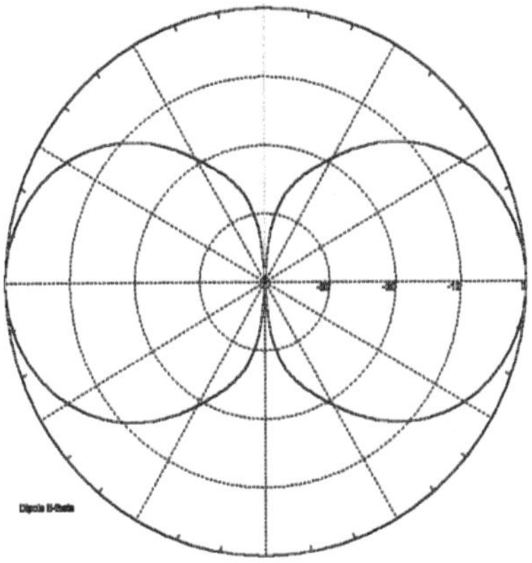

*Elevation Pattern*

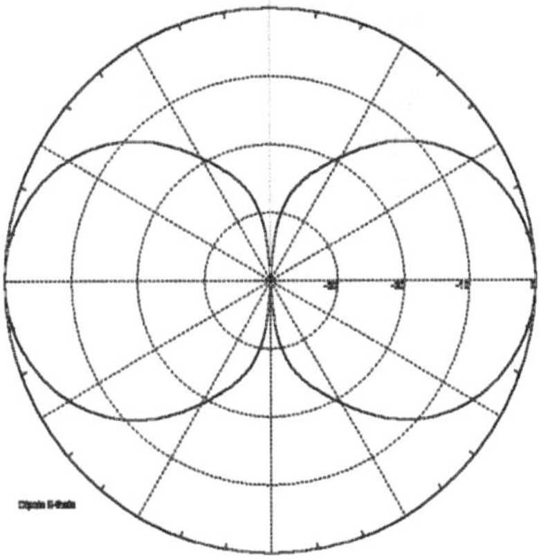

Special Considerations:

*Make Metal plate from good conductor, like aluminum or copper. Adjust width and length slightly to optimize match. Make plate large enough so that at least one width of space is between the edge of the plate and any open area of the slot. Center position of coaxial feed.*

Testing:

*For mid sized and small apertures use far field range with transmitter/receiver or network analyzer. Near field and compact ranges work also. Use calibrated reference antenna. Use axial ratio measurements to confirm quality of required polarization. For large apertures, use radio star techniques, holography or satellites with known EIRP. Check Appendix 6 for details.*

*Use VSWR bridge, Network Analyzer or similar device to optimize match.*

Type: ***Serpentine Slot***

Frequency of Operational: *5 Mhz to 5 GHz*
Gain Range: *0 to 2.12 dBi*
Bandwidth: *~ 5%*
Approximate Power Rating: *1 Kw*
Design Software: *Ensemble, IE3D, Moment Method*

Configuration:

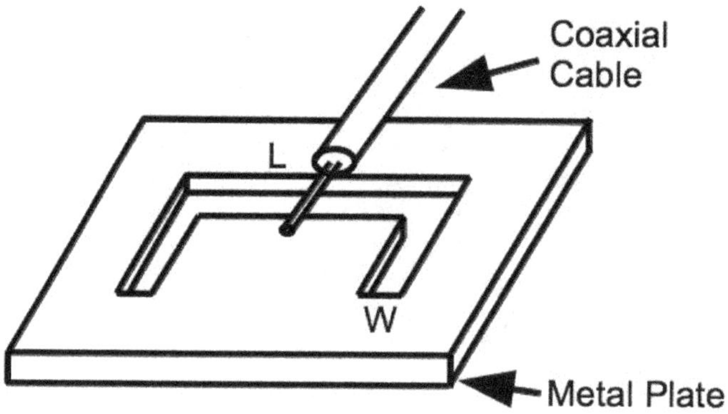


Design Approach:

*To calculate length:* *L = ~.5 lambda*
*W = ~.1 lambda*

Construction and Theory of Operation:

*This antenna is one of the simplest to design and build. It can be manufactured by cutting a slot on a metal ground plane and feeding it across the center with coax. The characteristics of the slot are very similar to a dipole in both gain and pattern. Obviously this antenna is one of the least expensive to build. It can be made into round or other conformal structures. It can also be made in the form of arrays as well as covered with radome material. Babinet's principle applies to the operation of these antennas; including the interesting fact that the primary polarization vector is 90 degrees from the orientation of the slot axis.*

*Azimuth Pattern (E)* *Elevation Pattern (H)*

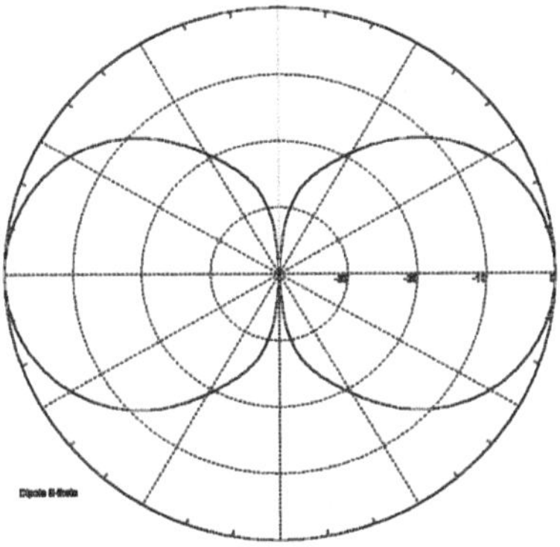

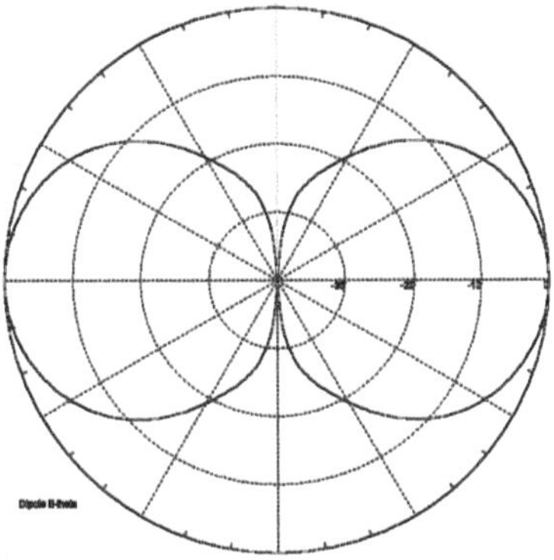

Special Considerations:

*Make Metal plate from good conductor, like aluminum or copper. Adjust width and length slightly to optimize match. Make plate large enough so that at least one width of space is between the edge of the plate and any open area of the slot. Center position of coaxial feed. Length includes all leg lengths, measure from center of slot.*

Testing:

*For mid sized and small apertures use far field range with transmitter/receiver or network analyzer. Near field and compact ranges work also. Use calibrated reference antenna. Use axial ratio measurements to confirm quality of required polarization. For large apertures, use radio star techniques, holography or satellites with known EIRP. Check Appendix 6 for details.*

*Use VSWR bridge, Network Analyzer or similar device to optimize match.*

Type: ***CP Slot***

Frequency of Operational: *5 Mhz to 50 MHz*
Gain Range: *0 to 2.12 dBi*
Bandwidth: *~ 5%*
Approximate Power Rating: *1 Kw*
Design Software: *Ensemble, IE3D, Moment Method*

Configuration:

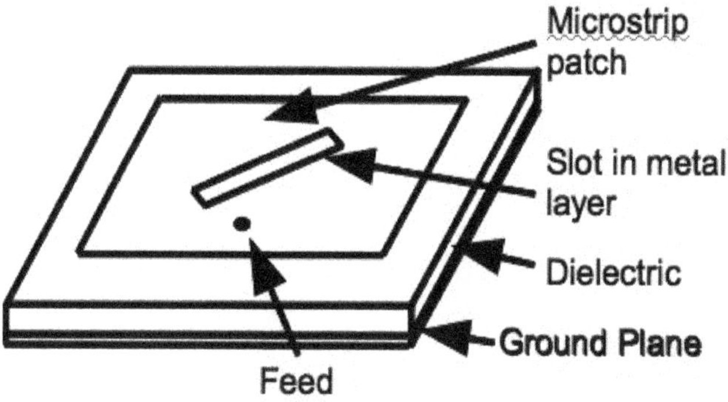


Design Approach:

*Use microstrip patch calculations, add slot to square patch, set at 45 degrees. Adjust width and length to optimize gain and minimize axial ratio. Bandwidth will be about 30% that of linear patch.*

Construction and Theory of Operation:

*Essentially a standard square microstrip patch with a ~1/4 wavelength long slot centered in the patch at 45 degrees. The fields generated by such a structure allow for a 90 degree phase shift between the vertical and horizontal polarizations, making the wavefront circular in nature. The angle of the slot is either +45 degrees for Left Hand Circular or -45 degrees for Right Hand Circular. The sample shown above is an example of Left Hand. Expect a loss of bandwidth, both in term of impedance as well as gain on the order of 70%.*

*Azimuth Pattern*

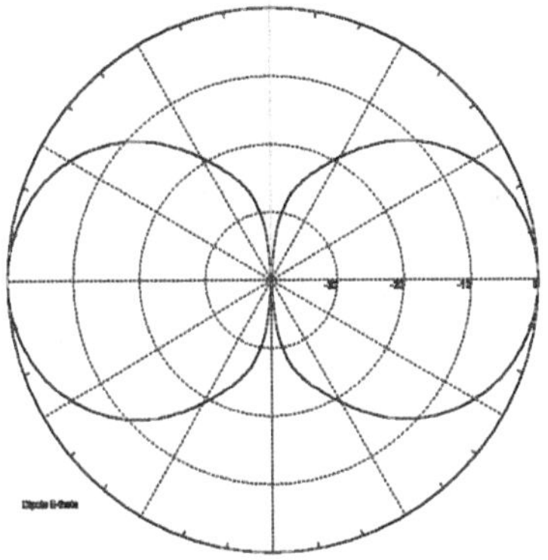

*Elevation Pattern*

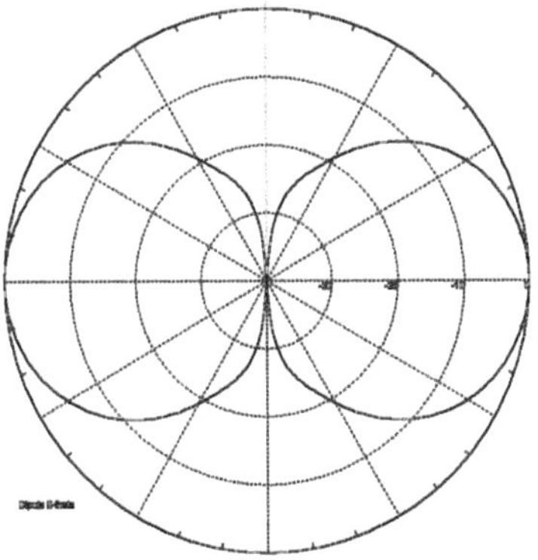

Special Considerations:

*Choosing the right dielectric material will effect bandwidth and beamwidth. In general the higher the dielectric constant the greater the bandwidth also, because the patch will be small as the dielectric constant is increased, expect lower gain as a consequence. The dielectric material's thickness also has a lot to do with bandwidth, generally, the thinner the lower the bandwidth. As noted above, the bandwidth for circularly polarization is about 30% that of a linear.*

Testing:

*For mid sized and small apertures use far field range with transmitter/receiver or network analyzer. Near field and compact ranges work also. Use calibrated reference antenna. Use axial ratio measurements to confirm quality of required polarization. For large apertures, use radio star techniques, holography or satellites with known EIRP. Check Appendix 6 for details. Use VSWR bridge, Network Analyzer or similar device to optimize match.*

| | |
|---|---|
| Type: | ***Slot Array*** |
| Frequency of Operational: | *500 Mhz to 50 GHz* |
| Gain Range: | *2 to 18 dBi* |
| Bandwidth: | *~ 5%* |
| Approximate Power Rating: | *10 Kw* |
| Design Software: | *FDTD, IE3D, Moment Method* |

Configuration:

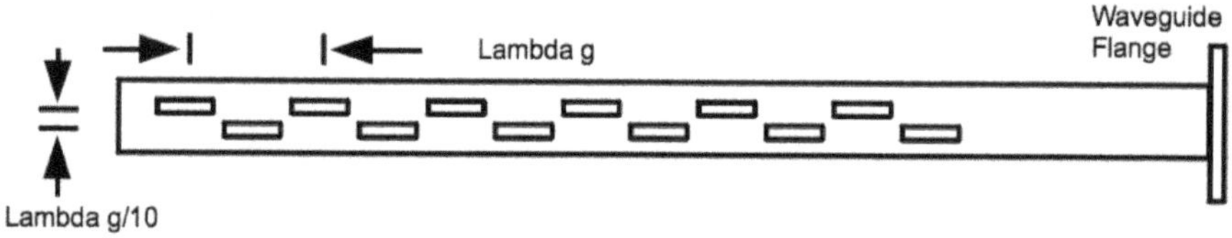


Design Approach:

*Use available software packages to determine optimal slot width, position and performance. Select waveguide for frequency of operation.*

Construction and Theory of Operation:

*Slot Arrays made of waveguide take advantage of the position of the fields insides the waveguide and place the slots over the peak energy positions. The offset between each successive slot is at guide lambda/2, compensating for the opposite polarity of current flow. If the waveguide is terminated with a load, this is known as a non-resonant design and will have wide band performance and as a consequence of the load, loosing some energy as well. A short on the waveguide makes the design a resonant, lower and higher Q approach. This type of antenna can operate with significant amounts of power and is commonly used in radar systems.*

Azimuth Pattern

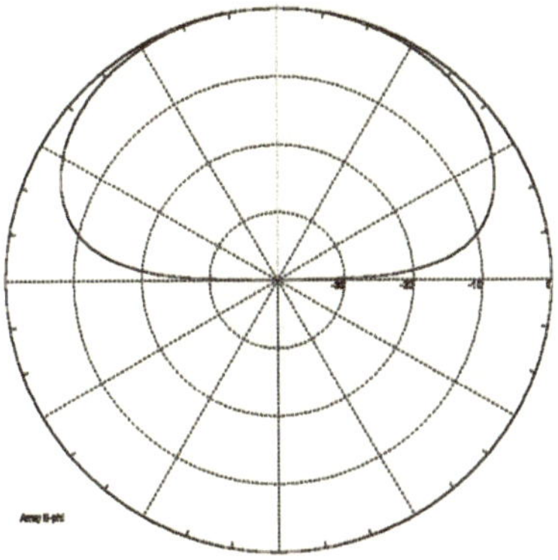

Elevation Pattern

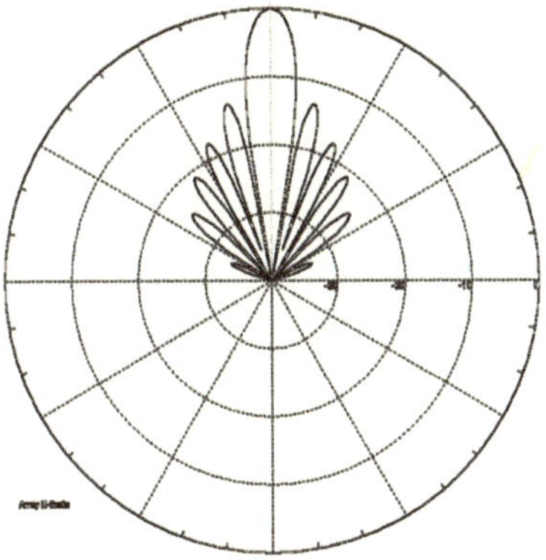

Special Considerations:

*Make sure the frequency of operation is not too close to the waveguide cutoff. In a non-resonant design, calculate the amount of average power to be used in the case of a transmitter, make sure the load in the termination end can handle the power. If several arrays are to be placed in a square configuration, be aware of the mutual coupling issues, more of a concern for short wall slots as apposed to broad wall slots.*

Testing:

*For mid sized and small apertures use far field range with transmitter/receiver or network analyzer. Near field and compact ranges work also. Use calibrated reference antenna. Use axial ratio measurements to confirm quality of required polarization. For large apertures, use radio star techniques, holography or satellites with known EIRP. Check Appendix 6 for details.*

*Use VSWR bridge, Network Analyzer or similar device to optimize match.*

**Questions:**

1. Suggest an antenna type that has the following characteristics:

   a. Frequency of 200 to 250 MHz
   b. Gain of 15 dBi
   c. Linear Polarization

2. Your boss tells you price is a huge factor in the design of antenna with the following characteristics:

   a. Frequency of 2000 MHz
   b. Gain of 20 dBi
   c. Linear Polarization

   Can waveguides be made of plastic?

3. Suggest an antenna design for the next generation rocket booster that is capable of going to the moon, with the following properties:

   a. Frequency of 2200 MHz
   b. Gain of 3 dBi
   c. Circular Polarization

4. Slot antennas are designed using the absence of metal, not the presence. How can you take advantage of this feature?

5. Research the term "annular ring" and design an antenna at 1575.43 MHz using this approach. What are the advantages and disadvantages of this design?

6. Design an Alford array at 1800 MHz with a gain of 12 dBi. What type of horizontal beam will it have?

7. How can you design a slot antenna with circular polarization ?

8. How can slot antennas be used as a filter?

# Chapter 8 – Helical Antennas

GPS Satellite using Helical Antennas

## Fundamentals

The helical antenna, first conceived of by John Kraus in 1947, consists of a conductive element wound in the form of a helix. This antenna type is normally mounted on a ground plane and consists of two basic types:

*Normal Mode* or Broadside beam

*Axial Mode* or End fire beam

In the normal mode, the dimensions of the helix are small compared to its wavelength of operation. These types of antenna act as circularly polarized monopoles are are useful for GPS receivers and other CP applications where broad or even hemispherical beam patterns are desired.

In the axial mode, the circumference is about one wavelength and the space between the coils is about .25 lambda. The length of the coil determines the gain, the longer the higher the gain. The sense of the circular polarized wave is determined by how the coil is wound. A right hand turn as seen from the transmitter to the receiver delineates right hand circular etc.

Adding the output of a left hand to a right hand circular pair of helixes yields a linear polarization. Using three helixes in close proximity and rotating the center antenna physically allows for the scanning of the resultant beam from the array.

Terminal impedance is between 100 and 200 ohms, with the resistive part of the impedance approximated by:

$$R = 140 * \left(\frac{C}{\lambda}\right) \quad [41]$$

where:

C is the circumference

lambda is the wavelength

The maximum directivity can be approximated by:

$$D = 15 * N * \left(\frac{(C^2 * S)}{\lambda^3}\right) \quad [42]$$

where:

N = Number of turns in the helix

C = Circumference

S = Spacing between turns

lambda = wavelength

The half-power beamwidth (3 dB points) can be approximated by:

$$HPBW\,(Degrees) = \frac{(52 * \lambda^{(\frac{3}{2})})}{(C * \sqrt{(N * S)})}$$ [43]

The beamwidth between nulls can be approximated by:

$$HPBW\,(Degrees) = \frac{(115 * \lambda^{(\frac{3}{2})})}{(C * \sqrt{(N * S)})}$$ [44]

**Applications**

Helical antenna have been used extensively in aerospace applications and terrestrial communications. For instance the Orbcomm constellation of satellites use collapsable VHF antennas that unfurl in space. In addition, the GPS constellation of satellites uses an array of helixes to transmit the spread spectrum information that allow earth bound receivers to determine time and location.

This type of antenna is also used for many applications on the ground, notably NASA communication sites for satellite uplinks and downlinks. Early in satellite history circular polarization was necessary to mitigate the effects of Faraday rotation, whereby a satellite broadcasting at lower frequency suffers from polarization rotation due to fluctuations in the ionosphere.

In addition, point to point communications are achieved easily with this antenna. Due to the very high cross pole isolation and wide bandwidth, these antenna can be ideal in an environment where hide data volume is required.

**Gain**

Like a yagi, the longer the antenna the higher the gain, after about 20 turns however it is more advantageous in terms of construction to array multiple antennas to achieve higher gain.

**Pattern**

Again, the pattern performance is like a yagi with the exception of the helix has a superior front to back ratio due to the ground plane on most designs.

**Polarization**

Helix antenna are inherently circular in polarization with the sense of rotation in construction being the same as the resultant polarization sense. In array designs if both right hand and left hand polarizations are required, placing opposite sense antennas in adjacent rows decreases the amount of mutual coupling effects that can detune an antenna and / or cause scan blindness.

**Assembly Techniques**

Helix antennas are normally held in place by forms, which might be poles in the case of large antenna and foam or other low loss material in the case of smaller varieties. Non form antennas are possible, necessitating strong wire or pipe for the helix element. The propensity to bend or otherwise detune the antenna is usually mitigated by the use of a form. In addition, using forms allow the helical element to be made from tape or wire, lowering the weight while keeping the strength.

As mentioned before, in an array of helical antennas, it is advantageous to use cups for reflectors and if the design requires both left hand and right hand polarization, placing opposite polarizations next to each other lowers the mutual coupling effects.

Tapering the helical element can increase operating bandwidth, these types are almost always wound on forms. Also, there is an advantage to tapering the helix from the bottom, near the ground plane to the top in terms of line thickness.

The quadrifilar helix is not a traveling wave helix in the sense of the axial mode helix, but two twisted loops (left-hand helix produces right hand polarization). Each twisted loop produces bidirectional circular polarization (same sense), but the feed causes cancellation in one direction and produces a unidirectional circularly polarized pattern.

The quadrifilar antenna is fed by running a coax through the center of the form to the top where it connects to the radiating elements. At the bottom of the antenna the elements are connected together and to ground, see diagram in following individual descriptions. These antennas have very good axial ratio over broad beamwidths. Model the proposed design to optimize gain, pattern and match.

**Design Options**

There are several interesting modifications of helix antenna designs to minimize mutual coupling effects in arrays or increase bandwidth. The following is a short list of some of the more popular modifications.

**Cups**

Helical elements can be placed in cups, then onto a ground plane. These cups are on the order of twice the diameter in width and one quarter the diameter in depth. These cups contain the fields at the base of the antennas and allow closer placement of element in an array setting.

**Mechanical rotation (beam steering)**

Kraus discovered that if three helixes are placed in an array, and the outside helices are rotated in opposite directions, the resultant combined beam will sweep from one position along the longitudinal axis of the array to the opposite side for every n rotations of the outboard helices.

**Array**

Arrays of helices can be made with good bandwidth and good gain characteristics. If, however the polarization is the same for all elements, mutual coupling effects have a dilitorious effect on beam steering and sidelobe performance. If the designer has the option to use both right hand and left hand polarizations, as in the case of consumer satellite reception, columns of right hand polarized elements can be closely placed to columns of left hand polarized elements with minimal mutual coupling effects.

### Log Periodic (conical)

An interesting style of helix was designed as a combination standard helix and log periodic. This style is best described as a conical helix and has very good wide band characteristics. A design example is included in the latter part of this chapter.

### Taper

There have been helical designs where the end 1/3 of the helix is tapered to about two thirds of its original diameter. There has come positive effects in beam symmetry and wide band performance.

### Counter wound for linear

If a helix of one polarization sense is combined in an array with another helix of the opposite sense, the resultant polarization sensitivity will be linear, not circular. Also, if a left hand polarized helix is added to the end of a right hand polarized helix of equal length, the resultant polarization will also be linear.

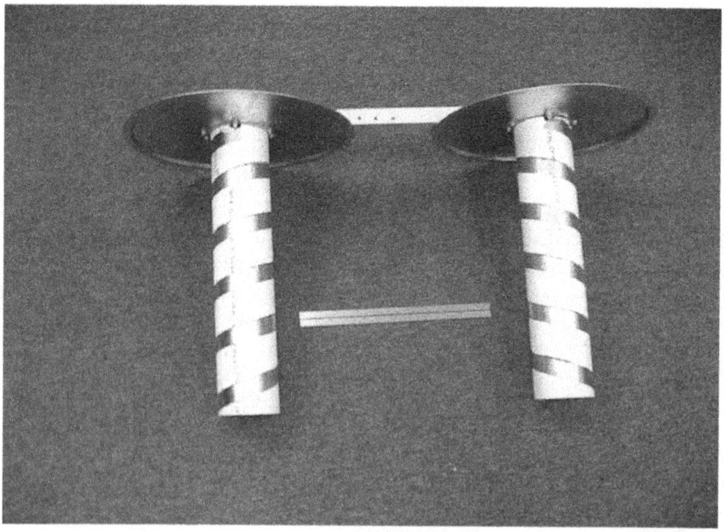

Example of left hand and right hand wound helixes (75)

John D. Kraus's Helix Array, used for Radio Astronomy

**Design Examples**

The following pages contain examples of antennas discussed in this chapter. These examples represent a first order design approach. For a more complete and optimized antenna solution, use the example closest to the requirement needs and optimize by using good simulation software, creating prototypes, testing and finally modifying prototypes to optimize performance.

| | |
|---|---|
| Type: | ***Standard Helix*** |
| Frequency of Operational: | *10 Mhz to 14 GHz* |
| Gain Range: | *0 to 30 dBic* |
| Bandwidth: | *~ 20%* |
| Approximate Power Rating: | *1 Kw* |
| Design Software: | *NEC-WIN Pro, IE3D, Moment Method* |

Configuration:

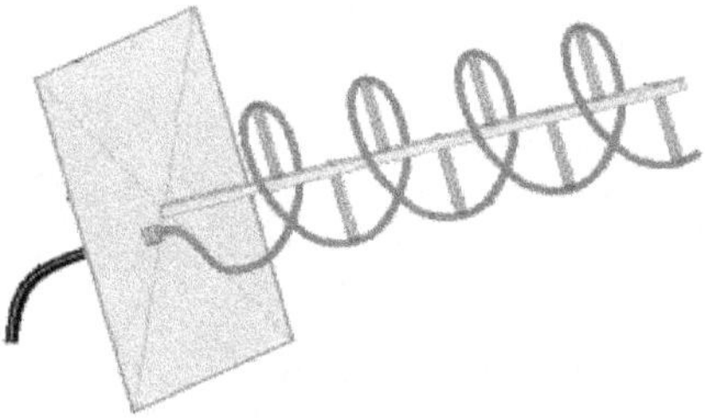

Design Approach:

*Circumference = 1 lambda*
*Spacing (S) = ~.1 lambda*
*Number of turns = ~10 for 14 dBic, 20 for 16 dBic, 30 for 17.5 dBic*
*L(meters) = 75/ f(MHz)*
*Feed Point = ~140 ohms at base of helix, move up conductor for 50.*

Construction and Theory of Operation:

*This antenna is a simple design with good performance characteristics. The helix can be made in array form, summed with and opposite turned helix for linear polarization. This antenna can also be tapered to increase broadband performance and improve axial ration. Bandwidths in excess of 30% can be achieved. Various other configurations including the reflector shape can be implemented to produce optimum antennas for specific applications.*

*Azimuth Pattern*

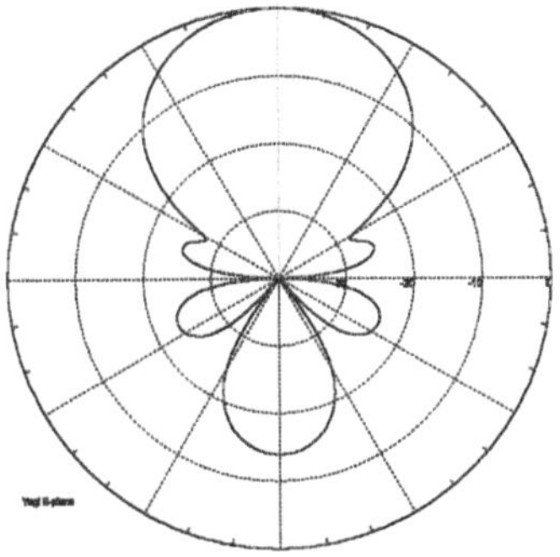

*Elevation Pattern*

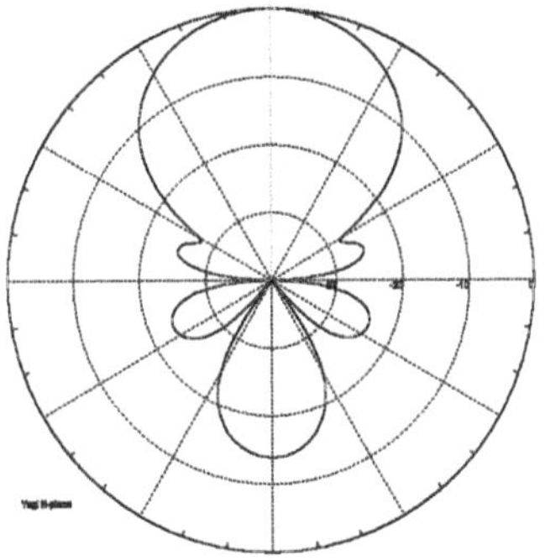

Special Considerations:

*Make a reflector (R ) out of aluminum mesh or plate, size at least 2 lambda per side. Support structure (B) can be made of fiberglass or aluminum tubing, though non conductive material is preferable. Cross supports (E) must be made of non-conductive material like fiberglass or PVC. Feed cable (C ) can be 50 ohm coax, matching can be accomplished by grounding end of helix and placing feed point away from the end. In this approach, a sliding strap is useful to find optimum position of feed.*

Testing:

*Far field range with transmitter/receiver or network analyzer. Near field and compact ranges work also. Use calibrated reference antenna. Use axial ratio measurements to confirm quality of circular polarization. Check Appendix 6 for details.*

| | |
|---|---|
| Type: | ***Tapered Helix*** |
| Frequency of Operational: | *5 Mhz to 5 GHz* |
| Gain Range: | *0 to 2.12 dBi* |
| Bandwidth: | *~ 5%* |
| Approximate Power Rating: | *1 Kw* |
| Design Software: | *NEC-WIN Pro, IE3D, Moment Method* |

Configuration:

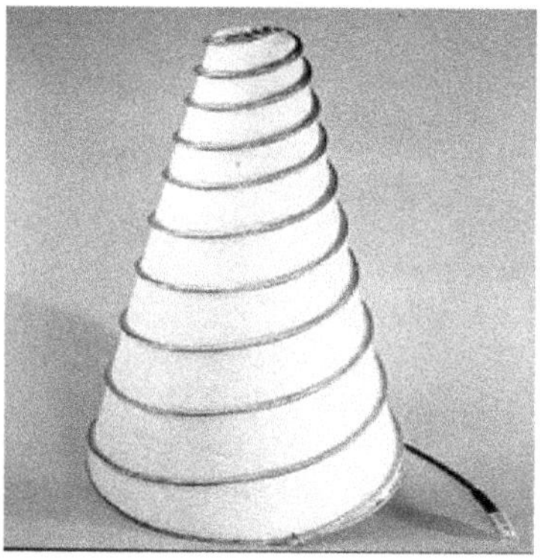

Design Approach:

*Circumference = 1 lambda at lowest frequency*
*Spacing (S) = ~.25 lambda*
*Number of turns = ~13 for 14 dBi, 23 for 16 dBi, 33 for 17.5 dBi*

Construction and Theory of Operation:

*This is a helix placed on a non conductive cone which enhances the bandwidth to a great degree. Used for wideband, multi-octave applications such as radio astronomy, surveillance and ultra-wideband radar. Made of 1,2 or 4 arms, with a feed at the center top of the structure. Usually requires a balun to match. Taper angles have been extensively evaluated and depend on required gain and bandwidth.*

*Azimuth Pattern*

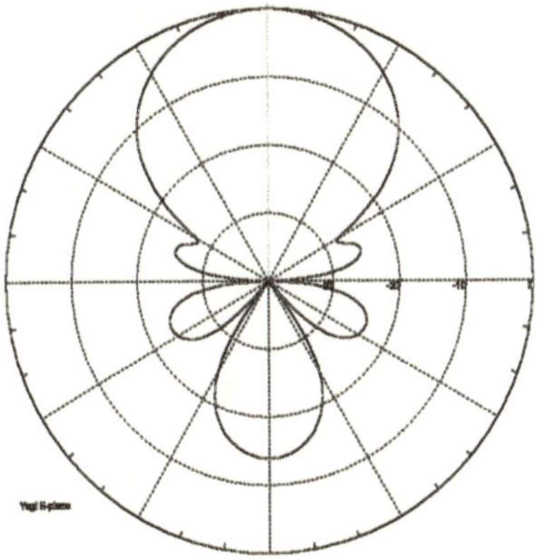

*Elevation Pattern*

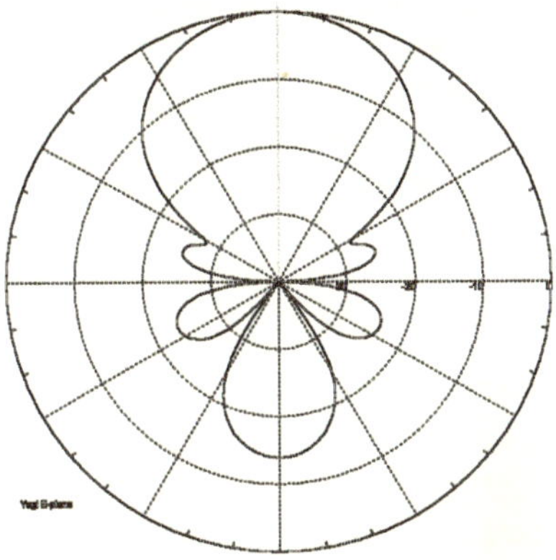

Special Considerations:

*For form this antenna is wound on cannot be conductive. Wideband balun necessary for multi arm designs. Ends of arms not terminated.*

Testing:

*Far field range with transmitter/receiver or network analyzer. Near field and compact ranges work also. Use calibrated reference antenna. Use axial ratio measurements to confirm quality of circular polarization. Check Appendix 6 for details.*

Type: ***Counter Wound Helices, Linear Polarized***

Frequency of Operational: *50 Mhz to 14 GHz*
Gain Range: 3 *to* 33 *dBi*
Bandwidth: *~ 15%*
Approximate Power Rating: *1 Kw*
Design Software: *NEC-WIN Pro, IE3D, Moment Method*

Configuration:

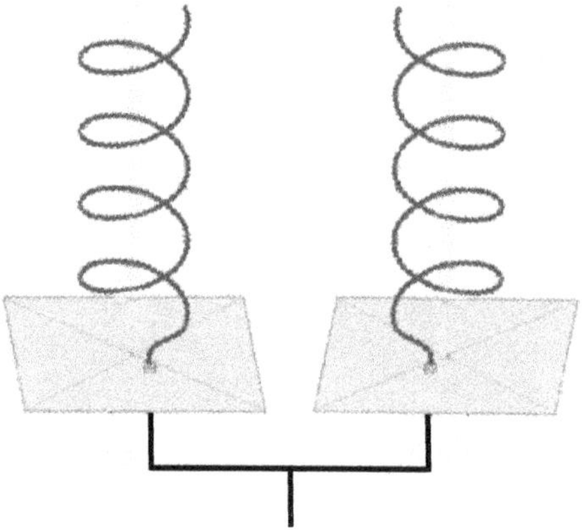

Design Approach:

*Circumference = 1 lambda*
*Spacing (S) = ~.25 lambda*
*Number of turns = ~13 for 14 dBi, 23 for 16 dBi, 33 for 17.5 dBi*
*L(meters) = 75 / f(MHz)*

*Feed Point = ~140 ohms at base of helix, move up conductor for 50.*

Construction and Theory of Operation:

*Counter wound helices when fed in phase with each other, produce a slant linear response in the far field. The gain increases based on the number of elements, in the case above, 3 dB. Also, little mutual coupling is realized in the above configuration, which is useful in the design of arrays.*

*Azimuth Pattern*

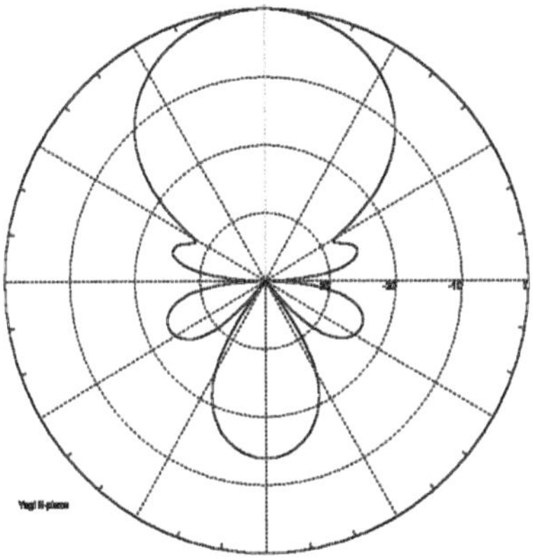

*Elevation Pattern*

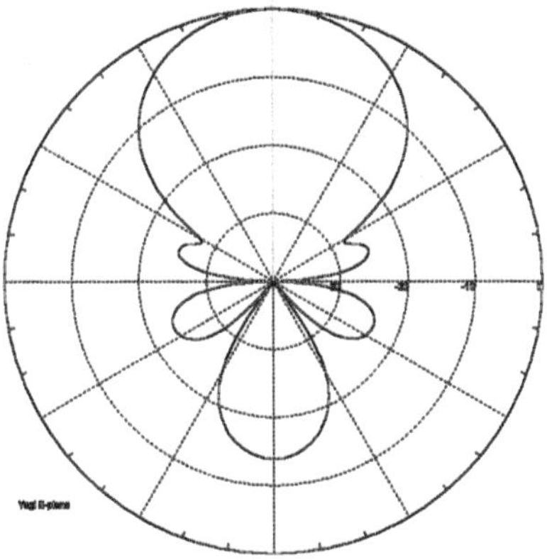

Special Considerations:

*Make sure impedances are matched and equal lengths of feed coax are used when combining helices. The polarization vector is 45 degrees relative to the line between antennas. Do not place antennas closer than .2 wavelength to minimize coupling. Placing similarly wound (same sense) helices requires greater distance to minimize mutual coupling.*

Testing:

*Far field range with transmitter/receiver or network analyzer. Near field and compact ranges work also. Use calibrated reference antenna. Use axial ratio measurements to confirm quality of circular polarization. Check Appendix 6 for details.*

Type: ***Short Helix for Wide Beam***

Frequency of Operational: *50 Mhz to 14 GHz*
Gain Range: *0 to 3.0 dBic*
Bandwidth: *~ 15%*
Approximate Power Rating: *1 Kw*
Design Software: *NEC-WIN Pro, IE3D, Moment Method*

Configuration:

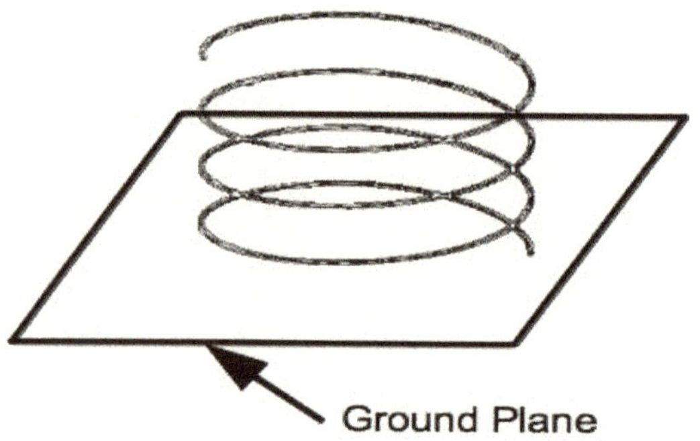


Design Approach:

*Circumference = 1 lambda*
*Spacing (S) = ~.1 lambda*
*Number of turns = ~1.5for 3 dBic, L(meters) = 75 / f(MHz)*
*Feed Point = ~140 ohms at base of helix, move up for 50.*

Construction and Theory of Operation:

*This is an axial mode helix design where the spacing between turns is small, compressing the antennas and thereby broadening the beam. Useful for GPS and other applications where hemispherical coverage is desired. Wind the conductive element on a low loss dielectric. Even smaller size (and less gain) will be obtained if the conductive element is wound on high dielectric constant material.*

*Azimuth Pattern*

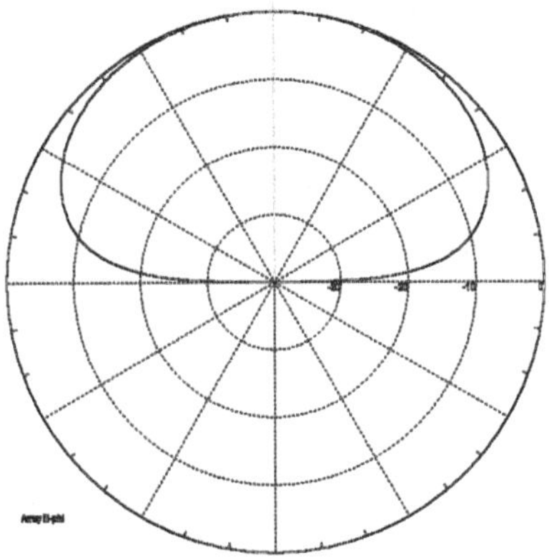

*Elevation Pattern*

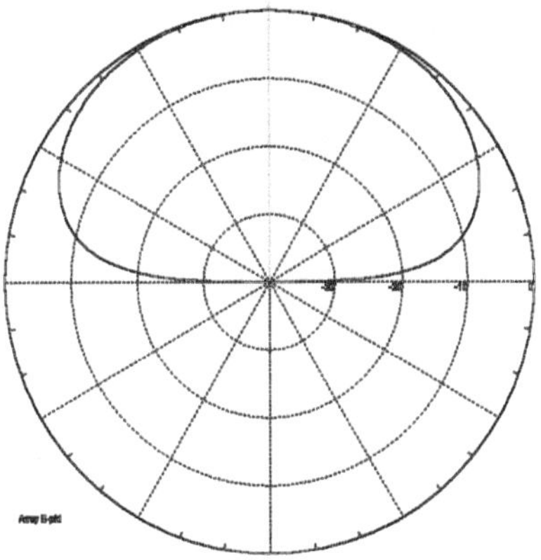

Special Considerations:

*Take into account any radome material, which can lower frequency of resonance. Also, test several versions of this type of antenna, fewer turns will yield wider beam and lower gain. Balance required gain with polarization purity.*

Testing:

*Far field range with transmitter/receiver or network analyzer. Near field and compact ranges work also. Use calibrated reference antenna. Use axial ratio measurements to confirm quality of circular polarization. Check Appendix 6 for details.*

| | |
|---|---|
| Type: | ***GPS Example, Quadra-Filar*** |
| Frequency of Operational: | *50 Mhz to 5 GHz* |
| Gain Range: | *0 to 3 dBic* |
| Bandwidth: | *~ 15%* |
| Approximate Power Rating: | *1 Kw* |
| Design Software: | *NEC-WIN Pro, IE3D, Moment Method* |

Configuration:

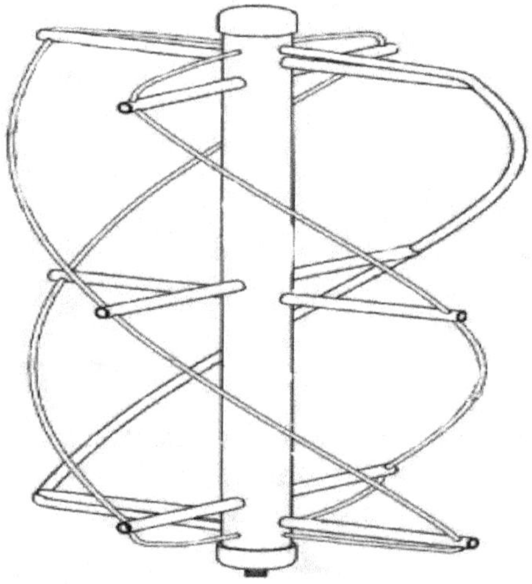

Design Approach:

*Circumference = 1 lambda*
*Spacing (S) = ~.3 lambda*
*Number of turns = ~1.5 for 3 dBic*

Construction and Theory of Operation:

*This antenna is fed at the top via a balun to each of two radiating elements. They are joined at the base to form a loop. The direction of twist with the elements dictates the polarization sense. The element can be made of stiff copper wire or small gauge tubing. The support structure can be made of PVC. Paint unit to protect from UV degradation.*

*Azimuth Pattern*

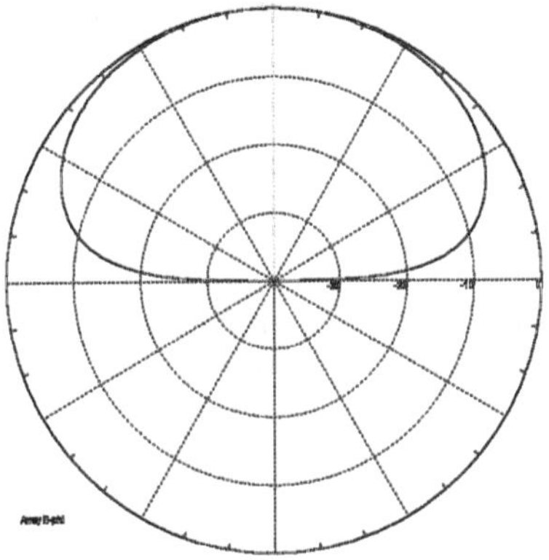

*Elevation Pattern*

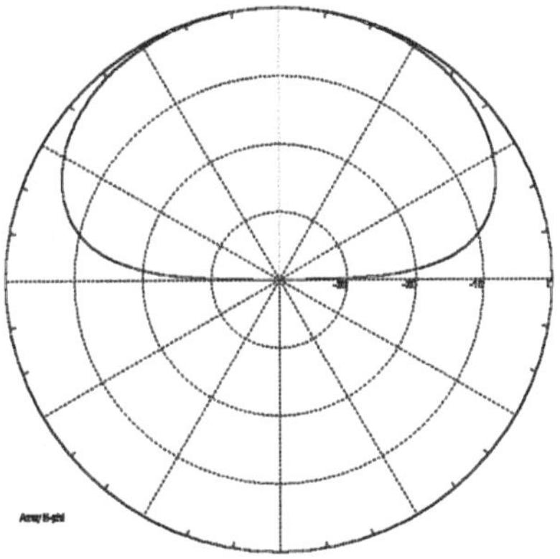

Special Considerations:

*The balun is critical and is necessary to combine essentially two loops with about 200 ohm impedances each to one 50 ohm output. These baluns can be made of coax, check literature for best option.*

Testing:

*Far field range with transmitter/receiver or network analyzer. Near field and compact ranges work also. Use calibrated reference antenna. Use axial ratio measurements to confirm quality of circular polarization. Check Appendix 6 for details.*

**Questions:**

1. Suggest an antenna type that has the following characteristics:

   a. Frequency of 200 to 400 MHz
   b. Gain of 15 dBi
   c. Linear Polarization

2. Your boss tells you price is a huge factor in the design of antenna with the following characteristics:

   a. Frequency of 144 MHz
   b. Gain of 20 dBic
   c. RHC Polarization

3. Suggest an antenna design for the next generation satellite that is capable of going to the moon, with the following properties:

   a. Frequency of 2200 MHz
   b. Gain of 3 dBic
   c. LHC Polarization

4. What is the minimum number of turns to create a helix antenna?

5. What happens when you feed the other end of the helix antenna?

6. How does the mutual coupling between helical antennas change from equal polarized antennas (like all RHC) to opposite (like RHC then LHC) ?

7. Can you couple to a waveguide with the feed end of a helix ?

8. What happens when you taper the upper third of a helix antenna ?

# Chapter 9 – Horn Antennas

Horn Antenna that discovered the afterglow of the Big Bang

## Fundamentals

Horn antennas are simple to design, efficient and easy to build. Essentially, they form a smooth transition from a waveguide to air. With a 50 ohm probe, this means a transition from 50 to 377 ohms. With standard waveguide, this provides a very efficient transition usable for high power as well as very good receiver applications. The aperture size dictates the gain of the antenna, sidelobes are low and front to back ratios are very good. Based on this and the simplicity of the design, these antennas tend to have very low noise temperatures, making their use as radio telescopes like the one pictured above popular.

Horns can be made circular, rectangular, sectoral (either E or H plane), exponentially tapered, ridged, corrugated or with the addition of exponentially tapered inserts, very wide band.

The circular horn can be fed with either a circular waveguide or with a rectangular waveguide plus a transition piece. The transition is several wavelengths long and adapts the rectangular waveguide to the circular input of the horn. In this case polarization is preserved. The circular horn has uniform pattern widths in both the E and H planes, which can be an advantage when total antenna efficiency is paramount. This quality is also advantageous for antenna systems that require good polarimetric properties, like meteorological radars and space communications facilities.

Rectangular horns are typically just an extension of the waveguide (assuming rectangular waveguide) in terms of the ratio of the broad wall to the short wall. These antennas have been used extensively in radar applications, standard gain antennas and in some radio astronomy applications. Flare angles for the horn can range from 15 to 50 degrees depending on application. The angles modify the beam parameters allowing for careful matching of the horn to the reflector or angle of radiation needed for particular applications. In some older designs, the walls of this type have been exponentially curved to allow a broader bandwidth. In more modern designs, wide band performance is achieved by exponential inserts or corrugations within the horn wall structure. This type of antenna can handle very high levels of power, even into the megawatt region, leading to its use in high power radars, oil reclamation equipment and microwave heaters. This horn design is almost always single pole and is sometimes moved physically to work at the orthogonal polarization.

Sectoral horns maintain the dimensions of one of the sides of the waveguide (either E or H) while flaring the opposite side. Very rectangular beam shapes can be achieved using this technique.

Corrugated horns achieve bandwidths of up to 50% and are used extensively in the satellite business, capable of high power and wide bandwidth in the C, Ku and Ka band commercial designs. Military applications also exist for this type of feed. Using this type has allowed many satellite design companies to standardize components leading to more competitive pricing. Below is a diagram of a typical commercial satellite antenna system, showing vertical and horizontal polarizations, each reflecting off of a different *shaped* main antenna, and the resulting beam footprint on the earth.

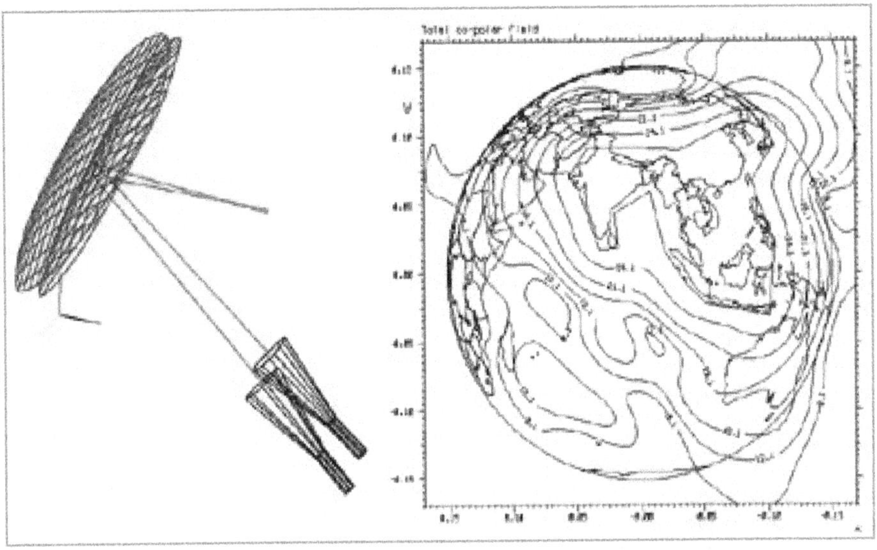

Example of horn feed dual grid reflector, used on satellites, showing shaped pattern

Horn apertures can also be square, being fed by square waveguide, which allows lower cost construction but minimizes bandwidth.

**Arrays**

Arrays of horns have been designed, taking advantage of low mutual coupling and the ability to have the horn openings touch each other leading to high total antenna efficiency and wide band performance. Construction can be expensive however due to large amounts of welding or machining.

**Calibration**

Horn antennas lend themselves nicely to use as calibration sources, due to their easily replicated designs. Most sophisticated antenna measurement facilities have a stable of calibration horns which are secondary standards to horns carefully calibrated at NIST. For the secondary standards, available from several manufacturers, gain over frequency plots are provided, usable to at least .1 dB accuracy.

In addition to being an amplitude standard, horn antenna have very good polarization qualities. Cross pole depths can exceed 40 dB for a well constructed standard gain horn. For circular polarization tests, these horn antennas are frequently spun physically to allow accurate measurements of axial ratio and other circular polarization properties. This is done in a antenna test range either indoor or outdoor. The horn antenna is used as a source, facing the antenna under test (AUT). The very predictable performance of these antennas allows for accurate and flexible use in antenna test facilities.

## Special Types of Horns

*Vivaldi*
This type is made with a square or rectangular aperture, but with inserts exponentially tapered to increase bandwidth. This type is available in either single or dual polarized variants.

*Coaxial feed*
This type of horn is actually a horn within a horn, the inner and outer horns are usually far apart in operating frequencies. For instance, C and Ku band feeds for commercial satellite uplink antennas. The beams from each horn overlap and have similar polarization properties.

*Horn with 90 degree Reflector*
As the picture at the beginning of this chapter shows, horns can be designed with large dimensions and with reflectors at their radiating apertures to simplify pointing and in the case of the great many used by the telephone companies from the 1960s to the 1990s, ease installation details. In the case of the horn antenna shown, this one verified theories on the residual temperature of the universe after the "big bang." Two scientists (Townes and Sholow) received the Nobel prize using this antenna for verifying the 2.7 degree Kelvin afterglow. The antenna is near Holmdel, New Jersey and is now a historic landmark.

## Design Considerations

There has been significant work done for the optimum dimensions for a horn antenna. Refer to T. Milligan's *Modern Antenna Design* for more information.

Example of Horn Fed Deep Space Network Antenna (Australia)

**Design Examples**

The following pages contain examples of antennas discussed in this chapter. These examples represent a first order design approach. For a more complete and optimized antenna solution, use the example closest to the requirement needs and optimize by using good simulation software, creating prototypes, testing and finally modifying prototypes to optimize performance.

Type: ***Pyramidal Horn***

Frequency of Operational: *100 Mhz to 500 GHz*
Gain Range: *5 to 35 dBi*
Bandwidth: *~ 15%*
Approximate Power Rating: *1 Mw*
Design Software: *Champ, GTD*

Configuration:

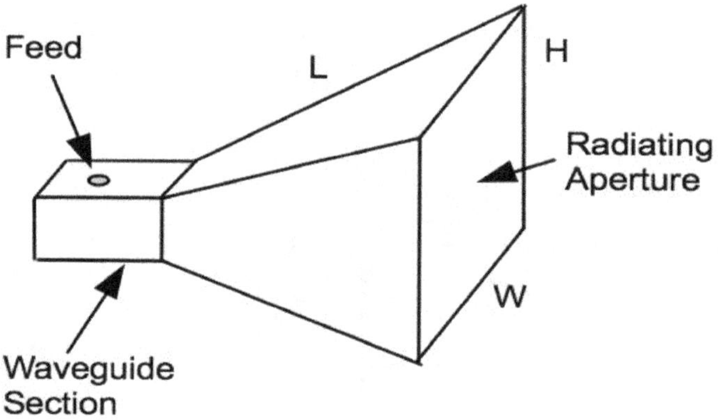


Design Approach:

*L = ~N lambda*
*W*H = Area to determine gain, ratio same as waveguide (~2:1)*
*Feed Point = ~0.25 lambda from end of waveguide section*
*Flare Angle = 20 to 60 Degrees, depending on beamwidth requirement*
*Waveguide Length = ~N lambda*

Construction and Theory of Operation:

*This antenna is useful for broadband and high power applications. It is simple to build with conductive components, usually aluminum or brass. Gain can be adjusted by adding length and aperture area. Use Bulkhead coaxial connector as feed, adjust length to optimize VSWR. This antenna is basically an extension of a waveguide, with a flare angle to meet impedance of air. The size of the aperture determines gain and beamwidth.*

*Azimuth Pattern*

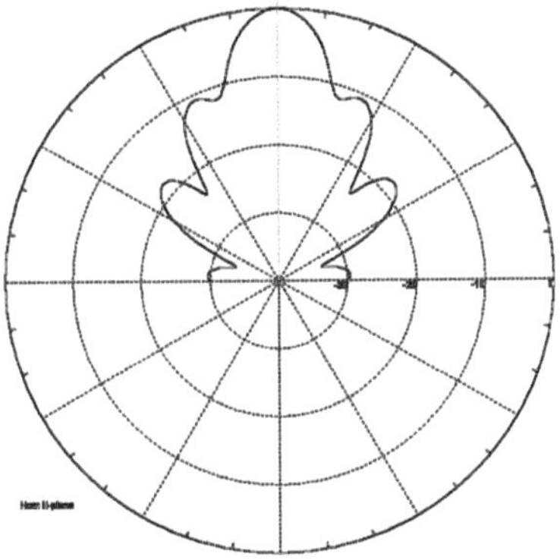

*Elevation Pattern*

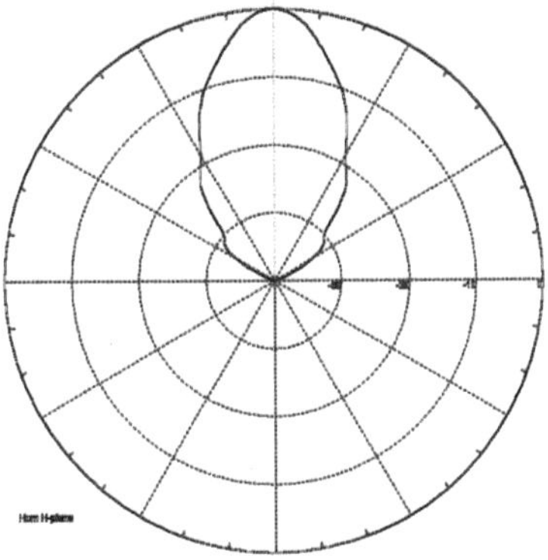

Special Considerations:

*Use brass, aluminum or other good conductor, weld or fully seal all joints. If this type of antenna is used for high power applications, make sure VSWR is optimized. If used outside, provide radome over aperture.*

Testing:

*For mid sized and small apertures use far field range with transmitter/receiver or network analyzer. Near field and compact ranges work also. Use calibrated reference antenna. Use axial ratio measurements to confirm quality of required polarization. For large apertures, use radio star techniques, holography or satellites with known EIRP. Check Appendix 6 for details.*

*Use VSWR bridge, Network Analyzer or similar device to optimize match.*

Type: ***Conical Horn***

Frequency of Operational: *50 Mhz to 50 GHz*
Gain Range: *0 to 35 dBic*
Bandwidth: *~ 20%*
Approximate Power Rating: *1 Mw*
Design Software: *Champ, GTD*

Configuration:

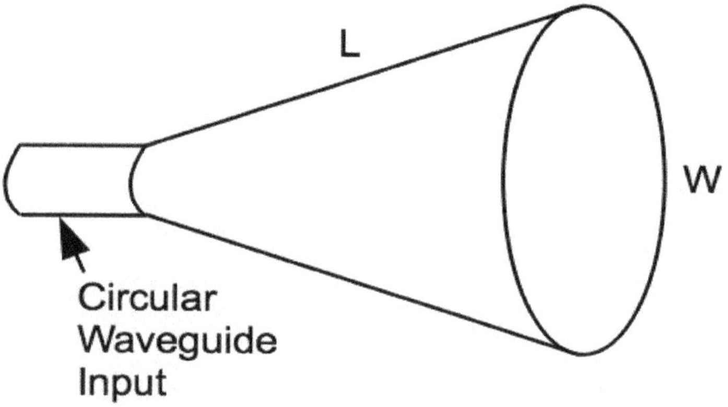


Design Approach:

*L = ~N lambda*
*W=pi * (R^2) = Area to determine gain*
*Feed Waveguide = TE00 mode*
*Flare Angle = 20 to 60 Degrees, depending on beamwidth requirement*
*Circular Waveguide Input Length = ~N lambda*

Construction and Theory of Operation:

*This antenna is useful for broadband and high power applications. It is simple to build with conductive components, usually aluminum or brass. Gain can be adjusted by adding length and aperture area. Use Bulkhead coaxial connector as feed, adjust length to optimize VSWR. This antenna is basically an extension of a waveguide, with a flare angle to meet impedance of air. The size of the aperture determines gain and beamwidth.*

*Azimuth Pattern* *Elevation Pattern*

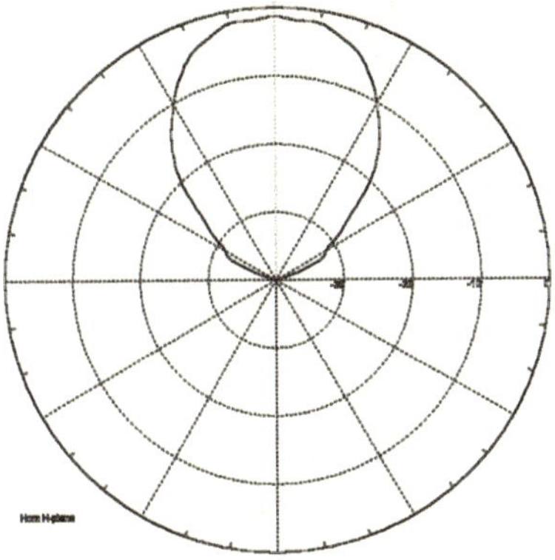

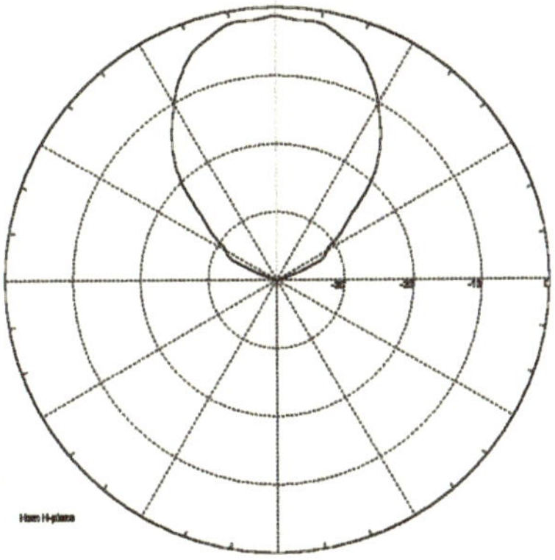

Special Considerations:

*Use brass, aluminum or other good conductor, weld or fully seal all joints. If this type of antenna is used for high power applications, make sure VSWR is optimized. If used outside, provide radome over aperture.*

Testing:

*For mid sized and small apertures use far field range with transmitter/receiver or network analyzer. Near field and compact ranges work also. Use calibrated reference antenna. Use axial ratio measurements to confirm quality of required polarization. For large apertures, use radio star techniques, holography or satellites with known EIRP. Check Appendix 6 for details.*

*Use VSWR bridge, Network Analyzer or similar device to optimize match.*

| | |
|---|---|
| Type: | ***Corrugated Horn*** |
| Frequency of Operational: | *5 0 Mhz to 50 GHz* |
| Gain Range: | *0 to 35 dBi* |
| Bandwidth: | *~ 20%* |
| Approximate Power Rating: | *1 Mw* |
| Design Software: | *Champ, GTD* |

Configuration:

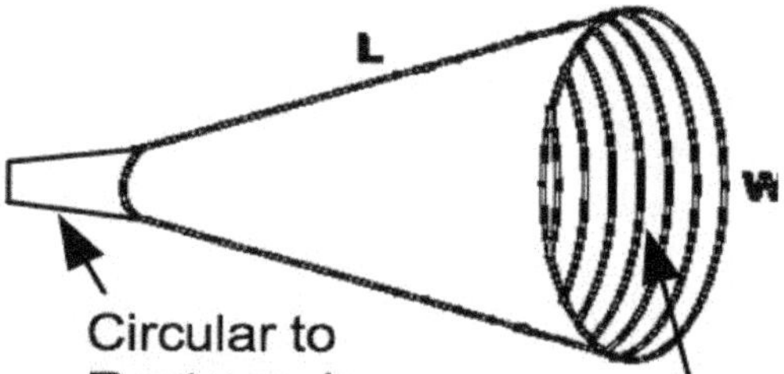


Design Approach:

*L = ~N lambda*
*W=pi * (R^2) = Area to determine gain*
*Feed Waveguide = TE00 mode*
*Flare Angle = 20 to 60 Degrees, depending on beamwidth requirement*
*Circular to Rectangular Adapter Length = ~4 lambda*
*Corrugation Depth ~0.25 lambda, separation ~0.25 lambda, use software*

Construction and Theory of Operation:

*This antenna is useful for broadband and high power applications. It is simple to build with conductive components, usually aluminum or brass. Gain can be adjusted by adding length and aperture area. This antenna can be fed by waveguide or waveguide to coax adapter, standard sizes. This antenna is basically an extension of a waveguide, with a flare angle to meet impedance of air. The size of the aperture determines gain and beamwidth.*

*Azimuth Pattern*

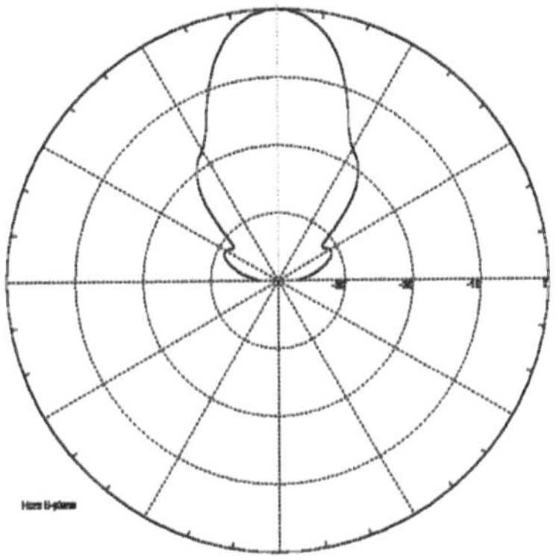

*Elevation Pattern*

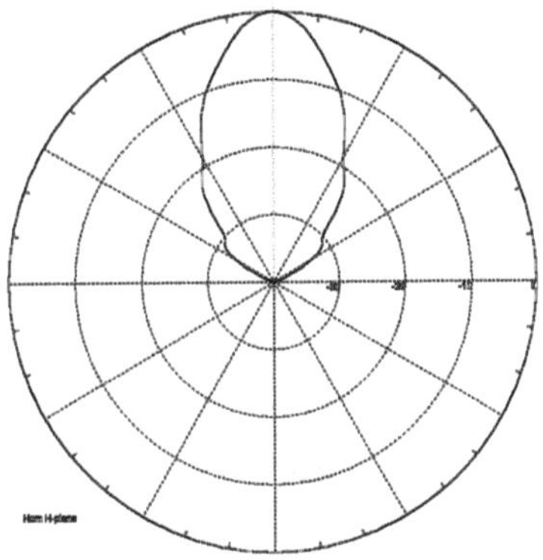

Special Considerations:

*Use brass, aluminum or other good conductor, weld or fully seal all joints. If this type of antenna is used for high power applications, make sure VSWR is optimized. If used outside, provide radome over aperture. Corrugations are distributed somewhat evenly but optimal performance comes from evaluating software models for particular application. Must have separate waveguide adapter due to machining constraints.*

Testing:

*For mid sized and small apertures use far field range with transmitter/receiver or network analyzer. Near field and compact ranges work also. Use calibrated reference antenna. Use axial ratio measurements to confirm quality of required polarization. For large apertures, use radio star techniques, holography or satellites with known EIRP. Check Appendix 6 for details.*

*Use VSWR bridge, Network Analyzer or similar device to optimize match.*

| | |
|---|---|
| Type: | ***Vivaldi*** |
| Frequency of Operational: | *50 Mhz to 40 GHz* |
| Gain Range: | *0 to 35 dBi* |
| Bandwidth: | *~ 25%* |
| Approximate Power Rating: | *100 Kw* |
| Design Software: | *Champ, GTD, HFSS* |

Configuration:

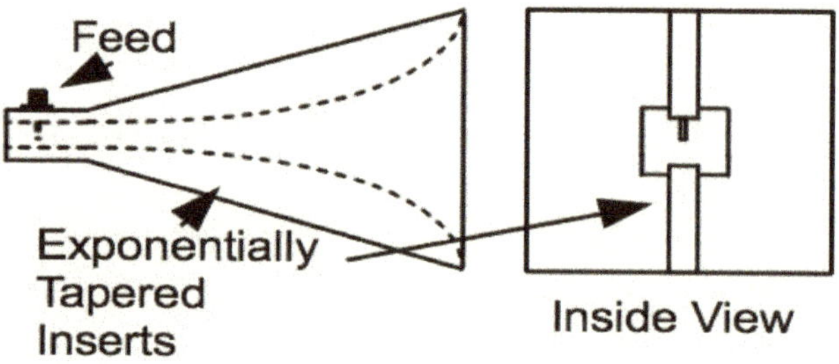


Design Approach:

*Length = ~N lambda*
*Width * Height = Area to determine gain*
*Feed = Coaxial*
*Flare Angle = 20 to 60 Degrees, depending on beamwidth requirement*
*Insert Width = ~0.5 lambda at center frequency*
*Insert Separation = ~0.5 lambda at center frequency*
*Distance from probe to end wall = 0.25 lambda*

Construction and Theory of Operation:

*This antenna is useful for broadband and high power applications. It is simple to build with conductive components, usually aluminum or brass. Gain can be adjusted by adding length and aperture area. This antenna is basically an extension of a waveguide, with a flare angle to meet impedance of air. The size of the aperture determines gain and beamwidth. The inserts allow very good performance over wide bandwidth. Inserts in both x and y dimensions allow for multipole operation.*

Azimuth Pattern

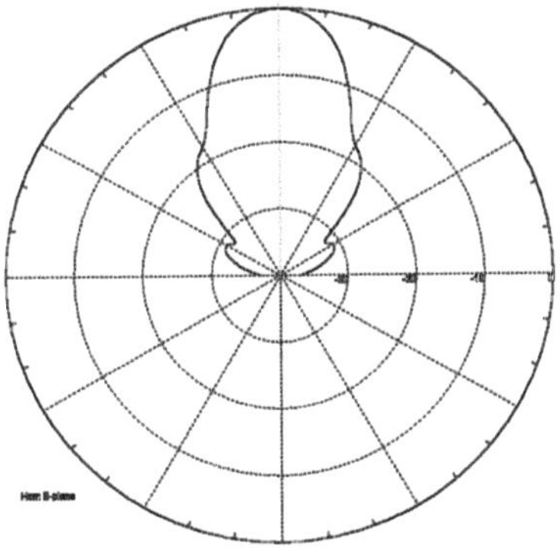

Elevation Pattern

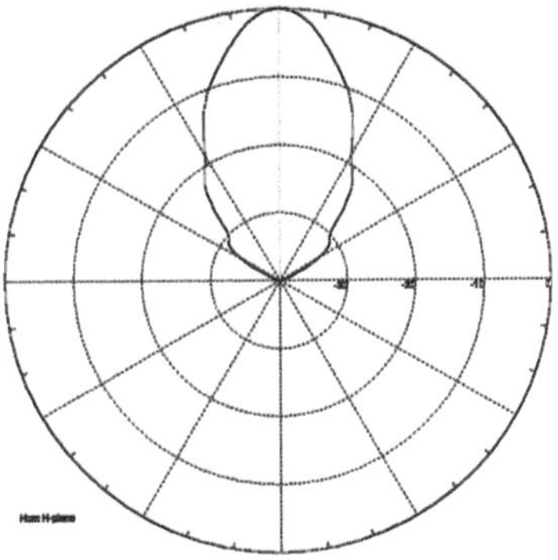

Special Considerations:

*Use brass, aluminum or other good conductor, weld or fully seal all joints. If this type of antenna is used for high power applications, make sure VSWR is optimized. If used outside, provide radome over aperture. Inserts are mirror images of each other but optimal performance comes from evaluating software models for particular application.*

Testing:

*Far field range with transmitter/receiver or network analyzer. Near field and compact ranges work also. Use calibrated reference antenna. Use axial ratio measurements to confirm quality of linear polarization. Check Appendix 6 for details.*

*2nd Edition*

| | |
|---|---|
| Type: | ***Dual Pole*** |
| Frequency of Operational: | *50 Mhz to 50 GHz* |
| Gain Range: | *0 to 35 dBi* |
| Bandwidth: | *~ 20%* |
| Approximate Power Rating: | *100 Kw* |
| Design Software: | *Champ, GTD* |

Configuration:

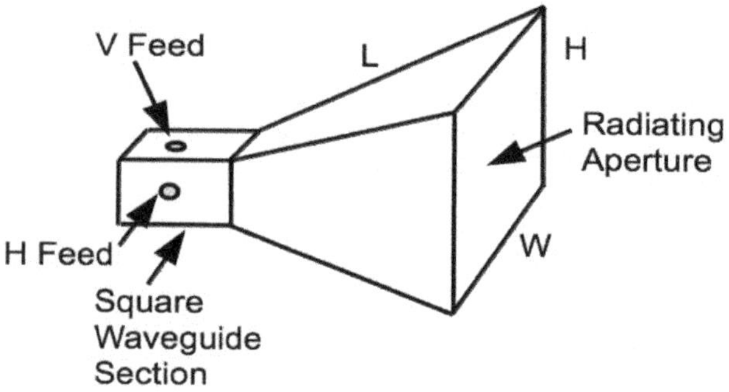


Design Approach:

*Length = ~N lambda*
*Width * Height = Area to determine gain, Width = Height*
*Feed = Coaxial or Waveguide*
*Flare Angle = 20 to 60 Degrees, depending on beamwidth requirement*
*Distance from probe to end wall = 0.25 lambda*

Construction and Theory of Operation:

*This antenna is useful for broadband and high power applications. It is simple to build with conductive components, usually aluminum or brass. Gain can be adjusted by adding length and aperture area. This antenna can be fed by waveguide or waveguide to coax adapter, standard sizes. This antenna is basically an extension of a waveguide, with a flare angle to meet impedance of air. The size of the aperture determines gain and beamwidth. The inserts allow very good performance over wide bandwidth. Inserts in both x and y dimensions allow for multipole operation.*

*Azimuth Pattern*

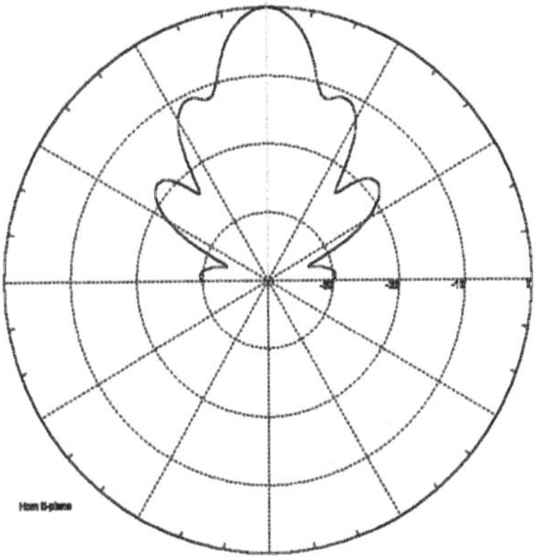

*Elevation Pattern*

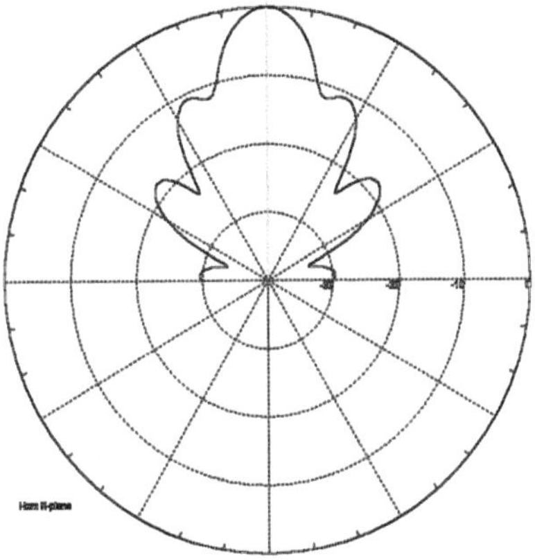

Special Considerations:

*Use brass, aluminum or other good conductor, weld or fully seal all joints. If this type of antenna is used for high power applications, make sure VSWR is optimized. If used outside, provide radome over aperture. Inserts are mirror images of each other but optimal performance comes from evaluating software models for particular application.*

Testing:

*Far field range with transmitter/receiver or network analyzer. Near field and compact ranges work also. Use calibrated reference antenna. Use axial ratio measurements to confirm quality of linear polarization. Check Appendix 6 for details.*

Type: ***Coaxial Fed Horn***

Frequency of Operational: *50 Mhz to 50 GHz*
Gain Range: *0 to 35 dBi*
Bandwidth: *~ 5%*
Approximate Power Rating: *1 Kw*
Design Software: *Champ, GTD*

Configuration:

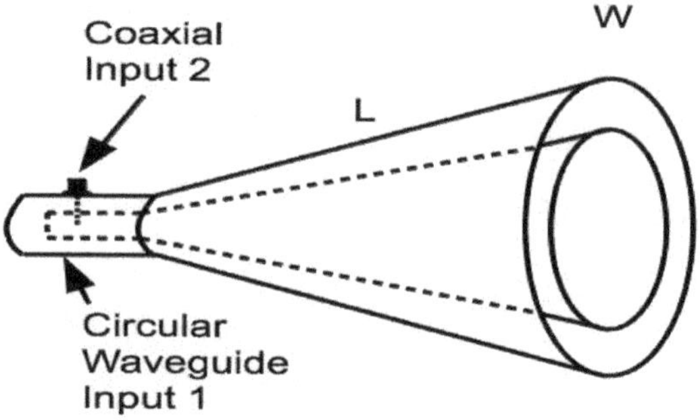


Design Approach:

*Length = ~N lambda*
*Width = Area to determine gain or Pi * R^2*
*Feed = Coaxial and Waveguide*
*Flare Angle = 20 to 60 Degrees, depending on beamwidth requirement*
*Distance from probe to end wall = 0.25 lambda*
*High band (Input 2) at least 2x Low Band (Input 1),e.g., C and Ku band*

Construction and Theory of Operation:

*This antenna is useful for dual band and high power applications. It is tricky to build with conductive components, usually aluminum or brass. Gain can be adjusted by adding length and aperture area. This antenna can be fed by waveguide or waveguide to coax adapter, standard sizes. This antenna is basically an extension of a waveguide, with a flare angle to meet impedance of air. The size of the aperture determines gain and beamwidth. The inserts allow very good performance over wide bandwidth. Inserts in both x and y dimensions allow for multipole operation.*

*Azimuth Pattern*

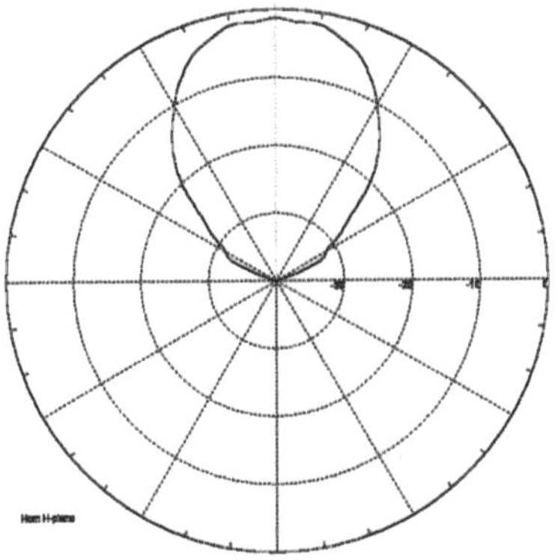

*Elevation Pattern*

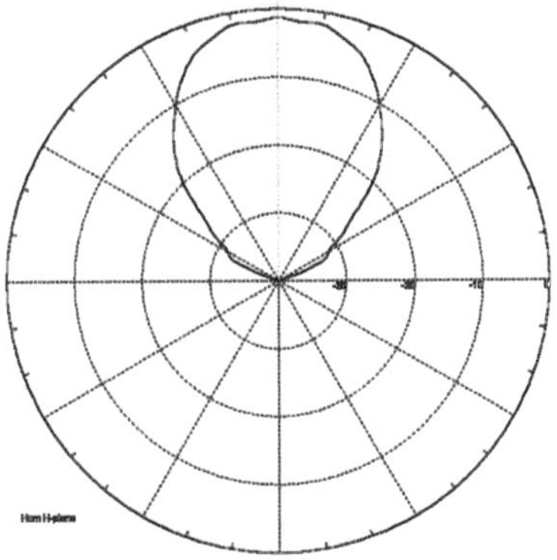

Special Considerations:

*Use brass, aluminum or other good conductor, weld or fully seal all joints. If this type of antenna is used for high power applications, make sure VSWR is optimized. If used outside, provide radome over aperture. Inserts are mirror images of each other but optimal performance comes from evaluating software models for particular application.*

Testing:

*Far field range with transmitter/receiver or network analyzer. Near field and compact ranges work also. Use calibrated reference antenna. Use axial ratio measurements to confirm quality of linear polarization. Check Appendix 6 for details.*

Type: ***Orthomode Feed***

Frequency of Operational: *50 Mhz to 50 GHz*
Gain Range: *0 to 35 dBic*
Bandwidth: *~ 15%*
Approximate Power Rating: *100 Kw*
Design Software: *Champ, GTD*

Configuration:

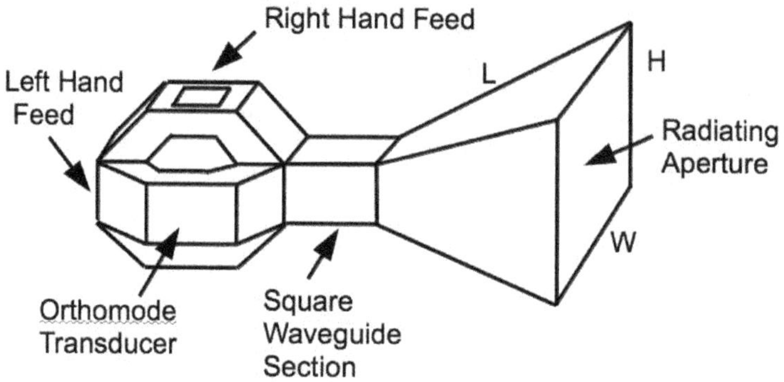


Design Approach:

*Length = ~N lambda*
*Width = Height = Area to determine gain*
*Feed = Waveguide*
*Flare Angle = 20 to 60 Degrees, depending on beamwidth requirement*

Construction and Theory of Operation:

*This antenna is useful for good circular polarization and high power applications. It is complex to build with conductive components, usually aluminum or brass. Gain can be adjusted by adding length and aperture area. This antenna can be fed by waveguide or waveguide to coax adapter, standard sizes. This antenna is basically an extension of a waveguide, with a flare angle to meet impedance of air. The size of the aperture determines gain and beamwidth. The inserts allow very good performance over wide bandwidth. Inserts in both x and y dimensions allow for multipole operation.*

*Azimuth Pattern*

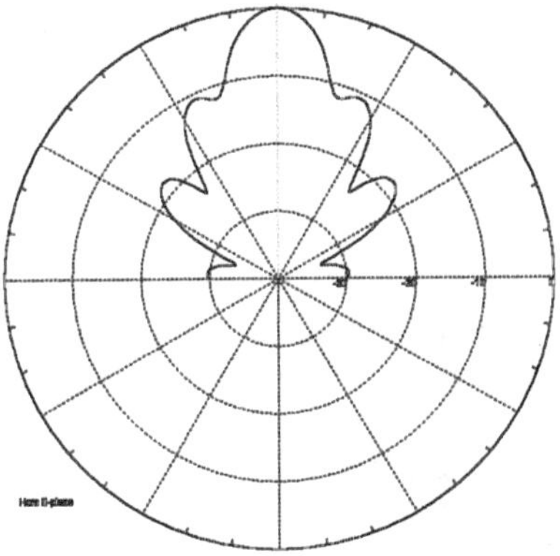

*Elevation Pattern*

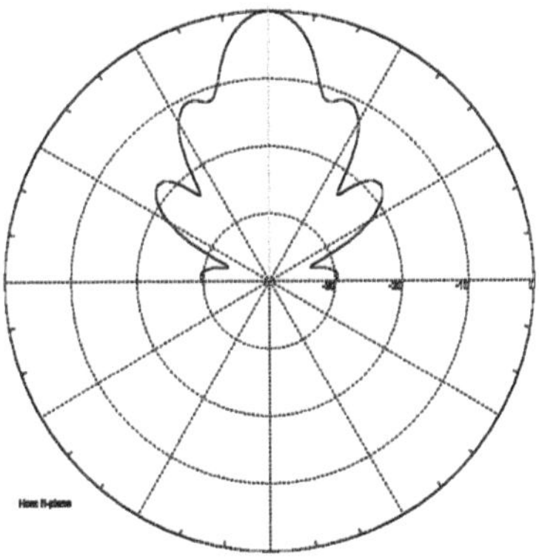

Special Considerations:

*Use brass, aluminum or other good conductor, weld or fully seal all joints. If this type of antenna is used for high power applications, make sure VSWR is optimized. If used outside, provide radome over aperture. Build orthomode transducer separately from horn antenna, optimize performance then add antenna, might require further tuning.*

Testing:

*Far field range with transmitter/receiver or network analyzer. Near field and compact ranges work also. Use calibrated reference antenna. Use axial ratio measurements to confirm quality of linear polarization. Check Appendix 6 for details.*

**Questions:**

1. Suggest an antenna type that has the following characteristics:

   a. Frequency of 200 to 400 MHz
   b. Gain of 15 dBi
   c. Linear Polarization

2. Your boss tells you price is a huge factor in the design of antenna with the following characteristics:

   a. Frequency of 20 MHz
   b. Gain of 20 dBi
   c. Linear Polarization

3. Does a linear polarized rectangular horn antenna require all sides to be filled in?

4. Suggest an antenna design for the next generation rocket booster that is capable of going to the moon, with the following properties:

   a. Frequency of 2200 MHz
   b. Gain of 13 dBi
   c. Circular Polarization

5. What is the relationship between the horn aperture and gain ?

6. How does the flare angle effect gain ?

7. How does the length effect bandwidth in a corrugated horn ?

8. Does a horn aperture have to be symmetrical ? If not, what are the consequences of the asymmetry ?

# Chapter 10 – Reflector Antennas

## Fundamentals

The first reference to parabolic surface is from Archimedes (287 – 212 BC) where in defense of the city of Syracuse on Sicily against the Roman fleet and army of Marcellus during the Punian war, he built large parabolic mirrors to concentrate the rays of the sun on the sails and wooden components of the fleet. Historians today feel that this story is probably not true, however numerous references and pieces of artwork depict the design and basic concept.

By strict definition, a parabolic dish a device that allows a bundle of parallel light or radio waves to impinge onto a reflector in the form of a paraboloid of revolution along its symmetry axis and those rays will be concentrated towards the focal point of the paraboloid.

These antennas are used extensively in radar, radio astronomy and communications. They are reasonably easy to construct and have been made in sizes beyond 100 meters in diameter. These surfaces can be stamped out, welded together, made in sections, formed with a wire grid in fiberglass and other techniques.

The mathematical definition of a parabola is as follows:

$$FP = PQ = \sqrt{(Xp^2)} + (f - Zp)^2 + f - Zp = 2 * f \qquad [45]$$

where:

FP = distance from the focal point to the surface

PQ = distance from the surface to the plane of the focal point

Xp = point on the x axis

f = distance from the center of the paraboloid to the focal point

Zp = distance from a point in the z axis (center of paraboloid through focal point) to the surface

in Cartesian coordinates:

$$X^2 = 4 * f * z \qquad [46]$$

in Spherical coordinates:

$$FP + PQ = p + (p * \cos(\phi)) = 2 * f \qquad [47]$$

or

$$p = \left( \frac{(2 * f)}{(1 + \cos(\phi))} \right) \qquad [48]$$

where:

p , ø are spherical coordinates

In many dish designs, a secondary reflector is used, typically a hyperboloid whose defining formula is:

$$\frac{x^2}{a^2}-\frac{x^2}{(c^2-a^2)}=1 \qquad [49]$$

where:

x,y = points in Cartesian space

z = point along axis of rotation

c = distance from reflector surface to the focal point of the main dish

In some cases a elliptical reflector is preferred, in this case the formula is:

$$\frac{z^2}{a^2}+\frac{x^2}{(c^2-a^2)}=1 \qquad [50]$$

From an electromagnetics standpoint the field from the feed induces surface currents on the reflector surface, which in turn are the sources of electromagnetic radiation. The field strength in a point P of space, outside the antenna, is found from the integration of the currents on the illuminated reflector surface. These lead to a representation of the radiated field as an integration of the field projected from the surface onto the aperture plane of the reflector, this is know as the Kirchhoff – Helmholtz aperture integration.

The aperture field and the farfield of the reflector antenna are related by a Fourier Transformation. Also it can be shown that the spatial field distribution in the focal plane is identical to the field distribution in the far field.

In fact the Inverse Fourier transform can be used after measuring the far field pattern to derive the quality of the radiating surface in terms of phase. In this way, a large aperture can be adjusted to optimize performance.

**Design**

Dish antennas come in a wide variety of forms. The classic round parabolic dish can be of a prime focus type, where the reflected radio wave come to a focal point where a horn or other suitable feed is suspended with legs to either transmit or receive as required. More advanced designs using a reflecting dish place a subreflector near the focal point, sending the radio waves to a feed point mounted on the surface. This has the effect of increasing the F/D ratio and increasing efficiency. In addition, complicated feeds, like those containing transmitters or cooled receiver components, can best be contained in a feed mechanism housed on the reflector surface. This type is known as a Cassegrain configuration and is used for the Deep Space Network of NASA as well a multitude of radio telescopes. Finally, beam waveguide systems have been developed to further increase F/D ratios by using a subreflector then sending the radio waves through the dish surface thence via reflectors down the mounting mechanisms into the pedestal or base of the antenna.

Subreflectors usually are made with a hyperbolic shape but have been made in elliptical shapes or spherical depending on the type of feed mechanism employed.

Some subreflectors nutate or move around to allow the movement of the focused beam to be placed at different locations, for instance multiple feeds.

Subreflectors can also be reflective at certain frequencies and transmit others to a feed mechanism behind the reflector, these types are known as dichroic and are made of reflective structures etched or otherwise placed on substrates shaped into the type of sub reflector needed. This allows for specific bands of interest to be selected for reflection or transmission; in most cases the lower frequencies are allowed to pass through to a focus in back of the reflectors and the high frequencies reflect onto feed mechanism fixed to the main dish.

Other dish configurations include offset paraboloidal sections, where there is no feed structure in front of the main reflector. This type is the most efficient due to the lack of aperture blockage.

Other parabolic sections are not offset but have been sliced in such a way as to modify the beam size for better coverage of the area of interest. A classic example of this is an air traffic radar, where a narrow beam is needed in azimuth and a wider beam needed in the vertical plane. In these cases the aperture of the antenna is longer in the horizontal plane and less so in the vertical plane.

**Aperture Efficiency**

Dish antennas are generally have the qualities of being frequency independent or broadband and being highly efficient. Offset parabolas, commercially available can reach 70% or more. The issue of efficiency is mainly dependent on several factors and can be mathematically presented thus:

$$\text{Aperture Efficiency} = E1 * E2 * E3 * E4 * E5 * E6 * E7 \qquad [51]$$

where:

E1 = illumination efficiency of the aperture by the feed function (taper)

E2 = spillover efficiency of the feed (and subreflector, if present)

E3 = radiation efficiency of the reflector surface (ohmic loss)

E4 = polarization efficiency of the feed-reflector combination

E5 = surface error efficiency ("Ruze loss"), also called scattering efficiency

E6 = focus error efficiency (both lateral and axial defocus)

E7 = blocking efficiency due to quadripod, subreflector, other obstructions

Generally speaking, reflector type antennas should be able to achieve aperture efficiencies in excess of 50%.

**Aperture Illumination**

A trade off is considered between low sidelobes and gain, where the illumination of the main dish by either the feed directly (as in prime focus) or a subreflector arrangement. Designers typically place the 5 to 9 dB points on the edge of the reflecting surface to minimize diffractive effects and lower the antennas noise temperature. This temperature is dependent on spillover effects, a lower edge taper number will increase the noise temperature of the antenna. Of course the more area illuminated by the feed, the greater the gain, a designer must be careful when making this decision and base it on expected signal strength form the far source.

**Surface Roughness**

Surface roughness has a significant effect on the beam forming properties of a reflecting dish. This roughness is defined as the deviation from the ideal mathematically derived curve, typically parabolic, but sometime spherical, hyperbolic or even elliptical. This deviation is characterized in either portion of a wavelength, inches or millimeters. The deviations occur from the manufacturing process, from sagging due to various elevation angles, temperature, wind loading, solar heating, deflection due to feed or subreflector structure deflections and other perturbations.

Generally, a beam will form when a dish roughness is better then $\lambda / 5$. It is much better however to achieve deviations from $\lambda / 25$ to $\lambda / 50$ for respectable performance. In some cases designs of l $\lambda$ / 100 are required to squeeze every bit of performance out of a reflector design. At low frequencies and small aperture sizes, making a good smooth surface is reasonably easy. At higher frequencies and large aperture sized, keeping good surface tolerances can be a significant challenge. As frequencies of operation tend to go up for communications and radio telescope need to be larger, several creative methods have been employed to maintain good surface specifications. These methods include the use of lasers to scan across the surface, and, using a retro reflector placed on the surface, a pulsed laser distance measuring apparatus can tell quite accurately the deviations of the reflector from ideal. Adjustments can then be made to the surface panels to optimize the operation of the antenna.

These errors are sometimes referred to as "RMS Losses" as they are the deviations from a perfect surface (be it paraboloidal or otherwise) and have an average RMS value across the surface. With this average value, one can calculate the gain loss thus:

$$Loss(dB) = 685 * \left( \frac{RMS}{\lambda} \right)^2 \qquad [52]$$

Another way to observe surface roughness is to use radio stars or satellites plus holographic techniques to plot the surface deviations from the ideal. This technique is dealt with in Appendix 5.

**Gain**

Gain for dish antennas is determined in the standard manner discussed in chapter two. The aperture in this case is determined by the area of the reflector minus any aperture blockage due to feed structures. In addition consideration must be given to the spillover and amplitude taper of the power impinged on the reflector.

**Patterns**

Patterns are also determined in a straight forward manner, based on aperture size and the effects of aperture blockage. In addition, the type of feed is important, with improvements in pattern performance when using offset feeds, Cassegrain and other sophisticated approaches.

**Diffraction**

Patterns for reflector antennas are influenced by diffractive effects of the reflector edges, structures to hold feed assemblies and other conductive components. Use a good modeling software tool that encompasses geometric optics to review the impact of these effects before committing to a final design.

**Effects of Struts**

When sidelobe performance is paramount, a three strut design is recommended. When gain symmetry is desired, a four strut arrangement is recommended, with the strut position in the +/- 45 degree positions relative to the vertical plane. This assumes linear polarization in both the vertical and horizontal planes. To optimize cross pole performance, and again assuming V and H polarizations, there should be four struts arranged in the vertical and horizontal planes. Scattering is the primary reason for the above effects and can be analyzed using a good physical optics and geometric optics modeling program.

It has also been found that for optimum performance, the struts should be placed at the edges of the reflector.

Blockage effects should also be considered with the struts and feed assembly, or subreflector assembly in the case of a Cassegrain system. This includes loss in gain and changes in sidelobe performance, especially parallel of the strut positions.

**Prime Focus**

This is the simplest of reflector designs whereby the feed assembly is placed near the center of the reflector, suspended at the focal point. Many if not most reflectors antennas systems are of this style. Use the standard parabolic design formulas to determine the dimensions and feed placement.

**Feed Subtended angle**

This can be calculated with the following formula:

$$\textit{Subtended Angle} = 4 * \tan^{-1}\left(\frac{1}{(4*(F/D))}\right) \qquad [52]$$

where,

F = Focal Distance, and

D = Diameter in same units as F

Generally, this means the angle from the prime focus point to the dish edges.

Typical feed tapers place the 13 dB points on these edges, this is a compromise between gain, antenna temperature (as spillover sees the ground beneath the dish), and sidelobe performance.

**Determining Focal Length**

A useful formula for determining the focal length of a parabola is:

$$F = \frac{D^2}{(16 * c)} \quad [53]$$

where,

D = Diameter, and

c = Depth of dish, same units as D

Example of Prime Focus Reflector, DSES 18 meter (31)

**Cassegrain**

This is a more efficient if not more expensive design for a reflector antenna. The prime focus position, where a feed normally would be found, is now the position for a smaller reflector assembly. These sub reflectors are usually shaped to hyperbolic curves, sometimes even elliptical. These secondary reflectors focus the energy onto a feed system, typically a horn, which is placed on the surface of the main reflector. NASA's Deep Space Network uses this style and has achieved efficiencies of up to 70% with this approach.

100 Meter Cassegrain Reflector, Effelsberg [37]

**Offset**

This is the most efficient of reflector geometries and is used in backyard dishes as well as spacecraft. It consists of a parabolic section, offset from the center with a feed system that is not in the field of view of the main reflector. Aperture efficiencies better than 70% have been achieved with this approach.

Example of an offset parabolic dish (38)

**Shaped**

Shaped reflectors are used to produce custom beam shapes for use in space (for instance) as communication satellites need to place the majority of their rf energy onto a particular country or region. The term 'shaped' means a departure from the standard parabolic or spherical geometries. Because the beam shape in the far field is related via the Fourier Transform to the phase and amplitude distribution on the radiating surface, this fact can be used to first define the far field pattern, then use the Fourier Transform to dictate the shape of the radiating surface. This approach can be controlled surprisingly well, as a very large number of geosynchronous satellites use antennas made in this way. For a standard offset shaped reflector, the surface resembles a potato chip, with waves and undulations defined by the design process. The use of geometric optics and physical optics is the most common approach to define the surface. First the contours are defined in the far field, taking note of the distance to the place the custom pattern must illuminate. In addition, for communications satellites, city tables are constructed with expected EIRP (essentially power on the ground) levels. Next the characteristics of the feed horn or assembly is taken into account. After a reasonable amount of computing time, a surface is defined and checked using compatible modeling software. The surface is build and physically measured over a range of operating temperature ranges. The resultant range of measurements are then re evaluated using the modeling software and the effects of temperature change and satellite movement is evaluated. If the surface is deemed good, over the range of temperatures, considering power levels at the surface in terms of contour and city requirements, the antenna will be integrated onto the satellite structure. After the satellite is completely built, tested in a thermo-vacuum oven and operated for a significant period of time, the antenna performance is analyzed using (typically) a large near field probe facility. The final patterns are derived and evaluated against customer requirements.

Example of dual gridded shaped satellite parabolic reflectors (39)

The satellite in the picture above consists of four main reflectors, the individual circular antenna assemblies shown are made up of two shaped reflectors, one with horizontal wires closely spaced imbedded in composite plastic, over another shaped reflector with vertical wires in similar composite material. The set of reflectors is feed by two horn antennas on the body of the satellite. In the center of the satellite bus there is another smaller offset, shaped reflector for use in a separate type of transmission. All five reflectors shown can have separate coverage areas.

**Dragonian**

For optimum performance, several steps need to be taken. First aperture blockage must be eliminated, this is accomplished for instance by designing an offset type geometry. A next good step is to under illuminate the main reflector, minimizing any diffractive effects and spillover. This lowers antenna noise temperature as well, in addition front to back numbers are very low. In the case of a Dragonian type geometry, the main dish can have edge illumination values in excess of 30 dB, as apposed to 10 dB or so for more standard applications. Performance for this type of antenna has been measured at 50+ dB for first sidelobe levels and at least 40 dB of cross pole performance. The downside of this antenna is a more complex construction as can be seen by the diagram that follows.

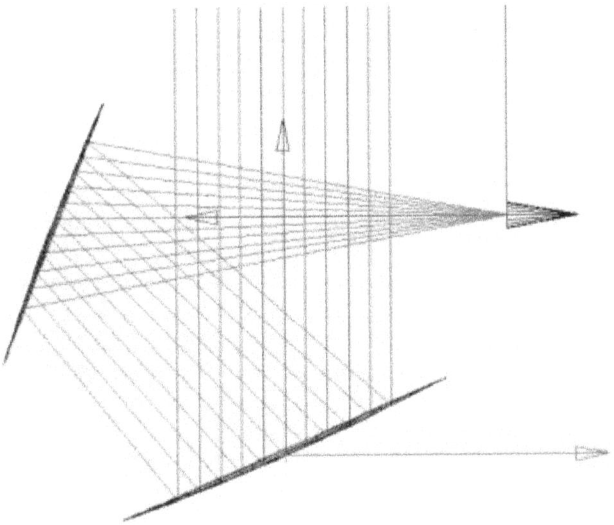

Dragonian Antenna Optics

**Adaptive (Correcting surface)**

For ground based reflector antennas that require the best efficiencies at any elevation angle, there are feedback mechanism that reshape the surfaces using pre-measured data or even real time approaches. The surface position is measured using laser probes and reflective references on the surface of the main reflective surface. In some cases these references are spots painted on the surface, in other cases there are moving mechanism with optical retroreflectors attached to them. In the case of the painted on spots, a camera mounted on the prime focus structure can accurately measure the positions of the various spots, then through a computer, can move the surface panels as necessary to optimize the shape of the reflector. In other cases a pre-measured set of surface movements over elevation is used to move panels as the dish moves.

**Arrays of Reflectors**

Arrays of reflectors have been assembled to improve gain and minimize beamwidth, primarily for use in radio astronomy. The first use of multiple reflectors was first achieved in the 1950s, in England, with the development of interferometers. Interferometers are made up of at least two antennas spaced many wavelengths apart. The beam pattern that results from such an arrangement is a series of "fringes" within the envelope of the individual antennas' pattern. The width of the fringes is dependent on the distance between elements. Using this arrangement can enable the fine measuring of the extent of a radio source, if large separations are used. In its ultimate configuration, the Very Long Baseline Array, images with resolutions in the few 100s of micro arc seconds are routinely accomplished. This array is made up of ten 90 foot Cassegrain reflectors positioned from California to the Virgin Islands. Using very stable local oscillators (based on Hydrogen maser clocks) the individual stations track an object of interest and record the receiver's output on video tape. The tapes are sent to a correlation and computation station in Socorro, New Mexico. Coincidently this is also the location of the Very Large Array, a 27 km wide array of 90 foot Cassegrain reflectors. The correlation and computation techniques used to reduce the raw data are similar in both cases, the resolutions are different however. Arrays of reflectors have been used in closer proximity for sending telemetry and communications from the Apollo spacecraft, this for redundancy. Arrays of reflectors have also been used FM CW Radar systems where there is a constant as apposed to pulsed radar transmission. In this case one antenna is used for transmission and the other for reception.

**Russian Multi Mirror (Ratan)**

This is a series of reflective surfaces in a circle of approximately 600 meters in diameter. The surfaces are parabolic in shape as if they were removed from a 600 meter dish edge. The circle of reflectors focuses its energy on a series of sub reflectors and then to specialized horn for reception of radio astronomical sources. This dish array has also been used to observe the movement of geosynchronous satellites and to observe the resultant variation in signal strength due to the troposphere.

**Coelestads**

These antenna arrays were also developed by the Russians in the 1950s and consist of two reflectors placed back to back, each pointed at a flat plate reflector some distance away. This is another form of interferometer and follows the same guidelines in terms of fringe development. The advantage is that only one receiver is needed, no long cables or other concomitant problems.

**Space Based**

Space based reflectors are unique in the sense that most have to be deployed using creative mechanisms. In the case of the dual gridded reflectors mentioned earlier, the reflectors are hinged and folded up against the satellite body during launch. Special release mechanisms made of paraffin plugs and redundant heater elements are used to start the process. Special springs and dampening mechanisms slowly (20 minutes or so) allow the reflectors to move to their final resting position.

Other space based reflectors have used inflatable structures or complex unfurling mechanisms. The main reflector for the Galileo spacecraft bound for Jupiter used the latter approach, however it hung up during the deployment phase, causing significant loss of bandwidth and science. Very large reflectors have been launched using fine mesh surfaces and composite superstructures, some of these antennas have exceeded 100 feet in diameter, again the unfurling mechanism take special attention are are almost always modeled using sophisticated software tools before they are launched. This is due in part to the fact that testing these light weight structures is very difficult in the gravity of earth.

**Tapered**

Due to the fact that many radar and communications facilities are in populated areas, enabling interference to enter the receivers, several schemes have been developed to mitigate the ability of signal other that the one of interest, to enter the feed horn or other feed mechanism. Careful analysis of the beam profile on the main reflector is important, in many cases the nine dB (or so) point of this feed is placed at the edges of the reflector. The positive effect is to reduce spillover and diffractive effects of the dish edges, the negative is a lowering of gain and broadening of the main beam pattern.

**Sidelobe Fences**

For reflectors and other types of arrays, fences made of reflective material are placed around the edges of a dish, parallel to the main beam direction. This reduces the effects of local interference; useful for both radar and communications systems.

**Edge Roll**

This is where the edge of the reflector rolled instead of abruptly stopping at the edge. This has the effect of reducing diffraction effects and therefore improving pattern performance.

**Csc squared**

This is a special shape applied to radar reflectors to best cover the ground (if in the air) or air (if on the ground). The pattern is widened in the direction of range, in other words, an airborne radar system has an elliptical pattern when viewing the ground, in the direction the radar antenna is pointed. This is very useful when mapping features on the ground or observing moving objects on the ground. Air traffic radar systems employ this same technique while looking for aircraft in the air, as it covers more elevation angles, compensating for aircraft at various altitudes.

Some useful formulas for a Csc squared antenna:

$$Range = h * Csc(\theta) \qquad [54]$$

where,

h = height of antenna (typically placed in an aircraft)
Theta = angle to ground

and,

$$Gain(dBi) = 10\log(Csc^2(\theta)) \qquad [55]$$

**Cylindrical**

This is a special type of reflector where one dimension of the reflector is elongated relative to the other. A classic example is the Vermillion River Observatory in Illinois, where a 600 foot long cylindrical shape was built on the ground, with a long array of feed placed at the focal "line" of the antenna. The feeds could be phased in such a way as to scan in the direction of the longitudinal axis of the cylinder. Scanning in the lateral axis of the reflector was accomplished with the motion of the earth over the course of a day. Extensive sky maps were made with this antenna at VHF frequencies leading to a better understanding of complex astrophysical phenomena.

**Spherical**

These types of reflectors are parabolic in one dimension and spherical in the other. In this way, multiple feeds can be placed in close proximity to each other and depending on their angle relative to the reflector, are used to receive emissions from multiple satellites. This is the most common use of this type of antenna and can be found at hotels and other venues where many channels are required for in room entertainment.

**Arecibo**

Another application of the spherical reflector shape is in radio and radar astronomy best exemplified by the 300 foot reflector in Arecibo, Puerto Rico. This antenna has an extensive history mapping pulsars, the Milky Way and taking radar images of numerous planets and asteroids. The reflector is a wire mesh sitting in a natural bowl depression in the mountains of Puerto Rico. Normally a suspended mesh would assume the shape of a caternary curve, but a multitude of adjustable ground supports have reshaped the reflector into a spherical shape. This is done to allow a moveable trolley assembly suspended over the antenna, to move a significant amount to cover more sky. This facility has the most powerful planetary radar on earth and has been used to send signals to our interstellar neighbors.

### Inside Balloon

Another interesting application of a spherical reflecting surface is for ground penetrating radar systems, using the inside surface of a large balloon as the reflector. Proposed for a Mars mission by Southwest Research Institute, this antenna has significant gain and with a modest transmitter, floating above the Martian surface, would be able to map the subsurface details with great resolution and depth. The surface of the reflector portion of the balloon can be coated with a very thin layer of conductive aluminum or silver, minimizing weight impact. Current designs for Martian balloon experiments predict floating above the surface for extended periods of time, rising high during the daytime and floating lower during the nighttime.

### Inflatable (F-111)

One interesting antenna design was that used on the Grumman F-111 aircraft. The radar system was capable of mapping object (moving and non-moving) on the ground, mapping other aircraft in the air, and identifying friend or foe. The antenna for this systems consisted of a parabolic dish with a reflective bladder on the upper portion. The bladder was inflated to produce a cosecant squared beam pattern, for ground mapping. The bladder could then be deflated to produce a more uniform beam for mapping the positions of other aircraft. In addition, several dipoles were suspended above the reflector dish that were used as a phase array to determine if the tracked airborne target was friend or foe, based on the transponder replies. This dish was steerable mechanically in both the elevation and azimuth angles, leading to significant area coverage.

### Sectioned

The Russian Ratan Radio Telescope is a good example of a sectioned parabolic reflector. Other examples are the reflectors used for air traffic radars and backyard satellite dishes. There are other interesting examples, such as:

### Ohio State Radio Telescope

The Ohio State Radio Telescope, also known as "The Big Ear", was designed in the 1950s to catalog radio sources in the Northern Hemisphere. The antennas was designed using a large flat movable reflector to scan

North and South. The reflected radio waves then progress to a parabolic section which then focuses the energy on a series of horn feed antennas placed near the ground. The first, flat reflector was 340 feet long and 100 feet tall, the parabolic section was 360 by 70 feet and the whole telescope took up over three football fields of space. This telescope cataloged thousands of radio sources over its career, the catalog is known appropriately as the Ohio Catalog. One interesting tidbit story about this antenna is, when it was completed, the designer (John Kraus) and other engineers walked into the control room, turned on the chart recorder and "started taking data." Normally, this is how all projects of this nature are run, but in this case, previously unknown radio sources started appearing on the chart paper, one after another

**Marseilles Radio Telescope**

Similar to the Kraus design, the flat reflector panel is 131 feet tall by 65 feet wide and the reflector is similar in size. The two reflectors are separated by 1,300 feet; the frequency of use in the decametric or HF bands. This telescope is located near Nancay, France and is still in use mapping the radio universe.

**Fresnel Reflectors**

During the C band backyard dish boom in the 1980s, a series of flattened reflectors based in part on the Fresnel equations. Essentially, these were designed by cutting a regular parabolic dish in concentric sections and lowering the three of four circular sections onto the same level, separated by one wavelength. This had the effect of producing a flatter, less obtrusive antenna. On consequence of such a design is a restricted bandwidth, in the case of the C band or Ku band design, this was not a factor.

**Caternary**

Reflectors in the early 1950s were designed in built very inexpensively in England by setting up a ring of poles and suspending wires between them in a caternary curve, created by gravity. Dishes in excess of 200 feet in diameter were made using this approach and were able to have a minimal amount of scanning capability by moving the focal point off of vertical. These telescopes were able to detect radio emissions from the Andromeda Galaxy as well as make detailed maps of the Milky Way.

**Beam Waveguide**

This type of reflector design is used by NASA in the Deep Space Network, tracking satellites in the far reaches of the solar system. These antennas, although more complex, have several advantages, like:

The relocation of sensitive electronics from a feed cone in the center of the main reflector to the pedestal equipment room

There is easier access for maintenance and modification such as adding new electronics for transmitting and receiving at additional radio frequency bands when needed to support future deep space missions.

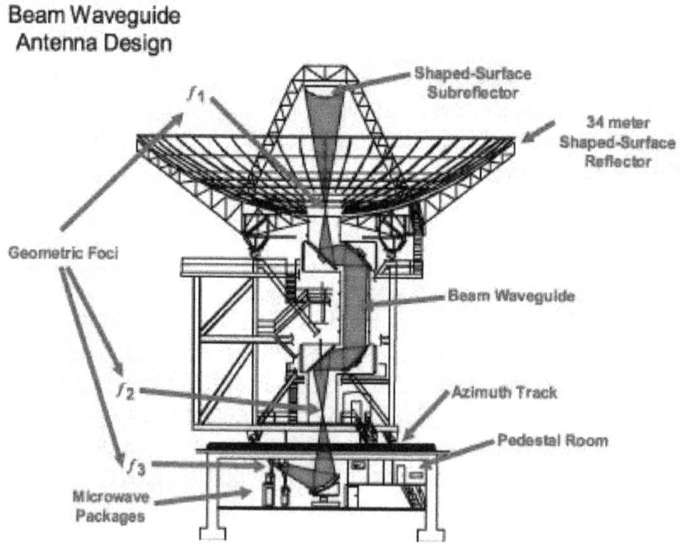


Example of Beam Waveguide Reflector design (40)

**Deep Space Network (DSN)**

NASA operates three large aperture antenna sites for the purpose of communicating with deep space probes, some beyond the solar system. They have facilities in Goldstone, California, Madrid, Spain and Alice Springs, Australia. With these facilities they can maintain a 24 hour stream of communications with any satellite. These antennas (the largest at each site is 73 meters) are carefully designed for efficiency, low noise and high power. Typically running in the S and X bands, efficiencies in excess of 70% have been achieved by careful geometric and physical optic approaches.

NASA Deep Space Network 73m and Optics

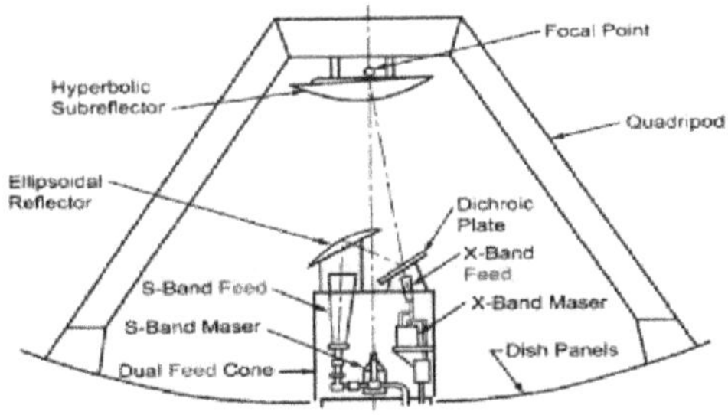


**Design Examples**

The following pages contain examples of antennas discussed in this chapter. These examples represent a first order design approach. For a more complete and optimized antenna solution, use the example closest to the requirement needs and optimize by using good simulation software, creating prototypes, testing and finally modifying prototypes to optimize performance.

Type: ***Prime Focus Parabola***

Frequency of Operational: *100 Mhz to 500 GHz*
Gain Range: *5 to 85 dBi*
Bandwidth: *~ 50+%*
Approximate Power Rating: *10+ Mw*
Design Software: *GTD, Physical Optics, Geometric Optics*

Configuration:

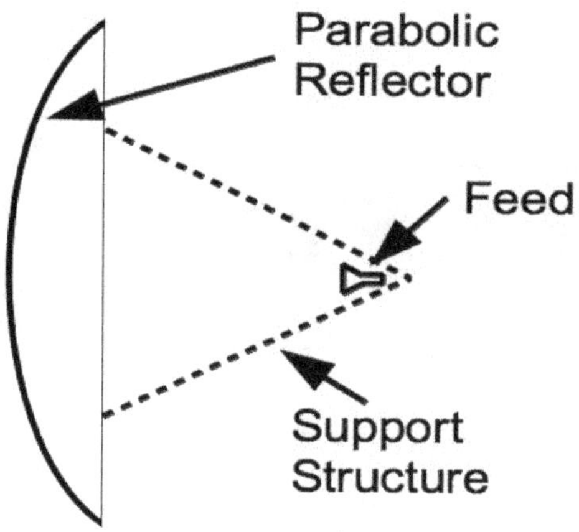


Design Approach:

*Use good physical or geometric optic simulation software*
*Use formula for parabola*

Construction and Theory of Operation:

*Essentially a paraboloid mirror, this antenna focusses the energy to a single spot where a horn or other suitable feed apparatus gathers or distributes the combined energy. Minimum diameter should be 5 lambda, the higher the number of lambda across the diameter will increase complexity and cost.*

*Azimuth Pattern*

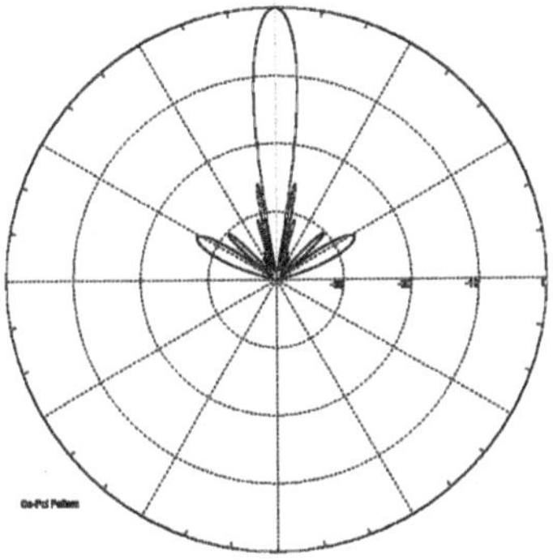

*Elevation Pattern*

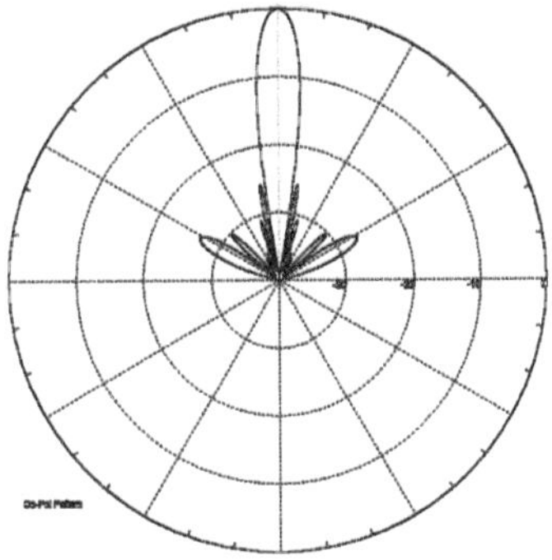

Special Considerations:

*Make sure that the surface roughness is as good as possible, with reason price wise. A roughness of 1/5 wavelength RMS is sufficient to form a beam but values better than 1/50 wavelength are much better for a good collimated beam and lower sidelobe structures. Satellite antennas for commercial use have roughness values better than 1/100 wavelength for instance. Reflecting structure can be made of steel with aluminum panels or all aluminum. For mid sized reflectors (< 3 meter diameter) wire mesh can be imbedded into fiberglass. Yet smaller apertures (<1 meter) can be made from stamped metal, preferably aluminum.*

*For larger reflectors keep in mind surface sag can be a problem when transitioning from high to low elevation angles. Also, wind loading can be a serious concern with very large structures. Counter weight large structures on the elevation angle.*

Testing:

*For mid sized and small apertures use far field range with transmitter/receiver or network analyzer. Near field and compact ranges work also. Use calibrated reference antenna. Use axial ratio measurements to confirm quality of required polarization. For large apertures, use radio star techniques, holography or satellites with known EIRP. Check Appendix 6 for details.*

Type: ***Cassegrain Parabola***

Frequency of Operational: *100 Mhz to 500 GHz*
Gain Range: *5 to 85 dBi*
Bandwidth: *~ 50+%*
Approximate Power Rating: *10+ Mw*
*Design Software:* *GTD, Physical Optics, Geometric Optics*

Configuration:

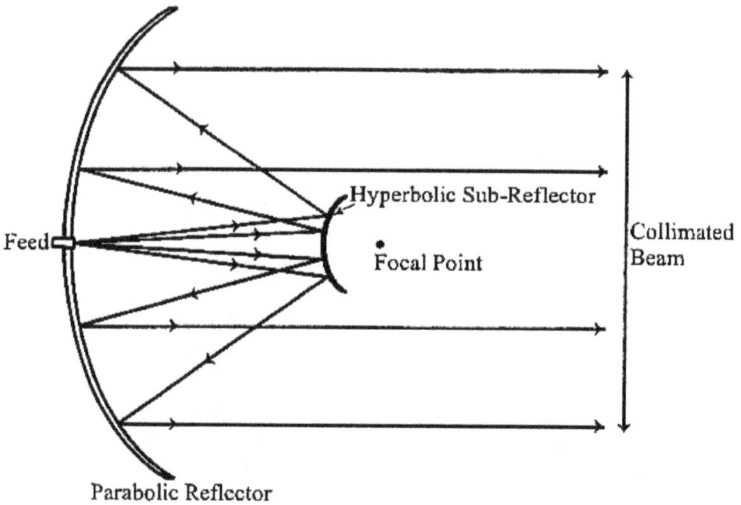


Design Approach:

*Use good physical or geometric optic simulation software*
*Use formula for parabola on main reflector and hyperbola (or elliptical) on sub reflector*

Construction and Theory of Operation:

*Essentially a paraboloid mirror, this antenna focusses the energy to a hyperbolic subreflector and then to a horn or other suitable feed apparatus gathers or distributes the combined energy. Minimum diameter should be 5 lambda, the higher the number of lambda across the diameter will increase complexity and cost.*

*Azimuth Pattern*

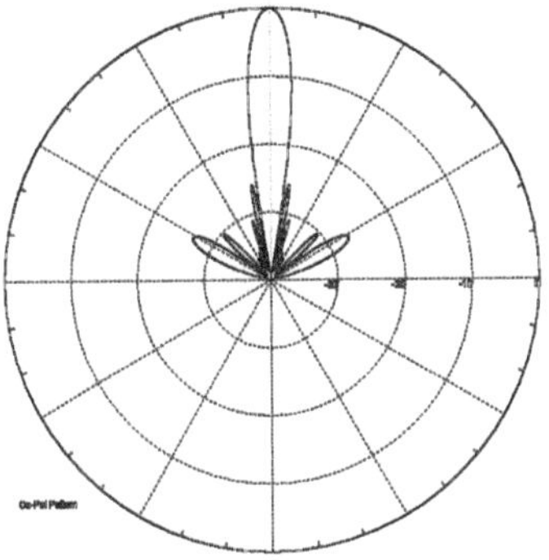

*Elevation Pattern*

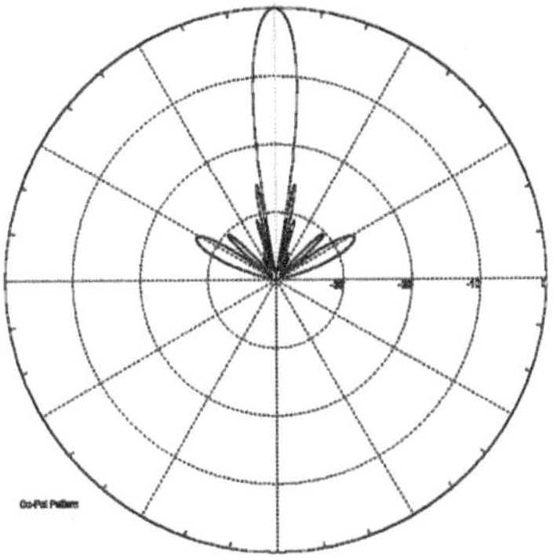

Special Considerations:

*Make sure that the surface roughness is as good as possible, with reason price wise. A roughness of 1/5 wavelength RMS is sufficient to form a beam but values better than 1/50 wavelength are required for a good collimated beam and lower sidelobe structures. Satellite antennas for commercial use have roughness values better than 1/100 wavelength for instance. Reflecting structure can be made of steel with aluminum panels or all aluminum. For mid sized reflectors (< 3 meter diameter) wire mesh can be imbedded into fiberglass. Yet smaller apertures (<1 meter) can be made from stamped metal, preferably aluminum.*

*For larger reflectors keep in mind surface sag can be a problem when transitioning from high to low elevation angles. Also, wind loading can be a serious concern with very large structures. Counter weight large structures on the elevation angle.*

Testing:

*For mid sized and small apertures use far field range with transmitter/receiver or network analyzer. Near field and compact ranges work also. Use calibrated reference antenna. Use axial ratio measurements to confirm quality of required polarization. For large apertures, use radio star techniques, holography or satellites with known EIRP. Check Appendix 6 for details.*

Type: ***Cassegrain Parabola-Polarization Twist***

Frequency of Operational: *100 Mhz to 500 GHz*
Gain Range: *5 to 85 dBi*
Bandwidth: *~ 50+%*
Approximate Power Rating: *10+ Mw*
*Design Software:* *GTD, Physical Optics, Geometric Optics*

Configuration:

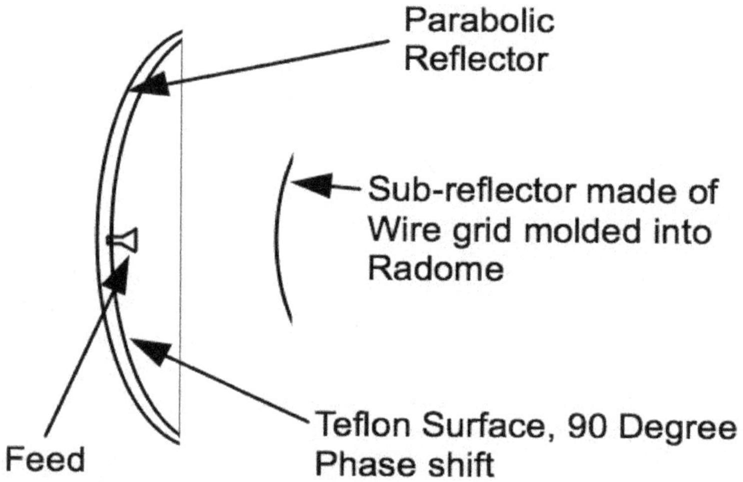


Design Approach:

*Use good physical or geometric optic simulation software*
*Use formula for parabola on main reflector and hyperbola (or elliptical) on sub reflector*

Construction and Theory of Operation:

*Essentially a paraboloid mirror, this antenna focusses the energy to a hyperbolic subreflector and then to a horn or other suitable feed apparatus gathers or distributes the combined energy. Minimum diameter should be 5 lambda, the higher the number of lambda across the diameter will increase complexity and cost. Main dish surface changes polarization by 90 degrees, allowing minimal feed blockage. Wire grid polarization opposite that of final beam.*

*Azimuth Pattern*

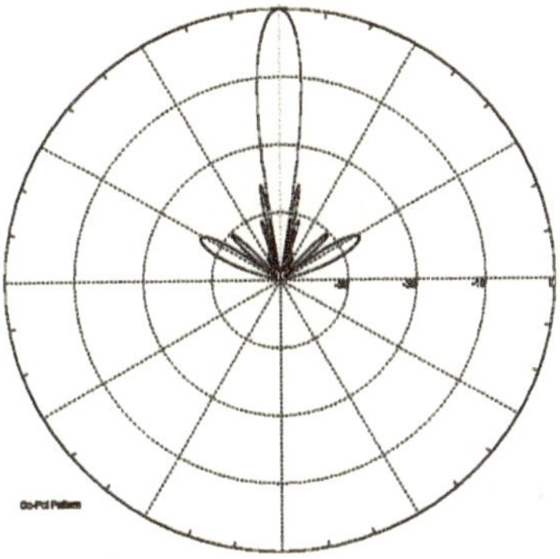

*Elevation Pattern*

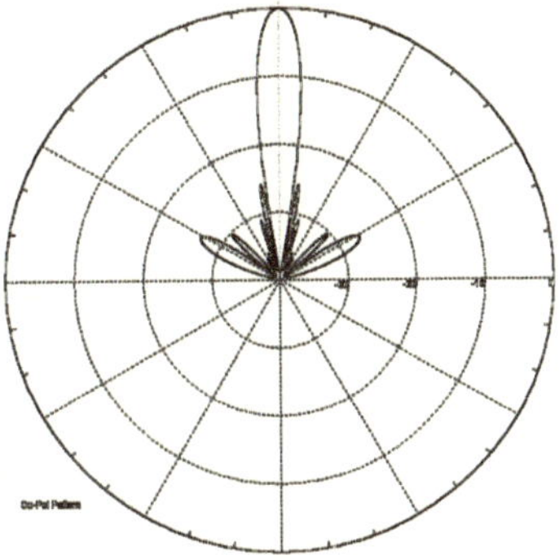

Special Considerations:

*Make sure that the surface roughness is as good as possible, with reason price wise. A roughness of 1/5 wavelength RMS is sufficient to form a beam but values better than 1/50 wavelength are required for a good collimated beam and lower sidelobe structures. Satellite antennas for commercial use have roughness values better than 1/100 wavelength for instance. Reflecting structure can be made of steel with aluminum panels or all aluminum. For mid sized reflectors (< 3 meter diameter) wire mesh can be imbedded into fiberglass. Yet smaller apertures (<1 meter) can be made from stamped metal, preferably aluminum.*

*For larger reflectors keep in mind surface sag can be a problem when transitioning from high to low elevation angles. Also, wind loading can be a serious concern with very large structures. Counter weight large structures on the elevation angle.*

Testing:

*For mid sized and small apertures use far field range with transmitter/receiver or network analyzer. Near field and compact ranges work also. Use calibrated reference antenna. Use axial ratio measurements to confirm quality of required polarization. For large apertures, use radio star techniques, holography or satellites with known EIRP. Check Appendix 6 for details.*

| | |
|---|---|
| Type: | ***Offset*** |
| Frequency of Operational: | *50 Mhz to 500 GHz* |
| Gain Range: | *0 to 85 dBi* |
| Bandwidth: | *~ 55+%* |
| Approximate Power Rating: | *1 Mw* |
| Design Software: | *GTD, Physical Optics, Geometric Optics* |

Configuration:

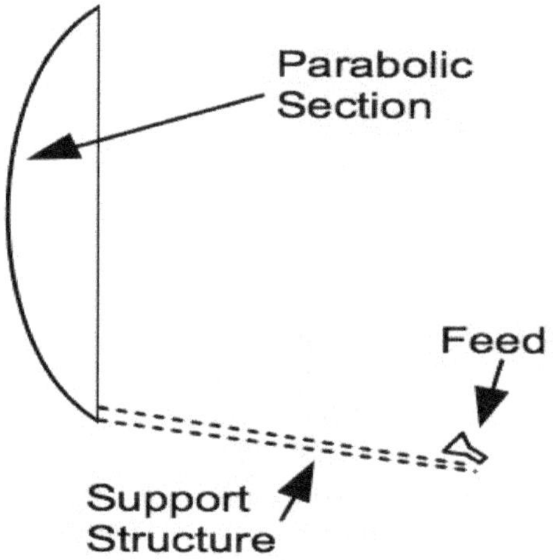


Design Approach:

*Use good physical or geometric optic simulation software*
*Use formula for parabolic section and offset*
*Reflector must be made elliptical to "present" circular aperture to far field*

Construction and Theory of Operation:

*Essentially a paraboloid mirror, this antenna focusses the energy to a single spot where a horn or other suitable feed apparatus gathers or distributes the combined energy. Minimum diameter should be 5 lambda, the higher the number of lambda across the diameter will increase complexity and cost. The focal point is not in the field of view of the reflector. As a consequence, the efficiency of these types of antennas are very high.*

*Azimuth Pattern*

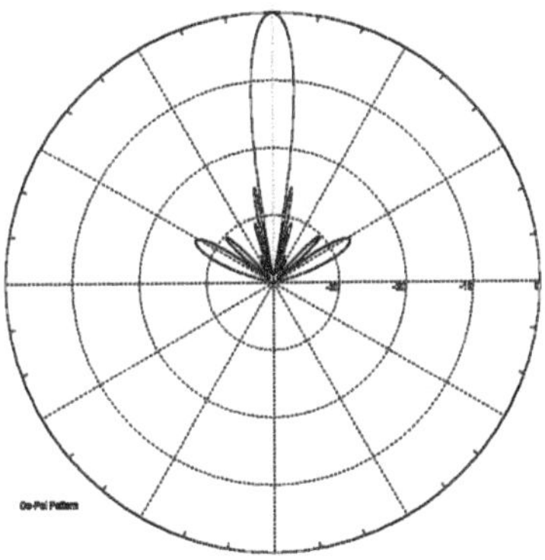

*Elevation Pattern*

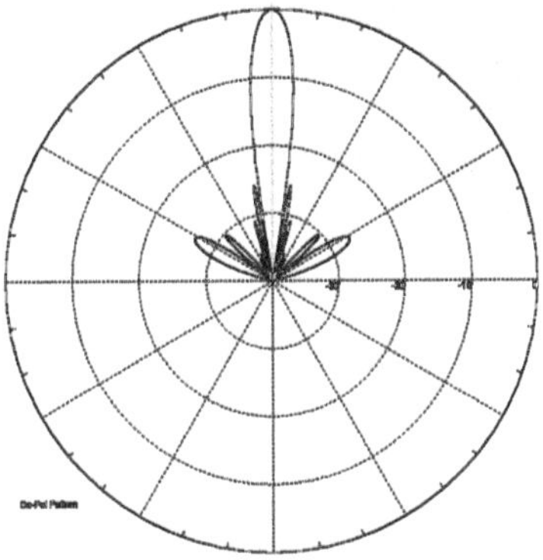

Special Considerations:

*Make sure that the surface roughness is as good as possible, with reason price wise. A roughness of 1/5 wavelength RMS is sufficient to form a beam but values better than 1/50 wavelength are required for a good collimated beam and lower sidelobe structures. Satellite antennas for commercial use have roughness values better than 1/100 wavelength for instance. Reflecting structure can be made of steel with aluminum panels or all aluminum. For mid sized reflectors (< 3 meter diameter) wire mesh can be imbedded into fiberglass. Yet smaller apertures (<1 meter) can be made from stamped metal, preferably aluminum.*

*For larger reflectors keep in mind surface sag can be a problem when transitioning from high to low elevation angles. Also, wind loading can be a serious concern with very large structures. Counter weight large structures on the elevation angle.*

Testing:

*For mid sized and small apertures use far field range with transmitter/receiver or network analyzer. Near field and compact ranges work also. Use calibrated reference antenna. Use axial ratio measurements to confirm quality of required polarization. For large apertures, use radio star techniques, holography or satellites with known EIRP. Check Appendix 6 for details.*

| | |
|---|---|
| Type: | ***Adaptive (self adjusting)*** |
| Frequency of Operational: | *5 Mhz to 500 GHz* |
| Gain Range: | *0 to 85 dBi* |
| Bandwidth: | *~ 55+%* |
| Approximate Power Rating: | *10 Mw* |
| Design Software: | *GTD, Physical Optics, Geometric Optics* |

Configuration:

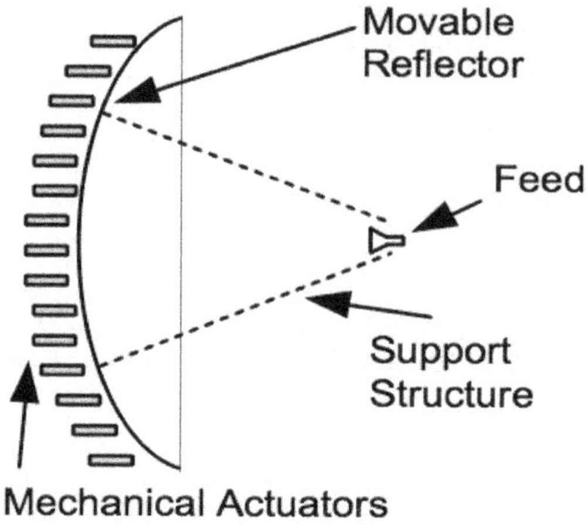


Design Approach:

*Use good physical or geometric optic simulation software*
*Use formula for parabola and use offset section*
*Design requires distance measuring system with full closed loop approach*

Construction and Theory of Operation:

*Essentially a paraboloid mirror, this antenna focusses the energy to a single spot where a horn or other suitable feed apparatus gathers or distributes the combined energy. Minimum diameter should be 5 lambda, the higher the number of lambda across the diameter will increase complexity and cost. Mechanical actuators allow for efficient operation at all elevation angles and allow higher frequency operation over temperature.*

*Azimuth Pattern* *Elevation Pattern*

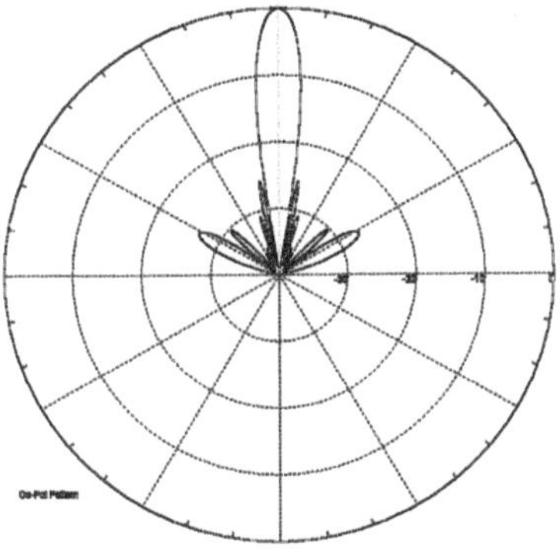

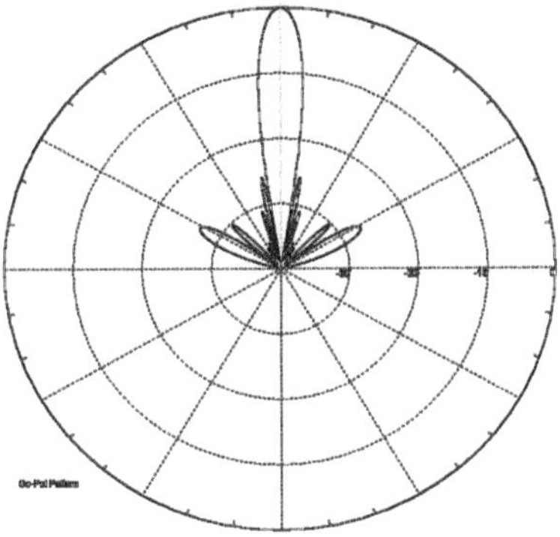

Special Considerations:

*Make sure that the surface roughness is as good as possible, with reason price wise. A roughness of 1/5 wavelength RMS is sufficient to form a beam but values better than 1/50 wavelength are required for a good collimated beam and lower sidelobe structures. Satellite antennas for commercial use have roughness values better than 1/100 wavelength for instance. Reflecting structure can be made of steel with aluminum panels or all aluminum. For mid sized reflectors (< 3 meter diameter) wire mesh can be imbedded into fiberglass. Yet smaller apertures (<1 meter) can be made from stamped metal, preferably aluminum.*

*For larger reflectors keep in mind surface sag can be a problem when transitioning from high to low elevation angles. Also, wind loading can be a serious concern with very large structures. Counter weight large structures on the elevation angle.*

Testing:

*For mid sized and small apertures use far field range with transmitter/receiver or network analyzer. Near field and compact ranges work also. Use calibrated reference antenna. Use axial ratio measurements to confirm quality of required polarization. For large apertures, use radio star techniques, holography or satellites with known EIRP. Check Appendix 6 for details.*

Type: ***Tapered – Sidelobe fence***

Frequency of Operational: *5 Mhz to 500 GHz*
Gain Range: *0 to 85 dBi*
Bandwidth: *~ 5%*
Approximate Power Rating: *1 Mw*
Design Software: *GTD, Physical Optics, Geometric Optics*

Configuration:

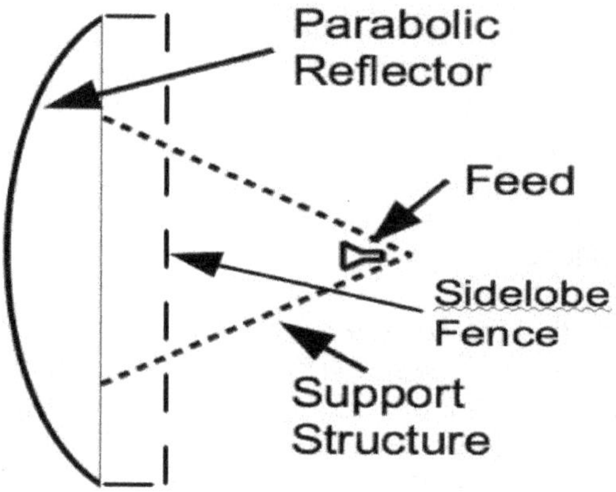


Design Approach:

*Use good physical or geometric optic simulation software*
*Use formula for parabola*
*Sidelobe fence should be at least 2 lambda long*

Construction and Theory of Operation:

*Essentially a paraboloid mirror, this antenna focusses the energy to a single spot where a horn or other suitable feed apparatus gathers or distributes the combined energy. Minimum diameter should be 5 lambda, the higher the number of lambda across the diameter will increase complexity and cost. A metal band is attached to the edge of the dish, several wavelengths long, to shield the dish from signals off axis from the main beam, in the sidelobes for instance. Tapering of the feed illumination also lowers sidelobe sensitivity, although it also lower gain and widens beamwidth.*

*Azimuth Pattern* *Elevation Pattern*

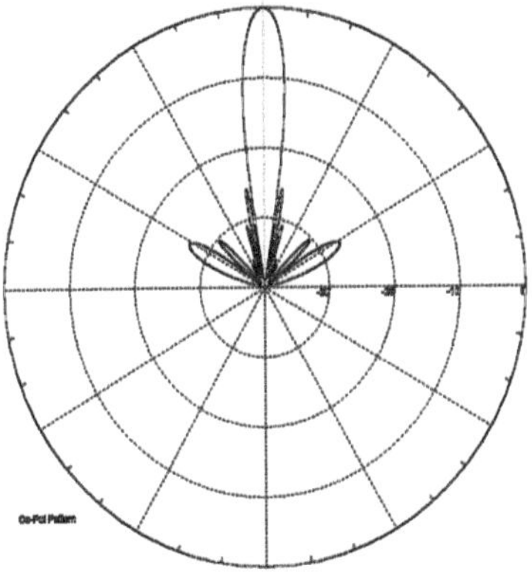

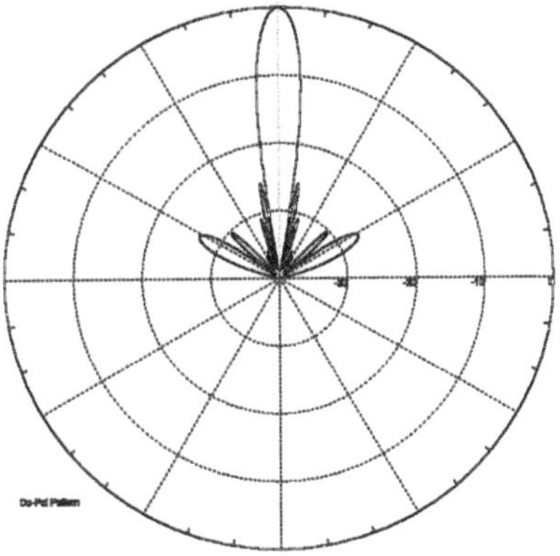

Special Considerations:

*Make sure that the surface roughness is as good as possible, with reason price wise. A roughness of 1/5 wavelength RMS is sufficient to form a beam but values better than 1/50 wavelength are required for a good collimated beam and lower sidelobe structures. Satellite antennas for commercial use have roughness values better than 1/100 wavelength for instance. Reflecting structure can be made of steel with aluminum panels or all aluminum. For mid sized reflectors (< 3 meter diameter) wire mesh can be imbedded into fiberglass. Yet smaller apertures (<1 meter) can be made from stamped metal, preferably aluminum.*

*For larger reflectors keep in mind surface sag can be a problem when transitioning from high to low elevation angles. Also, wind loading can be a serious concern with very large structures. Counter weight large structures on the elevation angle.*

Testing:

*For mid sized and small apertures use far field range with transmitter/receiver or network analyzer. Near field and compact ranges work also. Use calibrated reference antenna. Use axial ratio measurements to confirm quality of required polarization. For large apertures, use radio star techniques, holography or satellites with known EIRP. Check Appendix 6 for details.*

Type: ***Tapered – Rolled Edges***

Frequency of Operational: *5 Mhz to 500 GHz*
Gain Range: *0 to 85 dBi*
Bandwidth: *~ 55+%*
Approximate Power Rating: *1 Mw*
Design Software: *GTD, Physical Optics, Geometric Optics*

Configuration:

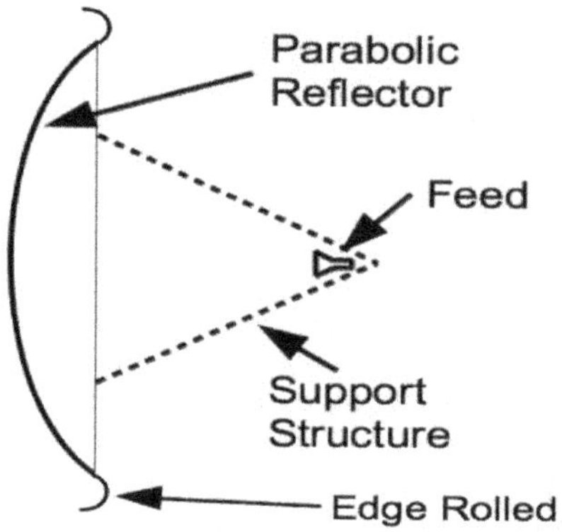


Design Approach:

*Use good physical or geometric optic simulation software*
*Use formula for parabola*
*Roll edges with at least 5 wavelengths of reflector material*

Construction and Theory of Operation:

*Essentially a paraboloid mirror, this antenna focusses the energy to a single spot where a horn or other suitable feed apparatus gathers or distributes the combined energy. Minimum diameter should be 5 lambda, the higher the number of lambda across the diameter will increase complexity and cost. A metal band is attached to the edge of the dish, several wavelengths long, to shield the dish from signals off axis from the main beam, in the sidelobes for instance. Tapering of the feed illumination also lowers sidelobe sensitivity, although it also lower gain and widens beamwidth.*

*Azimuth Pattern*

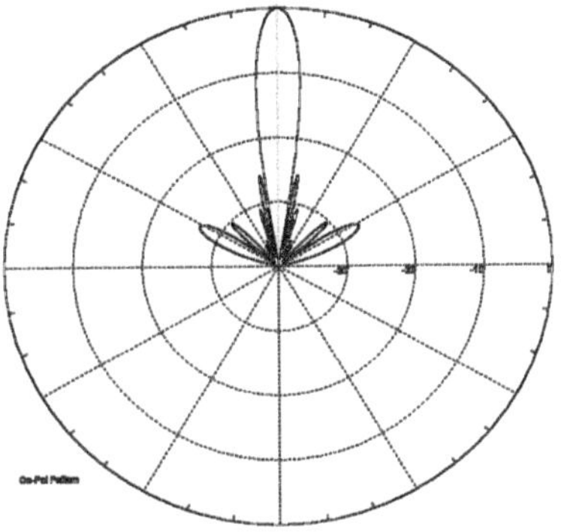

*Elevation Pattern*

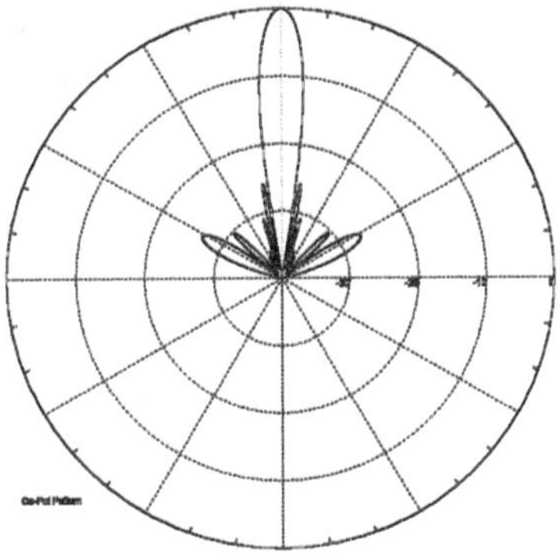

Special Considerations:

*Make sure that the surface roughness is as good as possible, with reason price wise. A roughness of 1/5 wavelength RMS is sufficient to form a beam but values better than 1/50 wavelength are required for a good collimated beam and lower sidelobe structures. Satellite antennas for commercial use have roughness values better than 1/100 wavelength for instance. Reflecting structure can be made of steel with aluminum panels or all aluminum. For mid sized reflectors (< 3 meter diameter) wire mesh can be imbedded into fiberglass. Yet smaller apertures (<1 meter) can be made from stamped metal, preferably aluminum.*

*For larger reflectors keep in mind surface sag can be a problem when transitioning from high to low elevation angles. Also, wind loading can be a serious concern with very large structures. Counter weight large structures on the elevation angle.*

Testing:

*For mid sized and small apertures use far field range with transmitter/receiver or network analyzer. Near field and compact ranges work also. Use calibrated reference antenna. Use axial ratio measurements to confirm quality of required polarization. For large apertures, use radio star techniques, holography or satellites with known EIRP. Check Appendix 6 for details.*

Type: ***CSC squared Paraboloid***

Frequency of Operational: *5 Mhz to 500 GHz*
Gain Range: *0 to 85 dBi*
Bandwidth: *~ 55+%*
Approximate Power Rating: *1 Mw*
Design Software: *GTD, Physical Optics, Geometric Optics*

Configuration:

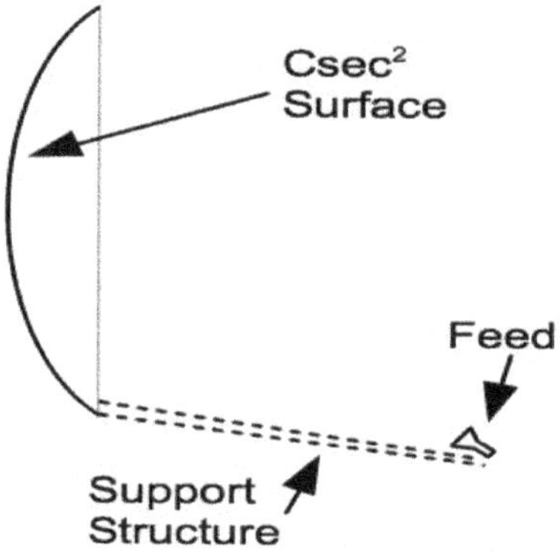


Design Approach:

*Use good physical or geometric optic simulation software*
*Use formula for Csec$^2$ and use offset section*
*Reflecting surface must be elliptical to "present" a circular aperture to the far field*
*For lower gain requirements, an offset microstrip patch works nicely, vary Csec$^2$ angle by adjusting distance from edge of patch to edge of ground plane.*

Construction and Theory of Operation:

*Essentially a Csec$^2$ shaped mirror, this antenna focusses the energy to a single spot where a horn or other suitable feed apparatus gathers or distributes the combined energy. Minimum diameter should be 5 lambda, the higher the number of lambda across the diameter will increase complexity and cost. This antenna produces a pattern used widely in radar work, that is wide on one side, useful for instance when "painting" the ground or in air craft surveillance modes.*

*Azimuth Pattern*

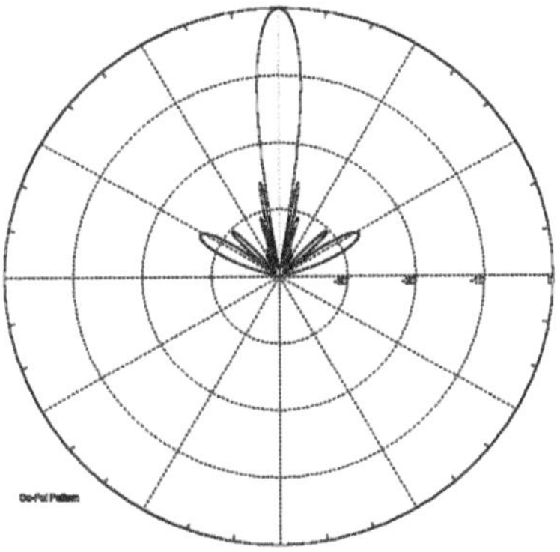

*Elevation Pattern*

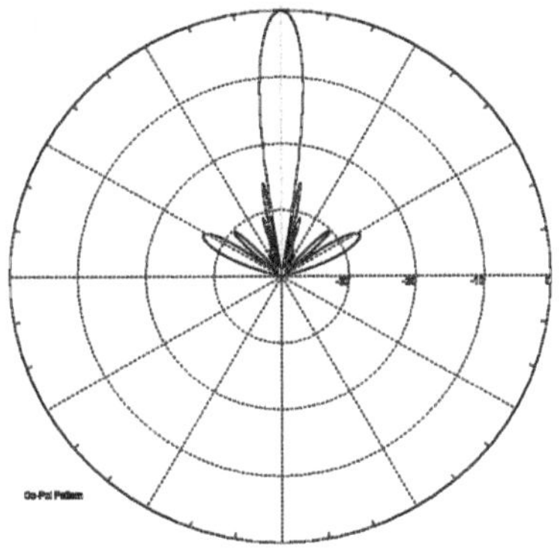

Special Considerations:

*Make sure that the surface roughness is as good as possible, with reason price wise. A roughness of 1/5 wavelength RMS is sufficient to form a beam but values better than 1/50 wavelength are much better for a good collimated beam and lower sidelobe structures. Satellite antennas for commercial use have roughness values better than 1/100 wavelength for instance. Reflecting structure can be made of steel with aluminum panels or all aluminum. For mid sized reflectors (< 3 meter diameter) wire mesh can be imbedded into fiberglass. Yet smaller apertures (<1 meter) can be made from stamped metal, preferably aluminum.*

*For larger reflectors keep in mind surface sag can be a problem when transitioning from high to low elevation angles. Also, wind loading can be a serious concern with very large structures. Counter weight large structures on the elevation angle.*

Testing:

*For mid sized and small apertures use far field range with transmitter/receiver or network analyzer. Near field and compact ranges work also. Use calibrated reference antenna. Use axial ratio measurements to confirm quality of required polarization. For large apertures, use radar techniques to verify sensitivity over altitude in the case of PPI operation, or ground returns in the case of airborne ground target acquisition. Check Appendix 6 for details.*

Type: ***Conical Paraboloid***

Frequency of Operational: *5 Mhz to 500 GHz*
Gain Range: *0 to 85 dBi*
Bandwidth: *~ 55+%*
Approximate Power Rating: *1 Mw*
Design Software: *GTD, Physical Optics, Geometric Optics*

Configuration:

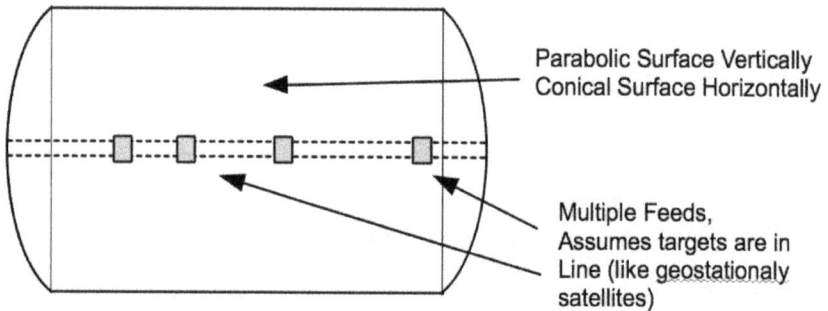

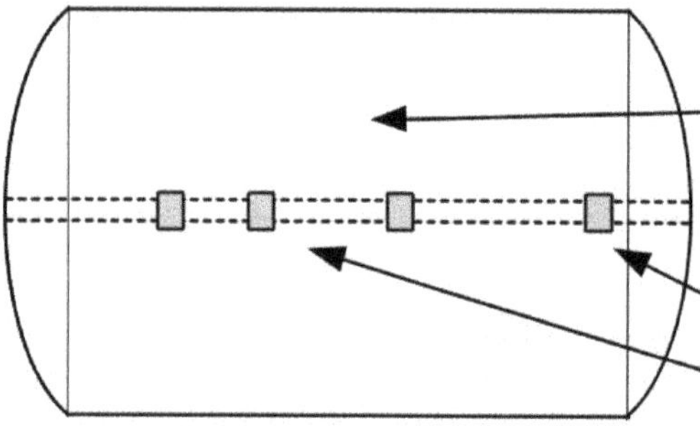


Design Approach:

*Use good physical or geometric optic simulation software*
*Use formula for parabola and conical surface*
*Expect fan beam performance with wider beam along short reflector axis*

Construction and Theory of Operation:

*Essentially a paraboloid mirror, this antenna focusses the energy to a single spot where a horn or other suitable feed apparatus gathers or distributes the combined energy. Minimum diameter should be 5 lambda, the higher the number of lambda across the diameter will increase complexity and cost. The antenna is conical in the horizontal plane to allow a feed "line" instead of "point." For the observation of multiple targets, as long as they are in a line, like geosynchronous satellites.*

*Azimuth Pattern*

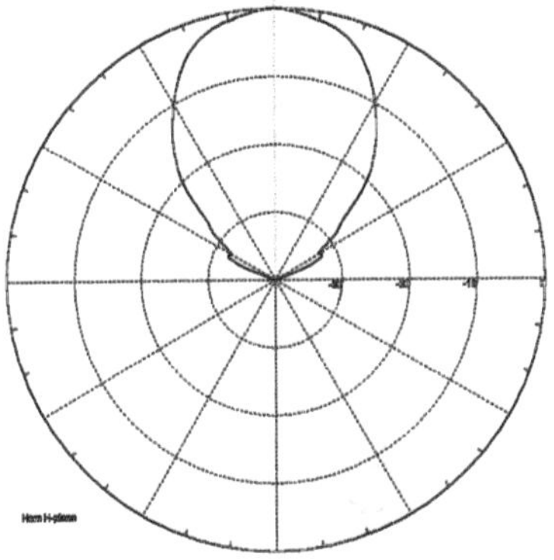

*Elevation Pattern*

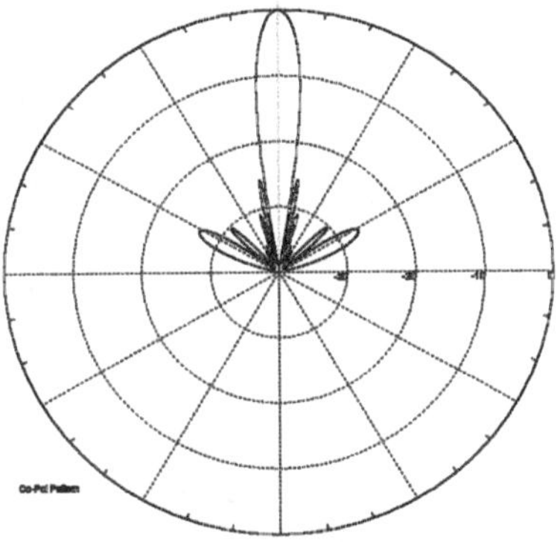

Special Considerations:

*Make sure that the surface roughness is as smooth as possible, within reason price wise. A roughness of 1/5 wavelength RMS is sufficient to form a beam but values better than 1/50 wavelength are required for a good collimated beam and lower sidelobe structures. Satellite antennas for commercial use have roughness values better than 1/100 wavelength for instance. Reflecting structure can be made of steel with aluminum panels or all aluminum. For mid sized reflectors (< 3 meter diameter) wire mesh can be imbedded into fiberglass. Yet smaller apertures (<1 meter) can be made from stamped metal, preferably aluminum.*

*For larger reflectors keep in mind surface sag can be a problem when transitioning from high to low elevation angles. Also, wind loading can be a serious concern with very large structures. Counter weight large structures on the elevation angle.*

Testing:

*For mid sized and small apertures use far field range with transmitter/receiver or network analyzer. Near field and compact ranges work also. Use calibrated reference antenna. Use axial ratio measurements to confirm quality of required polarization. For large apertures, use radio star techniques, holography or satellites with known EIRP. Check Appendix 6 for details.*

| | |
|---|---|
| Type: | ***Spherical*** |
| Frequency of Operational: | *5 Mhz to 50 GHz* |
| Gain Range: | *0 to 50 dBi* |
| Bandwidth: | *~ 55+%* |
| Approximate Power Rating: | *10 Mw* |
| Design Software: | *GTD, Physical Optics, Geometric Optics* |

Configuration:

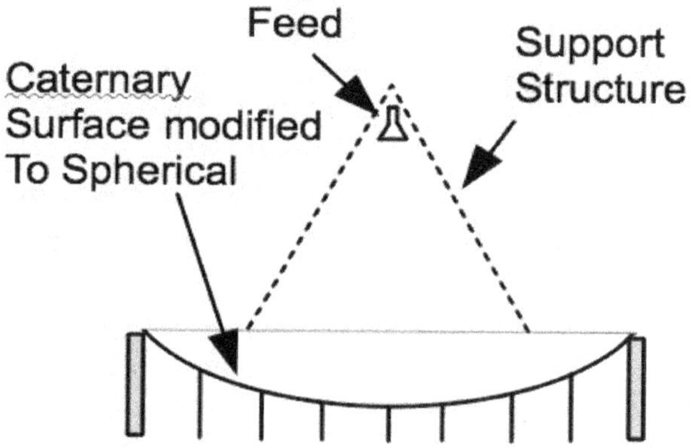


Design Approach:

*Use good physical or geometric optic simulation software*
*Use formula for sphere*
*Use distance measuring equipment to verify surface*
*Feed can be moved to give limited beam steering*

Construction and Theory of Operation:

*Essentially a spherical mirror, this antenna focusses the energy to a single spot where a horn or other suitable feed apparatus gathers or distributes the combined energy. Minimum diameter should be 5 lambda, the higher the number of lambda across the diameter will increase complexity and cost. The feed system needs to compensate for a non point focus, best approached with a linear feed parallel to the direction of the main beam.*

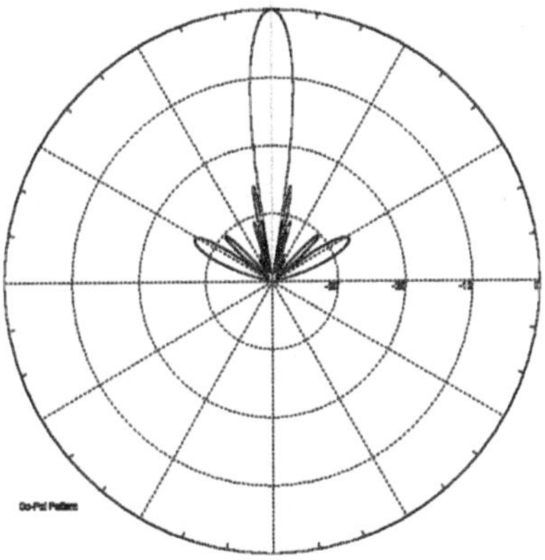

*Azimuth Pattern*

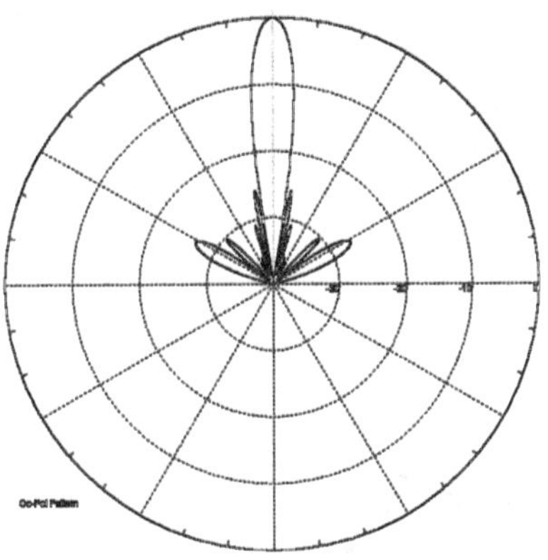

*Elevation Pattern*

Special Considerations:

*Make sure that the surface roughness is as good as possible, with reason price wise. A roughness of 1/5 wavelength RMS is sufficient to form a beam but values better than 1/50 wavelength are much better for a good collimated beam and lower sidelobe structures. Satellite antennas for commercial use have roughness values better than 1/100 wavelength for instance. Reflecting structure can be made of steel with aluminum panels or all aluminum. For mid sized reflectors (< 3 meter diameter) wire mesh can be imbedded into fiberglass. Yet smaller apertures (<1 meter) can be made from stamped metal, preferably aluminum.*

Testing:

*For mid sized and small apertures use far field range with transmitter/receiver or network analyzer. Near field and compact ranges work also. Use calibrated reference antenna. Use axial ratio measurements to confirm quality of required polarization. For large apertures, use radio star techniques, holography or satellites with known EIRP. Check Appendix 6 for details.*

Type: ***Sectioned Paraboloid***

Frequency of Operational: *5 Mhz to 500 GHz*
Gain Range: *0 to 45 dBi*
Bandwidth: *~ 55+%*
Approximate Power Rating: *100 Kw*
Design Software: *GTD, Physical Optics, Geometric Optics*

Configuration:

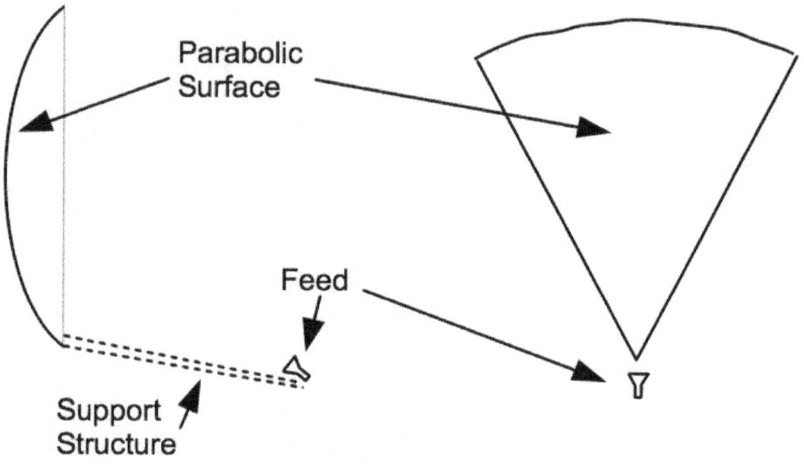


Design Approach:

*Use good physical or geometric optic simulation software*
*Use formula for parabola, use section*
*Illuminate very center of reflector to avoid asymmetrical beam*

Construction and Theory of Operation:

*Essentially a paraboloid mirror section, this antenna focusses the energy to a single spot where a horn or other suitable feed apparatus gathers or distributes the combined energy. Minimum diameter should be 5 lambda, the higher the number of lambda across the diameter will increase complexity and cost.*

*Azimuth Pattern*

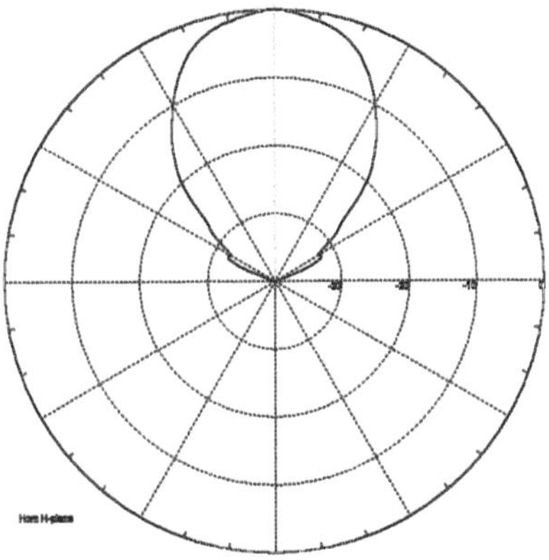

*Elevation Pattern*

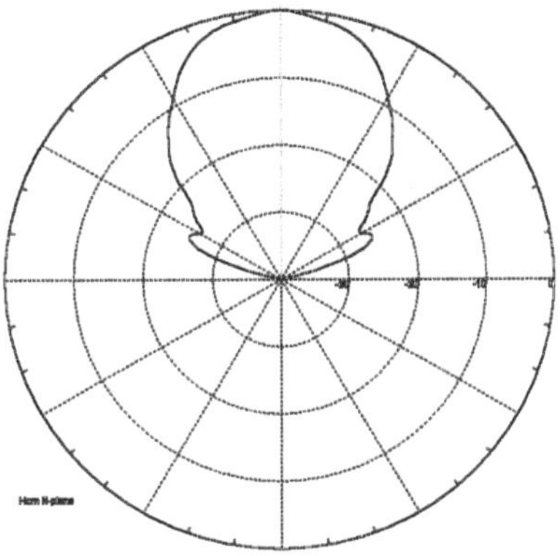

Special Considerations:

*Make sure that the surface roughness is as good as possible, with reason price wise. A roughness of 1/5 wavelength RMS is sufficient to form a beam but values better than 1/50 wavelength are required for a good collimated beam and lower sidelobe structures. Satellite antennas for commercial use have roughness values better than 1/100 wavelength for instance. Reflecting structure can be made of steel with aluminum panels or all aluminum. For mid sized reflectors (< 3 meter diameter) wire mesh can be imbedded into fiberglass. Yet smaller apertures (<1 meter) can be made from stamped metal, preferably aluminum.*

*For larger reflectors keep in mind surface sag can be a problem when transitioning from high to low elevation angles. Also, wind loading can be a serious concern with very large structures. Counter weight large structures on the elevation angle.*

Testing:

*For mid sized and small apertures use far field range with transmitter/receiver or network analyzer. Near field and compact ranges work also. Use calibrated reference antenna. Use axial ratio measurements to confirm quality of required polarization. For large apertures, use radio star techniques, holography or satellites with known EIRP. Check Appendix 6 for details.*

Type: ***Inflatable Paraboloid***

Frequency of Operational: *500 Mhz to 15 GHz*
Gain Range: *0 to 35 dBi*
Bandwidth: *~ 45+%*
Approximate Power Rating: *100 Kw*
Design Software: *GTD, Physical Optics, Geometric Optics*

Configuration:

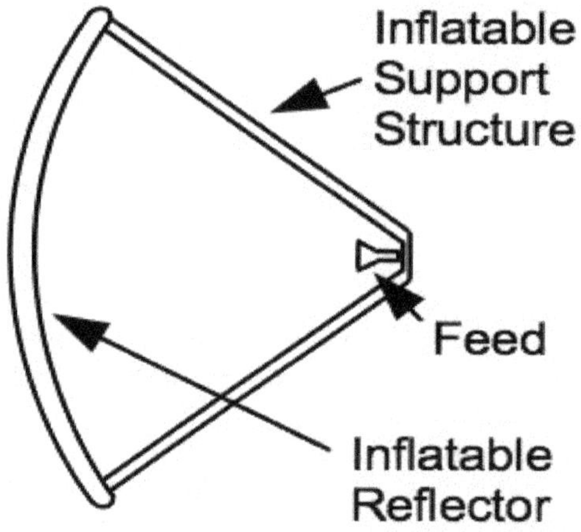


Design Approach:

*Use good physical or geometric optic simulation software*
*Use formula for parabola*
*Best used in space, use space qualified components*

Construction and Theory of Operation:

*Essentially a paraboloid mirror section, this antenna focusses the energy to a single spot where a horn or other suitable feed apparatus gathers or distributes the combined energy. Minimum diameter should be 5 lambda, the higher the number of lambda across the diameter will increase complexity and cost. This antenna is typically designed for use in space, where a large structure can be sent up in a small package, then deployed.*

*Azimuth Pattern* *Elevation Pattern*

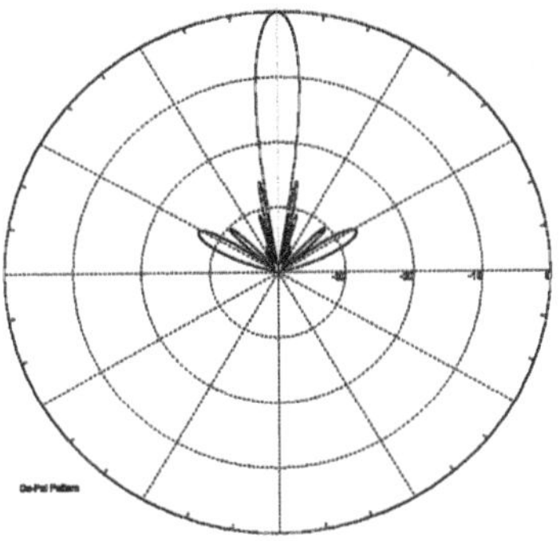

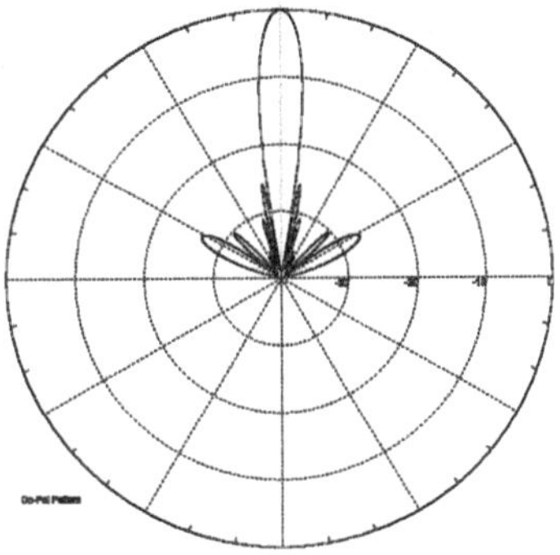

Special Considerations:

*Make sure that the surface roughness is as good as possible, with reason price wise. A roughness of 1/5 wavelength RMS is sufficient to form a beam but values better than 1/50 wavelength are much better for a good collimated beam and lower sidelobe structures. Satellite antennas for commercial use have roughness values better than 1/100 wavelength for instance. Reflecting structure can be made of with aluminum or gold coated inflatable materials, like plastic, rubber. For mid sized reflectors (< 3 meter diameter) wire mesh can be imbedded into the inflatable material. Yet smaller apertures (<1 meter) can be made from metals with memory, preferably aluminum.*

Testing:

*For mid sized and small apertures use far field range with transmitter/receiver or network analyzer. Near field and compact ranges work also. Use calibrated reference antenna. Use axial ratio measurements to confirm quality of required polarization. For large apertures, use radio star techniques, holography or satellites with known EIRP. Check Appendix 6 for details.*

Type: ***2 Sectioned Paraboloid, F-111***

Frequency of Operational: *500 Mhz to 20 GHz*
Gain Range: *0 to 35 dBi*
Bandwidth: *~ 55+%*
Approximate Power Rating: *100 Kw*
Design Software: *GTD, Physical Optics, Geometric Optics*

Configuration:

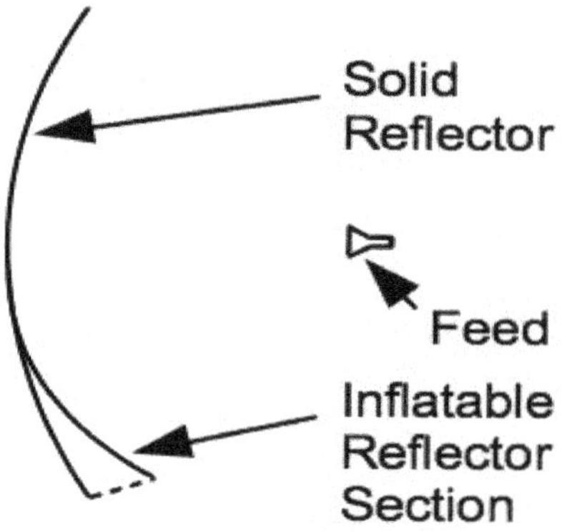


Design Approach:

*Use good physical or geometric optic simulation software*
*Use formula for parabola and $Csc^2$ design for each surface*
*Inflate pneumatically and maintain constant pressure*

Construction and Theory of Operation:

*Essentially a paraboloid mirror section, this antenna focusses the energy to a single spot where a horn or other suitable feed apparatus gathers or distributes the combined energy. Minimum diameter should be 5 lambda, the higher the number of lambda across the diameter will increase complexity and cost. An inflatable surface is attached to the main parabolic surface that, when inflated creates another shape, for instance a $Csc^2$ surface, useful in military aircraft applications.*

*Azimuth Pattern*

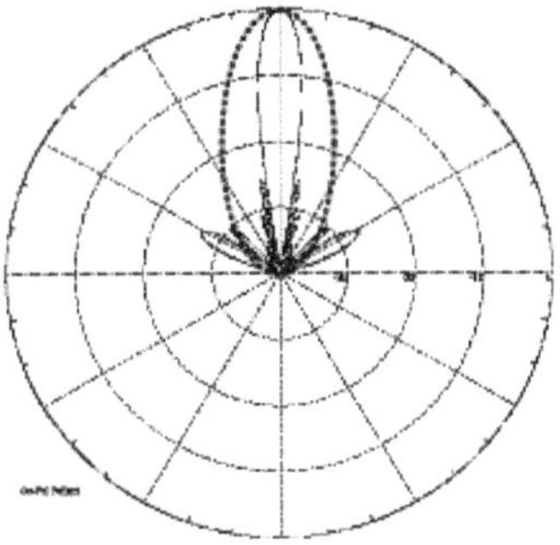

*Elevation Pattern*

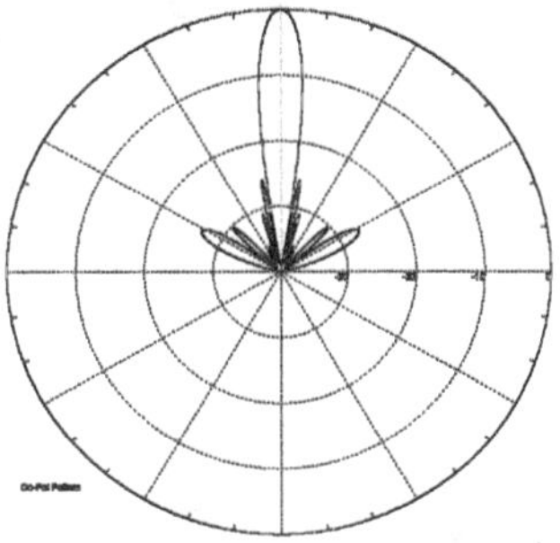

Special Considerations:

*Make sure that the surface roughness is as good as possible, with reason price wise. A roughness of 1/5 wavelength RMS is sufficient to form a beam but values better than 1/50 wavelength are required for a good collimated beam and lower sidelobe structures. Satellite antennas for commercial use have roughness values better than 1/100 wavelength for instance. Reflecting structure can be made of with aluminum or gold coated inflatable materials, like plastic, rubber. For mid sized reflectors (< 3 meter diameter) wire mesh can be imbedded into the inflatable material.*

Testing:

*For mid sized and small apertures use far field range with transmitter/receiver or network analyzer. Near field and compact ranges work also. Use calibrated reference antenna. Use axial ratio measurements to confirm quality of required polarization. For large apertures, use radio star techniques, holography or satellites with known EIRP. Check Appendix 6 for details.*

Type: ***Compressed Paraboloid, Fresnel Reflector***

| | |
|---|---|
| Frequency of Operational: | *50 Mhz to 50 GHz* |
| Gain Range: | *0 to 35 dBi* |
| Bandwidth: | *~ 15+%* |
| Approximate Power Rating: | *100 Kw* |
| Design Software: | *GTD, Physical Optics, Geometric Optics* |

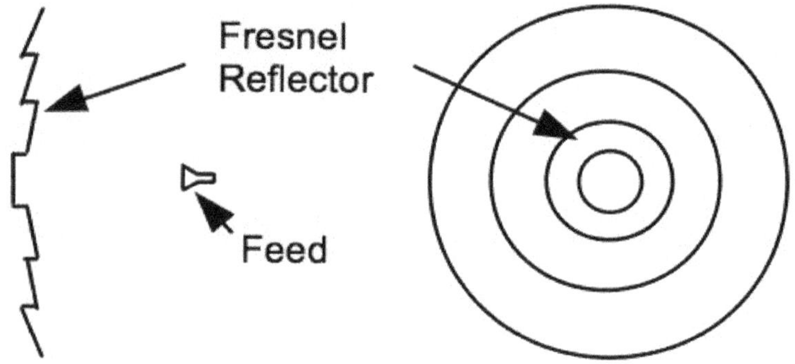


Configuration:

Design Approach:

*Use good physical or geometric optic simulation software*
*Use formula for parabola to design*
*Make sure steps are multiples of lambda*
*Reflector component can be imbedded in fiberglass or stamped*

Construction and Theory of Operation:

*Essentially a paraboloid mirror section, this antenna focusses the energy to a single spot where a horn or other suitable feed apparatus gathers or distributes the combined energy. Minimum diameter should be 5 lambda, the higher the number of lambda across the diameter will increase complexity and cost. Concentric circles are cut out and placed in n lambda depths to keep sections in phase. This reduces the depth of the reflector while maintaining the F/D ratio. This lowers the bandwidth and is typically used on single band operation.*

*Azimuth Pattern*

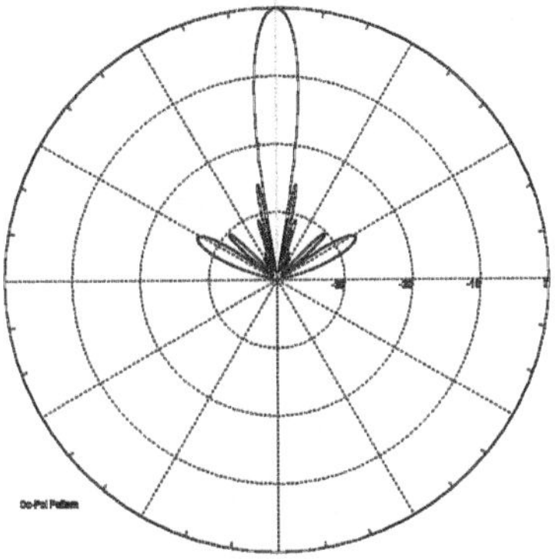

*Elevation Pattern*

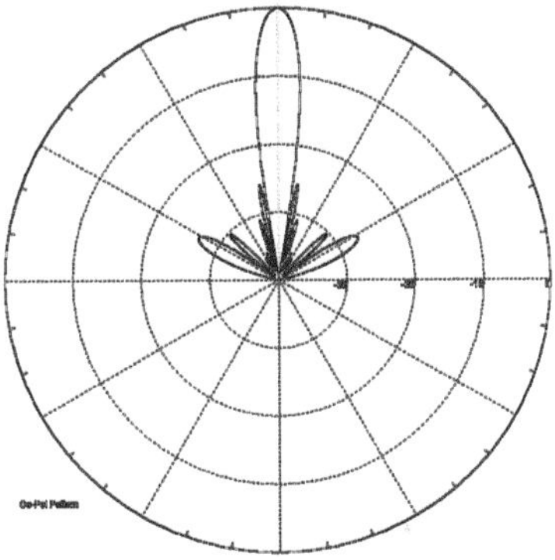

Special Considerations:

*Make sure that the surface roughness is as good as possible, with reason price wise. A roughness of 1/5 wavelength RMS is sufficient to form a beam but values better than 1/50 wavelength are much better for a good collimated beam and lower sidelobe structures. Satellite antennas for commercial use have roughness values better than 1/100 wavelength for instance. Reflecting structure can be made of steel with aluminum panels or all aluminum. For mid sized reflectors (< 3 meter diameter) wire mesh can be imbedded into fiberglass. Yet smaller apertures (<1 meter) can be made from stamped metal, preferably aluminum.*

*For larger reflectors keep in mind surface sag can be a problem when transitioning from high to low elevation angles. Also, wind loading can be a serious concern with very large structures. Counter weight large structures on the elevation angle.*

Testing:

*For mid sized and small apertures use far field range with transmitter/receiver or network analyzer. Near field and compact ranges work also. Use calibrated reference antenna. Use axial ratio measurements to confirm quality of required polarization. For large apertures, use radio star techniques, holography or satellites with known EIRP. Check Appendix 6 for details.*

Type: ***Gridded Paraboloid***

| | |
|---|---|
| Frequency of Operational: | *50 Mhz to 50 GHz* |
| Gain Range: | *0 to 35dBi* |
| Bandwidth: | *~ 25+%* |
| Approximate Power Rating: | *10 Kw* |
| Design Software: | *GTD, Physical Optics, Geometric Optics* |

Configuration:

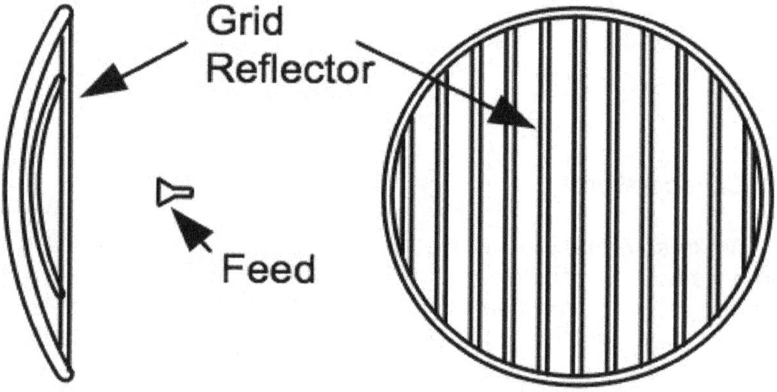


Design Approach:

*Use good physical or geometric optic simulation software*
*Use formula for parabola to design*
*Spacing of grid components must be less than lambda/5*
*Reflector is highly polarized in direction of grid components*

Construction and Theory of Operation:

*Essentially a paraboloid mirror section with reflecting surfaces made of tubes wires or etched lines, this antenna focusses the energy to a single spot where a horn or other suitable feed apparatus gathers or distributes the combined energy. Minimum diameter should be 5 lambda, the higher the number of lambda across the diameter will increase complexity and cost.*

*Azimuth Pattern*

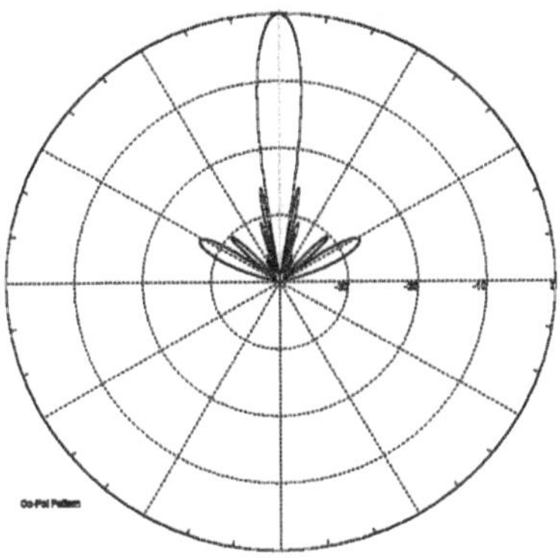

*Elevation Pattern*

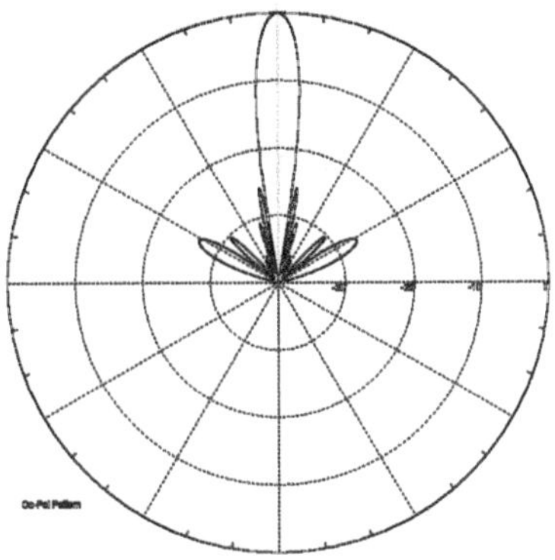

Special Considerations:

*Make sure that the surface roughness is as good as possible, with reason price wise. A roughness of 1/5 wavelength RMS is sufficient to form a beam but values better than 1/50 wavelength is required for a good collimated beam and lower sidelobe structures. Satellite antennas for commercial use have roughness values better than 1/100 wavelength for instance. Reflecting structure can be made of steel with aluminum panels or all aluminum. For mid sized reflectors (< 3 meter diameter) wire mesh can be imbedded into fiberglass. Yet smaller apertures (<1 meter) can be made from stamped metal, preferably aluminum.*

Testing:

*For mid sized and small apertures use far field range with transmitter/receiver or network analyzer. Near field and compact ranges work also. Use calibrated reference antenna. Use axial ratio measurements to confirm quality of required polarization. For large apertures, use radio star techniques, holography or satellites with known EIRP. Check Appendix 6 for details.*

Type: ***Concentric Paraboloid***

Frequency of Operational: *50 Mhz to 50 GHz*
Gain Range: *0 to 45 dBi*
Bandwidth: *~ 55+%*
Approximate Power Rating: *100 Kw*
Design Software: *GTD, Physical Optics, Geometric Optics*

Configuration:

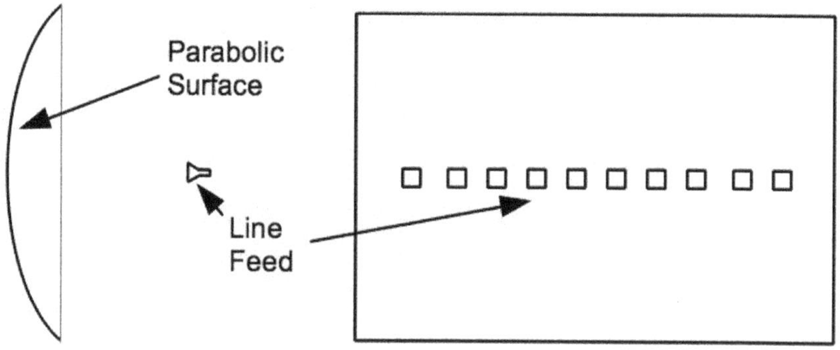


Design Approach:

*Use good physical or geometric optic simulation software*
*Use formula for cylinder and parabola to design*
*Fan beam forms with wider beam parallel to short axis*

Construction and Theory of Operation:

*Essentially a paraboloid mirror section, this antenna focusses the energy to a single spot where a horn or other suitable feed apparatus gathers or distributes the combined energy. Minimum diameter should be 5 lambda, the higher the number of lambda across the diameter will increase complexity and cost. One axis of the reflectors is elongated to create a fan beam. By properly phasing the feeds, the main beam can be moved.*

*Azimuth Pattern*

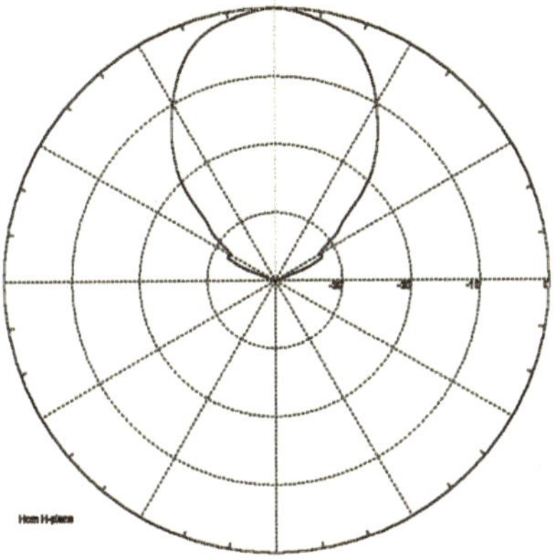

*Elevation Pattern*

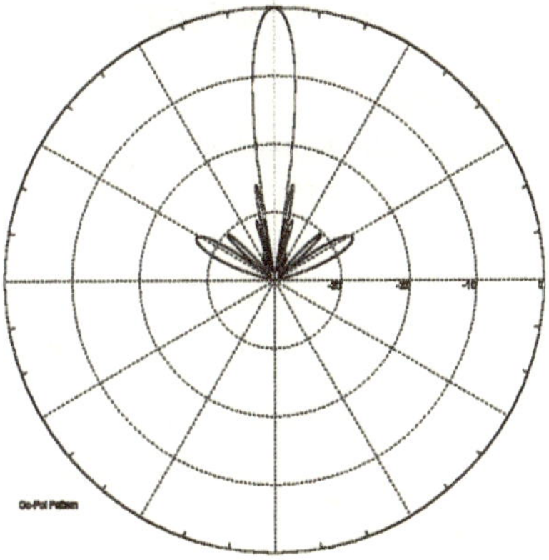

Special Considerations:

*Make sure that the surface roughness is as good as possible, with reason price wise. A roughness of 1/5 wavelength RMS is sufficient to form a beam but values better than 1/50 wavelength are much better for a good collimated beam and lower sidelobe structures. Satellite antennas for commercial use have roughness values better than 1/100 wavelength for instance. Reflecting structure can be made of steel with aluminum panels or all aluminum. For mid sized reflectors (< 3 meter diameter) wire mesh can be imbedded into fiberglass. Yet smaller apertures (<1 meter) can be made from stamped metal, preferably aluminum.*

Testing:

*For mid sized and small apertures use far field range with transmitter/receiver or network analyzer. Near field and compact ranges work also. Use calibrated reference antenna. Use axial ratio measurements to confirm quality of required polarization. For large apertures, use radio star techniques, holography or satellites with known EIRP. Check Appendix 6 for details.*

Type: ***Caternary Paraboloid***

Frequency of Operational: *5 Mhz to 20 GHz*
Gain Range: *0 to 45 dBi*
Bandwidth: *~ 55+%*
Approximate Power Rating: *1 Mw*
Design Software: *GTD, Physical Optics, GeometricOptics*

Configuration:

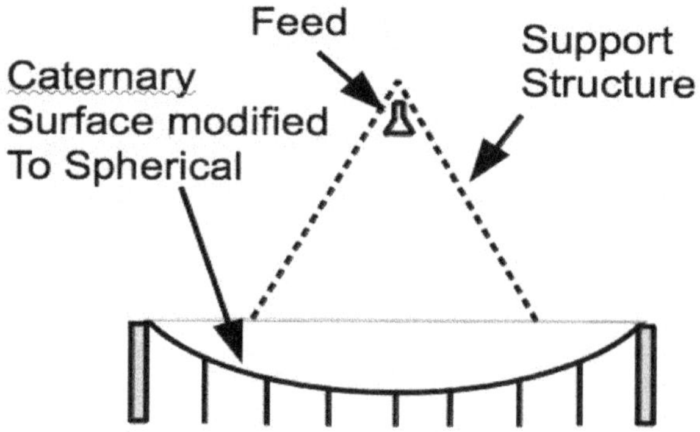


Design Approach:

*Use good physical or geometric optic simulation software*
*Expect an elongated feed point, consider traveling wave feed design*
*Feed position can be moved several degrees for scanning*
*Use distance measuring equipment from the feed to the surface to verify required surface.*
*Use formula for parabola as guide*

Construction and Theory of Operation:

*Essentially a paraboloid mirror section, this antenna focusses the energy to a single spot where a horn or other suitable feed apparatus gathers or distributes the combined energy. Minimum diameter should be 5 lambda, the higher the number of lambda across the diameter will increase complexity and cost. Wires beneath structure adjust surface to a parabolic shape.*

*Azimuth Pattern*

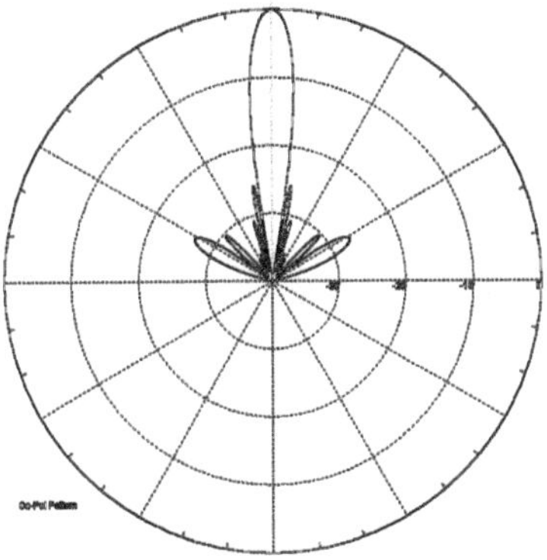

*Elevation Pattern*

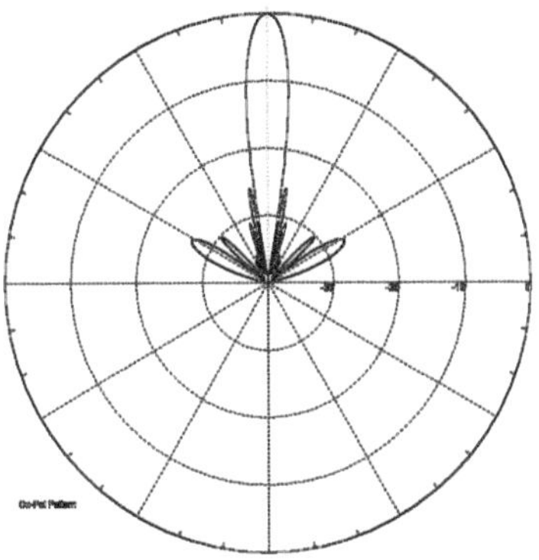

Special Considerations:

*Make sure that the surface roughness is as good as possible, with reason price wise. A roughness of 1/5 wavelength RMS is sufficient to form a beam but values better than 1/50 wavelength is required for a good collimated beam and lower sidelobe structures. Satellite antennas for commercial use have roughness values better than 1/100 wavelength for instance. Reflecting structure can be made of steel with aluminum panels or all aluminum. For mid sized reflectors (< 3 meter diameter) wire mesh can be imbedded into fiberglass. Yet smaller apertures (<1 meter) can be made from stamped metal, preferably aluminum.*

Testing:

*For mid sized and small apertures use far field range with transmitter/receiver or network analyzer. Near field and compact ranges work also. Use calibrated reference antenna. Use axial ratio measurements to confirm quality of required polarization. For large apertures, use radio star techniques, holography or satellites with known EIRP. Check Appendix 6 for details.*

Type: ***Beam Waveguide Paraboloid***

| | |
|---|---|
| Frequency of Operational: | *50 Mhz to 500 GHz* |
| Gain Range: | *0 to 85 dBi* |
| Bandwidth: | *~ 55+%* |
| Approximate Power Rating: | *1 Mw* |
| Design Software: | *GTD, Physical Optics, Geometric Optics* |

Configuration:

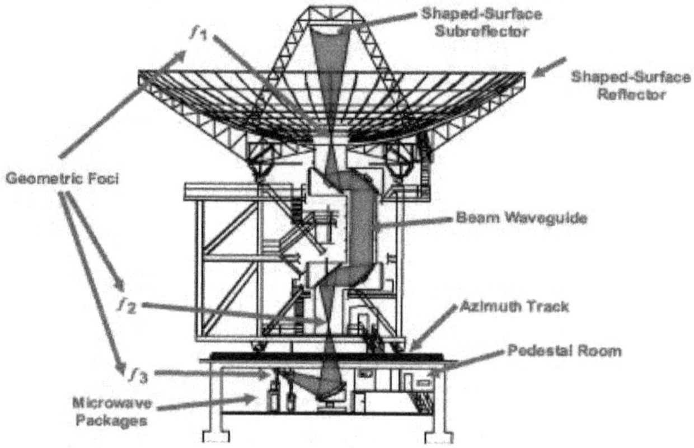


Design Approach:

*Use good physical or geometric optic simulation software*
*Use formula for parabolic section and hyperbolic (or elliptical) subreflector*
*Illuminate 80% of all reflective surfaces*
*Center all rotational axiis*

Construction and Theory of Operation:

*Essentially a paraboloid mirror system, this antenna focusses the energy to a single spot where a horn or other suitable feed apparatus gathers or distributes the combined energy. Minimum diameter should be 5 lambda, the higher the number of lambda across the diameter will increase complexity and cost. The beam extension allows for very high F/D ratios, good for spectroscopy.*

*Azimuth Pattern*

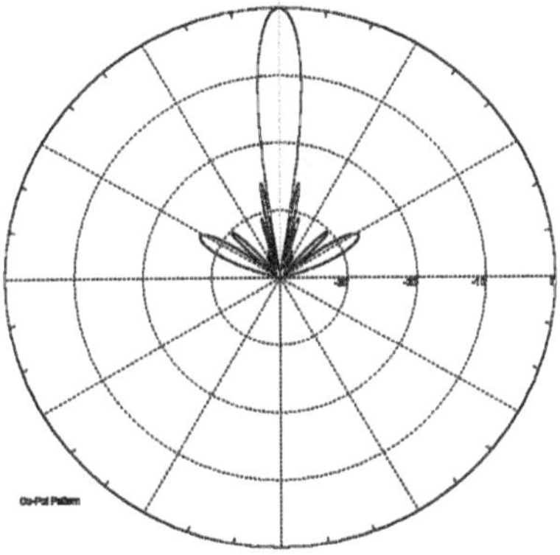

*Elevation Pattern*

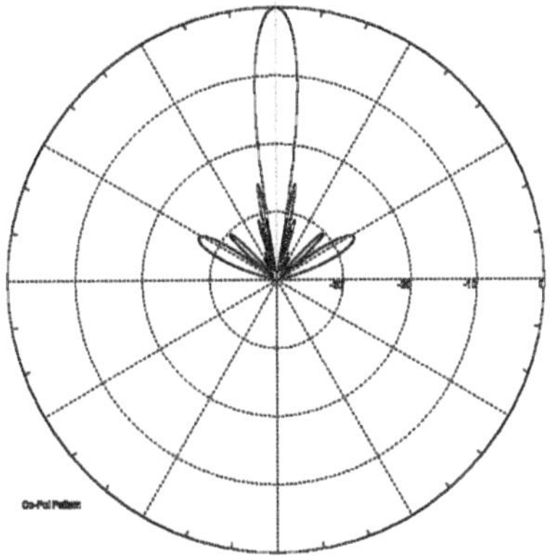

Special Considerations:

*Make sure that the surface roughness is as good as possible, with reason price wise. A roughness of 1/5 wavelength RMS is sufficient to form a beam but values better than 1/50 wavelength is required for a good collimated beam and lower sidelobe structures. Satellite antennas for commercial use have roughness values better than 1/100 wavelength for instance. Reflecting structure can be made of steel with aluminum panels or all aluminum. For mid sized reflectors (< 3 meter diameter) wire mesh can be imbedded into fiberglass. Yet smaller apertures (<1 meter) can be made from stamped metal, preferably aluminum.*

*For larger reflectors keep in mind surface sag can be a problem when transitioning from high to low elevation angles. Also, wind loading can be a serious concern with very large structures. Counter weight large structures on the elevation angle.*

Testing:

*For mid sized and small apertures use far field range with transmitter/receiver or network analyzer. Near field and compact ranges work also. Use calibrated reference antenna. Use axial ratio measurements to confirm quality of required polarization. For large apertures, use radio star techniques, holography or satellites with known EIRP. Check Appendix 6 for details.*

| | |
|---|---|
| Type: | ***Ratan Type*** |
| Frequency of Operational: | *50 Mhz to 50 GHz* |
| Gain Range: | *0 to 45 dBi* |
| Bandwidth: | *~ 55+%* |
| Approximate Power Rating: | *100 Kw* |
| Design Software: | *GTD, Physical Optics, Geometric Optics* |

Configuration:

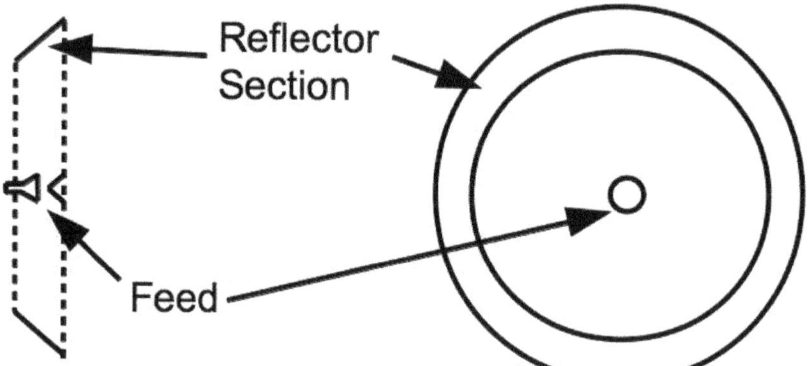


Design Approach:

*Use good physical or geometric optic simulation software*
*Use formula for parabolic section*
*Reflector section can be tilted slightly to steer main beam*

Construction and Theory of Operation:

*Essentially a paraboloid mirror section, this antenna focusses the energy to a single spot where a horn or other suitable feed apparatus gathers or distributes the combined energy. Minimum diameter should be 5 lambda, the higher the number of lambda across the diameter will increase complexity and cost. These systems are usually very large in extent to inexpensively create very small beamwidths, useful for radio astronomy.*

*Azimuth Pattern*

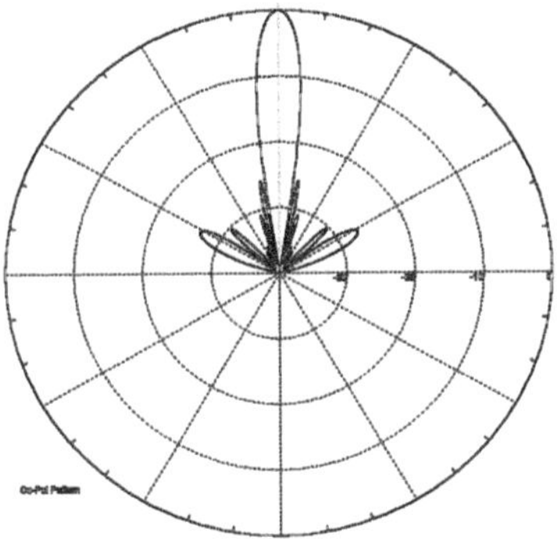

*Elevation Pattern*

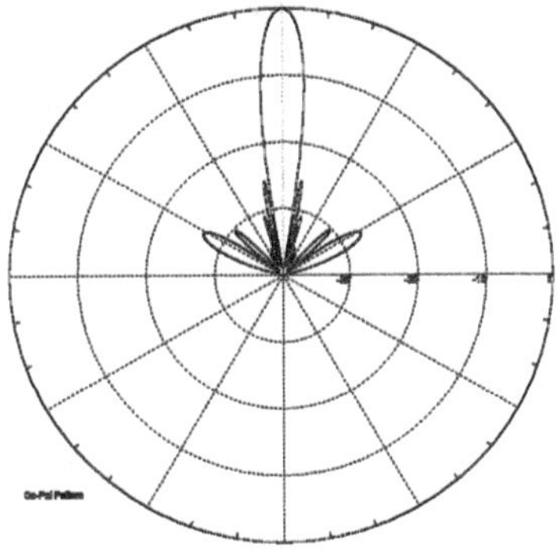

Special Considerations:

*Make sure that the surface roughness is as good as possible, with reason price wise. A roughness of 1/5 wavelength RMS is sufficient to form a beam but values better than 1/50 wavelength are much better for a good collimated beam and lower sidelobe structures. Satellite antennas for commercial use have roughness values better than 1/100 wavelength for instance. Reflecting structure can be made of steel with aluminum panels or all aluminum. For mid sized reflectors (< 3 meter diameter) wire mesh can be imbedded into fiberglass. Yet smaller apertures (<1 meter) can be made from stamped metal, preferably aluminum.*

Testing:

*For mid sized and small apertures use far field range with transmitter/receiver or network analyzer. Near field and compact ranges work also. Use calibrated reference antenna. Use axial ratio measurements to confirm quality of required polarization. For large apertures, use radio star techniques, holography or satellites with known EIRP. Check Appendix 6 for details.*

Type: ***Balloon Reflector***

Frequency of Operational: *50 Mhz to 15 GHz*
Gain Range: *0 to 45 dBi*
Bandwidth: *~ 55+%*
Approximate Power Rating: *100 Kw*
Design Software: *GTD, Physical Optics, Geometric Optics*

Configuration:

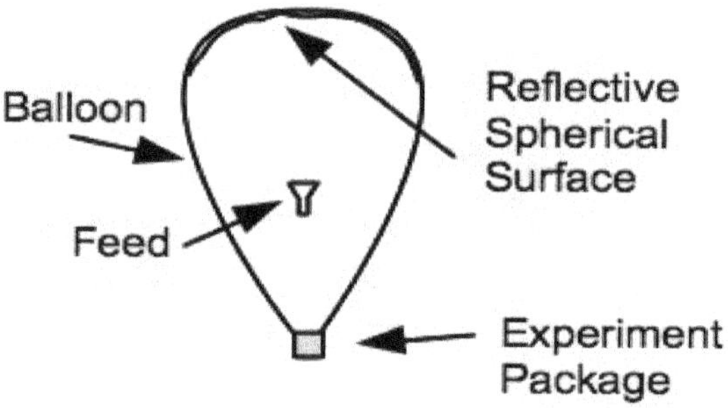


Design Approach:

*Use good physical or geometric optic simulation software*
*Use formula for spherical section for reflector*
*Design line feed for focal point, like traveling wave antenna*

Construction and Theory of Operation:

*Essentially a paraboloid or spherical mirror section, this antenna focusses the energy to a single spot where a horn or other suitable feed apparatus gathers or distributes the combined energy. Minimum diameter should be 5 lambda, the higher the number of lambda across the diameter will increase complexity and cost. The use of a reflective surface inside the balloon allows for a large aperture, useful for ground penetrating radar and other space research requirements.*

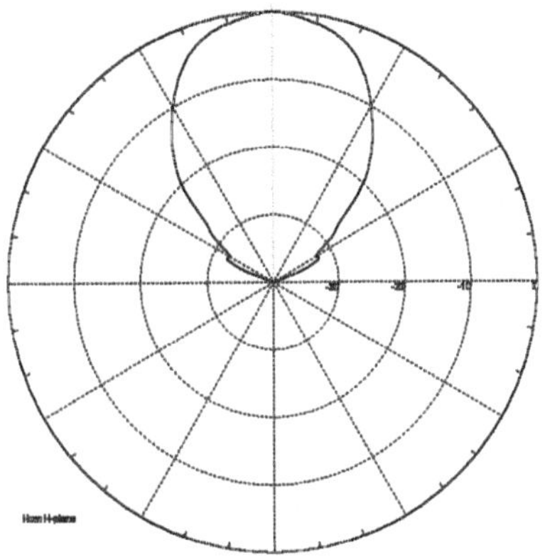


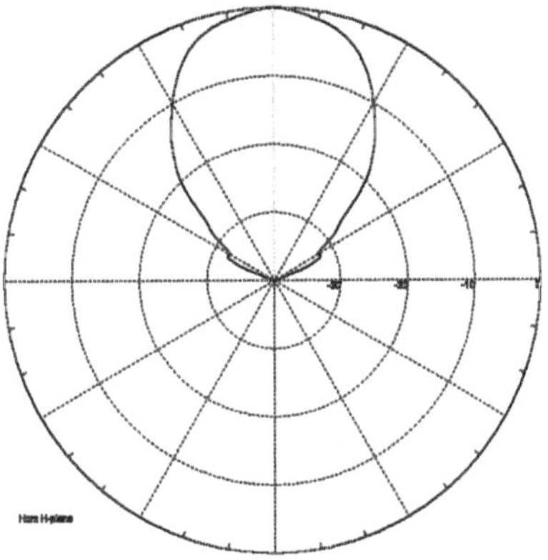


Special Considerations:

*Make sure that the surface roughness is as good as possible, with reason price wise. A roughness of 1/5 wavelength RMS is sufficient to form a beam but values better than 1/50 wavelength are much better for a good collimated beam and lower sidelobe structures. Reflecting structure can be made of aluminum or gold vapor deposited on mylar, polyethylene or other suitable balloon material. For mid sized reflectors (< 3 meter diameter) fine wire mesh can be imbedded into the balloon material. Yet smaller apertures (<1 meter) can be made from thin stamped metal, preferably aluminum.*

Testing:

*For mid sized and small apertures use far field range with transmitter/receiver or network analyzer. Near field and compact ranges work also. Use calibrated reference antenna. Use axial ratio measurements to confirm quality of required polarization. For large apertures, use radio star techniques, holography or satellites with known EIRP. Check Appendix 6 for details.*

Type: ***Gregorian Fed Reflector***

Frequency of Operational: *50 Mhz to 500 GHz*
Gain Range: *0 to 85 dBi*
Bandwidth: *~ 55+%*
Approximate Power Rating: *1 Mw*
Design Software: *GTD, Physical Optics, Geometric Optics*

Configuration:

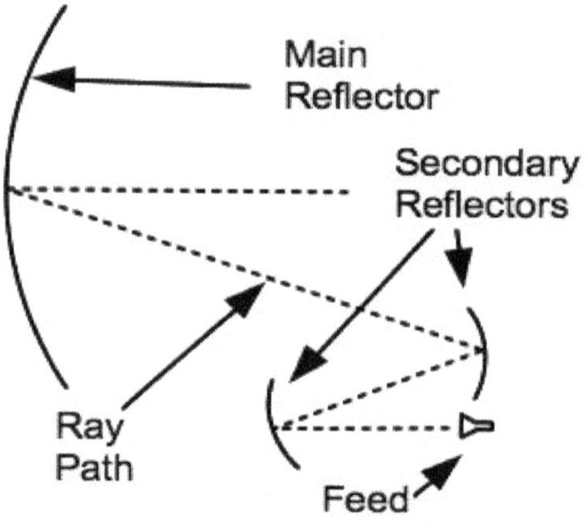


Design Approach:

*Use good physical or geometric optic simulation software*
*Use formula for parabolic section for main reflector, use Gregorian optics for secondary reflectors*

Construction and Theory of Operation:

*Essentially a paraboloid mirror section, this antenna focusses the energy to a single spot where a horn or other suitable feed apparatus gathers or distributes the combined energy. Minimum diameter should be 5 lambda, the higher the number of lambda across the diameter will increase complexity and cost.*

*Azimuth Pattern*

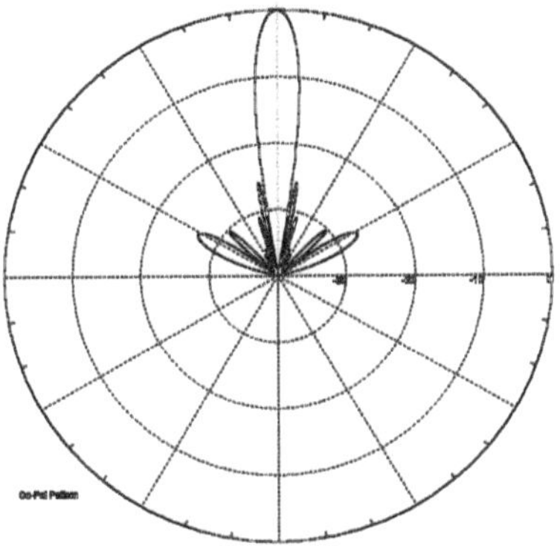

*Elevation Pattern*

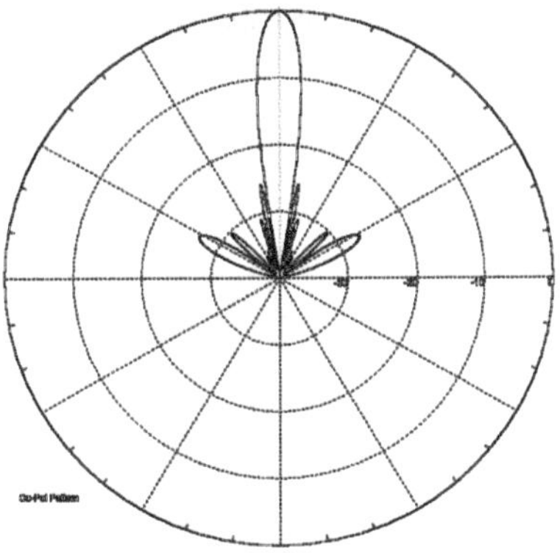

Special Considerations:

*Make sure that the surface roughness is as good as possible, with reason price wise. A roughness of 1/5 wavelength RMS is sufficient to form a beam but values better than 1/50 wavelength are much better for a good collimated beam and lower sidelobe structures. Satellite antennas for commercial use have roughness values better than 1/100 wavelength for instance. Reflecting structure can be made of steel with aluminum panels or all aluminum. For mid sized reflectors (< 3 meter diameter) wire mesh can be imbedded into fiberglass. Yet smaller apertures (<1 meter) can be made from stamped metal, preferably aluminum.*

*For larger reflectors keep in mind surface sag can be a problem when transitioning from high to low elevation angles. Also, wind loading can be a serious concern with very large structures. Counter weight large structures on the elevation angle.*

Testing:

*For mid sized and small apertures use far field range with transmitter/receiver or network analyzer. Near field and compact ranges work also. Use calibrated reference antenna. Use axial ratio measurements to confirm quality of required polarization. For large apertures, use radio star techniques, holography or satellites with known EIRP. Check Appendix 6 for details.*

| | |
|---|---|
| Type: | ***Corner Reflector*** |
| Frequency of Operational: | *10 Mhz to 5 GHz* |
| Gain Range: | *0 to 25 dBi* |
| Bandwidth: | *~ 55+%* |
| Approximate Power Rating: | *1 Mw* |
| Design Software: | *GTD, Physical Optics, Geometric Optics* |

Configuration:

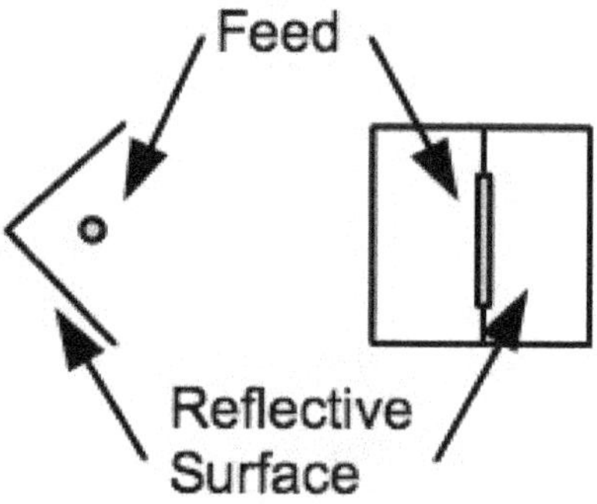


Design Approach:

*Use good physical or geometric optic simulation software*
*Use formula for flat reflective section for main reflector*
*Use dipole or biconical feed spaced .25 lambda from reflector*

Construction and Theory of Operation:

*Essentially a paraboloid mirror section, this antenna focusses the energy to a single spot where a horn or other suitable feed apparatus gathers or distributes the combined energy. Minimum diameter should be 5 lambda, the higher the number of lambda across the diameter will increase complexity and cost.*

Azimuth Pattern

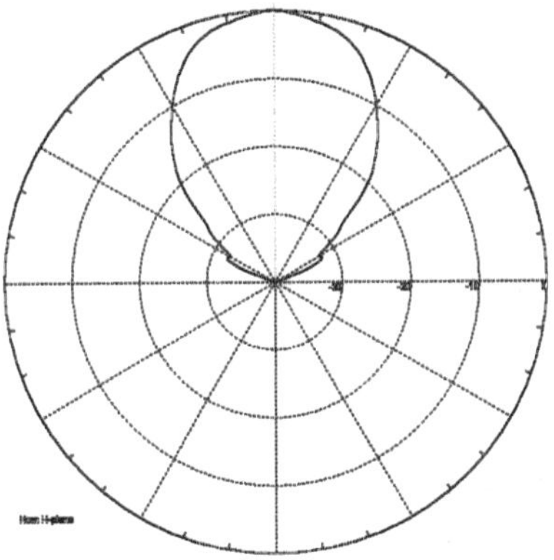

Elevation Pattern

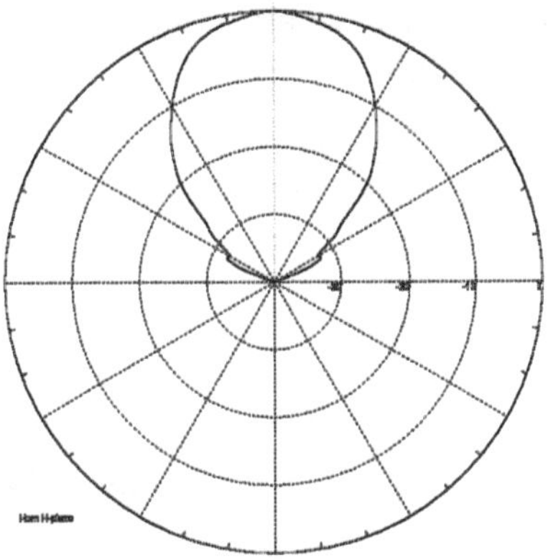

Special Considerations:

*Make sure that the surface roughness is as good as possible, with reason price wise. A roughness of 1/5 wavelength RMS is sufficient to form a beam but values better than 1/50 wavelength are much better for a good collimated beam and lower sidelobe structures. Satellite antennas for commercial use have roughness values better than 1/100 wavelength for instance. Reflecting structure can be made of steel with aluminum panels or all aluminum. For mid sized reflectors (< 3 meter diameter) wire mesh can be imbedded into fiberglass. Yet smaller apertures (<1 meter) can be made from stamped metal, preferably aluminum.*

*For larger reflectors keep in mind surface sag can be a problem when transitioning from high to low elevation angles. Also, wind loading can be a serious concern with very large structures. Counter weight large structures on the elevation angle.*

Testing:

*For mid sized and small apertures use far field range with transmitter/receiver or network analyzer. Near field and compact ranges work also. Use calibrated reference antenna. Use axial ratio measurements to confirm quality of required polarization. For large apertures, use radio star techniques, holography or satellites with known EIRP. Check Appendix 6 for details.*

Type: ***Short Backfire***

Frequency of Operational: *10 Mhz to 50 GHz*
Gain Range: *0 to 25 dBi*
Bandwidth: *~ 15+%*
Approximate Power Rating: *1 Mw*
Design Software: *GTD, Physical Optics, Moment Method*

Configuration:

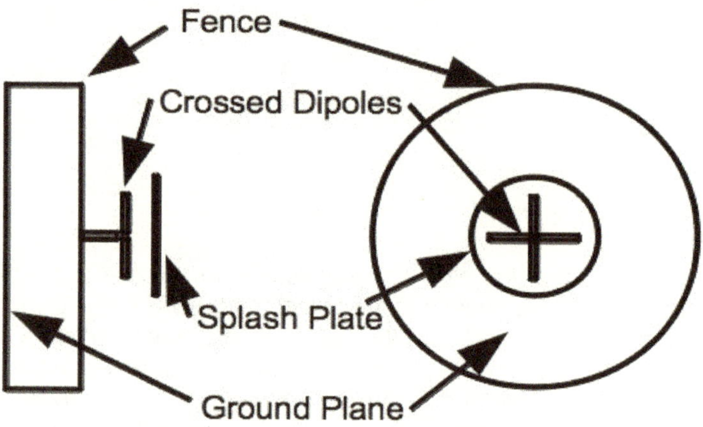


Design Approach:

*Dipole length = 0.5 lambda*
*Splash plate to dipole distance = 0.25 lambda*
*Ground Plane to Dipole distance = 0.25 lambda*
*Fence Depth = 0.5 lambda*
*Ground plane diameter = 2 lambda*

Construction and Theory of Operation:

*A very efficient antenna, these antennas are essentially a dipole with a flat ground reflective plate. Sidelobes are minimal and these antennas are useful in arrays. The fence acts to contain the sidelobes.*

*Azimuth Pattern*

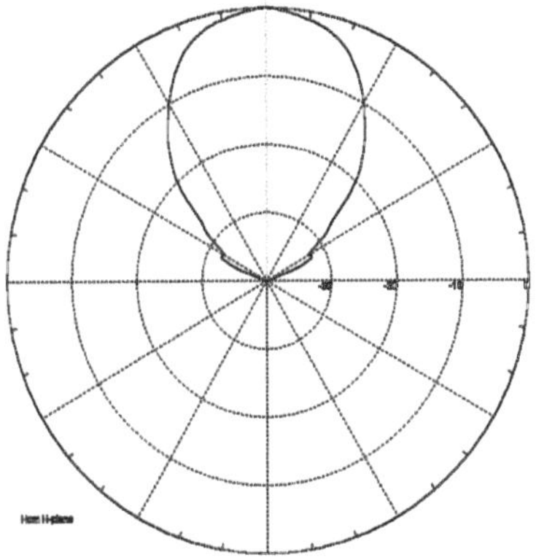

*Elevation Pattern*

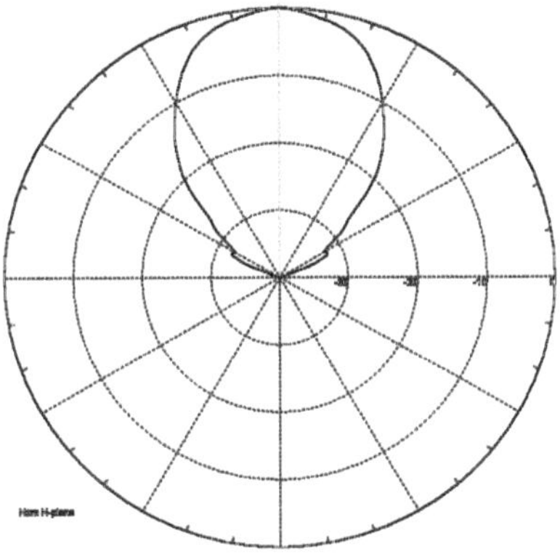

Special Considerations:

*Make sure that the surface roughness is as good as possible, with reason price wise. A roughness of 1/5 wavelength RMS is sufficient to form a beam but values better than 1/50 wavelength are much better for a good collimated beam and lower sidelobe structures. Satellite antennas for commercial use have roughness values better than 1/100 wavelength for instance. Reflecting structure can be made of steel with aluminum panels or all aluminum. For mid sized reflectors (< 3 meter diameter) wire mesh can be imbedded into fiberglass. Yet smaller apertures (<1 meter) can be made from stamped metal, preferably aluminum.*

*For larger reflectors keep in mind surface sag can be a problem when transitioning from high to low elevation angles. Also, wind loading can be a serious concern with very large structures. Counter weight large structures on the elevation angle.*

Testing:

*For mid sized and small apertures use far field range with transmitter/receiver or network analyzer. Near field and compact ranges work also. Use calibrated reference antenna. Use axial ratio measurements to confirm quality of required polarization. For large apertures, use radio star techniques, holography or satellites with known EIRP. Check Appendix 6 for details.*

**Questions:**

1. Suggest an antenna type that has the following characteristics:

   a. Frequency of 200 to 400 MHz
   b. Gain of 15 dBi
   c. Linear Polarization

2. Your boss tells you price is a huge factor in the design of antenna with the following characteristics:

   a. Frequency of 2000 MHz
   b. Gain of 20 dBi
   c. Linear Polarization

   Does the reflector have to be solid?

3. Suggest an antenna design for the next generation satellite that is capable of going to the moon, with the following properties:

   a. Frequency of 2200 MHz
   b. Gain of 3 dBi
   c. Circular Polarization

4. During the 50's and 60's telephone traffic was commonly sent over microwave links. There was a dish pointed up looking at a grid reflective plate which sent the signals horizontally. What was the advantage of such an arrangement ?

5. Saltellite antenns frequently have one shaped grid placed over another shaped grid. What advantage does this have ?

6. How far does the main beam move for a reflector if the feed is translated by one degree ?

7. How can an umbrella be used as a reflector?

# Chapter 11 – Lens Antennas

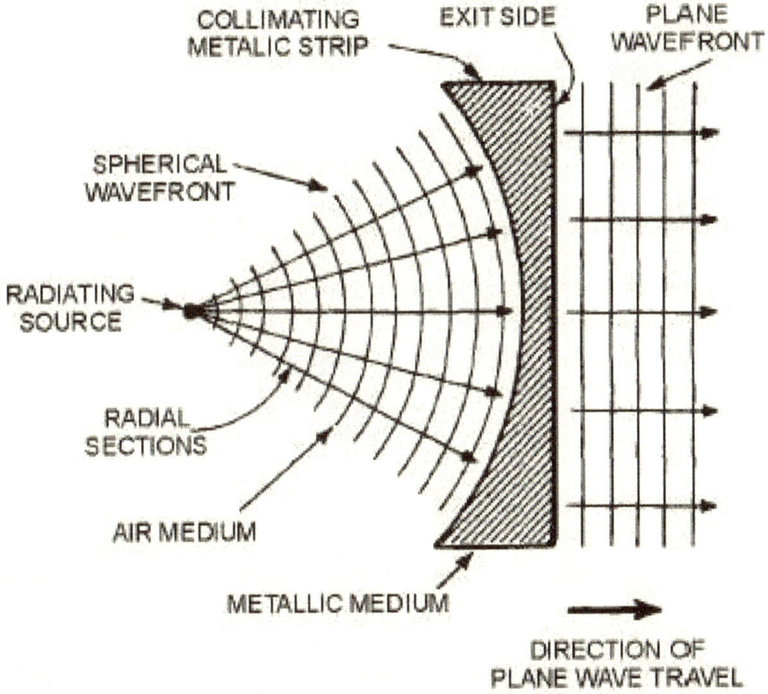


Example of Lens Operation (41)

## Fundamentals

### Metal Fresnel Lens

The metal type lens antenna is constructed of a series of thin metal plates with air or other dielectric between them. The curvature of the edges of the plates form the lens and the space between the metal plates form the "waveguide."

The lens antenna works in principle much like a lens in optics, with many of the same formula used to design and evaluate the structure. Various radii of curvature may be used to create a lens antenna. First, determine the desired gain for the lens by the familiar formula:

$$Gain(dBd) = 10 * \log(4.5 * Ae * Ah) \qquad [57]$$

where:

Ae = Aperture size in wavelengths in the E plane
Ah = Aperture size in wavelengths in the H plane

The gain in this case is dB above a dipole or 2.12 dBi.

Next calculate the focal length desired by:

$$F = \frac{(Lens\ Diameter)}{\left(2 * \tan\left(\frac{W}{2}\right)\right)} \qquad [58]$$

where:

Lens Diameter is in wavelengths and W is the width of the lens in wavelengths.

The spacing of the metal plates dictates the index of refraction thus:

$$Index = \sqrt{\left(1 - \left(\frac{\lambda}{2} * spacing\right)^2\right)} \qquad [59]$$

where:

lambda is operational wavelength
*spacing* is in terms of wavelengths

The index of refraction here is the ratio of the wavelength in the lens to the wavelength in free space. The radius of curvature is calculated next by the "lens makers formula" thus:

$$\frac{1}{f}=(n-1)*\left(\left(\frac{1}{R1}\right)-\left(\frac{1}{R2}\right)\right) \quad [60]$$

where:

R1 and R2 are the radii of curvature of the two lens surfaces (set one radius to infinity in the case of a single curved lens).

F = the focal length
n = the index of refraction

Note: a negative radius indicates a concave surface.

In the case of antenna the concave shape is equivalent to the convex shape of an optical lens. In the case of the optical component, waves are curved more at the outer edges than the center. Thus a convex lens can focus parallel light to a point, much like a telescope. In the case of a metal lens antenna the change in phase velocity between the metal plates performs the same function, however more material is needed at the edges as compared to the center of the lens to focus the radio waves to a point.

A zoned lens or Fresnel design can minimize the thickness of a metal lens antenna; the zones maintain the same radius of curvature however steps are added periodically to lower the profile of the antenna.

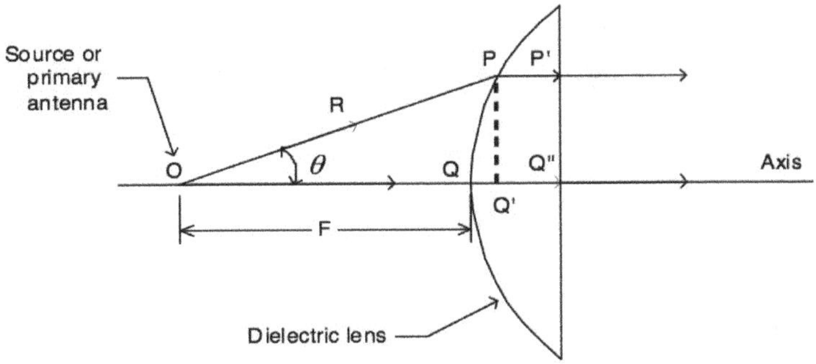


Example of a Dielectric Lens

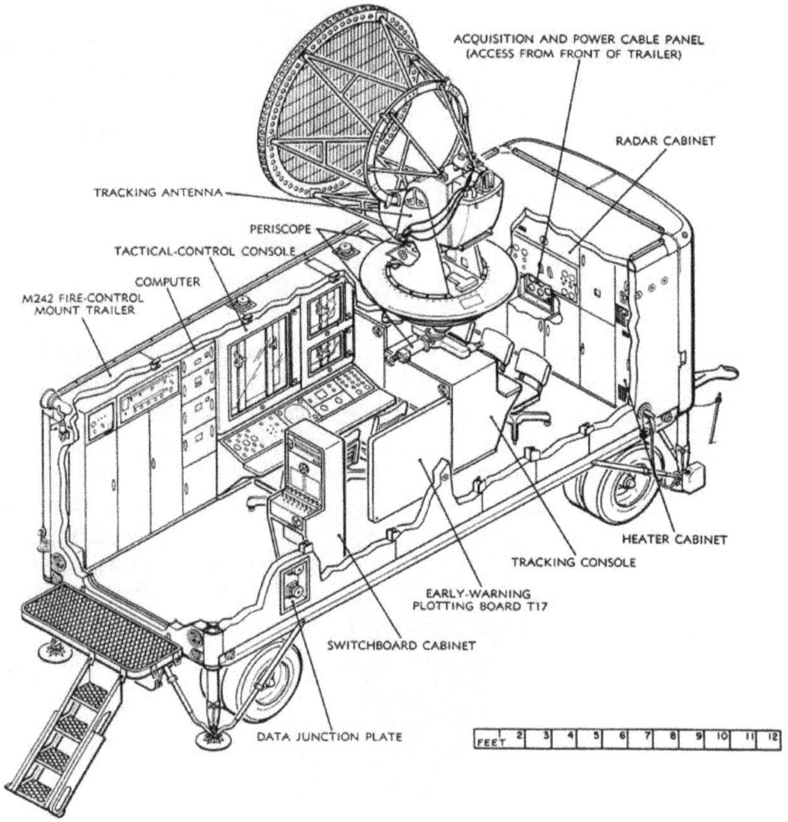


Radar System Using X-Band Lens Antenna (43)

Lens Antenna used on Nike Ajax Tracking Radar

**Luneberg Type**

An interesting type of lens antenna is composed of either a sphere or hemisphere of dielectric material that has an infinite number of focal points. Proposed by R. K. Luneberg in 1944, the lens is made of materials with relative dielectric constant Er, varying in spare distribution from 1 at the surface to 2 at the center thus:

$$Er = 2 - \left(\frac{r}{R}\right)^2 \qquad [61]$$

where:

Er = Relative Dielectric Constant
r = Radius
R = Radius of Lens

The Luneberg Lens can function as an antenna without movement of its spherical body. By movement of the small feed unit only towards the source of the electromagnetic beam. It is also capable of receiving and transmitting from and in multiple directions as the same time. [44]

Radio waves refract, just like light, at the border of two materials with different relative dielectric constants. An electromagnetic wave entering from a focus on the surface is refracted dat each border within the dielectric-material sphere, as the relative dielectric constant changes gradually from 1 to 2, and is eventually emitted as a plane wave from the opposite side of the sphere. [44]

A hemispherical type lens can be made with a reflecting ground plane at the "equator" thus simplifying design. The feed assembly is placed above the "equator" looking down at the reflector. Performance is reasonable with aperture efficiencies approaching 50%, sidelobe levels are usually around 20 dB down.

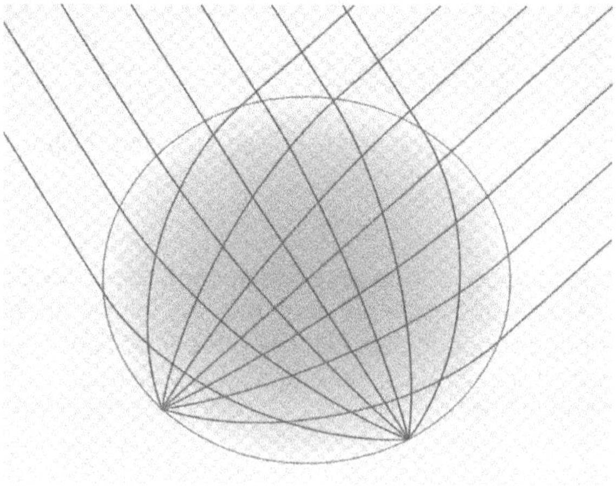

Ray Paths for a Luneberg Lens showing multiple sources

**Dielectric Lens**

Using the principle of changing thickness instead of dielectric constant, lens can be designed under the same algorithms as common optical lens. The choice of material is important, where low loss tangent is required for good aperture efficiency. Using the following formula, known as the Lens maker's equation, one can define the geometry of the lens antenna:

$$\frac{1}{f}=(n-1)*\left(\left(\frac{1}{R1}\right)-\left(\frac{1}{R2}\right)+\left(\frac{((n-1)*d)}{(n*R1*R2)}\right)\right) \qquad [62]$$

where:

f = focal length of the lens

n = refractive index of the lens material

R1 = is the radius of curvature of the lens surface closest to the feed

R2 = is the radius of curvature of the lens surface farthest from the feed

d = thickness of the lens

These types of antennas are used typically for millimeter and sub-millimeter designs. Applications include atmospheric radiometers and radar systems.

Applications

Lens designs have several distinct advantages over others. They can effectively operate with multiple simultaneous beams at different frequencies for instance. Typical uses include radar systems where multiple beams can be used in real time to locate targets. Other applications include the study of lightning propagation through clouds at VHF frequencies. Multiple feed in this case are connected to multiple receivers. As lightning moves very quickly, it is impractical to use a mechanically steering antennas.

The following picture shows a lens radar used in a fighter jet:

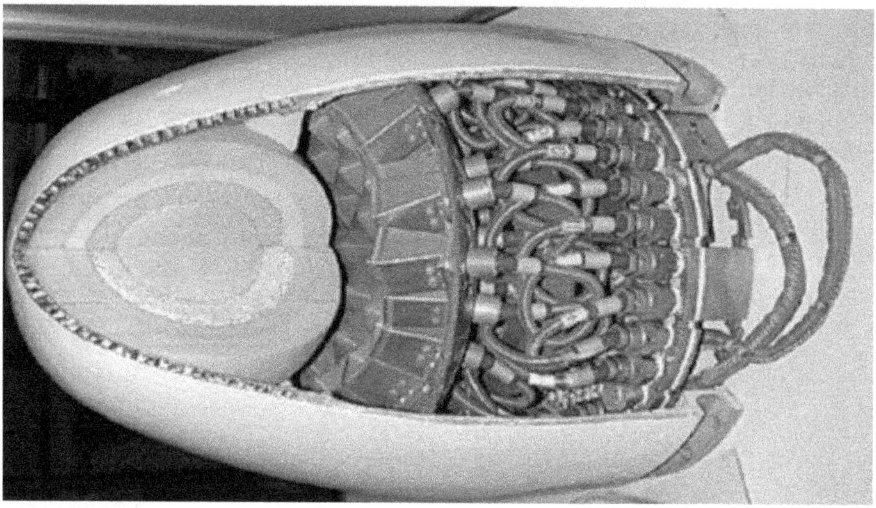

Notice the layers of dielectrics used in the lens portion. Also, the feeds are placed very closely to create complete coverage of a large portion of sky.

Other applications include use as a radar reflector, where a metal plate can be placed at the position of interest to create a highly reflective beam in one particular direction.

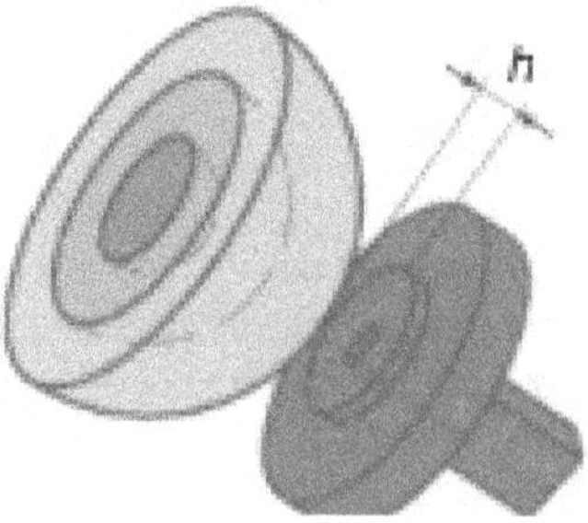


Maxwell's Half Fish Eye Lens

Large Luneberg Lens Radar at Kwajelein

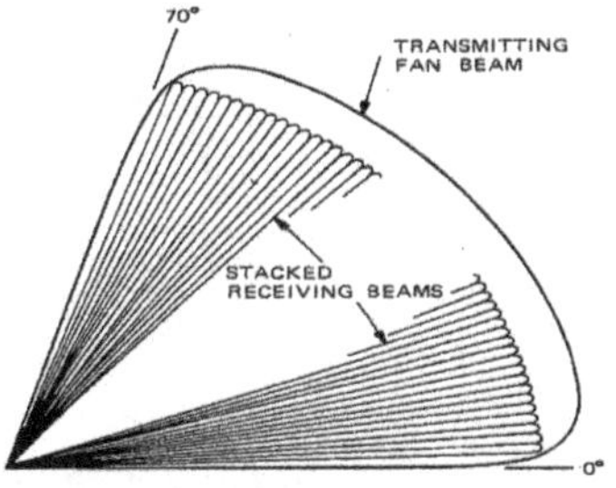


Multiple Beam Performance of Lens Radar

**Design Examples**

The following pages contain examples of antennas discussed in this chapter. These examples represent a first order design approach. For a more complete and optimized antenna solution, use the example closest to the requirement needs and optimize by using good simulation software, creating prototypes, testing and finally modifying prototypes to optimize performance.

| | |
|---|---|
| Type: | ***Metal Lens*** |
| Frequency of Operational: | *1 Ghz to 30 GHz* |
| Gain Range: | *5 to 50 dBi* |
| Bandwidth: | *~ 15%* |
| Approximate Power Rating: | *1 Mw* |
| Design Software: | *GTD, HFSS, Optics Software* |

Configuration:

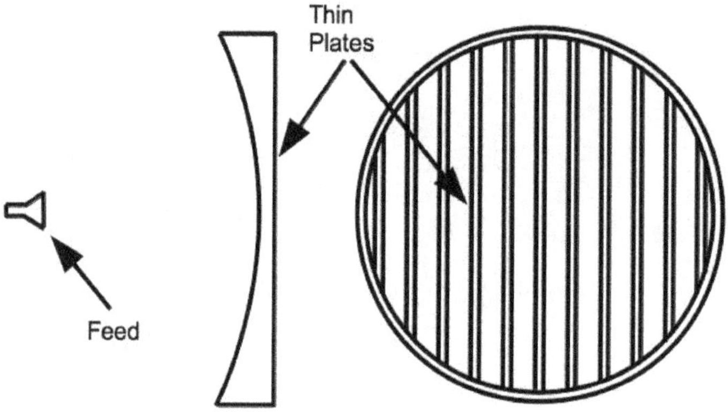


Design Approach:

*Use good electromagnetic surface simulation software, like HFSS*
*Use formula for lens starting with formulas at beginning of chapter*
*Use horn or similar directive feed*

Construction and Theory of Operation:

*Essentially a series of conductive strips in parallel with varying width to effect a change in propagation velocity, much like a glass lens used in optical telescopes. Metal must be conductive like aluminum.*

*Azimuth Pattern*

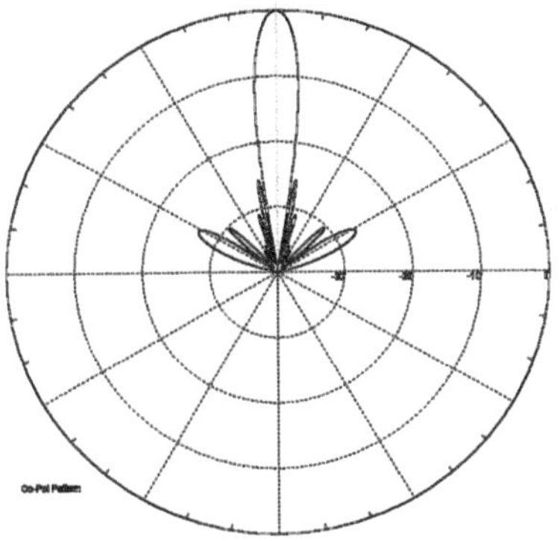

*Elevation Pattern*

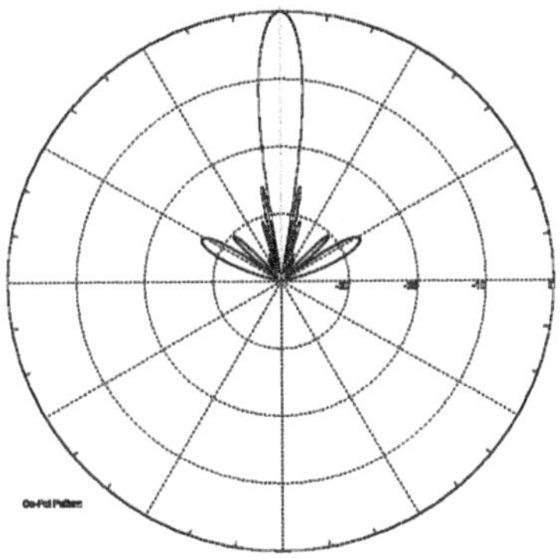

Special Considerations:

*Follow lens designers formula, specifically profiles of concave lens structures. Cut slices out with laser or other high quality milling apparatus. Design will not be too broadband.*

Testing:

*For mid sized and small apertures use far field range with transmitter/receiver or network analyzer. Near field and compact ranges work also. Use calibrated reference antenna. Use axial ratio measurements to confirm quality of required polarization. For large apertures, use radio star techniques, holography or satellites with known EIRP. Check Appendix 6 for details.*

Type: ***Dielectric Lens***

Frequency of Operational: *5 Ghz to 500 GHz*
Gain Range: *0 to 35 dBi*
Bandwidth: *~ 25%*
Approximate Power Rating: *1 Kw*
Design Software: *Physical Optics, Standard Optical software Ray tracing simulators*

Configuration:

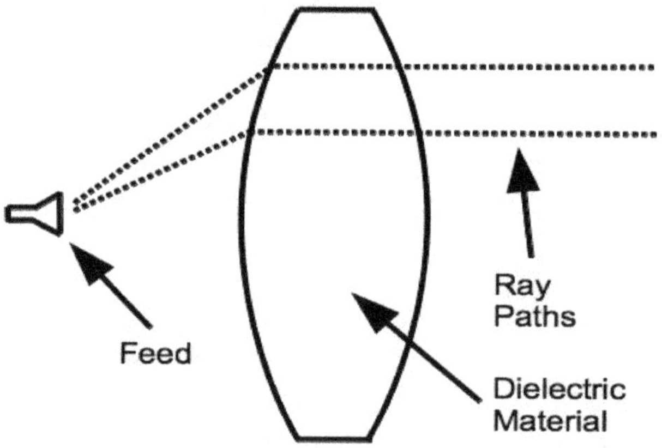


Design Approach:

*Use good ray tracing software and standard optical design rules*
*Use formula for lens starting with formulas at beginning of chapter*
*Use horn or similar directive feed*

Construction and Theory of Operation:

*This lens antenna is typically made of low loss dielectric materials, like teflon or even polycarbonate. The design is much like an optical lens design with similar equations. Useful for millimeter wave requirements up to and including sub-millimeter frequencies.*

*Azimuth Pattern*

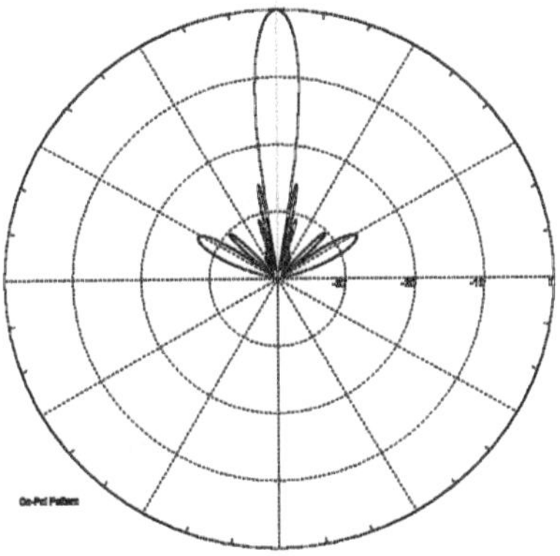

*Elevation Pattern*

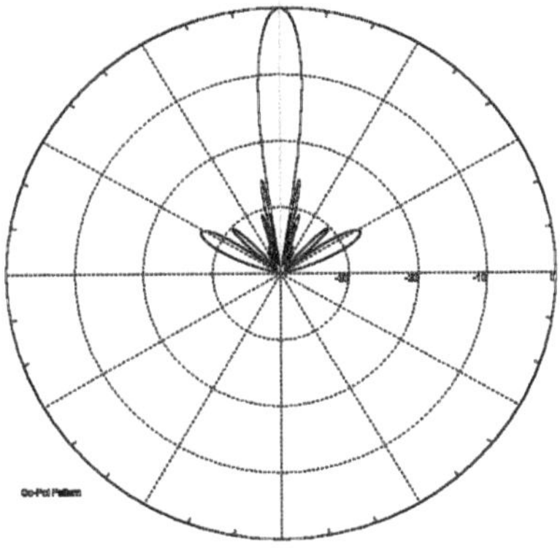

Special Considerations:

*Follow lens designers formula, specifically profiles of convex lens structures. Cut material out using lens grinding techniques, check optical lens design guides. Fresnel lens designs work as well.*

Testing:

*For mid sized and small apertures use far field range with transmitter/receiver or network analyzer. Near field and compact ranges work also. Use calibrated reference antenna. Use axial ratio measurements to confirm quality of required polarization. For large apertures, use radio star techniques, holography or satellites with known EIRP. Check Appendix 6 for details.*

Type: ***Luneburg Lens***

Frequency of Operational: *500 Mhz to 50 GHz*
Gain Range: *0 to 35 dBi*
Bandwidth: *~ 25%*
Approximate Power Rating: *10 Kw*
Design Software: *GTD, Physical Optics, Geometric Optics*

Configuration:

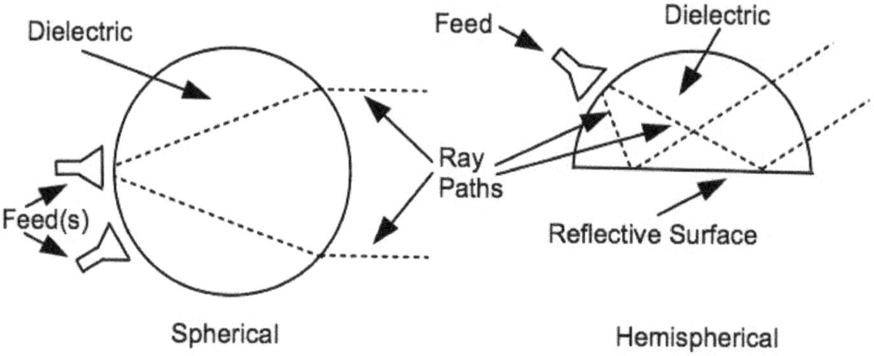


Design Approach:

*Use good ray tracing software and standard optical design rules*
*Use formula for lens starting with formulas at beginning of chapter*
*Use horn or similar directive feed*

Construction and Theory of Operation:

*This lens antenna is typically made of low loss dielectric materials, like teflon or even polycarbonate. The design is much like an optical lens design with similar equations. Multiple feeds can be used simultaneously. The lens is made of progressively lower dielectric constants starting from the center out. Many designs start with Epsilon 10 progressing to Epsilon 2.*

*Azimuth Pattern*

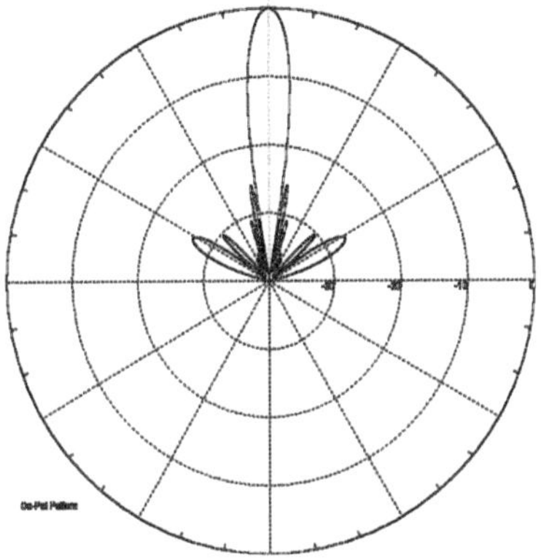

*Elevation Pattern*

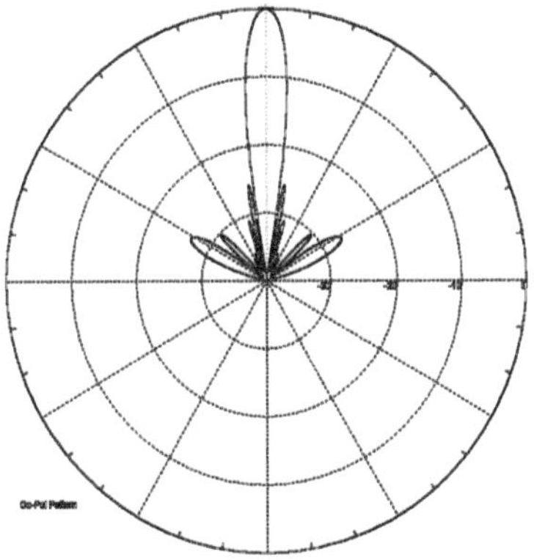

Special Considerations:

*Uniformity of materials is important when using this design. Concentric spheres of material work best. Be careful when using high power, not to overheat material. Feed can be moved physically or multiple feeds can be used at the same time.*

Testing:

*For mid sized and small apertures use far field range with transmitter/receiver or network analyzer. Near field and compact ranges work also. Use calibrated reference antenna. Use axial ratio measurements to confirm quality of required polarization. For large apertures, use radio star techniques, holography or satellites with known EIRP. Check Appendix 6 for details.*

**Questions:**

1. Suggest an antenna type that has the following characteristics:

   a. Frequency of 2 to 3 GHz
   b. Gain of 15 dBi
   c. Linear Polarization

2. Your boss tells you price is a huge factor in the design of antenna with the following characteristics:

   a. Frequency of 12 GHz
   b. Gain of 20 dBi
   c. Linear Polarization

   Could styrofoam and tin foil be used effectively?

3. Suggest an antenna design for the next generation rocket booster that is capable of going to the moon, with the following properties:

   a. Frequency of 22 GHz
   b. Gain of 13 dBi
   c. Circular Polarization

4. Is there an advantage to using variable dielectrics within a luneberg lens ?

5. What happens if a lens is shaped (e.i., not flat) ?

6. Would a phased array be a useful feed ? Why ?

7. How many feeds can a lens system use ?

8. If the feed is an omni, can you use multiple lenses ? How about multiple polarizations ?

9. Could you place a lens within a lens ? What would be the consequences in terms of gain and beam parameters ?

# Chapter 12 – Yagi Antennas

Yagi and his Famous Antenna (35)

## Fundamentals

Yagi antennas are made of a minimum of a radiating element, like a dipole, and a reflector. In most cases, additional elements or directors are added to increase gain. Elements can be crossed to create a dual pole design, which can be used for linear or circular polarization.

Typically, a main center boom is used to attach the antenna elements; because the currents are at zero for the directors and reflector, they can be attached directly to the center spar, even if its conductive. The driven element can be a dipole type or folded dipole (in the case of a wide band design). Matching the impedance is done with a 300 ohm input in the case of the folded dipole or a selection of several types of match assemblies in the case of a dipole and 50 ohm input impedance. For the latter match, common approaches are gamma, delta or similar designed configurations.

Manufacturing costs are low, wind load is low as well. Performance is reasonable with moderate sidelobe levels and moderate front to back ratios. Use of a good simulation package is very helpful allowing the designer to select optimum performance components. There is a need to compromise between best gain and best pattern. Several available simulators contain an iteration feature that can optimize performance depending on requirements.

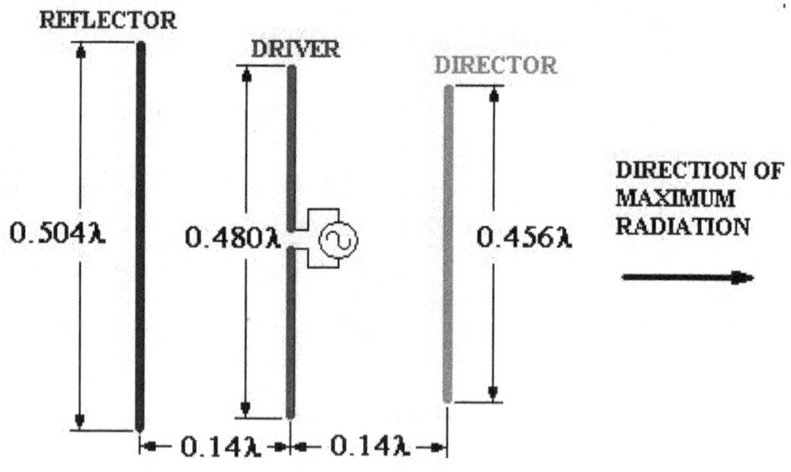


Typical Yagi Design (35)

**Applications**

Yagis are used extensively in communications applications, including public service, point-to-point terrestrial, ham radio, off-air television and radio astronomy. Along with monopoles, yagis are probably the most often used antenna.

**Arrays**

Like helical antennas, yagis can be arrayed to increase gain and allow for electrical steering. Care must be taken however to mitigate the effects of mutual coupling between elements, but in general, yagi arrays are easy to construct and have good electrical properties. There are many examples of these arrays including those used for radio astronomy, radar, and high gain amateur radio systems used (among other applications) for EME (Earth – Moon – Earth) communications.

**Construction Techniques**

The simplest form of yagi is the welding of the elements sans driven element to a spar, the driven element is electrically isolated by attached to the same boom. For the lowest cost, the basic construct can be made from stamped metal, typically aluminum. Low cost requirements dominated in several markets, especially off air television.

For more robust applications, larger diameter elements attached to a stronger spar will yield an antenna capable of taking high winds and ice loads. Recently, yagis have been placed inside of plastic radomes, this allows for an inexpensive construction inside while the radome protects the antenna from the elements.

When using aluminum construction, it is wise to use a coating like anodization, to minimize oxidation. Anodization comes in many colors, the most common are gold and black.

**Standard**

A "standard" yagi antenna is one that has the same length driven elements. This antenna is scalable and is the most common type of yagi.

**Tapered**

To increase bandwidth, tapering of the driven elements is a good option. This option is close to a log periodic in appearance with the exception of the more complicated feed mechanism needed for the log periodic.

**Feed Systems**

Matching a feed to a yagi is a relatively straight forward process. There are several techniques to choose from, depending on required bandwidth.

**Delta match**

In this approach, the driven element is not isolated from the center boom but is still ½ wavelength long. The feed line, typically a balanced twin lead is attached to a pair of feed conductors which are placed evenly on either side of the center of the driven element in such a way as to match the impedance to the element. In other words, the feed line feeds a "V" structure which attaches to the driven element, forming a triangle or Delta.

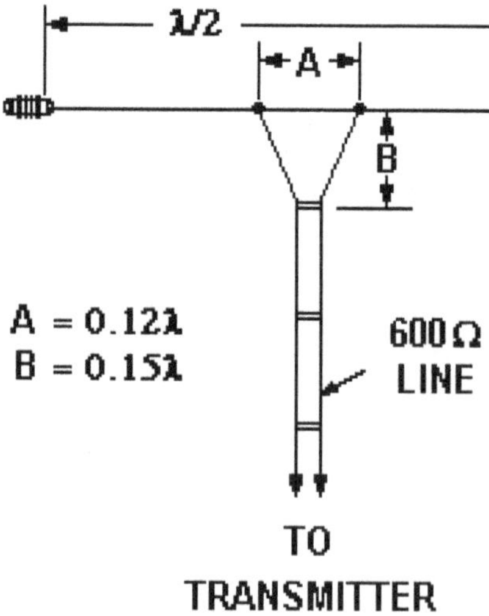


Example of Delta Match

**Gamma**

This is a simple match technique where the center of the feed coax is grounded and the center component of the coax is capacitively coupled to the driven element some specified distance from the center boom. Typically a slide mechanism is employed to tune the antenna optimally.

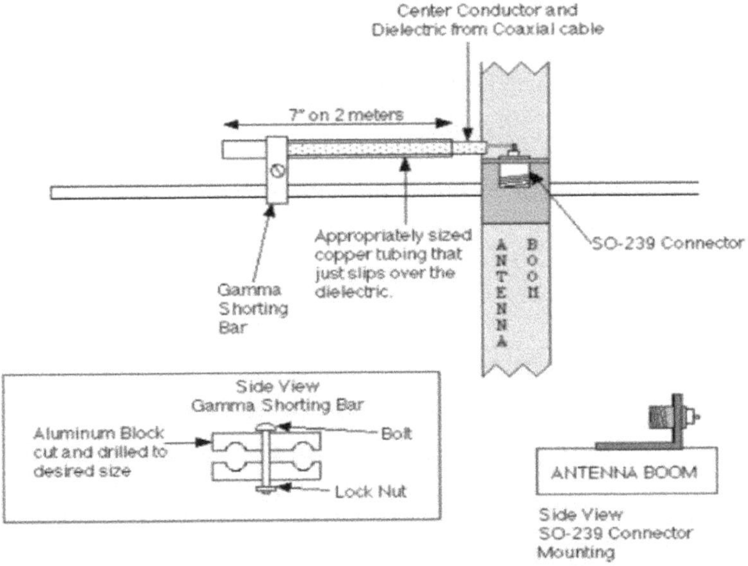


Example of Gamma Match

**T Match**

Another approach to matching the driven element in a yagi is to use a T match which applies equal currents to similar positions equally spaced on the element. In this example the dimension “C” is ½ wavelength. Dimension “A” is around ¼ wavelength and dimension “B” is around .1 wavelength. Optimal performance is found by experimentation.

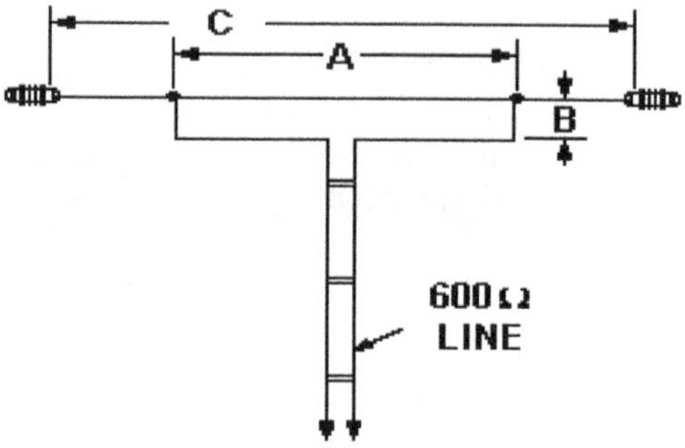


Example of “T” match

Where C = width of driven element, A and B are selected for best match.

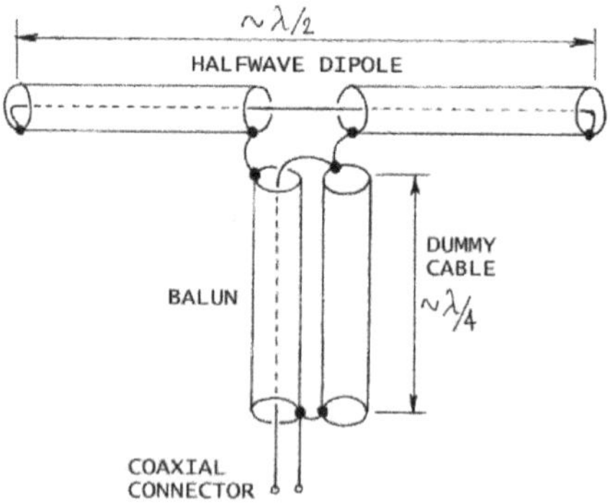


Dipole Match Circuit

**Capacitor and Balun**

For narrow band yagis, useful for filtering out interference, a capacitor can be placed across a dipole type driven element with a balun, in this case a few loops at an optimal distance from the capacitor. Fine tuning is accomplished by moving the loop in relationship to the coaxial attachment point.

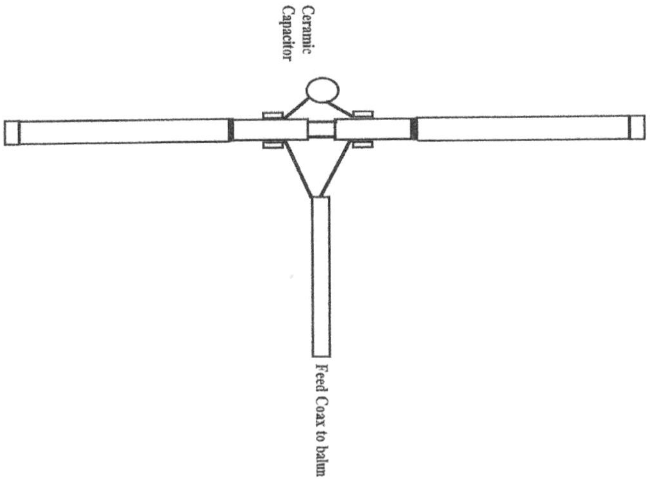


Example of cap and balun match for dipole fed yagi

**Folded or Bowtie Feed**

For Bowtie and folded type antennas, sometimes used in yagis to enhance bandwidth, transformers are typically used. The characteristic impedance of these types of antennas is around 300 ohms. Transformers for 300 to 75 and 300 to 50 are plentiful.

The designer should look for hardware that is used in the cable and television industry for these types of transformers.

**Design Examples**

The following pages contain examples of antennas discussed in this chapter. These examples represent a first order design approach. For a more complete and optimized antenna solution, use the example closest to the requirement needs and optimize by using good simulation software, creating prototypes, testing and finally modifying prototypes to optimize performance.

Type: ***Simple Yagi***

Frequency of Operational: *10 Mhz to 5 GHz*
Gain Range: *5 to 20 dBi*
Bandwidth: *~ 15 %*
Approximate Power Rating: *10 Kw*
Design Software: *Yagimax, Nec-Win Pro, Moment Method*

Configuration:

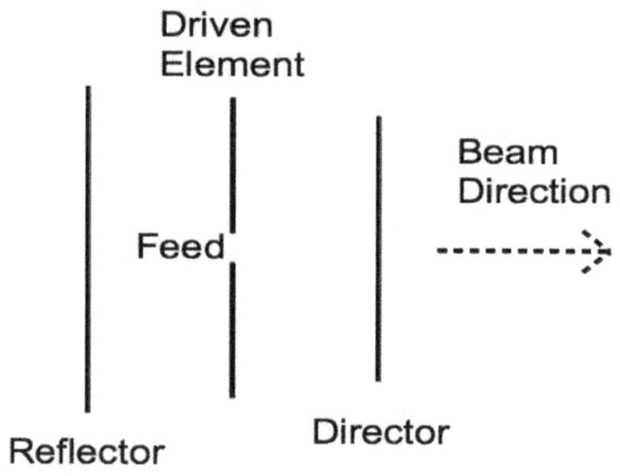


Design Approach:

*Reflector Length = 0.5 lambda; spacing to driven element = 0.125 lamda*
*Driven element length = 0.47 lambda; spacing to director = 0.37 lambda*
*Director length = 0.43 lambda; spacing between dir = 0.31 lambda*

Construction and Theory of Operation:

*This antenna is one of the simplest to design and build. Normally it is assembled on a supporting rod with the elements parallel to each other. Care must be taken to optimize the feed point to feed line matching. Several techniques are available including Gamma and Delta match techniques. The spacing between elements can be varied to optimize sidelobe or front to back performance. Large yagis have been constructed with wires and towers to work down to 10 MHz. Smaller yagis have been made by etching the elements on a substrate material. Polarization is parallel to the elements and if orthogonal elements are assembled as well, circular and orthogonal polarizations can be realized.*

*Azimuth Pattern*

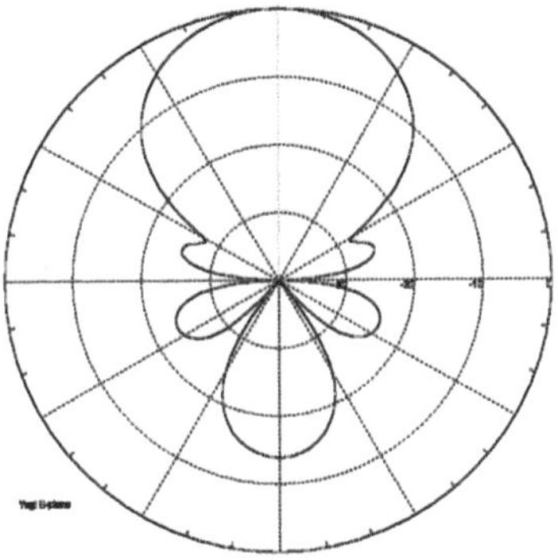

*Elevation Pattern*

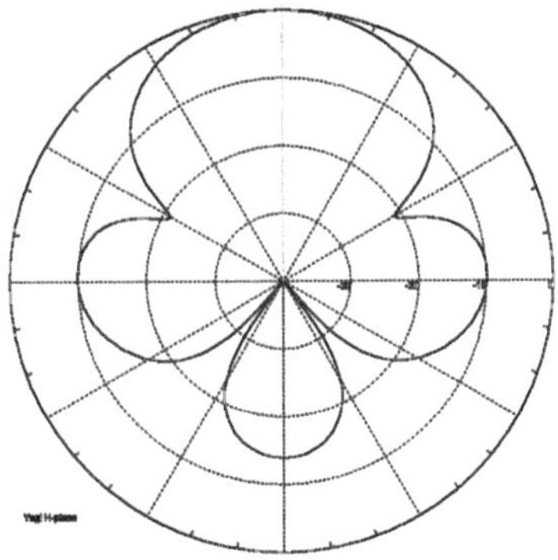

Special Considerations:

*Use folded dipole for wider band applications. Main support structure can be conductive and connected directly to the antenna elements, with the exception of the driven element. Impedance match with delta, gamma or balun systems.*

Testing:

*For mid sized and small apertures use far field range with transmitter/receiver or network analyzer. Near field and compact ranges work also. Use calibrated reference antenna. Use axial ratio measurements to confirm quality of required polarization. For large apertures, use radio star techniques, holography or satellites with known EIRP. Check Appendix 6 for details.*

Type: ***"Moxon" Yagi***

Frequency of Operational: *5 Mhz to 5 GHz*
Gain Range: *5 to 13 dBi*
Bandwidth: *~ 15%*
Approximate Power Rating: *10 Kw*
Design Software: *Nec-Win Pro, Moment Method*

Configuration:

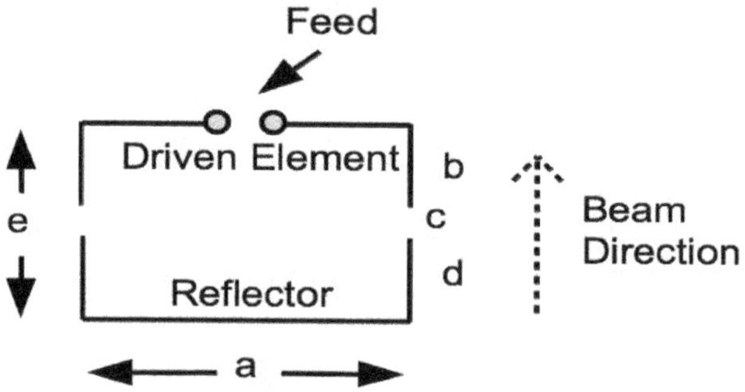


Design Approach:

*a = 0.37 lambda*
*b = 0.05 lambda*
*c = 0.02 lambda*
*d = 0.07 lambda*
*e = 0.14 lambda*

Construction and Theory of Operation:

*This antenna is one of the simplest to design and build. Normally it is assembled on a supporting rod with the elements parallel to each other. Care must be taken to optimize the feed point to feed line matching. The feed gap spacing can be varied for match optimization. The spacing between elements can be varied to optimize sidelobe or front to back performance. Large "Moxons" have been constructed with wires and towers to work down to 10 MHz. Smaller "Moxons" have been made by etching the elements on a substrate material. Polarization is parallel to the elements and if orthogonal elements are assembled as well, circular and orthogonal polarizations can be realized.*

*Azimuth Pattern*

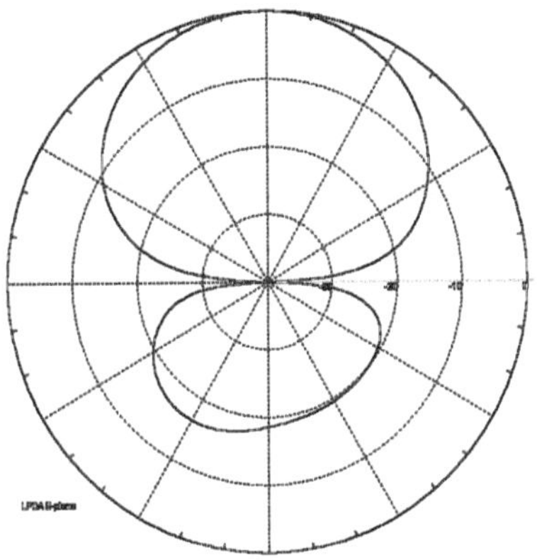

*Elevation Pattern*

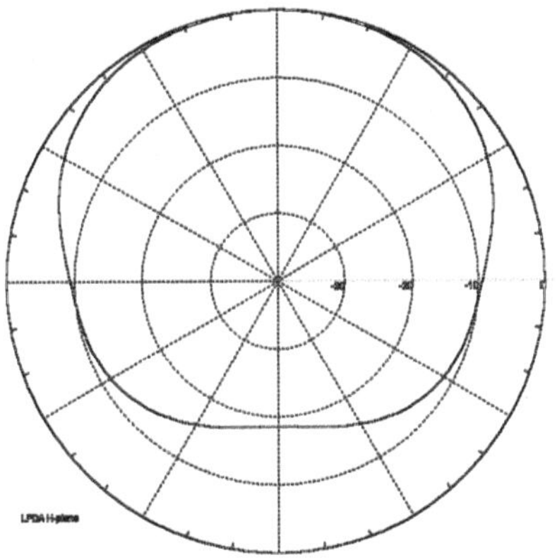

Special Considerations:

*Use aluminum tubing or for larger structures, wire. Adjust feed gap and/or gap "c" for best performance. Antenna is useful in confined areas or in mobile use.*

Testing:

*For mid sized and small apertures use far field range with transmitter/receiver or network analyzer. Near field and compact ranges work also. Use calibrated reference antenna. Use axial ratio measurements to confirm quality of required polarization. For large apertures, use radio star techniques, holography or satellites with known EIRP. Check Appendix 6 for details.*

Type: ***Tapered Yagi***

Frequency of Operational: *5 Mhz to 5 GHz*
Gain Range: *0 to 15 dBi*
Bandwidth: *~ 20%*
Approximate Power Rating: *10 Kw*
Design Software: *Yagimax, Nec-Win Pro, Moment Method*

Configuration:

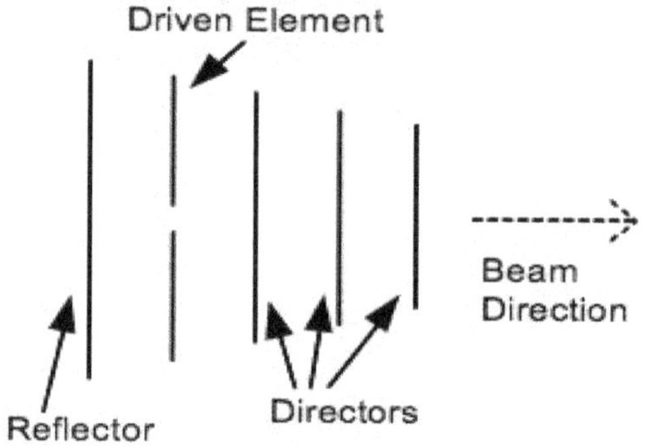


Design Approach:

*Reflector Length = 0.5 lambda; apex angle 20 to 50 degrees*
*Driven element length = 0.47 lambda;spacing between elements = 0.31 lambda*
*Director 1 length = 0.43 lambda*
*Director 2 length = 0.41 lambda*

Construction and Theory of Operation:

*This is a broader band version of a yagi. Normally it is assembled on a supporting rod with the elements parallel to each other. Care must be taken to optimize the feed point to feed line matching. Several techniques are available including Gamma and Delta match techniques. The spacing between elements can be varied to optimize sidelobe or front to back performance. Large yagis have been constructed with wires and towers to work down to 10 MHz. Smaller yagis have been made by etching the elements on a substrate material. Polarization is parallel to the elements and if orthogonal elements are assembled as well, circular and orthogonal polarizations can be realized.*

*Azimuth Pattern*

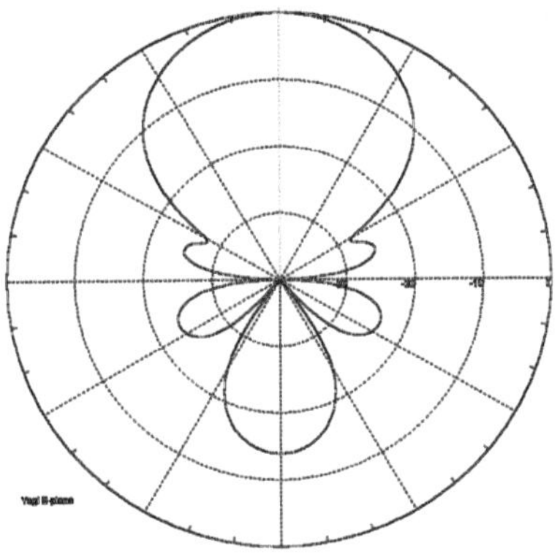

*Elevation Pattern*

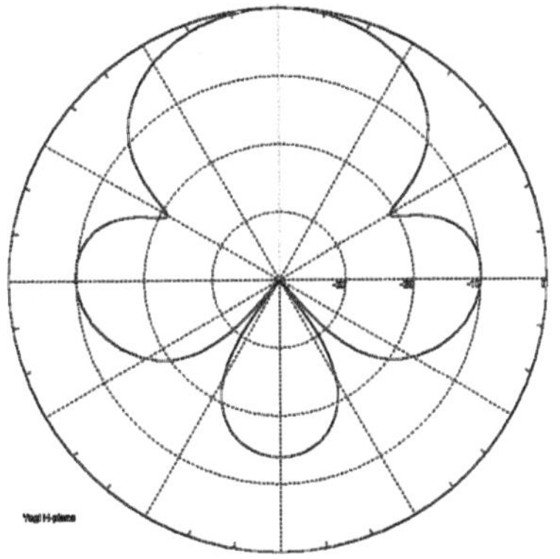

Special Considerations:

*Use folded dipole or bowtie driven elements for wider band applications. Main support structure can be conductive and connected directly to the antenna elements, with the exception of the driven element. Impedance match with delta, gamma or balun systems.*

Testing:

*For mid sized and small apertures use far field range with transmitter/receiver or network analyzer. Near field and compact ranges work also. Use calibrated reference antenna. Use axial ratio measurements to confirm quality of required polarization. For large apertures, use radio star techniques, holography or satellites with known EIRP. Check Appendix 6 for details.*

*2nd Edition*

| | |
|---|---|
| Type: | ***Cross Polarized*** |
| Frequency of Operational: | *5 Mhz to 5 GHz* |
| Gain Range: | *0 to 25 dBi* |
| Bandwidth: | *~ 15%* |
| Approximate Power Rating: | *10 Kw* |
| Design Software: | *Yagimax, Nec-Win Pro, Moment Method* |

Configuration:

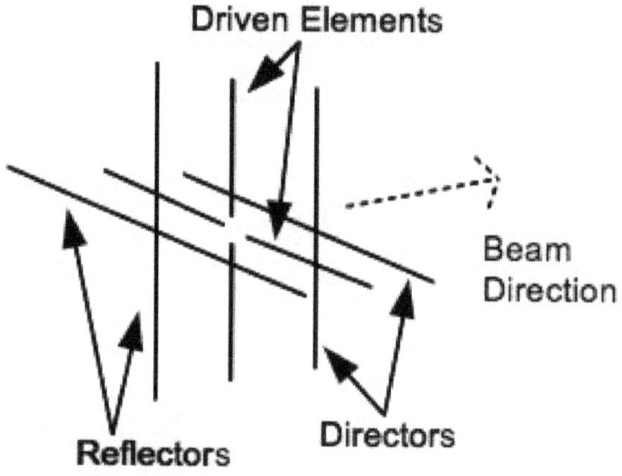


Design Approach:

*Reflector Length = 0.5 lambda; spacing to driven element = 0.125 lambda*
*Driven element length = 0.47 lambda; spacing to director = 0.37 lambda*
*Director length = 0.43 lambda; spacing between director = 0.31 lambda*

Construction and Theory of Operation:

*Essentially a dipole with a pair of inductors in parallel with capacitors resonant at the frequency of interest and held down on both ends with insulators and and guy wires. Impedance is 72 ohms, polarization is horizontal with a broad figure 8 azimuth pattern.*

*Azimuth Pattern*

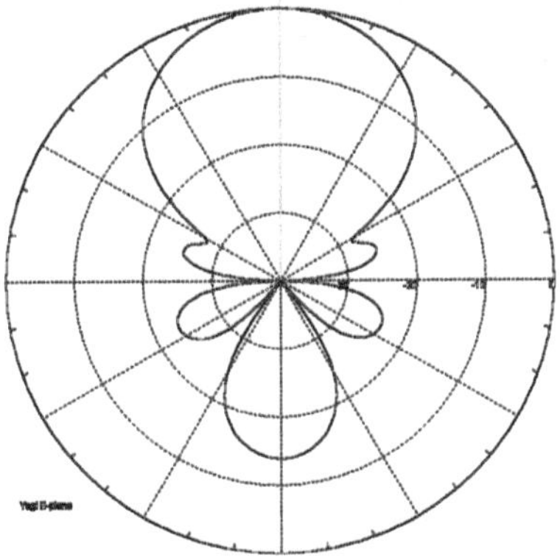

*Elevation Pattern*

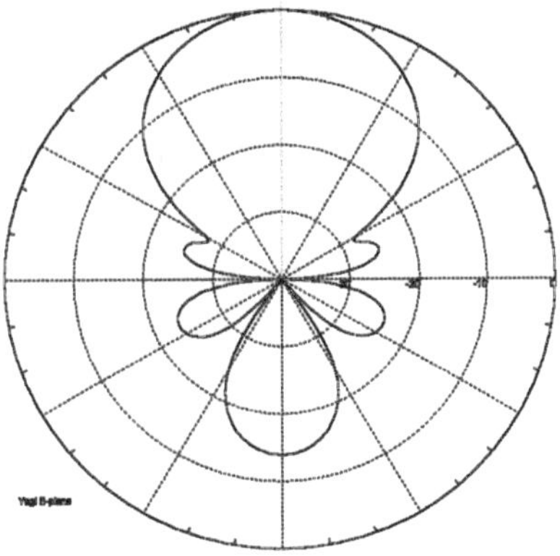

Special Considerations:

*Antennas can use common support structure, including conductive material. Individual antennas can be used separately or combined into right hand and left circular polarizations.*

Testing:

*For mid sized and small apertures use far field range with transmitter/receiver or network analyzer. Near field and compact ranges work also. Use calibrated reference antenna. Use axial ratio measurements to confirm quality of required polarization. For large apertures, use radio star techniques, holography or satellites with known EIRP. Check Antenna Testing Appendix for details.*

| | |
|---|---|
| Type: | ***Delta Matched Yagi*** |
| Frequency of Operational: | *5 Mhz to 500 MHz* |
| Gain Range: | *0 to 20 dBi* |
| Bandwidth: | *~ 5%* |
| Approximate Power Rating: | *10 Kw* |
| Design Software: | *Yagimax, Nec-Win Pro, Moment Method* |

Configuration:

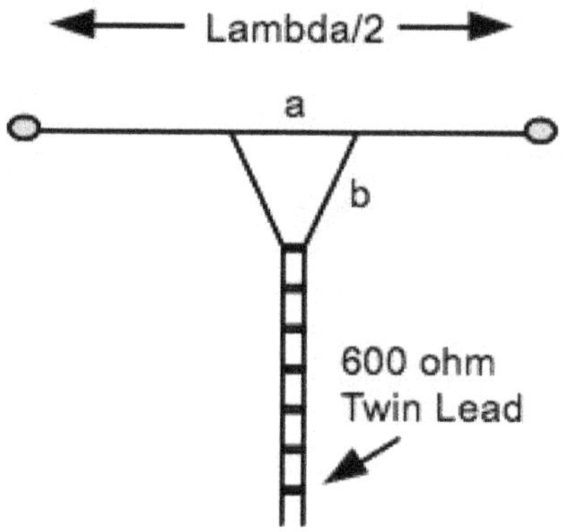


Design Approach:

*a = 0.5 lambda*
*b = 0.15 lambda*

*This is a driven element for a standard yagi with standard design formula mentioned previously*

Construction and Theory of Operation:

*This is the driven element portion of a standard yagi and is essentially a half wave monopole driven by a twin lead feed line. The delta portion changes the impedance (in this case, 600 ohms) to that required to nicely match the half wave radiator. This system can match a wide range of impedances but has a disadvantage of radiating slightly, modifying the antenna pattern.*

*Azimuth Pattern*

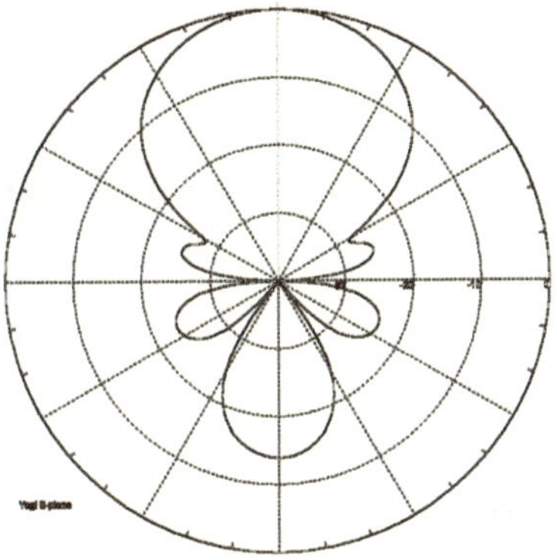

*Elevation Pattern*

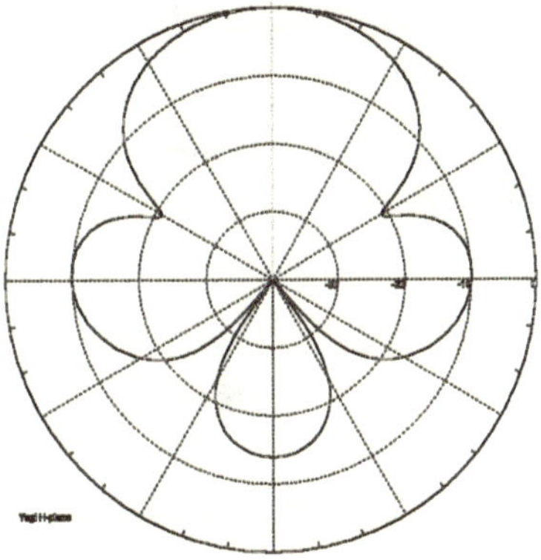

Special Considerations:

*Use aluminum pipe for delta section. The twin lead can be converted to coax with a balun or other suitable device. Use components that are weather resistant, for outdoor use.*

Testing:

*For mid sized and small apertures use far field range with transmitter/receiver or network analyzer. Near field and compact ranges work also. Use calibrated reference antenna. Use axial ratio measurements to confirm quality of required polarization. For large apertures, use radio star techniques, holography or satellites with known EIRP. Check Antenna Test Appendix for details.*

Type: ***Gamma Matched Yagi***

Frequency of Operational: *5 Mhz to 5 GHz*
Gain Range: *0 to 20 dBi*
Bandwidth: *~ 15%*
Approximate Power Rating: *10 Kw*
Design Software: *Yagimax, Nec-Win Pro, Moment Method*

Configuration:

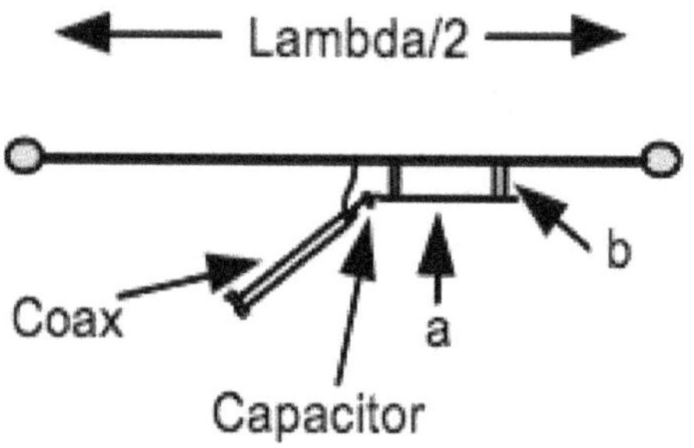


Design Approach:

*a = lambda/20*
*b = lambda/140*
*Diameter of tuning structure is 1/3 that of driven element*
*Capacitor value = 8 pF/Lambda in Meters*
*Coax shield attached to center of driven element*
*Short right side of match to driven element*
*Insulate left side of match from driven element*
*This is a driven element for a standard yagi with standard design formula mentioned previously*

Construction and Theory of Operation:

*This is the driven element portion of a standard yagi and is essentially a half wave monopole driven by a coaxial feed line. The gamma portion changes the impedance (in this case, 50 ohms) to that required to nicely match the half wave radiator. This system can match a wide range of impedances.*

*Azimuth Pattern*

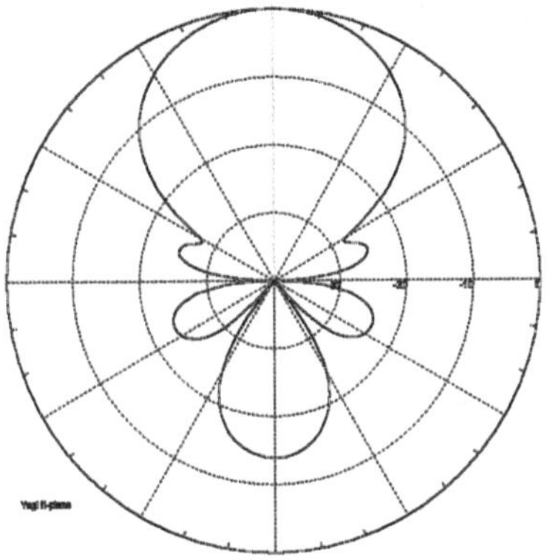

*Elevation Pattern*

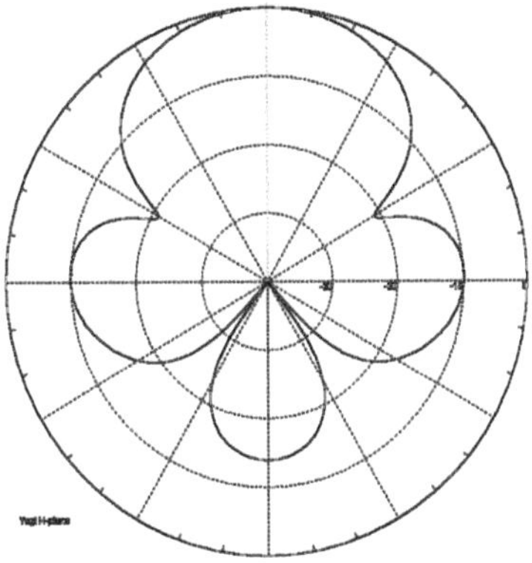

Special Considerations:

*Use aluminum pipe for gamma section. Use components that are weather resistant, for outdoor use.*

Testing:

*For mid sized and small apertures use far field range with transmitter/receiver or network analyzer. Near field and compact ranges work also. Use calibrated reference antenna. Use axial ratio measurements to confirm quality of required polarization. For large apertures, use radio star techniques, holography or satellites with known EIRP. Check Appendix 6 for details.*

*Use VSWR bridge or similar device to optimize match. Slide shorting section to center frequency of interest.*

Type: ***Cap and Balun***

Frequency of Operational: *5 Mhz to 500 MHz*
Gain Range: *0 to 15 dBi*
Bandwidth: *~ 15%*
Approximate Power Rating: *10 Kw*
Design Software: *Yagimax, Nec-Win Pro, Moment Method*

Configuration:

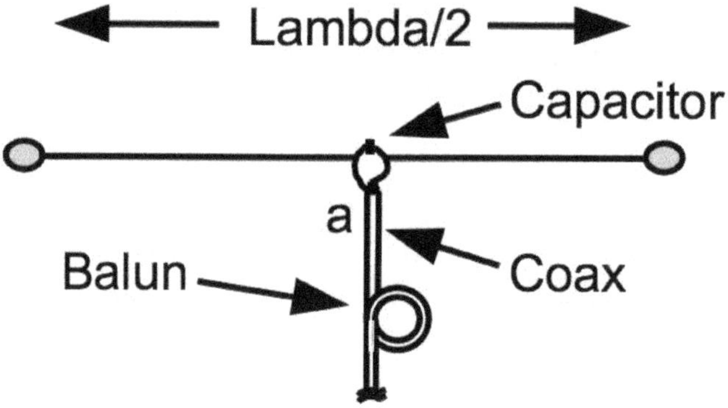


Design Approach:

*a = lambda/16*
*Number of turns on Balun - 2*
*Diameter of Balun - lambda/16*
*Capacitor value = 8 pF/Lambda in Meters*
*This is a driven element for a standard yagi with standard design formula mentioned previously*

Construction and Theory of Operation:

*This is the driven element portion of a standard yagi and is essentially a half wave dipole driven by a coaxial feed line. The balun portion changes the impedance (in this case, 50 ohms) to that required to nicely match the half wave radiator. This system can match a wide range of impedances. "Rolling" the balun along the coax effects the resonant frequency.*

*Azimuth Pattern* *Elevation Pattern*

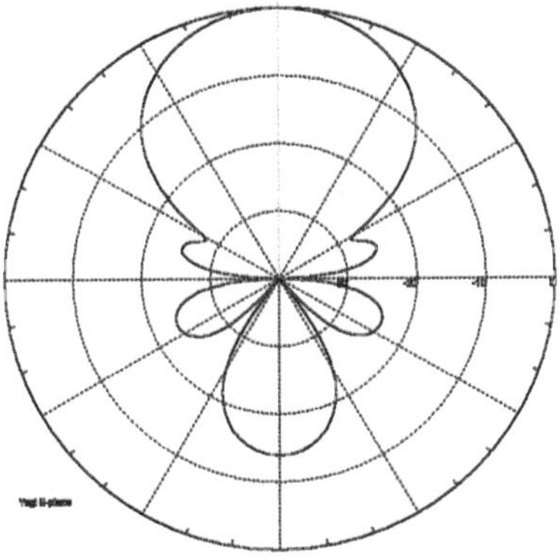

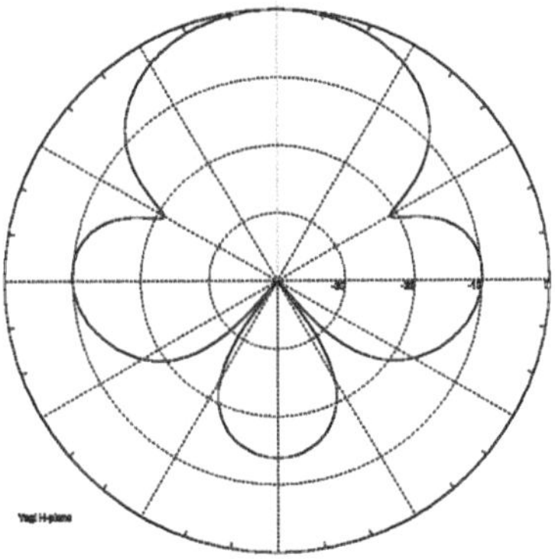

Special Considerations:

*Use aluminum pipe for gamma section. Use components that are weather resistant, for outdoor use.*

Testing:

*For mid sized and small apertures use far field range with transmitter/receiver or network analyzer. Near field and compact ranges work also. Use calibrated reference antenna. Use axial ratio measurements to confirm quality of required polarization. For large apertures, use radio star techniques, holography or satellites with known EIRP. Check Appendix 6 for details.*

*Use VSWR bridge, Network Analyzer or similar device to optimize match.*

Type: ***Folded Dipole or Bowtie Fed Yagi***

Frequency of Operational: *5 Mhz to 5 GHz*
Gain Range: *0 to 20 dBi*
Bandwidth: *~ 25%*
Approximate Power Rating: *10 Kw*
Design Software: *Yagimax, Nec-Win Pro, Moment Method*

Configuration:

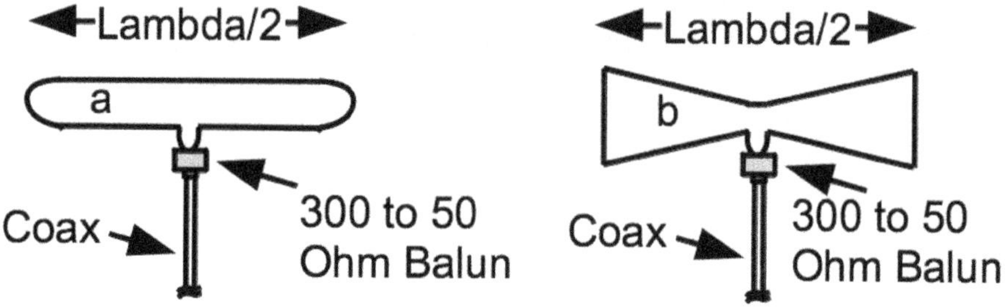


Design Approach:

*a = 0.125 lambda*
*b = 0.125 lambda to 0.25 lambda*
*This is a driven element for a standard yagi with standard design formula mentioned previously*

Construction and Theory of Operation:

*This is the driven element portion of a standard yagi and is essentially a half wave dipole driven by a coaxial feed line. In both cases, wider bandwidths are realized by using loop or bowtie configurations. Baluns must be used to match to 50 ohm coax systems, otherwise ladder line of 300 ohm impedance can be used.*

*Azimuth Pattern* *Elevation Pattern*

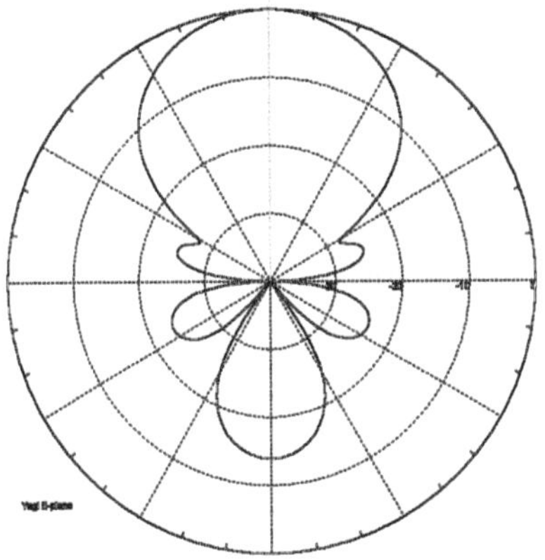

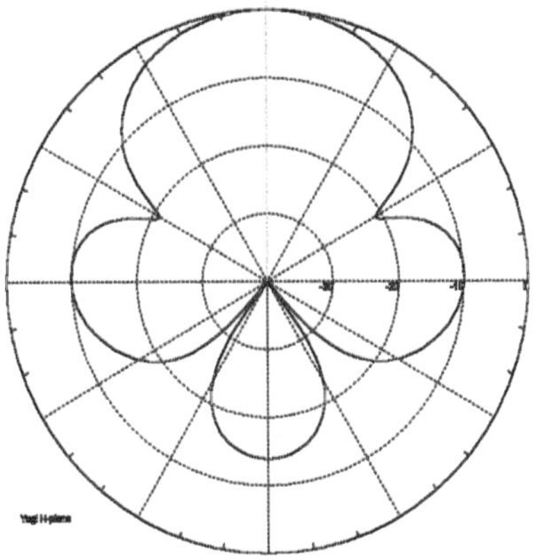

Special Considerations:

*Use aluminum pipe for gamma section. Use components that are weather resistant, for outdoor use.*

Testing:

*For mid sized and small apertures use far field range with transmitter/receiver or network analyzer. Near field and compact ranges work also. Use calibrated reference antenna. Use axial ratio measurements to confirm quality of required polarization. For large apertures, use radio star techniques, holography or satellites with known EIRP. Check Appendix 6 for details.*

*Use VSWR bridge or similar device to optimize match.*

**Questions:**

1. Suggest an antenna type that has the following characteristics:

   a. Frequency of 200 to 300 MHz
   b. Gain of 15 dBi
   c. Linear Polarization

2. Your boss tells you price is a huge factor in the design of antenna with the following characteristics:

   a. Frequency of 20 MHz
   b. Gain of 20 dBi
   c. Linear Polarization

   Can wire be used in this application?

3. Suggest an antenna design for the next generation rocket booster that is capable of going to the moon, with the following properties:

   a. Frequency of 2200 MHz
   b. Gain of 2 dBic
   c. Circular Polarization

4. Can a yagi be imbedded in plastic or fabric ? How would that affect the resonant frequency ?

5. How do you add the outputs of a crossed yagi to create circular polarization ? Slant linear ?

6. How does varying the size of the reflector affect performance ?

7. What happens when you add many directors to performance ?

8. What happens when you use a folded dipole as the driven element ? Specifically bandwidth ?

# Chapter 13 - Frequency Independent Antennas

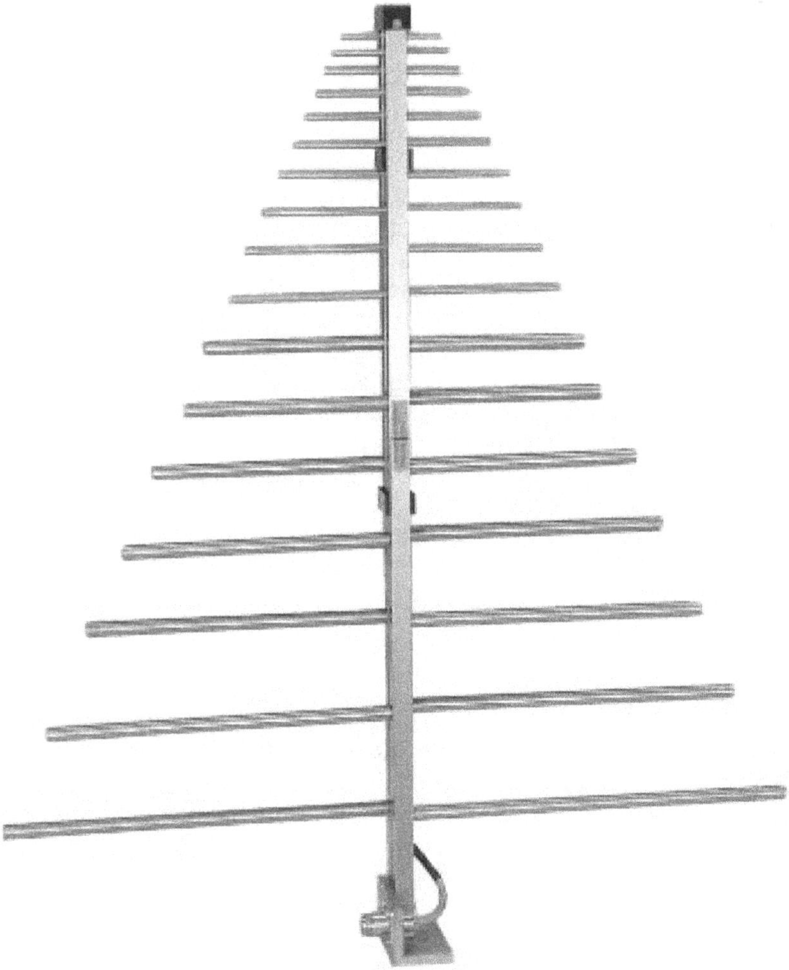

Example of Log Periodic (32)

**Fundamentals**

**Log Periodic**

The first type of a frequency independent antenna is a log periodic, these designs are wide band in nature, some times up to several octaves. The basic construction is to have every other element tied together in a zig zag pattern with a feed point near the front of the antenna. See the diagram below for details. Invented in the 1950s, this type of antenna has been used extensively in military and aerospace applications. Polarizations are typically linear with some versions designed with orthogonal elements to allow either Vertical or Horizontal sense.

One very popular application is with off air Television, where channel 2, near 56 MHz needs to be received as well as channel 83 (old UHF) near 800 MHz. The mass production of this type of antenna has lowered the cost significantly for this application. Another application is with overseas aircraft flights where the mode of communication is in the shortwave bands, with frequencies from 5 to 20 MHz required to allow for diurnal propagation variations. In this case the FAA will have a large log periodic placed about 100 feet in the air with a rotating mast to orient the antenna correctly. This style of antenna also lends itself nicely to broadband emissions testing in the areas of EMC (Electro Magnetic Compatibility) and EMI (Electro Magnetic Interference) applications. Also, this type of antenna is found in many antenna test ranges, as it can be traced back to NIST standards easily and covers a significant amount of bandwidth, minimizing the number of standard gain antennas needed at these test ranges.

**Gain**

The gain on these types of antenna is usually quite good, at least 15 dBi. The front to back ratios also are impressive as is the cross pole quality. Depending of frequency of operation, the antenna is either rotated with a known transmitter in the far field or tested with a field strength meter in the case of low frequency fixed installations.

**Bandwidth**

As mentioned earlier, these antennas are inherently broad banded, sometimes exceeding several octaves in span. Less broad designs are made as well for applications not requiring wide band performance. From a theoretical standpoint, bandwidths from "DC to Daylight" are achievable however matching constraints and manufacturing difficulties restrict research in this area.

**Patterns**

Patterns are generally straight forward in nature and depend on frequency being tested and number of elements. E and H patterns are similar in nature with low sidelobe and backlobe amplitudes. Depending on size, patterns can be taken in the far field or near field. In some cases, like low frequency fixed applications, the use of field strength meters is required to sample the energy are many azimuth angles and plotting of the radiation pattern.

**Polarization**

Polarization is linear in nature due to the fact that the broad band performance makes it difficult to combine vertical and horizontal antenna together with a 90 degree offset (in phase) over wide frequency ranges.

**Applications**

**Off-Air**

The most common application of this type of antenna, quickly being replaced by satellite and cable systems.

**OTH-B Radar (Over The Horizon - Backscatter)**

During the 1960's, this type of antenna was used in large vertical and horizontal arrays to created HF (2 to 30 MHz) radar systems capable of using the ionosphere to reflect energy and achieving great ranges (sometimes in the thousands of miles) to track aircraft and ships over the oceans. A similar radar system is in current use but does not require the low frequencies of operation or the bandwidth. In the case of the first operational research radars, very large geographic areas could be covered, including whole states.

**DF (Direction Finding)**

A very common application for this type of antenna is in the area of direction finding, where the source can be at most any frequency across a broad band. The FCC uses a log periodic to find illegal transmitters, other "special" government agencies do the same.

**SWL**

One of the very best antennas for listening to shortwave broadcast is the log periodic. The FAA, Military, DEA, and Intelligence departments of the government use this antenna to locate, monitor, or communicate with aircraft, ships, and distant broadcasts to determine location and frequency of operation.

**HAM**

Like shortwave operations, HAM radio operators use this type of antenna to cover large amounts of bandwidth, eliminating multiple antennas for specific bands.

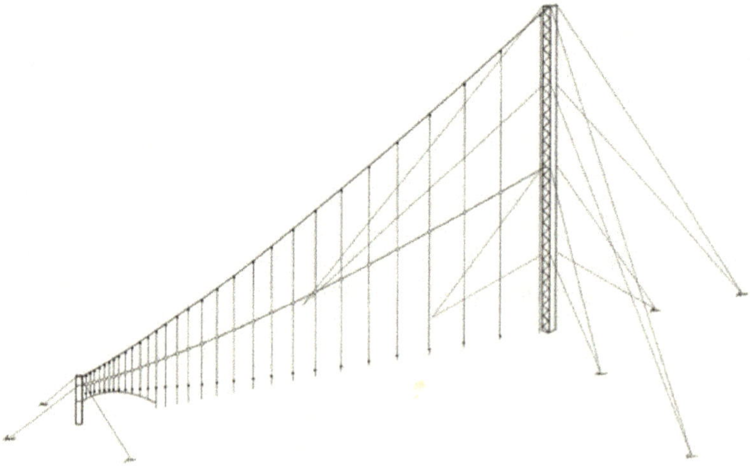

Example of HF Log Periodic (32)

**Construction Techniques**

Log Periodic antennas are constructed using several methods, depending on application. For a typical wire or tube design, every other radiating element is placed on one longitudinal element. A mirror image is placed on another longitudinal element and the two are mated in close proximity using a loop or balun construction technique. Refer to details in antenna papers and representative examples for the latest design approach. Other techniques include the stamping of metal in trapezoidal shapes with (again) every other element represented longitudinally. Two such metal plates can be added together in a "V" shape with a divergence angle around 45 degrees and used for wideband dish feeds as well as surveillance applications. Another approach is to etch the longitudinal element on one side of a double sided PC board. The mirror image of every other radiating element is placed on the opposite side. Using a feed through connector such as a SMA bulkhead connector enables the construction of a fairly inexpensive wide band antenna. Dual polarization is achieved by using four longitudinal elements as apposed to two, with one pair 90 degrees from the other.

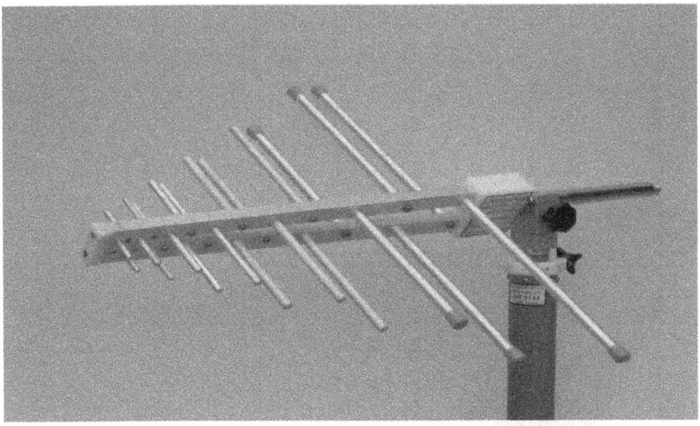

Laboratory Standard Log Periodic

**Spirals and Conical Spirals**

There has been significant recent work on the optimization of spiral and conical spiral antennas. These types are discussed in a later chapter but are being commented on to illuminate their wide band properties. These designs can be many octaves in bandwidth with any polarization required. They can also be used for direction finding in both azimuth and elevation planes. They are used commonly for wide band spectrum analysis and have been used in aircraft and spacecraft. Constructed of etched spirals on either a ground plane or cavity, they are polarized in the same way as the turns when fed from the center or oppositely polarized when feed at the edge. A right hand circular spiral can be coupled with a left hand spiral to create linear polarizations. In addition, a four or more armed spiral can be used for direction finding by sampling both the phase and amplitude of each arm.

Example of a Wide Band two arm spiral antenna

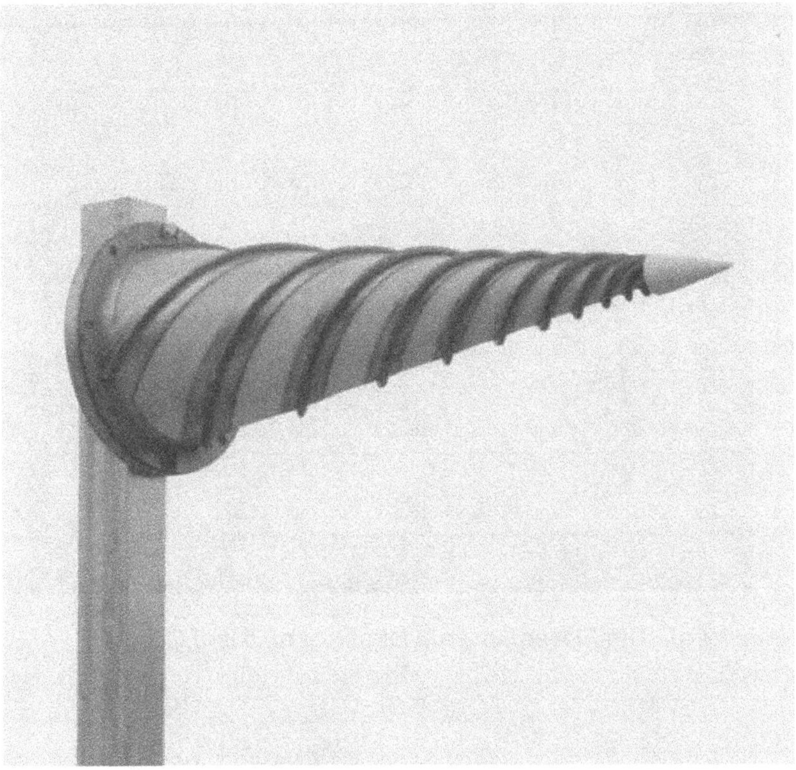

Example of a Wide Band single arm Conical Spiral

Conical spirals are spirals with better control of beamwidth. In other words unlike spirals, their beams can be very directional while retaining broadband performance. They can also be considered like helix antennas but with a conical shape. Helicals have well defined beams with good front to back ratios. Conical Spirals share these traits.

**The Eleven Antenna**

Originally designed for use in radio astronomy, this antenna can have up to a 10:1 bandwidth. The feed impedance is around 200 ohms on four arms which can be used as a monopulse array if desired. The phase center stays very stable which is an advantage in a reflector application.

The Eleven Antenna for use from .5 to 5 GHz

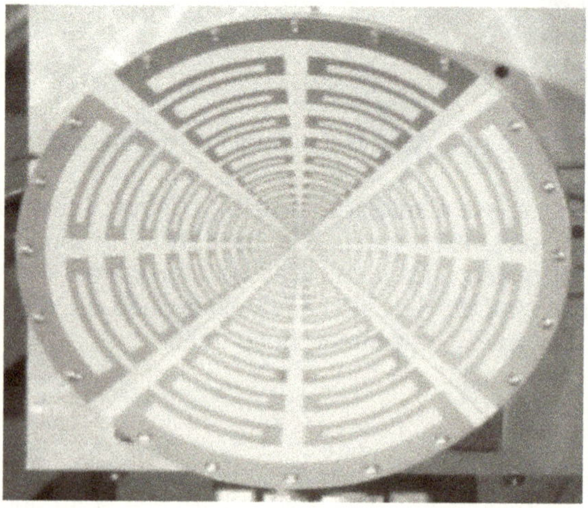

The Eleven Antenna for use from 2 to 10 GHz

**Current Sheet Antenna**

Suggested by Wheeler and developed by Munk and Harris Aerospace, this antenna can have in excess of 10:1 bandwidth and combines array performance with wideband performance. Munk realized that one of the great issues with arrays is the impedance changes as the main beam is moved, he solved the problem by capacitively coupling the radiating arms of a dipole array. Typically an array is constructed using both polarizations with multiple feeds combined beneath the array. This approach has been used for arrays that can broadband spans up to upper frequencies of 50 GHz and beyond using semiconductor manufacturing techniques.

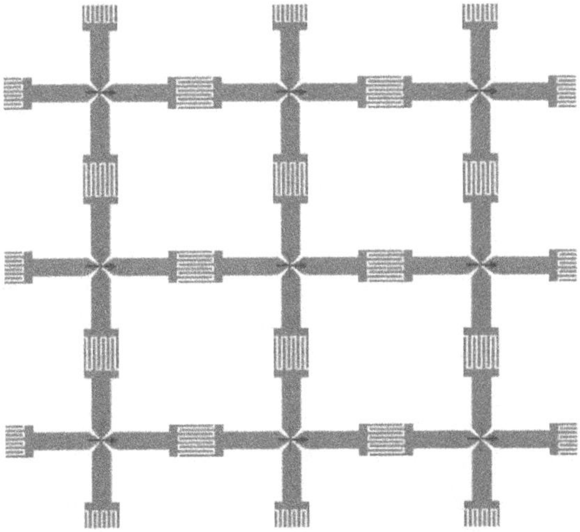

Current Sheet Layout

Current Sheet antennas have feed impedances near 50 ohm, useful for using commonly available components. Some designs have included a second array imbedded into the first to create two independent wide band antennas, each with separate steering functions. These arrays can be conformal and have been used successfully for both radar and communications. Substrates can be low loss foams for light weight and efficient operation.

The following picture shows the wideband performance compared to an idealized 8 element array.

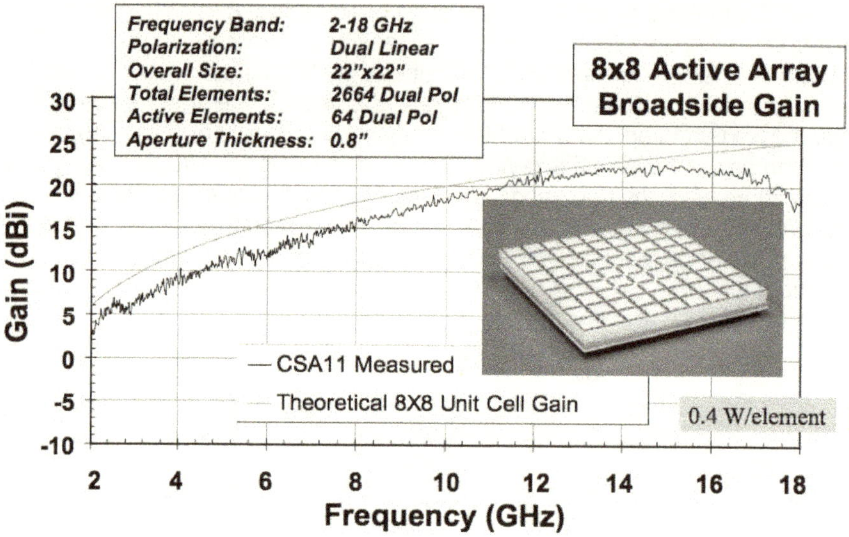


**QSC Antenna**

This antenna is designed around a log-periodic structure within a conical shape. It has very good performance and can be easily used for wideband communications as well as monopulse direction finding.

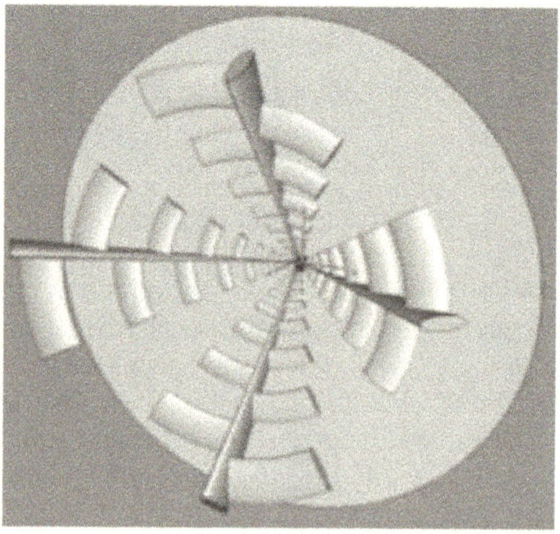

**Inverted Conical Spiral (ICS)**

This antenna has great wideband performance and is another combination of conical shape and broadband element designs. The element is a sinuous spiral shape used initially in flat spiral applications. The addition of the conical shape enhances performance. Here are two examples:

## Spiral Antennas

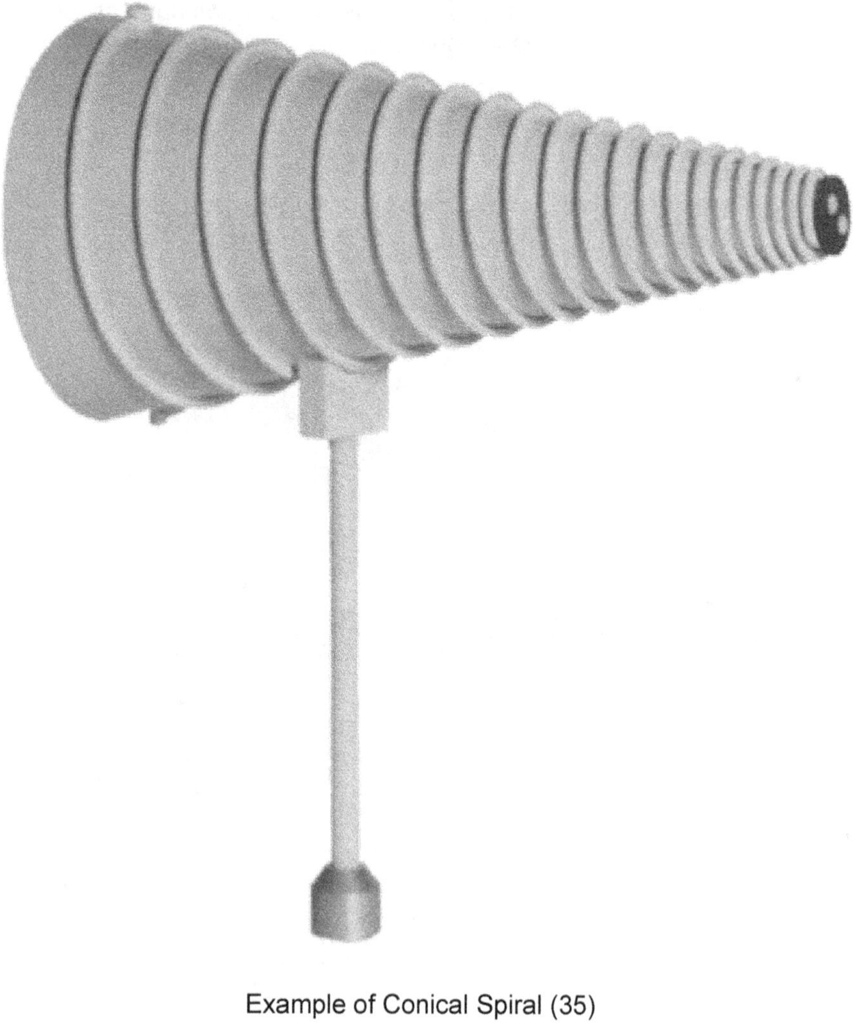

Example of Conical Spiral (35)

**Fundamentals**

A spiral antenna is one that is either etched on a substrate and placed over a cavity or wound in a conical fashion over a suitable form. The main characteristics of this type of antenna is frequency independence or wide band operation. Cavity spirals commercially available run from 2 - 18 GHz or several other very wide band slices of spectrum.

Of the etched variety, there are several types, including Archimedean, Sinuous and Equiangular. Archimedean maintain the same distance between conductive elements as they wrap around a center feed point. Equiangular spirals increase the spacing at a constant rate as they unwrap. Sinuous spirals wrap back and forth in a symmetrical pattern as they open up towards the perimeter.

Conical spirals, like the one shown above, exhibit wide band performance as well but with a more constrained beam and as a result, better gain.

Conical spirals, sometimes called log-spiral antennas have several discrete properties, namely:

1. The angle between a radial line to the spiral and a tangent line to the spiral is a constant value for all point on the spiral.

2. At least one and a half turns need to be used for effective radiation.

3. It is necessary to continue the spiral beyond the maximum radius of RF radiation for the antenna to operate properly at the low frequency range. The magnitude of the currents reflecting from the truncated low frequency end of the spiral arms must be a small fraction of that allowed to radiate into space. The spiralcan be continued with an appropriate expansion coefficient, and a suitable radius can be found to allow a good impedance match.

4. The maximum radius of the spiral corresponds to the lowest frequency of operation, thus:

$$Rmax=\left(\frac{\lambda}{2}\right)*\pi \qquad [58]$$

The equation of a conical or log spiral is:

$$R=Ro*e^{A*\theta} \qquad [59]$$

where:

Ro is the minimum value of the radius,
A is the expansion coefficient

The conical antenna can be fed from the bottom, much like a helix or by the top with an even number of radiating elements, matched with a balun.

For an equilateral or Archimedian spiral, there are similar but distinct attributes. Spirals can be thought of as a traveling wave antenna, where currents in neighboring turns which flow in the same direction add in phase and as a result, radiate.

For an equiangular type spiral antenna the equation is the same as that of a conical spiral, only the antenna is made on a flat surface.

Alternatively, an Archimedian spiral can be designed with the following formula:

$$R=Ro*\phi \qquad [60]$$

where: ø is the angle of the turn.

In this case, although the Archimedian spiral has less bandwidth, it does have superior beam symmetry.

Each arm of an Archimedean spiral is linearly proportional to the angle ø and is described by the following formulas:

$$R=(Ro*\phi)+R1 \qquad [61]$$

$$R=(Ro*(\phi-\pi))+R1 \qquad [62]$$

where:

*R1* = inner radius of circle
*Ro* = outer radius of circle

The spacing between each turn is given by:

$$Ro = \frac{(2 * W)}{\pi} \quad [63]$$

or

$$Ro = \frac{(s + w)}{\pi} \quad [64]$$

where:

s = spacing
w = width of line

In a self complimentary structure the spacing or width can be determined by:

$$s = w = \left( \frac{(R2 - R1)}{4} \right) * N \quad [65]$$

where:

R2 = overall radius
N = number of turns

In the case of a four arm spiral the arm width becomes:

$$w = \left( \frac{(R2 - R1)}{8} \right) * N \quad [66]$$

For these antennas, the lower frequency of operation can be found by:

$$F(low) = \frac{c}{(2 * \pi * R2)} \qquad [67]$$

and the upper frequency of operation can be found by:

$$F(high) = \frac{c}{(2 * \pi * R1)} \qquad [68]$$

Another type of spiral antenna is a sinuous form, where the radiating elements change direction as they radiate outward.

Sinuous antennas have the advantage of having good linear properties, if fed with the proper hybrids. In addition, these antennas can be used effectively as monopulse type antennas for direction finding.

**Gain**

For cavity backed spirals, the gain over the wide band of operation is around -10 to +3 dBi. For conical or log spirals, the maximum gain is dependent on the taper angle and can be as high as 10 dBic, with a corresponding narrow beam. Although spirals of this type can be used without a ground counterpoise, yielding a dipole like pattern, a more prevalent design philosophy is to use a cavity, which is larger than the radiating element and approximately one quarter of a wavelength in depth at the lowest frequency of operation.

**Polarization**

Polarization is typically circular, just like a helix antenna, with the sense (left hand or right hand) following the turn direction of the elements. Just as in helixes, opposite circular polarization can be added together to achieve linear polarization. With a spiral antenna the central feed connects to the polarization (L or R) that follows the mechanical design of the radiating element. Placing a feed at the end opposite the center feed allows for use of the opposite polarization. Thus both pols are available on one spiral.

**Applications**

Cavity backed spiral antennas find many uses in the Military, for wide band signal detection and direction finding. The log spiral varieties are useful in wide band dish feed designs as well as direction finding. Many radio astronomy dish antennas use this type of antenna to cover many bands of interest. In addition, multi arm spirals can bed used as a monopulse antenna by sampling phase and amplitude from each arm at the frequency of interest. Doing so allows for the location of an emitter in the case of a surveillance application.

**Types**

**Equi-angular**

This style of spiral antenna is characterized by a multitude of arms that spiral outwards from a center point. As they do so, they also increase in separation in a linear way. Large versions of this antenna have been designed and constructed for use in the HF regions of the electromagnetic spectrum.

Refer to the literature for applicable impedance matching components, as this is sometime the most challenging characteristic of this type of antenna.

An example of an Equi-angular spiral

**Archimedian**

This style of spiral antenna is characterized by a multitude of arms that spiral outwards from a center point. As they do so, they remain constant in separation. Version of this antenna are often used in direction finding and wide spectral observations. Military and civilian agencies use this type to locate communications and radar emissions.

An example of an Archimedian spiral

**Tapered**

This style of spiral antenna is characterized by a multitude of arms that spiral outwards from a center point. As they do so, they also increase in separation in a linear way. In addition, the width of the arms increases linearly. This type of antenna is simple to construct and has been formed into a condical shape for use as a wide band reflector feed.

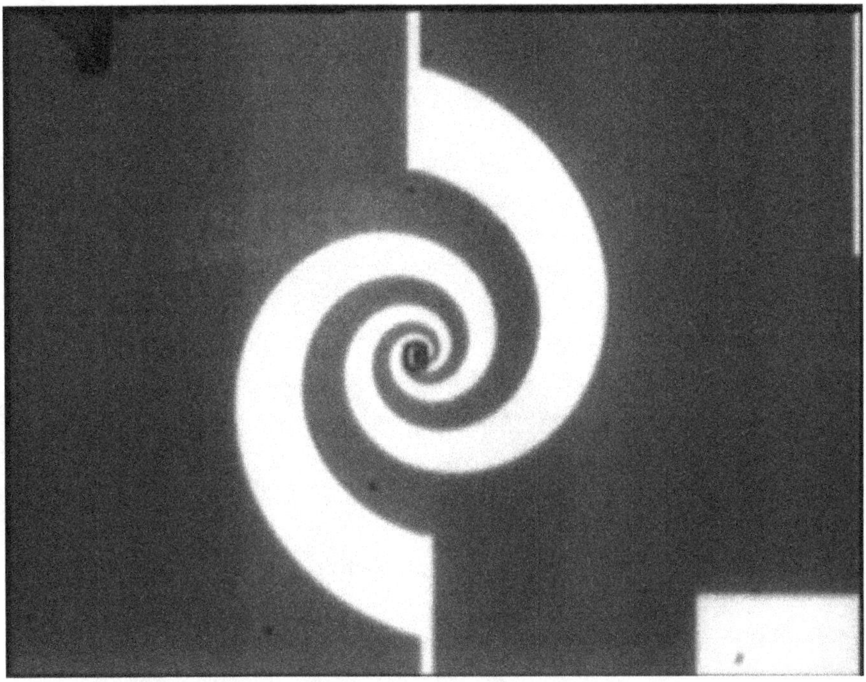

An example of a Tapered Spiral

**Sinuous**

This style of spiral antenna is characterized by a multitude of arms that spiral outwards from a center point. As they do so, they reverse direction every 45 degrees or so. In addition, the width of the arms increases linearly.

This style of antenna can be used for linear polarization applications and for direction finding applications if used in a monopulse configuration.

Also, a sinuous antenna can be used in a conical shape which enhances performance, either with or without a ground plane.

Example of Sinuous Spiral antenna with four arms

**Modal Decomposition**

Spiral antennas can be used in a variety of applications. One of the more innovative approaches to spiral antenna use is the application of modal decomposition for the purposes of direction finding using a single spiral antenna. Different modes are sampled on the spiral to achieve this operation. One set of modes operates like monopulse (Peak and Null at boresite). Another mode rotates phase twice in one azimuthal spin. In addition, another mode can be sampled for altitude determination.

**Construction Techniques**

The cavity back examples of spiral antennas are usually etched on a low loss substrate and placed over a conductive cavity for best results.

Log or conical spiral antennas are usually placed on a form using wire or coax as the radiating element. In some cases foil or even etched elements are used, depending on size.

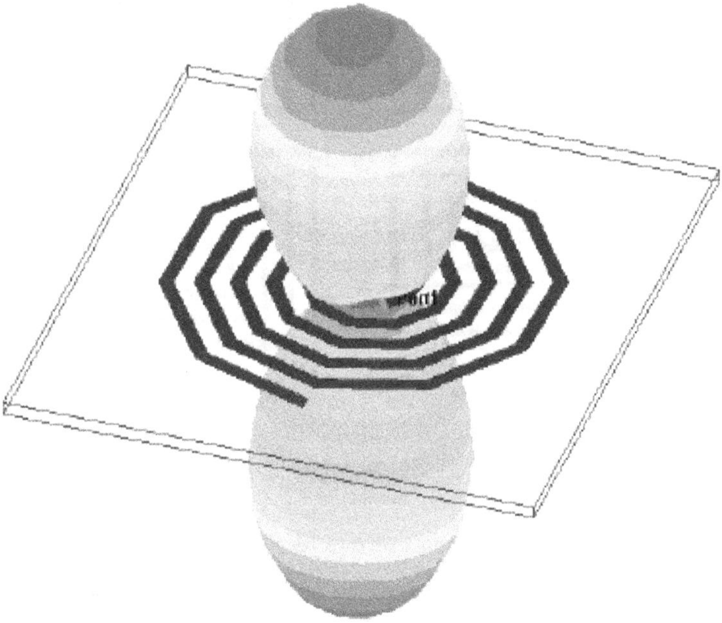

Radiation Pattern of Spiral Antenna

**Spiral Arrays**

Arrays of these types of antennas allow wideband agile beam operations. With a capable feed network, both LHC and RHC polarizations can be used. Combining the two obtains linear polarization. Again these arrays can be conformal, lightweight and reasonably priced.

Spiral Array

**Vivaldi Arrays**

Another array type that produces very good broadband performance is typically used in the Radio Astronomy community. It consists of a multitude of Vivaldi elements (discussed in the horn antenna section). These elements are a derivation of wide band horn inserts and use an exponential taper to allow for wideband operation. Fed with a small balun the impedance can be set to 50 ohms allowing for readily available amplifiers, phase shifters and splitters to be attached. Typically these antennas are constructed using printed circuit techniques keeping the costs reasonable. The following pictures show an individual element as well as an array.

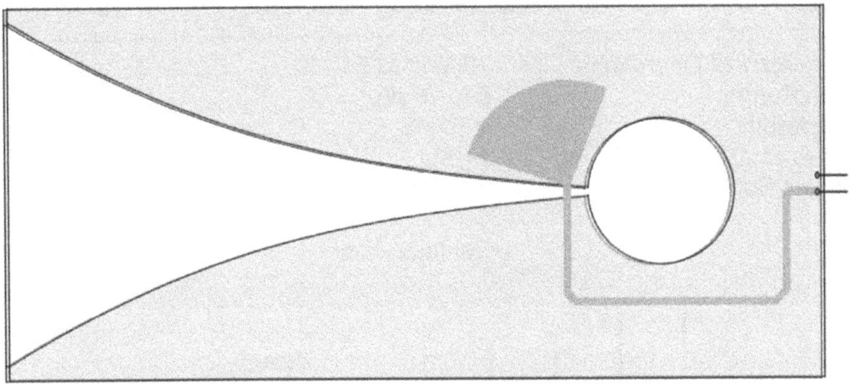

Vivaldi Element and Feed

Vivaldi Array

**Design Examples**

The following pages contain examples of antennas discussed in this chapter. These examples represent a first order design approach. For a more complete and optimized antenna solution, use the example closest to the requirement needs and optimize by using good simulation software, creating prototypes, testing and finally modifying prototypes to optimize performance.

Type: **_Standard Log Periodic_**

Frequency of Operational: *10 Mhz to 5 GHz*
Gain Range: *6 to 15 dBi*
Bandwidth: *~ 30+%*
Approximate Power Rating: *1+ Kw*
Design Software: *Yagimax, Nec-Win Pro, Moment Method*

Configuration

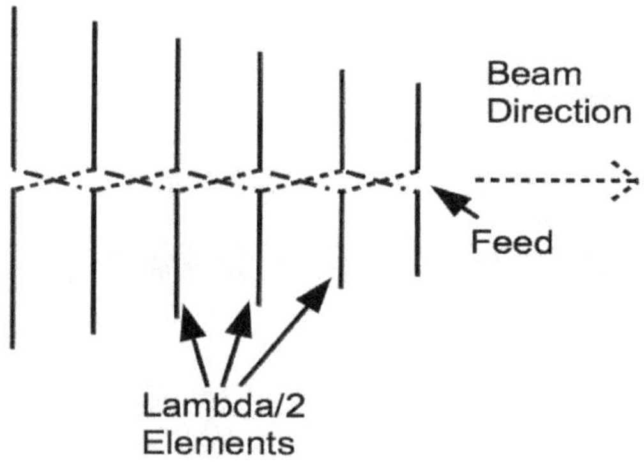


Design Approach:

*Length of elements = 0.5 lambda*
*Spacing of elements = 0.14 lambda*
*Number of elements = 5+*
*Apex angle = 10 to 50 Degrees*

Construction and Theory of Operation:

*Log Periodics are very useful for bandwidths of 2:1 or greater. When planning out this type of antenna, start with the longest element, then place the next longest .14 lambda away. Continue the procedure until done, then check to see that the apex angle is correct. The length of the elements will be dictated by the highest and lowest frequencies, then divide the remaining elements between them For instance, to design an antenna to operate between 100 and 500 MHz, design a dipole for 100,200,300,400, and 500 MHz. When assembling, cross over each element of the dipole to the next to make a 180 phase shift in adjacent dipoles (see diagram) feed with a 4:1 balun.*

*Azimuth Pattern*

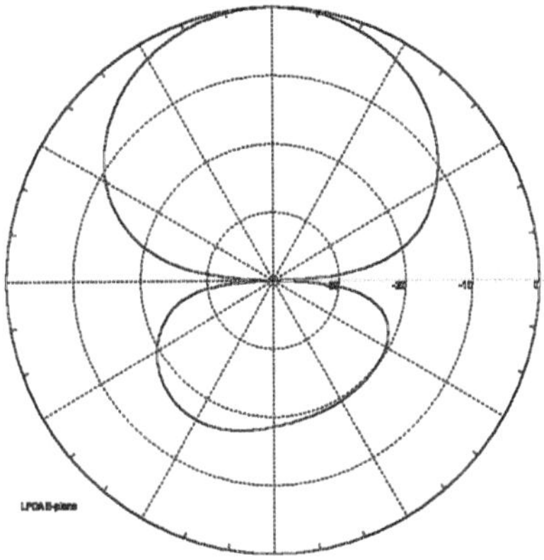

*Elevation Pattern*

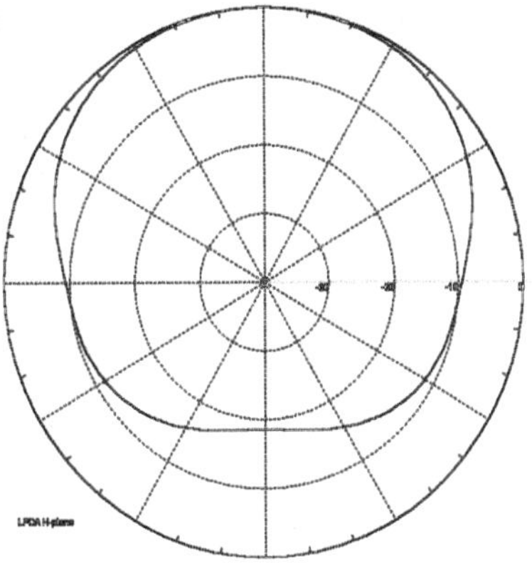

Special Considerations:

*Can be fed with coax, carefully consider match by using 4:1 balun. Every other element on same side is fed in phase, cross over wires must be kept separate. For instance separate both wire paths by affixing one path on top of tubular element and the other on the bottom. Central boom (not shown) can be conductive.*

Testing:

*For mid sized and small apertures use far field range with transmitter/receiver or network analyzer. Near field and compact ranges work also. Use calibrated reference antenna. Use axial ratio measurements to confirm quality of required polarization. For large apertures, use radio star techniques, holography or satellites with known EIRP. Check Appendix 6 for details.*

*Use VSWR bridge or similar device to optimize match.*

Type: ***Trapezoidal Log Periodic***

Frequency of Operational: *5 Mhz to 50 GHz*
Gain Range: *0 to 18 dBi*
Bandwidth: *~ 25%*
Approximate Power Rating: *10 Kw*
Design Software: *IE3D, Nec-Win Pro, Moment Method*

Configuration:

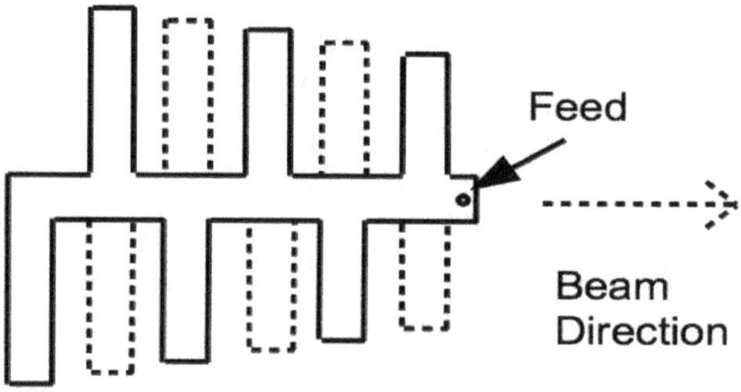


Design Approach

*Length of elements = 0.5 lambda*
*Spacing of elements = 0.14 lambda*
*Width of elements = 0.1 lambda*
*Number of elements = 5+*
*Apex angle = 10 to 50 Degrees*
*Separate conductive elements by 0.1 lambda*

Construction and Theory of Operation:

*Log Periodics are very useful for bandwidths of 2:1 or greater. When planning out this type of antenna, start with the longest element, then place the next longest .14 lambda away. Continue the procedure until done, then check to see that the apex angle is correct. The length of the elements will be dictated by the highest and lowest frequencies, then divide the remaining elements between them For instance, to design an antenna to operate between 100 and 500 MHz, design a dipole for 100,200,300,400, and 500 MHz. When assembling ,place conductive element .1 lambda apart and feed with a 4:1 balun.*

*Azimuth Pattern*

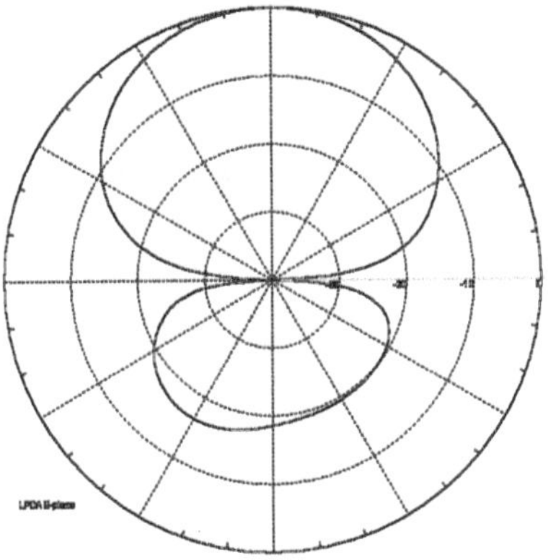

*Elevation Pattern*

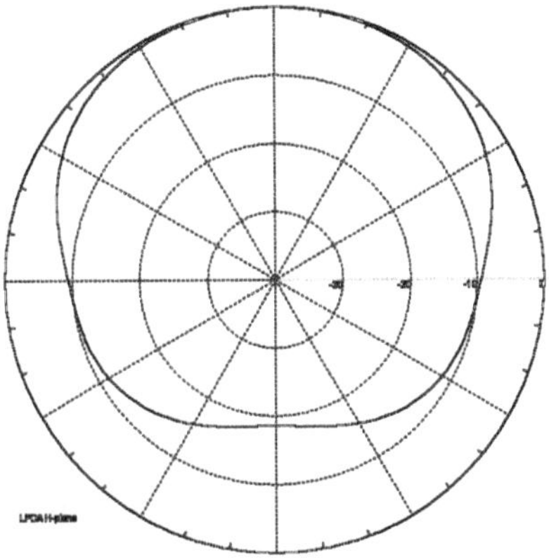

Special Considerations:

*Can be fed with coax, carefully consider match by using 4:1 balun. Every other element on same side is fed in phase, parallel, conductive elements must be kept separate. For instance separate paths by using insulators. Central boom (not shown) must not be conductive.*

Testing:

*For mid sized and small apertures use far field range with transmitter/receiver or network analyzer. Near field and compact ranges work also. Use calibrated reference antenna. Use axial ratio measurements to confirm quality of required polarization. For large apertures, use radio star techniques, holography or satellites with known EIRP. Check Appendix 6 for details.*

*Use VSWR bridge or similar device to optimize match.*

Type: **V shaped Log Periodic**

Frequency of Operational: *5 Mhz to 50 GHz*
Gain Range: *0 to 16 dBi*
Bandwidth: *~ 35%*
Approximate Power Rating: *10 Kw*
Design Software: *IE3D, Nec-Win Pro, Moment Method*

Configuration:

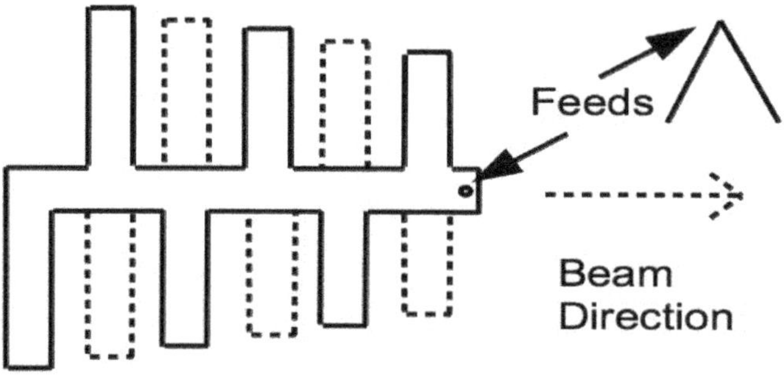


Design Approach:

*Length of elements = 0.5 lambda*
*Spacing of elements = 0.14 lambda*
*Width of elements = 0.1 lambda*
*Number of elements = 5+*
*Apex angle = 10 to 50 Degrees*
*Separate conductive elements by 0.1 lambda at feed end, up to 0.5 at the wide end*

Construction and Theory of Operation:

*Log Periodics are very useful for bandwidths of 2:1 or greater. When planning out this type of antenna, start with the longest element, then place the next longest .14 lambda away. Continue the procedure until done, then check to see that the apex angle is correct. The length of the elements will be dictated by the highest and lowest frequencies, then divide the remaining elements between them For instance, to design an antenna to operate between 100 and 500 MHz, design a dipole for 100,200,300,400, and 500 MHz. When assembling place conductive element .1 lambda apart and feed with a 4:1 balun.*

*Azimuth Pattern* *Elevation Pattern*

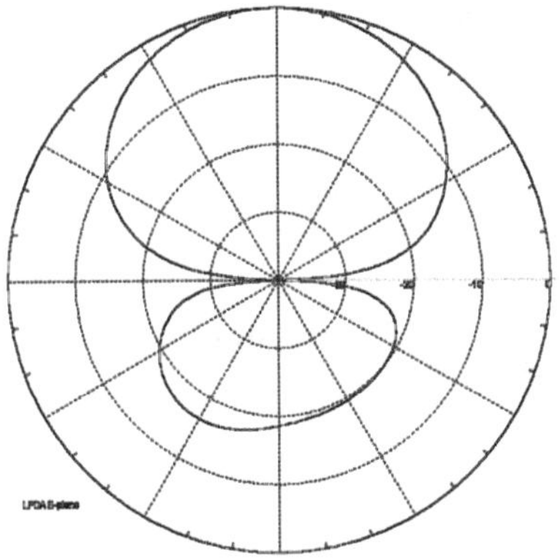

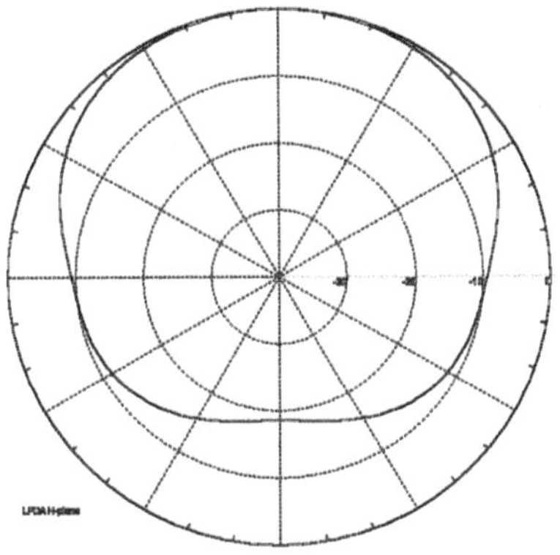

Special Considerations:

*Can be fed with coax, carefully consider match by using 4:1 balun. Every other element on same side is fed in phase, with a taper, conductive elements must be kept separate. For instance separate paths by using insulators. The feed end separation of .1 lambda is opposite the larger end and at up to .5 lambda.*

Testing:

*For mid sized and small apertures use far field range with transmitter/receiver or network analyzer. Near field and compact ranges work also. Use calibrated reference antenna. Use axial ratio measurements to confirm quality of required polarization. For large apertures, use radio star techniques, holography or satellites with known EIRP. Check Appendix 6 for details.*

*Use VSWR bridge or similar device to optimize match.*

Type: ***Etched Log Periodic***

Frequency of Operational: *5 Mhz to 50 MHz*
Gain Range: *0 to 16 dBi*
Bandwidth: *~ 5%*
Approximate Power Rating: *1 Kw*
Design Software: *Yagimax, Nec-Win Pro, Moment Method*

Configuration:

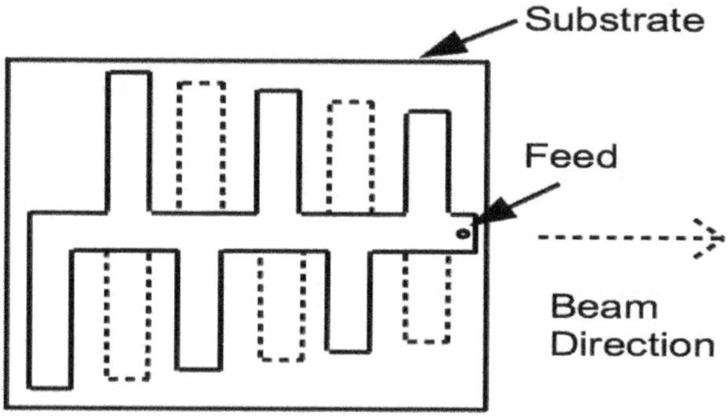


Design Approach:

*Length of elements = 0.5 lambda*
*Spacing of elements = 0.14 lambda*
*Width of elements = 0.1 lambda*
*Number of elements = 5+*
*Apex angle = 10 to 50 Degrees*
*Separate conductive elements by 0.1 lambda*

Construction and Theory of Operation:

*Log Periodics are very useful for bandwidths of 2:1 or greater. When planning out this type of antenna, start with the longest element, then place the next longest .14 lambda away. Continue the procedure until done, then check to see that the apex angle is correct. The length of the elements will be dictated by the highest and lowest frequencies, then divide the remaining elements between them For instance, to design an antenna to operate between 100 and 500 MHz, design a dipole for 100,200,300,400, and 500 MHz. When assembling place conductive elements on either side of a dielectric sheet and feed with a 4:1 balun.*

*Azimuth Pattern*

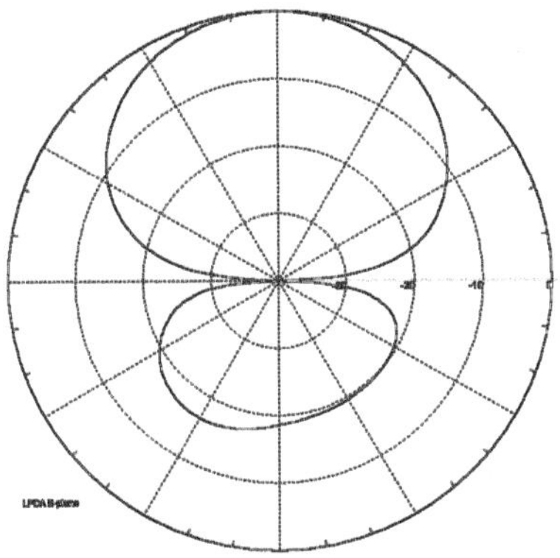

*Elevation Pattern*

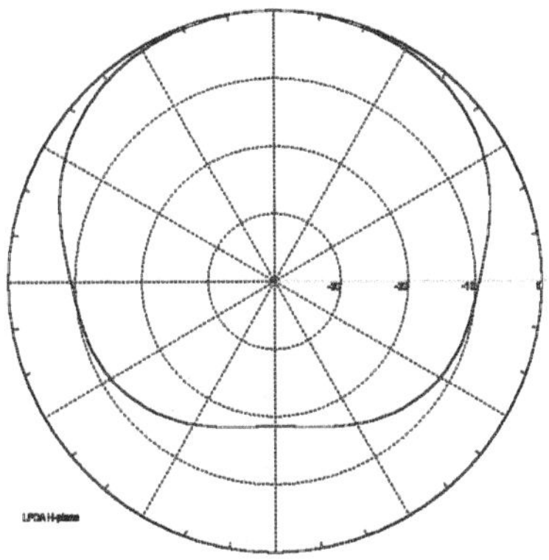

Special Considerations:

*Can be fed with coax, carefully consider match by using 4:1 balun. Every other element on same side is fed in phase, conductive elements must be kept separate. Use low loss dielectric like foam or teflon-fiberglass.*

Testing:

*For mid sized and small apertures use far field range with transmitter/receiver or network analyzer. Near field and compact ranges work also. Use calibrated reference antenna. Use axial ratio measurements to confirm quality of required polarization. For large apertures, use radio star techniques, holography or satellites with known EIRP. Check Appendix 6 for details.*

*Use VSWR bridge or similar device to optimize match.*

Type: ***Crossed Log Periodic***

Frequency of Operational: *5 Mhz to 50 MHz*
Gain Range: *0 to 16 dBi*
Bandwidth: *~ 5%*
Approximate Power Rating: *1 Kw*
Design Software: *Yagimax, Nec-Win Pro, Moment Method*

Configuration:

Design Approach:

*Length of elements = 0.5 lambda*
*Spacing of elements = 0.14 lambda*
*Width of elements = 0.1 lambda*
*Number of elements = 5+*
*Apex angle = 10 to 50 Degrees*

Construction and Theory of Operation:

*Log Periodics are very useful for bandwidths of 2:1 or greater. When planning out this type of antenna, start with the longest element, then place the next longest .14 lambda away. Continue the procedure until done, then check to see that the apex angle is correct. The length of the elements will be dictated by the highest and lowest frequencies, then divide the remaining elements between them For instance, to design an antenna to operate between 100 and 500 MHz, design a dipole for 100,200,300,400, and 500 MHz. When assembling place conductive elements on either side of a dielectric sheet and feed with a 4:1 balun.*

*Azimuth Pattern* *Elevation Pattern*

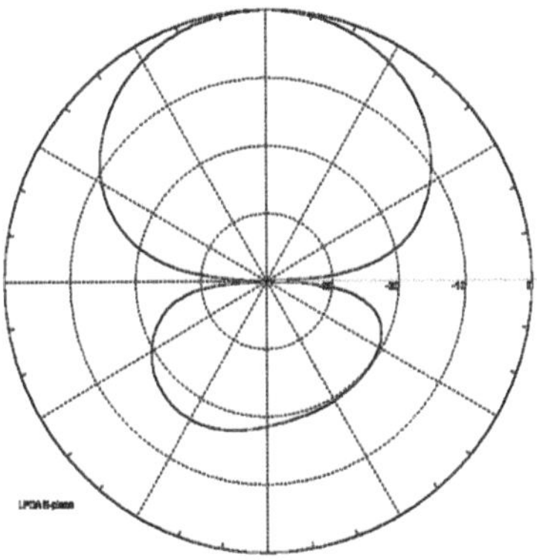

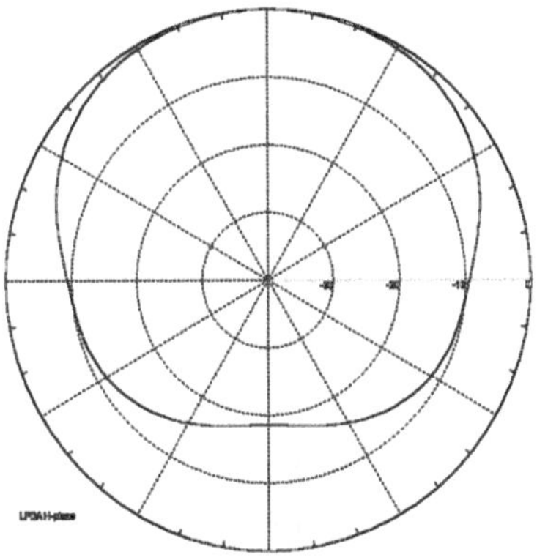

Special Considerations:

*Can be fed with coax, carefully consider match by using 4:1 balun. Every other element on same side is fed in phase, cross over wires must be kept separate. For instance separate both wire paths by affixing one path on top of tubular element and the other on the bottom. Central boom (not shown) can be conductive.*

Testing:

*For mid sized and small apertures use far field range with transmitter/receiver or network analyzer. Near field and compact ranges work also. Use calibrated reference antenna. Use axial ratio measurements to confirm quality of required polarization. For large apertures, use radio star techniques, holography or satellites with known EIRP. Check Appendix 6 for details.*

*Use VSWR bridge or similar device to optimize match.*

Type: ***Simple Spiral***

Frequency of Operational: *100 KHz to 50 GHz*
Gain Range: *-5 to 3 dBic*
Bandwidth: *~ 20+%*
Approximate Power Rating: *10+ Watts*
Design Software: *IE3D, Nec-Win Pro, Moment Method*

Configuration:

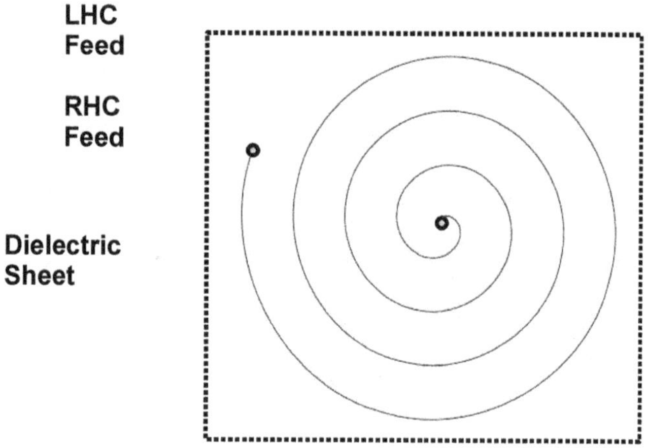


Design Approach:

*Number of turns – 3 to 100*
*Space between turns – at least one sheet thickness*
*Treat antenna like an inductor in a resonant circuit*
*Dielectric sheet is of low loss material*

Construction and Theory of Operation:

Spiral antennas are most useful for wide bandwidth applications. They can be conformal, using etching techniques. This antenna should be considered an inductor within a resonant circuit. Useful for omnidirectional applications, like rf tags and handheld devices. This antenna is probably best used in lower frequency applications where lower gain is acceptable. This antenna can be made in mass using inexpensive materials and large volume etching techniques.

*Azimuth Pattern*

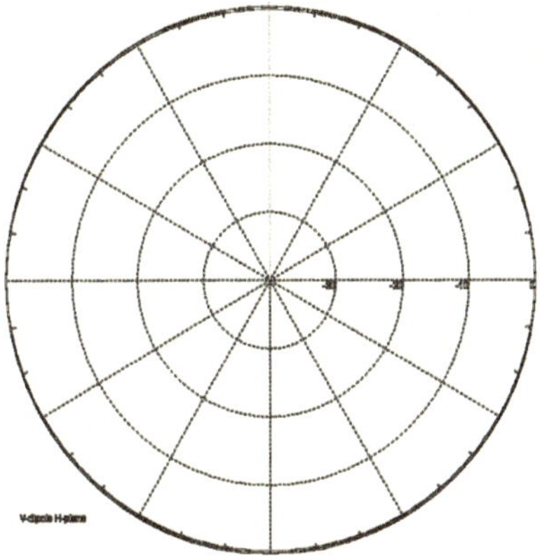

*Elevation Pattern*

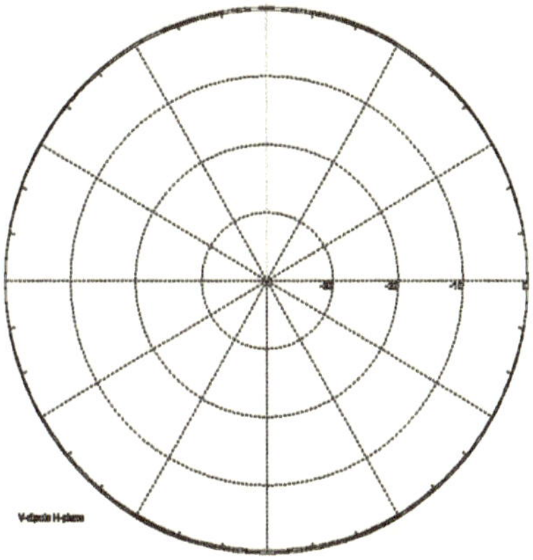

Special Considerations:

*Use low loss material for dielectric sheet with thickness of at least . 005 inches thick. Assume a higher Q (or lower bandwidth) operation as it needs to be part of a resonant circuit.*

Testing:

*For mid sized and small apertures use far field range with transmitter/receiver or network analyzer. Near field and compact ranges work also. Use calibrated reference antenna. Use axial ratio measurements to confirm quality of required polarization. For large apertures, use radio star techniques, holography or satellites with known EIRP. Check Appendix 6 for details.*

*Use VSWR bridge or similar device to optimize match.*

Type: ***Multi pole Spiral***

Frequency of Operational: *5 Mhz to 50 GHz*
Gain Range: *0 to 3 dBic*
Bandwidth: *~ 55%*
Approximate Power Rating: *1 Kw*
Design Software: *IE3D, Moment Method*

Configuration:

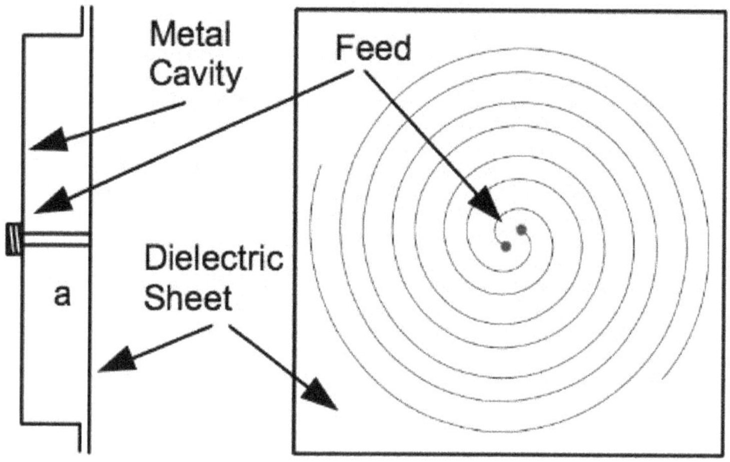


Design Approach:

*Number of turns – 2 to 5*
*Space between turns – at least one sheet thickness*
*Dielectric sheet is of low loss material*
*Distance "a" = 1 lambda at highest frequency*

Construction and Theory of Operation:

*Spiral antennas are most useful for wide bandwidth applications. They can be conformal, using etching techniques. This antenna has excellent circularly polarized radiation properties. Also, it can be used for very wide bandwidth applications like spectrum surveillance. Without the cavity the antenna is mostly omnidirectional, with the cavity the antenna has a hemispherical beam.*

*Azimuth Pattern*

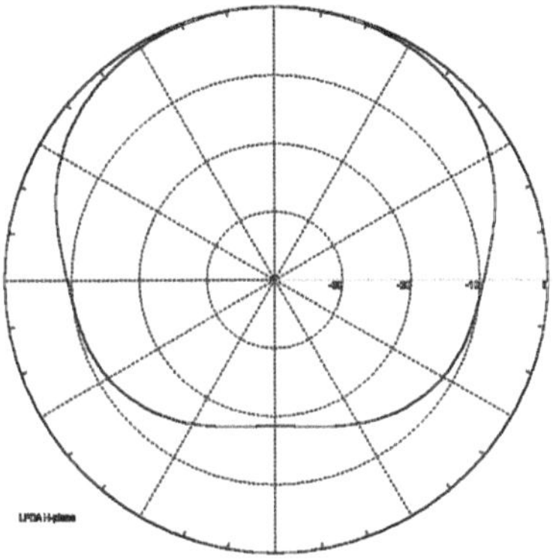

*Elevation Pattern*

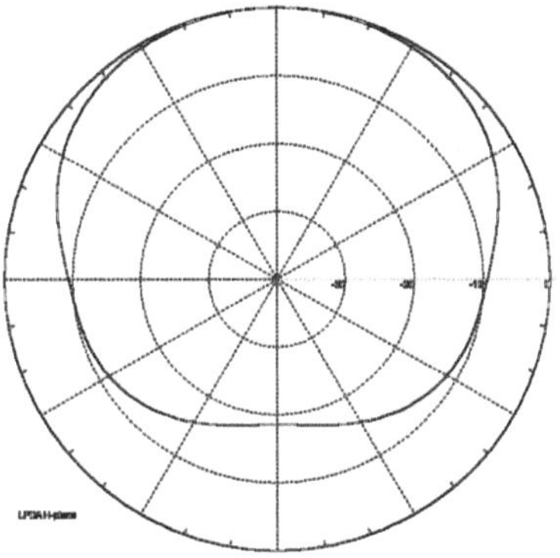

Special Considerations:

*Use low loss material for dielectric sheet with thickness of at least . 005 inches thick. Add resonant cavity to increase directivity by 3 dB. Cavity should be made of aluminum or other non corrosive material. Feed consists of center conductor on one let of spiral and shield on other side. Four arm spirals can also be made, requiring a balun for best match to 50 ohm line.*

Testing:

*For mid sized and small apertures use far field range with transmitter/receiver or network analyzer. Near field and compact ranges work also. Use calibrated reference antenna. Use axial ratio measurements to confirm quality of required polarization. For large apertures, use radio star techniques, holography or satellites with known EIRP. Check Appendix 6 for details.*

*Use VSWR bridge or similar device to optimize match.*

Type: ***Archimedian Spiral***

Frequency of Operational: *5 Mhz to 50 GHz*
Gain Range: *0 to 3 dBic*
Bandwidth: *~ 85%*
Approximate Power Rating: *1 Kw*
Design Software: *IE3D, Nec-Win Pro, Moment Method*

Configuration:

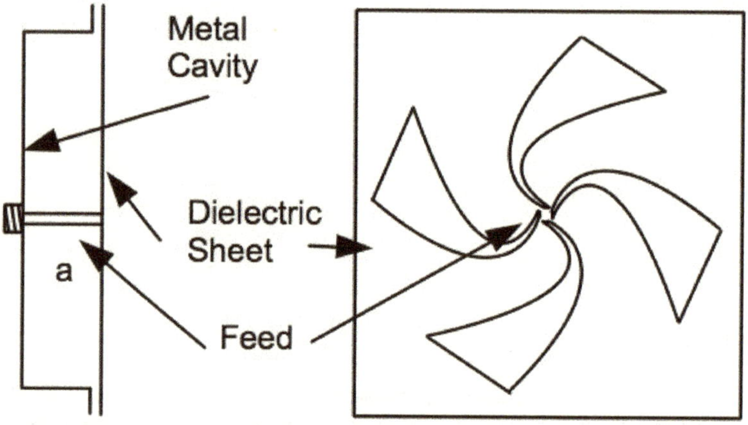


Design Approach:

*Number of turns – 2 to 5*
*Space between turns – at least one sheet thickness*
*Dielectric sheet is of low loss material*
*Distance "a" = 1 lambda at highest frequency*

Construction and Theory of Operation:

*Spiral antennas are most useful for wide bandwidth applications. They can be conformal, using etching techniques. This antenna has excellent circularly polarized radiation properties. Also, it can be used for very wide bandwidth applications like spectrum surveillance. Without the cavity the antenna is mostly omnidirectional, with the cavity the antenna has a hemispherical beam.*

*Azimuth Pattern*

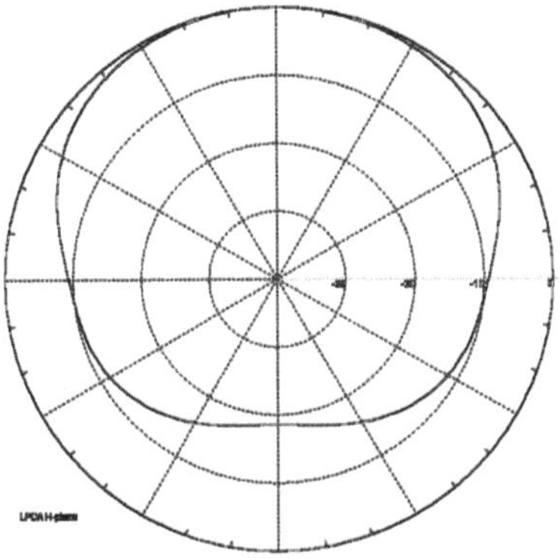

*Elevation Pattern*

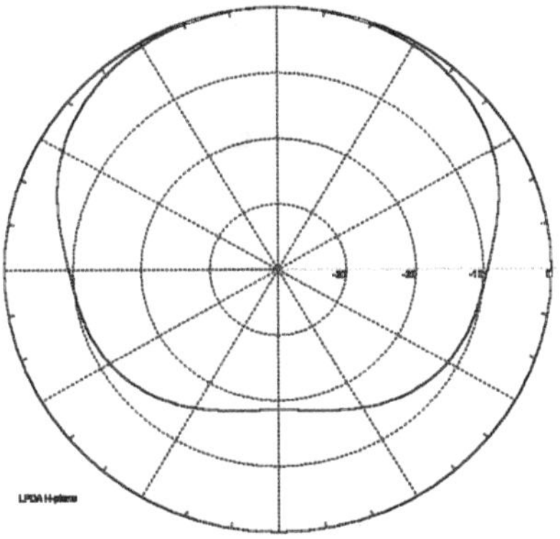

Special Considerations:

*Use low loss material for dielectric sheet with thickness of at least .005 inches thick. Add resonant cavity to increase directivity by 3 dB. Cavity should be made of aluminum or other non corrosive material. Feed consists of center conductor on one let of spiral and shield on other side. Four arm spirals can also be made, requiring a balun for best match to 50 ohm line.*

Testing:

*For mid sized and small apertures use far field range with transmitter/receiver or network analyzer. Near field and compact ranges work also. Use calibrated reference antenna. Use axial ratio measurements to confirm quality of required polarization. For large apertures, use radio star techniques, holography or satellites with known EIRP. Check Appendix 6 for details.*

*Use VSWR bridge or similar device to optimize match.*

Type: ***Tapered Spiral***

Frequency of Operational: *5 Mhz to 50 GHz*
Gain Range: *0 to 3 dBic*
Bandwidth: *~ 85%*
Approximate Power Rating: *1 Kw*
Design Software: *IE3D, Nec-Win Pro, Moment Method*

Configuration:

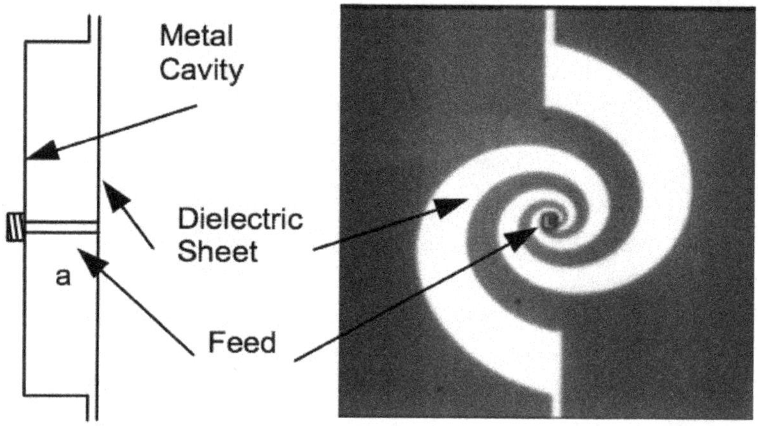


Design Approach:

*Number of turns – 2 to 5*
*Space between turns – at least one sheet thickness*
*Dielectric sheet is of low loss material*
*Distance "a" = 1 lambda at highest frequency*

Construction and Theory of Operation:

*Spiral antennas are most useful for wide bandwidth applications. They can be conformal, using etching techniques. This antenna has excellent circularly polarized radiation properties. Also, it can be used for very wide bandwidth applications like spectrum surveillance. Without the cavity the antenna is mostly omnidirectional, with the cavity the antenna has a hemispherical beam.*

*Azimuth Pattern*

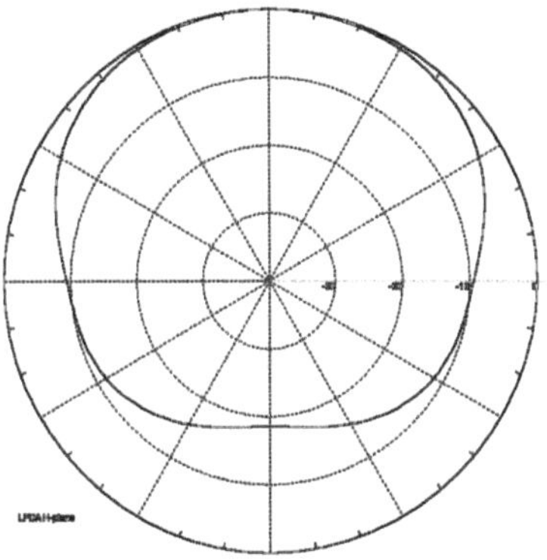

*Elevation Pattern*

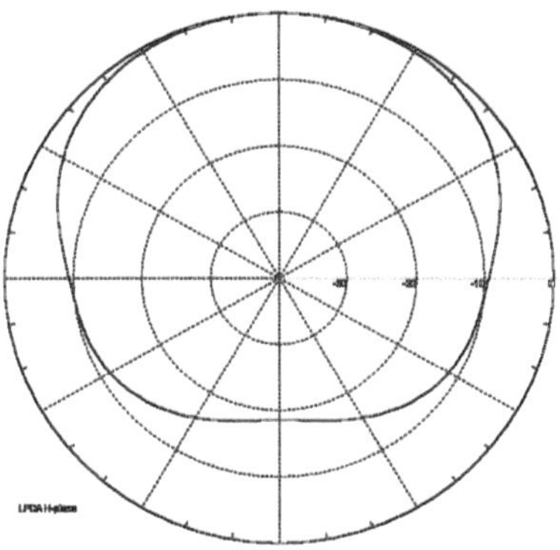

Special Considerations:

*Use low loss material for dielectric sheet with thickness of at least .005 inches thick. Add resonant cavity to increase directivity by 3 dB. Cavity should be made of aluminum or other non corrosive material. Feed consists of center conductor on one let of spiral and shield on other side. Four arm spirals can also be made, requiring a balun for best match to 50 ohm line.*

Testing:

*For mid sized and small apertures use far field range with transmitter/receiver or network analyzer. Near field and compact ranges work also. Use calibrated reference antenna. Use axial ratio measurements to confirm quality of required polarization. For large apertures, use radio star techniques, holography or satellites with known EIRP. Check Appendix 6 for details.*

*Use VSWR bridge or similar device to optimize match.*

| | |
|---|---|
| Type: | ***Sinuous*** |
| Frequency of Operational: | *5 Mhz to 50 GHz* |
| Gain Range: | *0 to 3 dBi* |
| Bandwidth: | *~ 85%* |
| Approximate Power Rating: | *1 Kw* |
| Design Software: | *IE3D, Nec-Win Pro, Moment Method* |

Configuration:

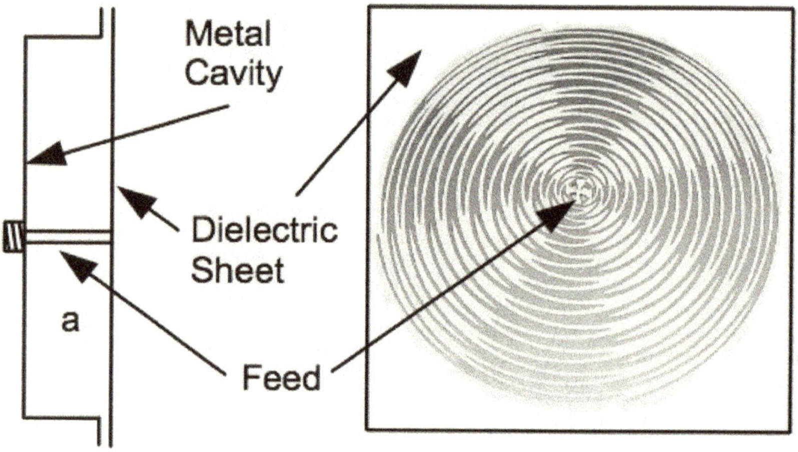


Design Approach:

*Number of turns – 2 to 50*
*Space between turns – at least one sheet thickness*
*Dielectric sheet is of low loss material*
*Distance "a" = 1 lambda at highest frequency*

Construction and Theory of Operation:

*Spiral antennas are most useful for wide bandwidth applications. They can be conformal, using etching techniques. This antenna has excellent circularly polarized radiation properties. Also, it can be used for very wide bandwidth applications like spectrum surveillance. Without the cavity the antenna is mostly omnidirectional, with the cavity the antenna has a hemispherical beam.*

*Azimuth Pattern*

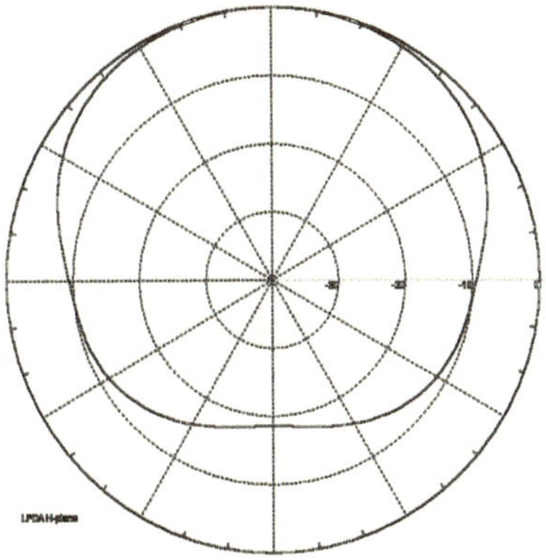

*Elevation Pattern*

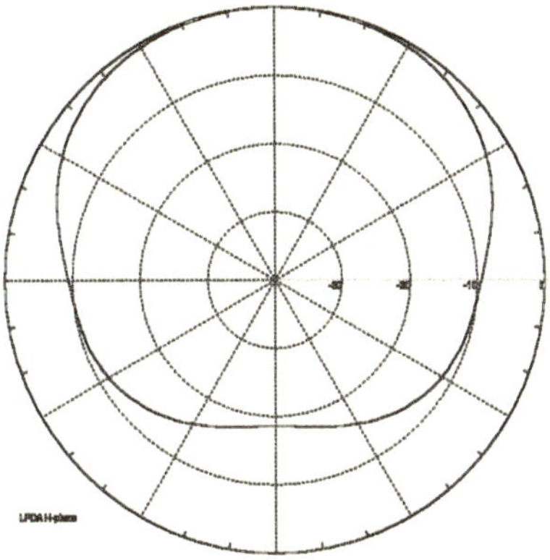

Special Considerations:

*Use low loss material for dielectric sheet with thickness of at least .005 inches thick. Add resonant cavity to increase directivity by 3 dB. Cavity should be made of aluminum or other non corrosive material. Feed consists of center conductor on one let of spiral and shield on other side. Four arm spirals can also be made, requiring a balun for best match to 50 ohm line.*

Testing:

*For mid sized and small apertures use far field range with transmitter/receiver or network analyzer. Near field and compact ranges work also. Use calibrated reference antenna. Use axial ratio measurements to confirm quality of required polarization. For large apertures, use radio star techniques, holography or satellites with known EIRP. Check Appendix 6 for details.*

*Use VSWR bridge or similar device to optimize match.*

**Questions:**

1. Suggest an antenna type that has the following characteristics:

   a. Frequency of 200 to 4000 MHz
   b. Gain of 3 dBi
   c. Linear Polarization

2. Your boss tells you price is a huge factor in the design of antenna with the following characteristics:

   a. Frequency of 20 - 200 MHz
   b. Gain of 2 dBic
   c. RHC Polarization

   Can the cavity be made of plastic?

3. Suggest an antenna design for the next generation rocket booster that is capable of going to the moon, with the following properties:

   a. Frequency of 2200 to 10 GHz
   b. Gain of 3 dBic
   c. Circular Polarization

4. Design a planar array with two distinct frequency bands, separated by several GHz. For instance could you place a 5 GHz microstrip patch on a 1 GHz patch and use the later as a ground plane ?

5. What happens to the polarization when you add a RHC antenna output to a LHC output from a set of spiral antennas ?

6. Does a sinuous antenna exhibit circular polarization characteristics ?

7. If you design a low frequency spiral antenna without a cavity, what happens to the polarization in front of vs. behind the radiating element ?

# Chapter 14 – Phased Array Antennas

Cobra Dane Phased Array Radar (41)

## Fundamentals

Phased arrays are defined as antennas which are made up of a plurality of active or passive elements. These elements can be made of other antennas that are connected in series, in parallel, have additional electronics like phase and amplitude control and, can in fact be built of a multitude or reflectors.

The advantages of phased arrays are numerous, a sample of the inherent qualities follows:

1. Can be conformal
2. Electrically Steerable in millisecond to microsecond speeds

3. Can be thin
4. Can have multiple beams and / or multiple nulls
5. Can survive significant damage

The initial stages of design include those covered in Chapter 2, where the necessary gain, bandwidth and polarization requirements are considered. A phased array is typically not as broadband as reflectors, and if the design requirements include broad frequency coverage, careful consideration must be paid to inter element interaction such as mutual coupling. In addition it is best to have elements approximately ½ lambda apart for best array performance, this is not always practical however; from a physical standpoint, certain geometries can have overlapping elements, from an electrical standpoint, the closer the proximity of an adjacent element, the more the electrical interaction. This latter effect can include detuning of the preferred frequency of operation and the concomitant loss of VSWR performance.

Always model the array before "cutting metal", use a simulation program that can include the characteristics of the elements as well as the entire array. The following formula characterizes the performance of an array:

$$AF = \frac{(\sin(N * \partial/2))}{(\sin(\partial/2))} \qquad [71]$$

where,

$$\partial = (k * d * \cos(\o))$$

$$k = \frac{2 * N}{\lambda}$$

$$\lambda = Wavelength$$

$$N = Number\ of\ Elements$$

$$\o = Polar\ Angle$$

and to estimate gain:

$$Gain(dBi) = Element\ Gain + (n * \log(Element\ Gain)) \qquad [72]$$

where,

Element Gain is in dBi

n = number of elements

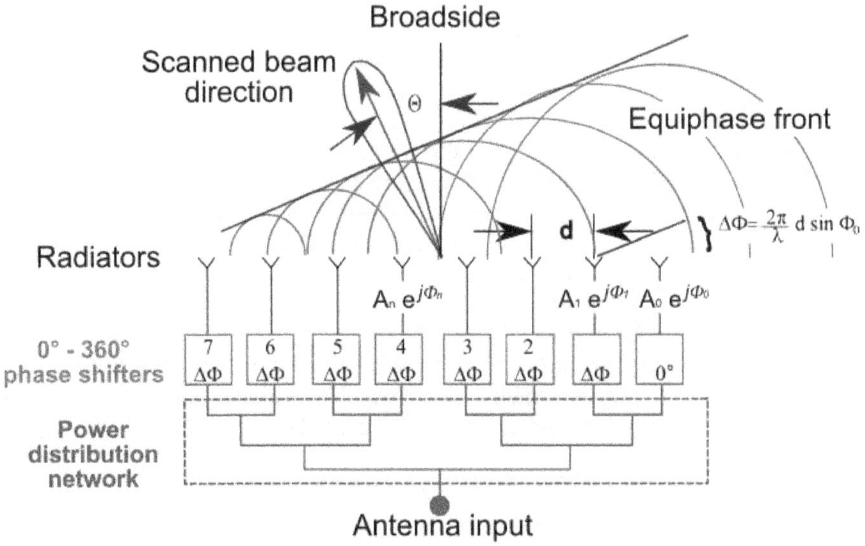


Generalized Phased Array

There are two main groups of phased arrays, the first which manifests a single beam, the second where the beam (or beams) can be moved about relative to the plane of the array. The first type is used widely for cellular, satellite, and many other consumer, commercial, scientific and military applications. In this case the array is physically pointed to the source of interest. The advantages here are there are simple and inexpensive methods to produce the elements and array. These include etching, stamping, or even silk screen techniques. The feed system for the first type can be incorporated on the same plane as the elements or just behind. Microstrip techniques tend to dominate this design philosophy. The second type of phased array incorporated more sophisticated hardware, typically circuitry connected to the output of each element.

The circuitry at the very least includes phase control and in some sophisticated systems, amplitude control as well. With phase control, the beam is shifted by adding delays of various magnitudes to the elements in such a way as to cause the plane of incoming radiation to be in phase at an angle other than perpendicular to the physical structure of the array. The number of phase shift increments is important to evaluate, as a minimum number is surely less expensive. Start with 4 bits, or 180, 90, 45 and 22.5 degree delays. The design of phase shifters is discussed below. Evaluate the performance of the array (gain, beam steering limits, sidelobe magnitude, etc.) and try more and less bits to maximize performance to cost ratio.

Consider the basis layout of the array early on in the design cycle as there are many. The simplest is probably the square array of rows and columns. Consider next the diamond shaped array where the physical portion of the array is turned 45 degrees and the individual element are turned back 45 degrees (see diagram). The advantage of such a layout is lower sidelobes and more efficient performance.

The array can be sparsed to minimize element count and allow for an amplitude taper that will lower sidelobe. The overall gain of the array goes down a bit as a consequence of this approach.

Finally, arrays can be very sparsed or even randomly placed over large distances. The consequences of this approach is the elements become interferometers, having a multitude of beams. Dealing effectively with this problem requires a reasonable amount of computing power.

**Element Selection**

Careful attention should be paid to the element selection of the phased array.

Several factors should be considered:

1. Does the array need to be conformal?
2. Will there be emphasis on the sidelobe performance?
3. To what degree, if any, does the main beam need to be

steered?
4. Are there cost considerations?
5. Is the array dual polarized?
6. Will the array be used to transmit high power?
7. What environmental concerns are there?

## Conformal Designs

If the array needs to be conformal, consider microstrip, slot or spiral elements. The slots and spirals will need a cavity or open area behind them. If the allowable thickness is less than an inch, a microstrip approach will work best. In the simplest form, an array of elements and feed lines can be etched onto a substrate less than ¼ inch thick for frequencies above 1 GHz. Below that frequency, the substrate needs to be thicker. The bandwidth will also drive the thickness. Use the antenna designer software discussed in appendix 10 to get a feel for required thickness.

If the array does not need to be conformal, many alternatives are available. At this point, the designer needs to consider beam steering angle, power ratings, cost etc.

## Sidelobe Performance

For good sidelobe performance, the designer should model several types of elements set at the anticipated spacing for the array. If the main beam is not steered or steered very little, consider reasonably high gain yagis or helixes. Also amplitude taper helps a lot here, where the elements are attenuated, starting low near the center and increasing in attenuation going to the perimeter. Consider Taylor taper, Chebycheff or array thinning (sparsing) here. Expect a slightly lower gain and a slightly larger beamwidth when tapering. Taper levels better than 30 dB down from the main beam can be achieved. As the beam is moved farther away from the bore sight, expect sidelobe degradation to increase, gain to diminish and beamwidth to increase.

## Beam Steering Magnitude

The magnitude of the required beam steer (assuming an "active" array) will drive such factors as element selection and element location. The main beam will only be able to be steered within the element's beam profile.

Steering beyond this point will cause the gain to drop off significantly. For a small area of beam coverage, for instance +/- 15 degrees, element gains of 10 dBi or less can be considered. For wider coverage areas, near omni antennas should be considered, for instance a dipole or wide beam microstrip design. In these cases the beamwidth of the element can be 90 + degrees and as a result, array beams can be steered by 90+ degrees. In some cases back to back arrays vertically oriented, can steer 360 in azimuth, with degraded sidelobe performance. Advanced radars and communication arrays have been designed using this technique. Inmarsat satellite arrays on transoceanic aircraft are an example of this approach.

**Cost**

If cost is an issue remember that a multitude of elements, with the exception of microstrip, will have cost reductions due to large number of identical components. For microstrip, the larger the array, the more etching is required and the volume discounts are less considerable. Using off the shelf components, for instance yagis purchased from an antenna vendor can be lowered in price due to the volume. As with anything else, if custom elements are required, the price will be higher.

**Polarization Considerations**

Arrays can be single or multiply polarized, in other words, vertical, horizontal, right hand circular or left hand circular. There are some exceptions for instance slant linear, however in the case of the feed network there is only a single feed required. In most cases for circularly polarized designs, the phase relationship between vertical and horizontal feed is taken care of at the element and a single feed results. In the case of separate vertical and horizontal or separate left hand and right hand circular polarizations, 2 feeds per element might be required. The resultant feed network doubles in size as a result.

**High Power**

For high power applications, consider the average power as well as the peak power. In simple terms average power will relate to the heating of elements and peak power will influence arcing potential, mulipaction and PIM product development. Choose the element carefully, consider horns and microstrip before dipole structures. Again, is possible use a comprehensive modeling tools to examine the fields around the element as well as the array. If possible use a sidelobe fence for radars with vertical beams, such as wind profiling systems. On search or meteorological radars, place the system on a tower or other elevated platform, in addition develop a safety plan that both predicts and measures RF radiation in the near field. Do not let personnel near an operating radar and include safety features like turning of the transmit power if doors near the radar are opened.

**Environmental Considerations, e.g. Space**

Radars can be used in unusual environments, like space or in aircraft, which require the designer to consider additional factors during the design phase. Space applications require that the designer consider temperature, vacuum, shock, vibration, high energy particles among other things. Generally speaking the following specifications must be considered:

| | | |
|---|---|---|
| 1. | Temperature | -40 to +150 C |
| 2. | Pressure | Local pressure (~30 inches at sea level) to 0 inches |
| 3. | Shock | Up to 20 g during launch and deployment |
| 4. | Vibration | Use profile supplied by launch vehicle manufacturer |
| 5. | Charged Particles | High energy electrons and protons, depends on mission |

The temperature is defined by how long an antenna and its components are in view of the sun. Special paint and covers can mitigate much of the variation in temperature. Consider the output power of the antenna system (assuming radar or communications applications) as well for a source of heating. Always test the antenna and associated components for temperature effects, inside a chamber with multiple cycles to simulate the expected environment. For instance if the antenna is to be used on a geostationary satellite, perform at least 8 cycles between expected cold and expected hot limits, then include a survival cold and hot level. These cycles should be set for a reasonable rate of change, around 2 degrees per hour, with a plateau of at least an hour. The antenna should be tested for VSWR at the very least during these tests.

Pressure variations are important insofar as during launch the pressure can be around 30 inches (assuming ground or sea launch) and within (typically) 8 minutes can change to 0. If antenna components are designed to outgas it is important to consider the rate of change. Also, adhesives, paints, substrate materials and foams can outgas and change electrical characteristics while in space, thereby detuning the antenna. Always test for pressure effects in a chamber during the design cycle.

During launch, significant g loads are experienced followed by g loads from stage separations and deployment operations. These should be considered and tested during the design phase. Normally, the deployment of the various satellites components including the antenna is done last and at slow speeds to minimize g loads. Excessive g loads can deform or break antenna components and as a result, significant portions of a mission can be lost. Consider the antenna on the Jupiter probe Galileo, the high gain antenna deployment mechanism malfunctioned, causing a significant loss of data throughput. The only reliable method of sending and receiving data was through an omni antenna, a difference of at least 30 dB in gain, the only recourse was to use much lower data rates, compromising the quantity of science completed.

Vibrations occurs from the launch, staging and insertion phases of the mission. Vibration can also occur during satellite re-orientation, for instance during the use of reaction control wheels, which spin like gyros in several different orientations. These wheels cause "rumble", which is not high in magnitude, but can cause resonance problems if the array is used at very high frequencies, due to the movement of a ground plane or element position.

Another consideration is with high energy particles in space. High energy electrons and protons are constantly streaming through space. This is due in great part to the solar wind, which varies in intensity over an 11 year cycle. Examine the N.O.A.A space environment laboratories archives for intensity and type of particle. When designing an array for use in space, consider the degrading abilities of these particles. They effect electronics, for instance the phase shifters and associated digital components. This mandates rad hard (or radiation hard) components. Fewer qualified components exist as a result and usually they are not the state of the art in capability. If a phase array is designed using substrate material, make sure the manufacturer has tested and qualified the material for use in space.

## Element Options

### Microstrip

Composed of resonant surfaces, usually 0.25 or 0.5 wavelength long and 0.6 wavelength wide, these structures are normally etched and placed over a ground plane. The ground plane and element are separated by a substrate, sometimes air, whose properties define the beamwidth and bandwidth. See the microstrip antenna section for more details. The feeds for these elements can be co-planar or fed from beneath the ground plane. These types of elements can take a reasonable amount of power, under certain circumstances up to 1 Kilowatt per element. Their bandwidth can be up to 10% of the frequency of interest. Phased arrays of this type are typically etched but can in cases of large volume, be die cut or even silk screened with conductive ink. The feed system is a series of bisecting lines, thus equal in phase to thee center of the array, if any of the connections are moved from the center, the main beam will be move as well. In some cases this is done on purpose to affect a down or uptilt.

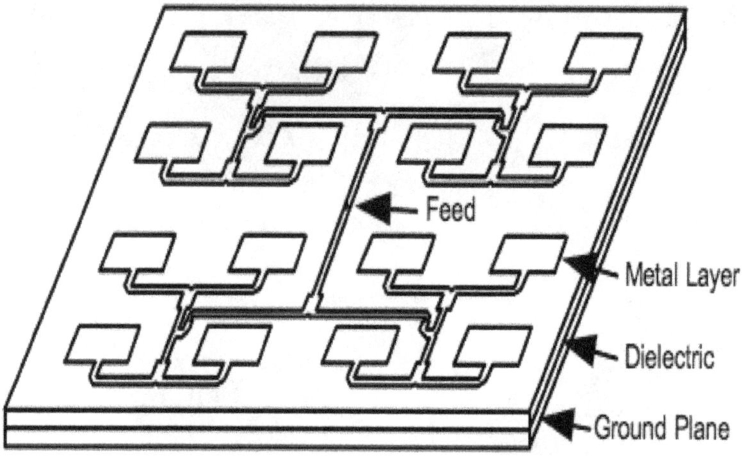


Microstrip Phased Array, 19 dBi Gain

**Horn**

One of the advantages of a horn element is its inherent broadband performance. There are several types of horns to be considered with some approaching 50% bandwidth. In addition, horns can operated at very high power levels. The disadvantages are with the more complicated mechanical details and thick structures required, as well as the added weight.

**Yagis**

Yagis have been used successfully in many phased array designs, typically those used for boundary layer radars and radio astronomy antennas. They are a bit more complex to manufacture and should be used over a ground plane to avoid high antenna temperatures. In concert with the other antenna elements, yagi arrays can have good sidelobe performance, 30 Db down from the main beam in the case of a sparsed array.

**Helix**

Helixes are reasonably easy to manufacture and have an inherent circularly polarized operation. They can be reasonably broadband and in the case of left hand and right hand polarized designs, the helixes can be placed in close proximity as long as the neighbor element is oppositely wound.

**Short Backfire**

Short Backfire antennas have good gain characteristics and considering their normal size, can be used in lower numbers to create a phased array. Each element can have up to 20 dBi and low mutual coupling attributes.

**Dipole / Crossed Dipole / Drooped Dipole**

One of the more common types of elements is the dipole over a ground plane. These elements have bandwidths of around 5% and good overall beam coverage. These elements have been used successfully for phased arrays that require large angular fields of view. They are reasonably easy to manufacture with inherent impedances of 72 ohms. Crossed dipoles allow for dual polarization, including vertical, horizontal, left hand and right hand polarizations. Drooped dipoles have additional advantages in terms of beam steerability and mutual coupling. To examine these advantages use a good simulator package with the individual element factors entered.

**Co-Co**

These elements are actually referred to as coaxial collinear elements and are made up of a several radiating elements in a linear orientation. Usually made up of ½ wavelength sections of coax attached in series with every other element placed so that a shield of one element contacts the center conductor of the next element and the center conductor of the first element contacts the shield of the second. Refer to Chapter 4 for more details. These elements can be simple to manufacture and simple to feed. They do however exhibit frequency dependent beam steering as a result of the linear feed arrangement. If these elements are fed in the center the beam can oblate or even split in two if too much bandwidth is required. Also, the amount of steering the array of co-co can achieve is minimal, say no more than 20 degrees. Sidelobe performance is average.

**Element Spacing**

In the best of worlds, the element spacing should be ½ wavelength, but the size of elements and the close proximity sometimes conspire to cause significant problems with this distance. More commonly, .6 to .8 wavelengths are used with the concomitant increase in sidelobes and grating lobes. At 1 wavelength, an end-fire condition occurs which creates havoc for most applications. Ground based boundary layer radars see significant ground returns. In general, the spacing will reduce the efficiency of the array and increase the sidelobe magnitude.

That being said, there are some interesting exceptions to the rules mentioned above. Sparsed arrays, when the adjacent elements are between .5 and .8 wavelengths work well, and interferometers definitely are useful, if the designer can correlate out the un-necessary beams. Use the software attached to this book to get a better picture of the spacing sensitivity.

**Mutual Coupling**

Mutual coupling occurs when elements of a phased array are interacting with one another. The fields associated with the element design can cause detuning and loss of gain in extreme cases. Expect the resonant frequency of an array to go down when the elements are tightly packed. Therefore, it might be prudent to design higher frequency resonant elements before building the array. In addition, the mutual coupling is a complex number, consisting of a real and imaginary part which convert to an amplitude and phase change. Examining the VSWR of each element (while the other elements are active in receive mode or terminated) will reveal the magnitude of the mutual coupling. Typically the loading of the element in proximity to other elements tends to lower the frequency of resonance. If the array is electronically steered, examine the mutual coupling with each bit of phase shift, in some cases a combination of mutual coupling from adjacent elements at certain phase offsets could be significant, leading to losses in gain in that particular beam direction. Again, the use of a good software simulator can help to predict the performance of the array in all intended beam directions. Munk of Ohio State's Electroscience Laboratory observed that problems found in phased array beam movement such as scan blindness and VSWR changes is due to mismatches at each element. Once this is addressed the performance improves greatly.

**Grating Lobes**

As mentioned above, grating lobes can occur when the element spacing is at or above 1 wavelength. In the case of an interferometer, this phenomenon can occur but the correlation techniques can minimize the effects. If it necessary to space elements at this distance, select the type which has minimal sidelobe structure 90 degrees from the main beam. For instance, short backfire or selectively designed microstrip elements. The following formula predicts the position of the first expected grating lobe of an array:

$$\O gl = \sin^{-1}\left(\frac{1}{Dmax}\right) * 57.3 \qquad [73]$$

where:

*Dmax* = number of wavelengths across aperture

**Patterns**

Patterns are taken the same way for arrays as well as single elements, keep in mind that the patterns will most likely be more complex due to the distributed radiators. For element spacing approaching 1 lambda, expect grating lobes as mentioned above. Always look at the E plane as well as the H plane. Radiation Distribution Plots or RDPs should be taken to observe the complete operation of the array.

**Gain**

Gain on a phased array is determined in the same method as mentioned in the introduction part of this book as well as the antenna test section in Appendix 5. The following formula applies:

$$Gain(dBi) = 10 * \log\left(\frac{(4 * \pi * Ae)}{\lambda^2}\right) \qquad [74]$$

where,

$$Ae = Effective\ Aperture\ of\ the\ Array$$

$$\lambda = wavelength\,(same\ units\ as\ Ae)$$

This number of course does not include losses due to feed lines, power combiners and phase shifters, as the case may be. These components conspire to reduce the true gain of an array. Also, in the case of electrically steered arrays, changes in beam position can present varying amounts of loss due the fact that some phase shifters are in a low degree position and some in a high degree position. For instance, phase shifters set to 180 degrees will have losses a bit higher than those set to 22.5 degrees. Finally, if the array is sparsed or includes amplitude taper modules, the efficiency or overall gain will be lower.

The exception to the above rule is a linear fed array, microstrip or coaxial collinear for instance, where the beam is moved by virtue of changing the frequency of operation. In these cases the efficiency can be very high. There have been interesting two dimensional linear array designs, see the microstrip section for more details.

**Phasing**

A phased array is steered by adding a certain amount of consecutive phase shifts along the line of the angle needed to move the beam. In other words, consider a line of elements, .6 lambda spacing; if all phase shifts are set to 0, the beam will be boresite or perpendicular to the line of elements. Now suppose we set element 1 to 0 phase shift, element 2 to 10 degree phase shift and so on until element n is n*10 degrees phase shifted. Now draw a line from the first element to the last that includes the 10 degrees times the lambda/360 in physical space. This line represents the beam offset plane and therefore the direction of best gain.

In real life, the 10 degree or other low increment is difficult to achieve and can in certain cases be very lossy to implement. More often, 3 to 6 bits of phase shift is used, meaning 180, 90, 45, 22.5, 11.25 etc. degree increments are employed. In fact, these low amounts of phase shift can work quite well, use a good design package to evaluate (or do a sensitivity analysis) of high few phase shift bits are needed to create an acceptable beam over the complete field of view of intended operations.

To calculate the phase shift required for an individual element given a particular angular offset use:

$$Phase\ Delay = \left( \frac{((n-1)*(2*\pi)*Sp)}{\lambda} \right) * \sin(Steering\ Angle) \quad [75]$$

where:

$$N = Number\ of\ Elements$$

$$Sp = Spacing\ (same\ units\ as\ \lambda)$$

Note: Phase Delay and Steering Angle are both in degrees

Reverting to first principles, consider an array of four elements placed horizontally, looking vertically. Consider D as the the distance between elements, and Theta as the angle off of vertical the array needs to be steered in. Then:

$$x = D * \sin(Theta) \quad [76]$$

x is therefore the distance from the a plane wave perpendicular to the steering angle to the element. Normally this would be the second element as the first is usually set at zero degrees. This formula can be repeated for each element or one can use the generalized formula for an arbitrary element:

$$\emptyset = (n-1)*(2\pi)*\left( \frac{x}{\lambda} \right) \quad [77]$$

Where:

$$\emptyset = Element\ phasing\ for\ Element\ n$$

$$n = Element\ Number$$

$x$ = *Extra distance wave has to travel*

$$\lambda = Wavelength$$

At least two schools of thought exist for where, on a large phased array, the zero phased element is placed. One school set this element on an edge of the array, another school sets the zero phased element in the center of the array. Depending on application and magnitude of beam steering, the design engineer should model the array and apply both approaches before making final decision on the steering technique. In the case of the zero phase element being placed in the center of the array, minimal insertion losses will be realized and of course the edges most likely will have some level of phase shift. In this scenario, a slight amplitude taper will be realized lowering the sidelobes a bit.

There are several types of phase shifters, the most common is a set of bits that are switched in and out by PIN diode switches. PIN diodes work by applying a reverse voltage (against the natural current flow of the diode) to turn the switch off, this takes low current but high voltage to make a good attenuated setting. To turn the diode on, requires a low voltage, higher current setting in the direction of current flow to make a low loss "on" condition. The shifter assemblies are usually two port devices with a straight through, 0 degree phase, "off" setting. Additional lines are added to switch in for instance 180 degrees of phase as the frequency of interest and so one. There are two basic types of delay lines, one that is a full longer line and one that uses reflection properties to affect the shift. A good four bit PIN diode phase shifter should have 1 dB of less insertion loss. The time to make a phase shift change can be in the microseconds.

Other types of phase shifters include sliding trombone, ferrite and varactor to name a few. In the case of a ferrite, a magnetic field is applied to create a phase shift, this can be done in small increments, these shifters can be lossy however.

Trombone type phase shifters are the least lossy but require a mechanical mechanism to move and as a result are very slow.

**Steering Losses**

Steering losses are caused by the aspect area of the array changing from perpendicular to the array to an offset position. In other words, the amount of available area for the gain calculation mentioned above diminishes as the angle to the "target" changes off of boresite.

The following formula allows the designer to evaluate this loss.

$$Steering\ Loss = Gain(dBi) * \cos(offset\ angle) \qquad [78]$$

Expect to see twice the steering loss when moving in both x and y directions relative to the plane of the array.

**Scan Blindness**

Mutual coupling between closely spaced element can conspire to lower the gain in particular beam directions in a phased array. It is important to model the array with a good software tool to detect the possibility of this phenomenon. On an electrically scanned phased array, if the mutual coupling is examined (or the effects in VSWR) there will be different values per beam. In some cases the proximity of elements viewed from a certain look angle can in fact cause significant mutual coupling and therefore losses in the gain of the array.

**Interferometers**

As mentioned before, when antenna elements go beyond 1 wavelength in separation, a multitude of antenna "lobes" are generated in the pattern. The interferometric patterns are the convolved pattern of the individual element and the full array. The pattern of the array depends on the projected baseline as viewed from the "target" or area the array is pointed at. In radio astronomy, where this technique is used most often used, the projected baseline changes as the "target" moves across the sky. Imagine seeing a large array from the moon, placed on the earth, as the earth moves, so does the distance and orientation of the array. There can be an advantage to this as the interferometric lobes will not flatten out or disappear until the source of the radio wave (or target) is larger than the beamwidth of the lobe. So the extent of the source can be determined by this method, better known as earth rotational aperture synthesis. Two excellent examples of this are the Very Large Array (VLA) in Socorro, New Mexico and the Very Large Baseline Array (VLBA), situated from California to the Virgin Islands. In both cases, the earth rotational aperture synthesis techniques is used. In the case of the VLBA, made up of 10 90 foot dishes operating over 10 bands of frequencies, the resolving capabilities are in the milli to micro-arcsecond range. In the case of the VLA, made up of 27 90 foot dishes, the resolving power is lower, however the coverage is more complete. The VLA and VLBA produce a multitude of interferometric beams or fringes, using correlators and "cleaning" algorithms, these instruments produce highly detailed maps of the radio sources. Because there are so many combinations of baselines, each producing a separate interferometer, the operation of the array can be understood by holding a colander over one's head and rotating it. The holes in the colander represent the many beams or fringes created by the array. As one rotates, a complete picture of the radio source can be created.

In the 1990s the Japanese designed and launched an orbiting segment of the VLBA, further increasing the resolution of the VLBA array. Day by day changes in the dynamics of the radio source could be observed, thus resolving several theories of the physical processes that energize parts of our universe.

**Coelestads**

An adaptation of an optical instrument, Russian radio astronomers in the 1950s designed in interesting interferometer by placing two flat panels or mirrors at a particular distance and between the pair, back to back dishes with a receiver. In this way separate steerable dishes, being more expensive, were not required. The instrument could track to a limited degree and by changing the distance to the radio source, a measurement of the angular extent could be measured.

**Tapering**

Phased arrays can be tapered in order to reduce the magnitude of the sidelobes and thus improve performance. The improvements come at a price, however and the designer needs to understand that the gain will be reduced slightly and the main beam will become oblate slightly as well. In the field of satellite communications, tapering is usually necessary to avoid receiving unwanted signals from closely space satellites, or worse, sending signals to several satellites instead of one. The particular regulation that applies to 2 degree orbital separation is:

$$29-(25*\log(\theta)) \qquad [79]$$

Several tapering approaches are described below, this list is not exhaustive however and a good designer will do their best to match the requirements of the antenna performance to the appropriate design technique.

**Amplitude**

In general, there are several ways to amplitude taper an array to reduce sidelobes, use a good simulation program to examine the performance of each before making a final decision on the design.

**Taylor**

Based on the Taylor series of polynomials, this is a very popular approach to amplitude weighting an array. An array is usually tapered symmetrically about its center. The Taylor taper applies amplitude shaping to the individual elements, with very low near the center and much higher (depending on the magnitude of the taper) at the ends. The minor sidelobes decrease as:

$$\frac{1}{(\sin\theta)} \qquad [79]$$

with increasing angle $\theta$ outside the angular region delimited by:

$$\left|\left|\left(\frac{D}{\lambda}\right)\sin(\theta)\right|\right| < \bar{n} \qquad [80]$$

where:

D = antenna dimensions

$\lambda$ = wavelength

$\bar{n}$ is an integer that denotes level of taper

Taylor taper coefficients can be found in math reference books like the CRC.

**Chebycheff**

Another popular approach is to use Chebycheff polynomials, depending on the order, the taper can be mild to severe, the latter affecting the gain to a significant degree.

**Sparsing**

One interesting technique to minimize sidelobe levels is to sparse the array. This lowers cost as well but the tradeoff is gain. Imagine a Taylor taper curve with the minimum attenuation in the center. This is how the sparsing works as well, by removing the un-needed elements in a random way relative to azimuth from the center of the array. Use the following approach for a more concise definition of the sparsing.

Random number generator times aperture distribution time scale factor k, which delineates level of thinning required. When the random number is higher than the product of k time the aperture distribution, the element is removed and therefore a lower k produces greater thinning. Thinning reduces gain by the ratio of the remaining elements to the initial total, but array size determines beamwidth.. The number of elements determines the average sidelobe level at 10 log(N) dB [Milligan].

**Multiple Beam**

In some military applications (and soon in commercial), multiple beams are required. When a designer uses a good antenna performance software package, they will discover how the proper placement of phase across the array can allow for multiple beams to be formed, this is helpful for several applications, like communicating with several entities at once. The cost of course is gain and sidelobe performance. For optimum control both phase and amplitude should be adjusted as needed, using a (usually) complex feedback adaptive array algorithm.

**Multiple Null**

In similar fashion to multiple beams, multiple nulls or specific null placement can be had by using the adaptive array algorithm. Null placement is sometimes necessary to minimize interference. A case in point is when a transmitter is trying to interfere with the proper operation of a radar. The effect of this jamming can be minimized by placing a null in the direction of the transmitter. For more information, refer to the bibliography.

**Optical / RF Hybrids**

There have been some radar applications where the power

amplifier, phase shifters, antenna elements and receiver front ends have been integrated into a single unit. These units are usually mass produced. In many of these cases optical fibers connect to these modules to provide input and output connections to other components of the radar. In cases where the coaxial cables to connect to the antenna would be too heavy (for instance along a tall tower), fiber optics can be used to send the signals closer to the antenna / amplifier combination. Fiber optics can be very broadband and very low loss, useful qualities in many antenna array systems.

**Optical Phased Arrays**

In the 1970s, a Canadian inventor used integrated circuit fabrication technology to make an optical phase array. He used a multitude of dipoles, set to resonate at optical frequencies. Each dipole was connected to a diode capable of rectifying the optical signals into direct current. The purpose was to make an efficient solar cell but mass production was problematic and the invention did not go very far. The array was directional, however and had performance was predictable using the equations presented in this book. It is probable therefore that further developments will be made to take advantage of the unique properties of phased arrays in the optical regime. The ability of a phased array to move the beam around and map an area of interest will have applications in many fields.

**SAR**

Synthetic Aperture Radars or SARs are a very interesting application of phased array technology. Although the technique does not require a phased array, many applications include this design type. Essentially, an array or other antenna is used in a radar mode while physically moving from one place to another. While the antenna is moved, the output of the array is recorded, sometimes on film, sometimes on high speed digital media. Phase and amplitude of the scans are used to create a map of great detail that utilized the combination of the resolution of the array and the distance the array traveled. In other words, long strips of radar imagery have been created which show amazing detail from an aircraft or the Space Shuttle. The very first Shuttle flight included a SAR systems that produced enough detail while in flight to allow archeologists to discover numerous subterranean roads and structures in the Middle East.

SAR technology is also used as a highly advanced radar surveillance approach to mapping aircraft positions and ground details. In many cases the rotating dome of an Airborne Warning And Coordination (AWAC) platform has been replaced by a phased array in SAR mode along the longitudinal axis of the aircraft. Resolution has been greatly improved by this technology.

**Beam Formers**

One of the best approaches to forming a beam is in the form of matrices. These can use discrete devices and/or be made with PC board microstrip lines and components. Typical of these components are splitters and hybrids but in some interesting applications these can also be ultrasonic transducers, fiber optics, lenses and water tanks. Matrices are useful if there are a low number of fixed beams required. Normally there are no or very few active devices making these arrays very reliable. In addition, because they usually have a port per beam, different radios (for instance) can be placed on these ports giving significantly more bandwidth for signals. In a cellular phone environment this is a constant challenge which can be easily addressed by a matrix.

Following is a brief description of several popular types of beam forming matrices.

**Butler**

A Butler matrix is a design which enables the multiple selection of beams with a combination of antenna elements, cables and hybrid couplers. The advantages are the ability to receive or transmit from multiple directions simultaneously. The disadvantages are that only a few beam positions can be utilized before the circuit complexity increases beyond a reasonable size. Also, a loss in gain and thus efficiency occurs with the cable and coupler losses. 1R, 4L etc denote the beam position relative to boresight in the following diagram which illustrates an eight beam application:

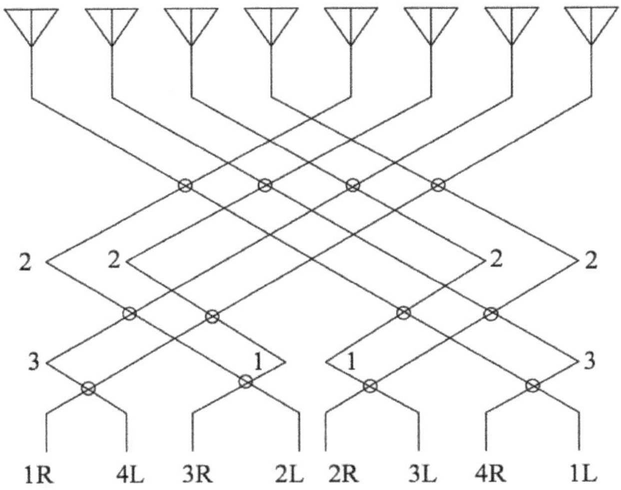

Butler Matrix, the circles are 90 degree hybrids

**Adler**

Another method to rapidly create several stationary beam positions is to use an Adler matrix as shown below:

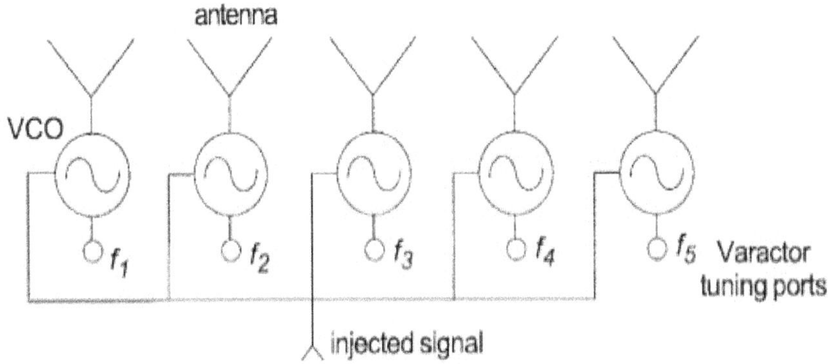


This circuit creates beams by setting the voltages for the varactors, which define the phase of the signal being injected. This can be done rapidly allowing the beams to scan a large amount of space in a short period of time. This array can be used for reception and low power transmission. Varactors are diodes whose capacitance varies with applied voltage. The amount of phase shift created by a change in capacitance follows this formula:

$$\Delta\phi = \omega * \sqrt{(L * C)} \quad [81]$$

where:

$\Delta\phi$ = phase change

$\omega$ = frequency

L = inductance

C = capacitance

**Nolan Matrix**

This array is a made of a combination of directional couplers and phasing lines to produce an efficient matrix that, like a Butler arrangement, has multiple beam outputs and multiple antenna inputs. This matrix can be used for receive or transmit. The layout is as follows, circles are phasing lines, rectangles are hybrids (as shown on right):

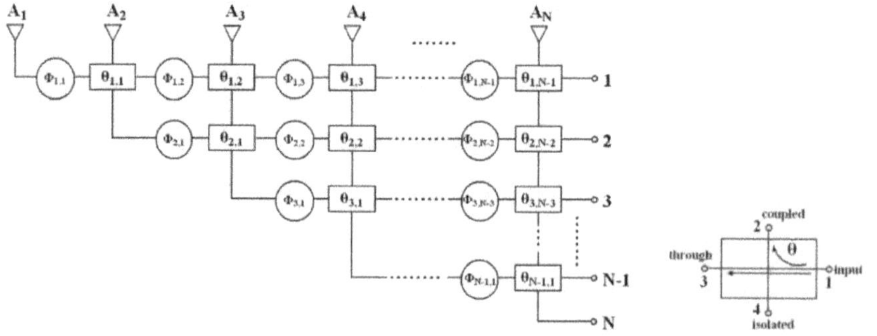


Nolan Matrix

**Blass Matrix**

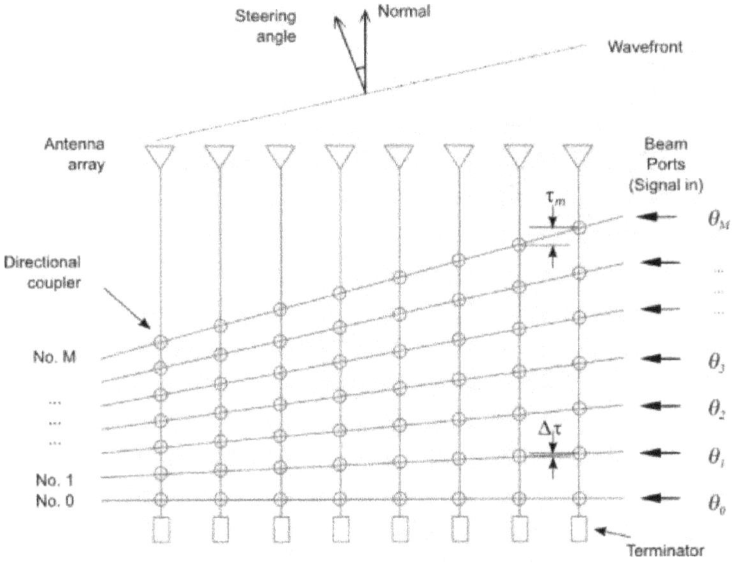


**Rottman Lens Matrix**

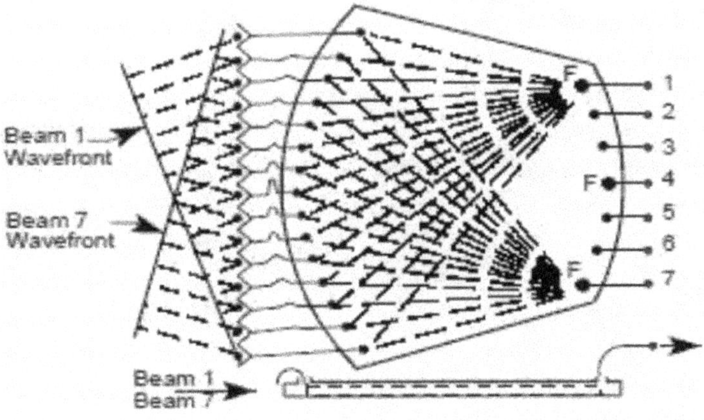


**Van Atta Retrodirective Array Matrix**

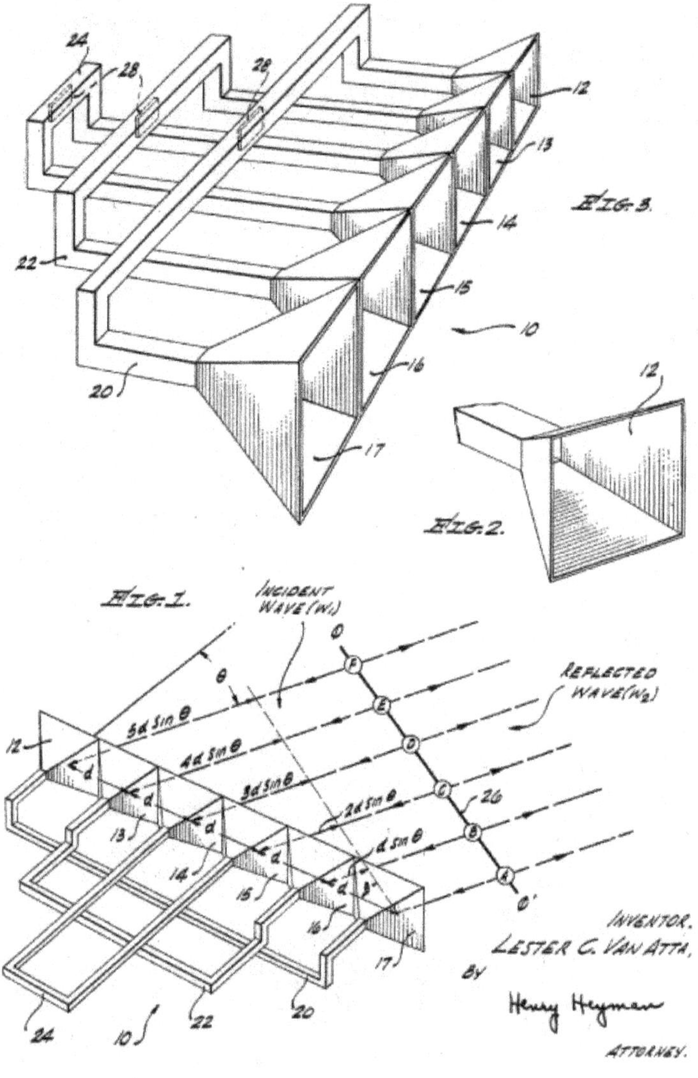


**Frequency Dependent Beam Steering**

Using a series fed array like a microstrip or coaxial collinear offers an interesting alternative to external phase shifters and / or directional couplers. As one element feeds another through a phasing line in a series fed arrangement, a phase relationship is established (typically at broadside) at the frequency of operation. If this frequency is varied, the phase relationship to the next element will vary and so on until the end of the array. The sequential adjustment in phase move the beam as a result. The formula for calculating beam shift vs frequency is:

$$\varphi=\left(\frac{360}{\lambda}\right)*d*\sin(\theta) \quad [82]$$

where:

$\varphi$ = phase shift between two successive elements

$\lambda$ = wavelength

$d$ = distance between the radiating elements

$\theta$ = beam steering in degrees

An advantage to this type of phase shifting is in the simplicity, no phase shifters and very good efficiency. Also, very small beam position adjustments can be made. The disadvantage is that the frequency must change to effect the beam movement, meaning that this technique is useful for radar systems but seldom used in communication systems.

In some military radar applications, azimuth scanning is done be rotating a phased array fed with a series type serpentine. The bottom line is fed to the next above it and so on until the top. A serpentine waveguide is used to create the phase shifter and series fed array. So as the radar is changed in frequency the elevation of the beam is changes, this in combination with the mechanical rotating allow for significant sky coverage.

**Amplitude Controllers**

Amplitude controllers are used typically in adaptive phased array applications, where optimum beam, sidelobe and null control is required.

They are used in conjunction with phase shifters to present a particular phase and amplitude to an array element. Normally, the attenuators are composed of several "bits" just like their phase shifter counterparts, but delay lines are replaced by attenuators. Review the specific literature in Appendix 9 for more information on adaptive phased array technology.

**Reflectennas**

These phased arrays are actually individual element on a (typically but not necessarily) flat ground plane. The array is fed by a single element like a horn, but the elements on the reflecting plane each have a feed point that is terminated by a shorted line of varying length to create a reflectors each with a specific reflection phase. Doing this allows the forming of a near field phase response that can be transposed to a far field antenna pattern. Patterns other than the smallest beam can be created, for instance if a design requires the coverage of a country from a geo-stationary satellite, this can be done by use of this technique. Also, construction is simplified as this array does not need to be formed into a paraboloid or hyperboloid. The drawback of this type of design is its inherent narrow bandwidth. In fact the more elements required, the narrower the bandwidth (or higher Q).

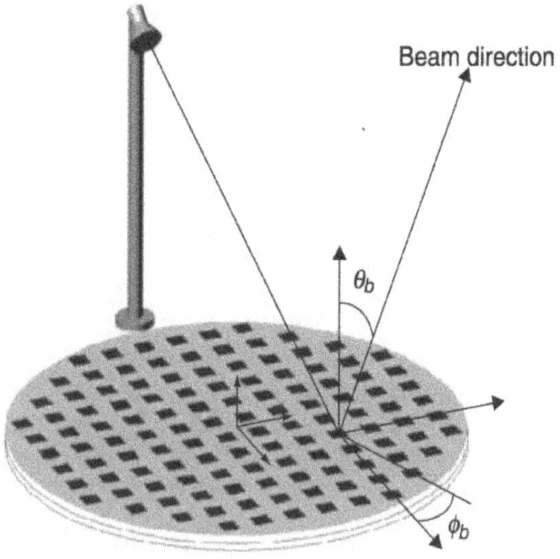


Reflectenna with microstrip reflectors

Patriot Radar system showing multiple phased arrays

**Design Examples**

The following pages contain examples of antennas discussed in this chapter. These examples represent a first order design approach. For a more complete and optimized antenna solution, use the example closest to the requirement needs and optimize by using good simulation software, creating prototypes, testing and finally modifying prototypes to optimize performance.

Type: ***Square Phased Array***

Frequency of Operational: *10 Mhz to 500 GHz*
Gain Range: *6 to 35 dBi*
Bandwidth: *~ 10 %*
Approximate Power Rating: *1+ Mw*
Design Software: *IE3D, Ensemble, Moment Method*

Configuration:

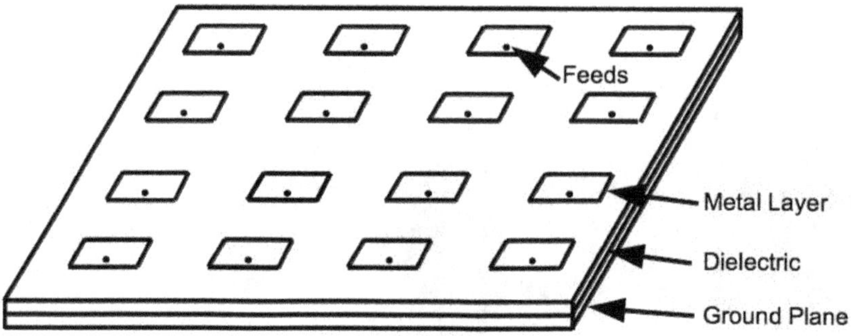


Design Approach:

*Spacing of Elements = 0.5 to 0.9 lambda*
*Number of elements = 2 to 256*

Construction and Theory of Operation:

*Phased arrays are made up of a multitude of elements in a line, square or other shape. The elements can be made up of dipoles, horns, patches, yagis, log periodics or other suitable antennas. Typically, the elements all "look" in one direction only. It is important not to exceed 0.9 lambda spacing between elements, otherwise grating lobes will occur and compromise beam coverage and gain. The elements can be fed in phase for a broadside beam, or sequentially out of phase to move the beam. The array beam can be moved within the dimensions of the single element beam; any further movement will cause a loss of gain. Amplitude tapers can also be applied to the elements to minimize sidelobes. For instance, a 0.5 0.7 0.9 1 1 0.9 0.7 0.5 taper (in terms of power) across an eight element array will reduce sidelobes from approximately -14 dB down to -18 dB.*

*Azimuth Pattern*

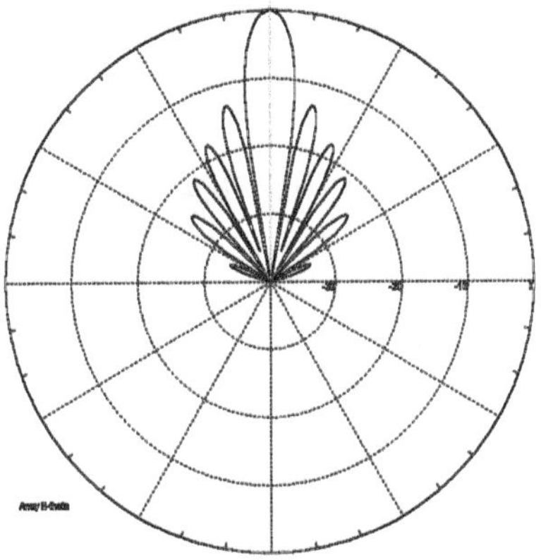

*Elevation Pattern*

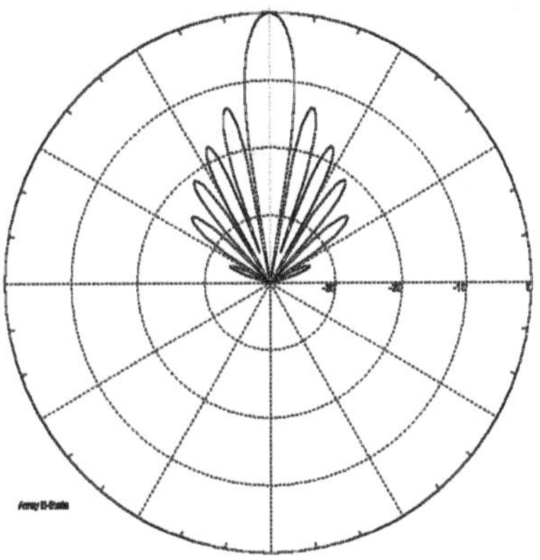

Special Considerations:

*Care must be taken in the design of the feed network, for boresite arrays, each element must be fed in phase, within a few degrees. Amplitude distribution is also important, equal amplitudes will create highest gain with higher sidelobes. Tapering amplitude will lower gain slightly but improve sidelobe performance. Test individual elements first, then small subarrays, careful to correct mutual coupling influences.*

Testing:

*For mid sized and small apertures use far field range with transmitter/receiver or network analyzer. Near field and compact ranges work also. Use calibrated reference antenna. Use axial ratio measurements to confirm quality of required polarization. For large apertures, use radio star techniques, holography or satellites with known EIRP. Check Appendix 6 for details.*

*Use VSWR bridge or similar device to optimize match.*

*$2^{nd}$ Edition*

| | |
|---|---|
| Type: | ***Diamond Phased Array*** |
| Frequency of Operational: | *50 Mhz to 50 GHz* |
| Gain Range: | *0 to 35 dBi* |
| Bandwidth: | *~ 5%* |
| Approximate Power Rating: | *10 Kw* |
| Design Software: | *IE3D, Ensemble, Moment Method* |

Configuration:

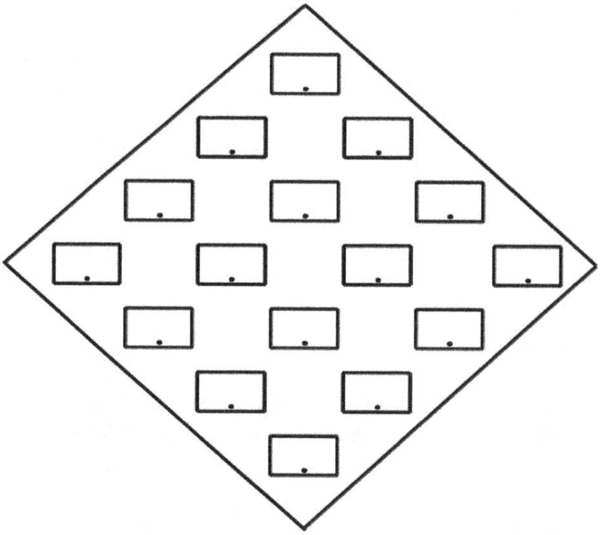

Design Approach:

*Spacing of Elements = 0.5 to 0.9 lambda*
*Number of Elements = 2 to 256*

Construction and Theory of Operation:

*Having the same gain as the square array, the advantage of a diamond array is lower sidelobes. This is due to the fact that there is less energy in the principle planes of the array where the sidelobes are. Spacing is the same between elements, also, the feed network can be co-planar but is usually on a lower layer with pins from the feed point of the individual elements down to the feed system.*

*Azimuth Pattern*

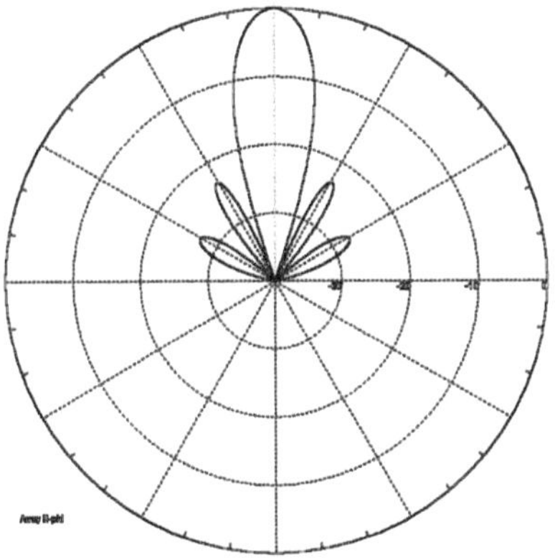

*Elevation Pattern*

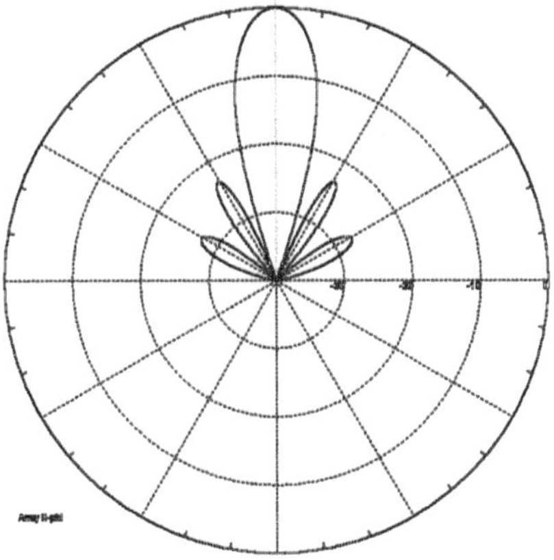

Special Considerations:

*Care must be taken in the design of the feed network, for boresite arrays, each element must be fed in phase, within a few degrees. Amplitude distribution is also important, equal amplitudes will create highest gain with higher sidelobes. Tapering amplitude will lower gain slightly but improve sidelobe performance. Test individual elements first, then small subarrays, careful to correct mutual coupling influences.*
*Feed from lower layer.*

Testing:

*For mid sized and small apertures use far field range with transmitter/receiver or network analyzer. Near field and compact ranges work also. Use calibrated reference antenna. Use axial ratio measurements to confirm quality of required polarization. For large apertures, use radio star techniques, holography or satellites with known EIRP. Check Appendix 6 for details.*

*Use VSWR bridge or similar device to optimize match.*

Type: ***Sparsed Array***

Frequency of Operational: *50 Mhz to 50 GHz*
Gain Range: *0 to 35 dBi*
Bandwidth: *~ 5%*
Approximate Power Rating: *100 Kw*
Design Software: *IE3D, Ensemble, Moment Method*

Configuration:

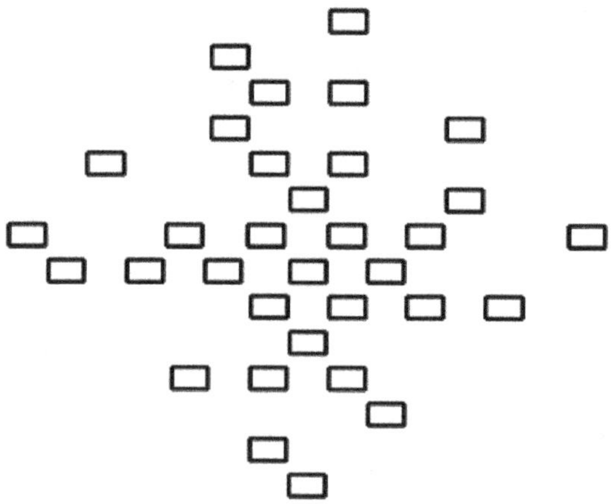

Design Approach:

*Use Random number generator multiplied by Taylor series to define which element are removed. Center of array filled, perimeter very sparse.*

*Spacing of Elements = 0.5 to 0.9 lambda*
*Number of elements = 2 to 256*

Construction and Theory of Operation:

*Having the same gain as the square array, the advantage of a diamond array is lower sidelobes. This is due to the fact that there is less energy in the principle planes of the array where the sidelobes are. Spacing is the same between elements, also, the feed network can be co-planar but is usually on a lower layer with pins from the feed point of the individual elements down to the feed system.*

*Azimuth Pattern*

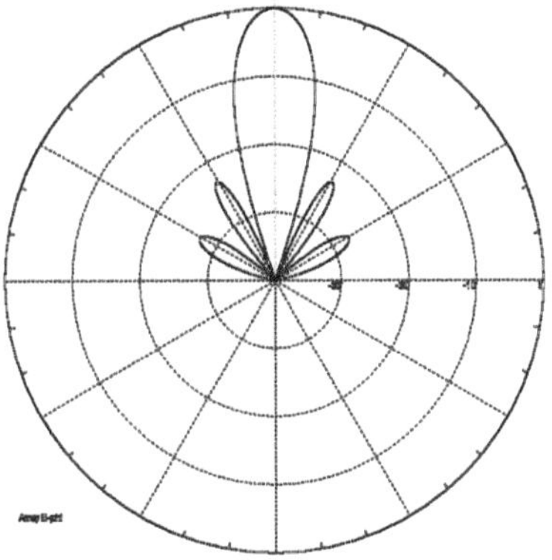

*Elevation Pattern*

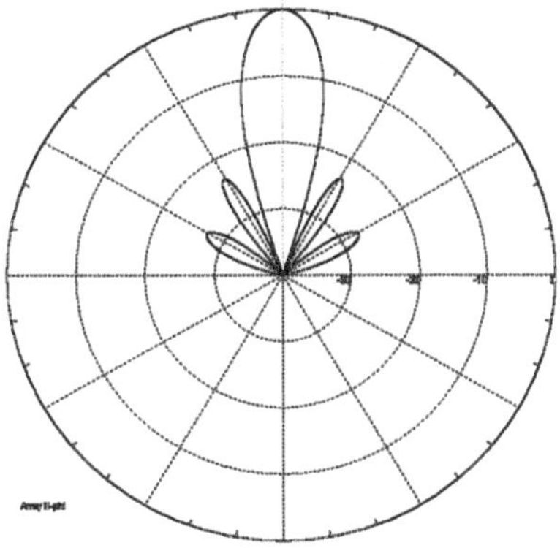

Special Considerations:

*Care must be taken in the design of the feed network, for boresite arrays, each element must be fed in phase, within a few degrees. Amplitude distribution is also important, equal amplitudes will create highest gain with higher sidelobes. Tapering amplitude will lower gain slightly but improve sidelobe performance. Test individual elements first, then small subarrays, careful to correct mutual coupling influences.*
*Feed from lower layer.*

Testing:

*For mid sized and small apertures use far field range with transmitter/receiver or network analyzer. Near field and compact ranges work also. Use calibrated reference antenna. Use axial ratio measurements to confirm quality of required polarization. For large apertures, use radio star techniques, holography or satellites with known EIRP. Check Appendix 6 for details.*

*Use VSWR bridge or similar device to optimize match.*

| | |
|---|---|
| Type: | ***Conformal Array*** |
| Frequency of Operational: | *50 Mhz to 50 GHz* |
| Gain Range: | *0 to 3 dBi* |
| Bandwidth: | *~ 5%* |
| Approximate Power Rating: | *10 Kw* |
| Design Software: | *IE3D, Ensemble, Moment Method* |

Configuration:

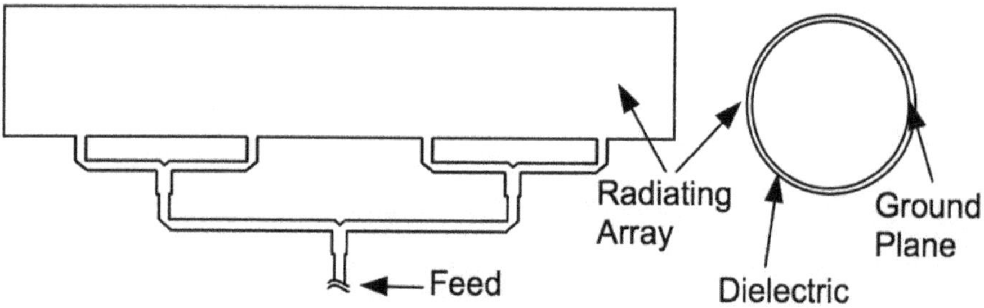


Design Approach:

*Spacing of Edge Feeds = 1 lambda*
*Number of elements = 2 to 32*

Construction and Theory of Operation:

*This phased array is made up of a single very wide microstrip element. Feed every wavelength, in phase and wrap around a curved surface yields an antenna with wide beamwidth, good gain and very low profile. This antenna is used commonly in rockets and satellites, where omni beams are required. All surfaces are conductive, typically etched from copper clad dielectric material. Feed with a corporate, in phase feed.*

*Azimuth Pattern* *Elevation Pattern*

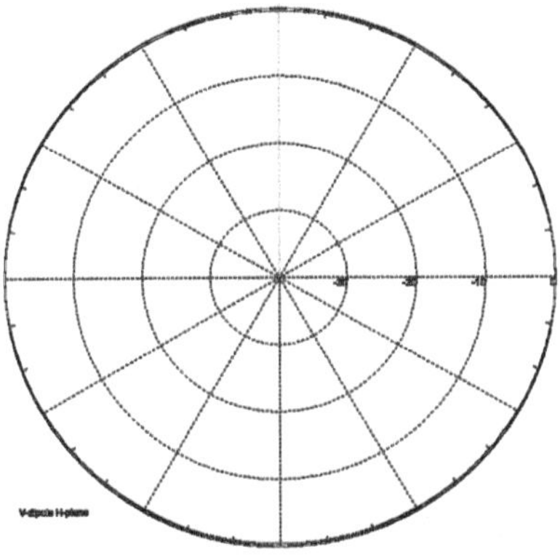

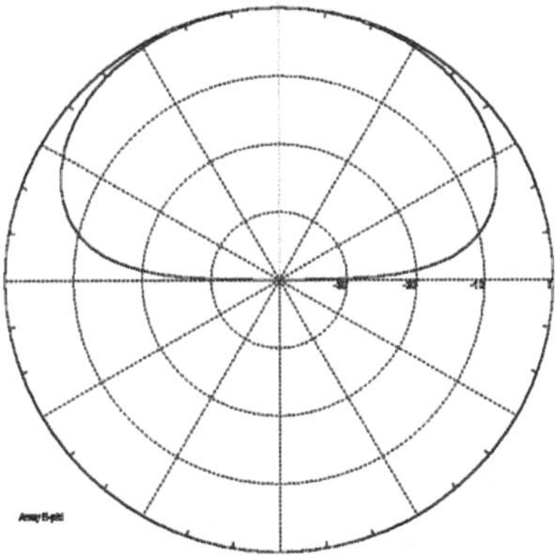

Special Considerations:

*For harsh environments, a radome can be attached onto the surface of the radiating array. This will lower the frequency of resonance, so design the resonant element to compensate. If used on a rocket, expect the plume to attenuate the signal.*

Testing:

*For mid sized and small apertures use far field range with transmitter/receiver or network analyzer. Near field and compact ranges work also. Use calibrated reference antenna. Use axial ratio measurements to confirm quality of required polarization. For large apertures, use radio star techniques, holography or satellites with known EIRP. Check Appendix 6 for details.*

*Use VSWR bridge or similar device to optimize match.*

Type: ***Interferometer***

Frequency of Operational: *5 Mhz to 500 GHz*
Gain Range: *0 to >65 dBi*
Bandwidth: *~ 5%*
Approximate Power Rating: *1 Kw*
Design Software: *IE3D, Ensemble, Moment Method*

Configuration:

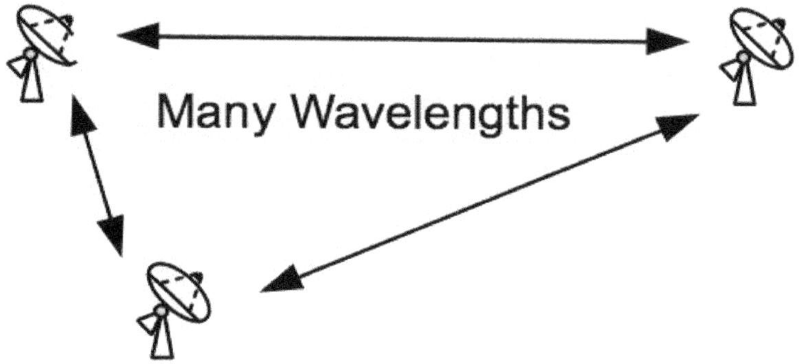


Design Approach:

*Spacing between elements = >10 lambda*
*Number of elements = 2 to N*

Construction and Theory of Operation:

*This antenna array is made up of a multitude of elements with significant spacing between them. The resultant antenna pattern is a combination of the element pattern times the pattern of an antenna the size of the baseline between the two elements. The finer structure is useful for observing the minute structure of radio astronomical sources. Arrays of this type can be spread of hundreds of feet to thousands of miles. Resolutions on the order of tens of microarcseconds are achievable. Good examples of this type of antenna are the Very Long Baseline Array (across the U.S.) and the Very Large Array (Socorro, New Mexico).*

*Azimuth Pattern*

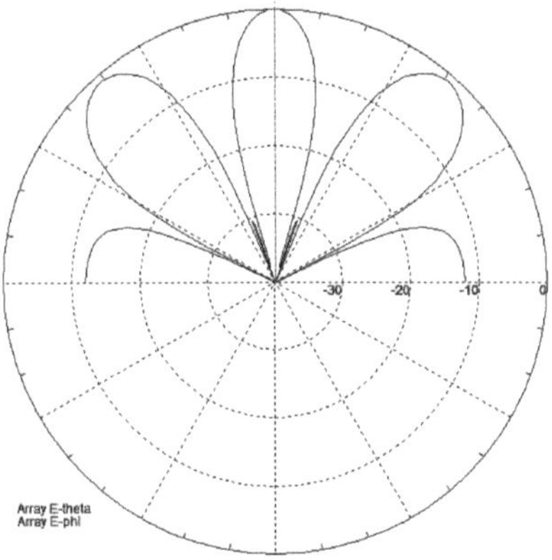


*Elevation Pattern*

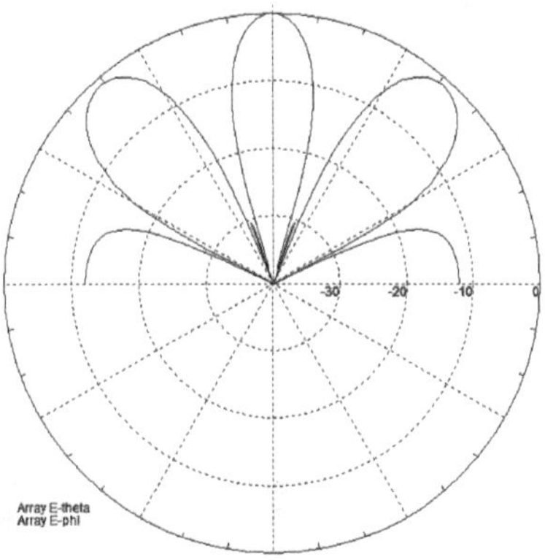


Special Considerations:

*Phase must be controlled to a high degree, either by using one frequency reference and connecting the antennas with coax or RF links. The other option is to use very stable oscillators (like Cesium Beam or Hydrogen Maser) at each element to maintain a very stable phase reference.*

Testing:

*Use calibrated reference antenna. Use axial ratio measurements to confirm quality of required polarization. For large apertures, use radio star techniques, holography or satellites with known EIRP. Check Appendix 6 for details.*

Type: ***Binomial Spaced Array***

Frequency of Operational: *5 Mhz to 500 GHz*
Gain Range: *0 to 35 dBi*
Bandwidth: *~ 5%*
Approximate Power Rating: *10 Kw*
Design Software: *IE3D, Ensemble, Moment Method*

Configuration:

Design Approach:

*Spacing between elements = Binomial spacing between elements*

*Number of elements = 2 to 64+*

Construction and Theory of Operation:

*This antenna array minimizes the number of elements and retains much of the antenna beam performance by illuminating the aperture in a tapered way. The beamwidth is slightly oblated by this technique and the gain is slightly lower. This style of array was used by radio astronomers to achieve high resolution over large baselines without prohibitive costs.*

*Azimuth Pattern*

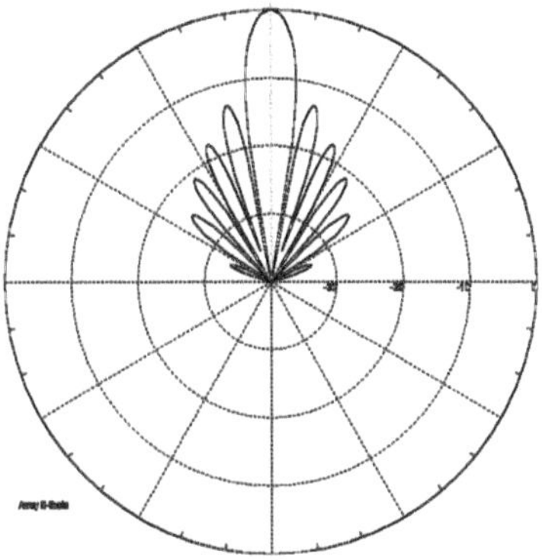

*Elevation Pattern*

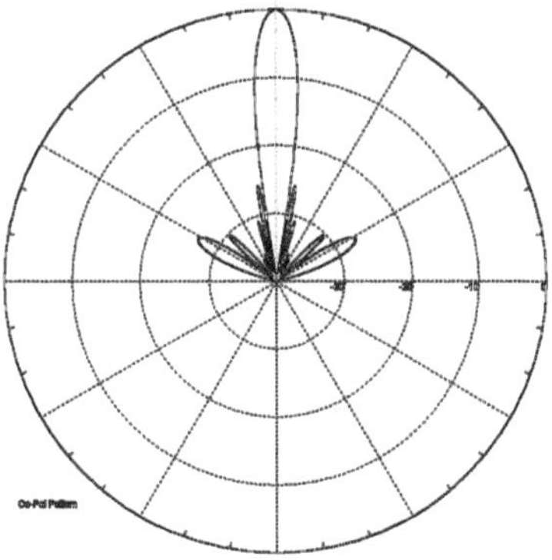

Special Considerations:

*Model this antenna and compare to other layouts before committing to this design. For large elements, this works well, especially if the elements are placed on tracks to make different baseline layouts.*

Testing:

*For mid sized and small apertures use far field range with transmitter/receiver or network analyzer. Near field and compact ranges work also. Use calibrated reference antenna. Use axial ratio measurements to confirm quality of required polarization. For large apertures, use radio star techniques, holography or satellites with known EIRP. Check Appendix 6 for details.*

Type: ***OTH-B (Over the Horizon - Backscatter)***

Frequency of Operational: 1 *Mhz to 500 MHz*
Gain Range: *0 to 35+ dBi*
Bandwidth: *~ 5%*
Approximate Power Rating: *10 Kw*
Design Software: *IE3D, Moment Method*

Configuration:

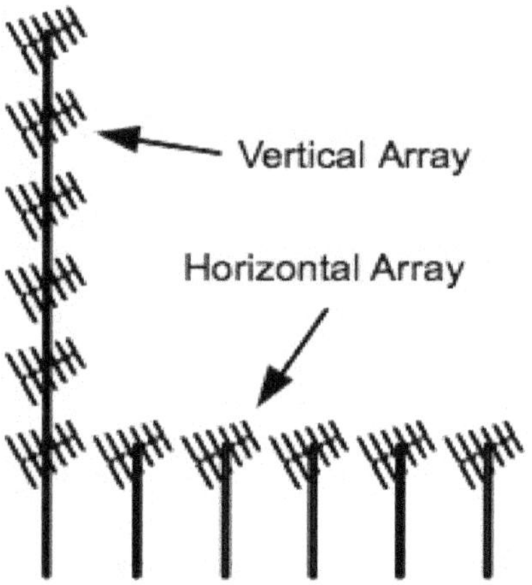


Design Approach:

*Spacing between elements = 0.5 to 0.8 Lambda*
*Number of elements = 2 to 16*

Construction and Theory of Operation:

*This is a HF array operating in the range of 1 to 500 MHz, the first versions of this antenna worked exclusively in the shortwave or 1 to 30 MHz range and took advantage of the reflective properties of the ionosphere. Modern examples of this type of antenna are used as a radar component to track ships and aircraft over distances far exceeding line of sight.*

*Azimuth Pattern* *Elevation Pattern*

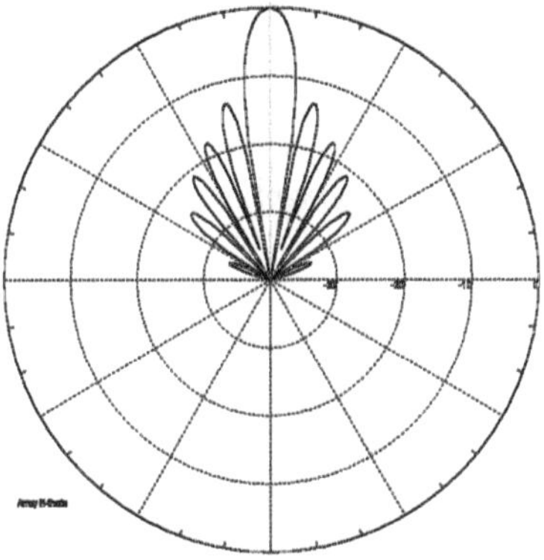

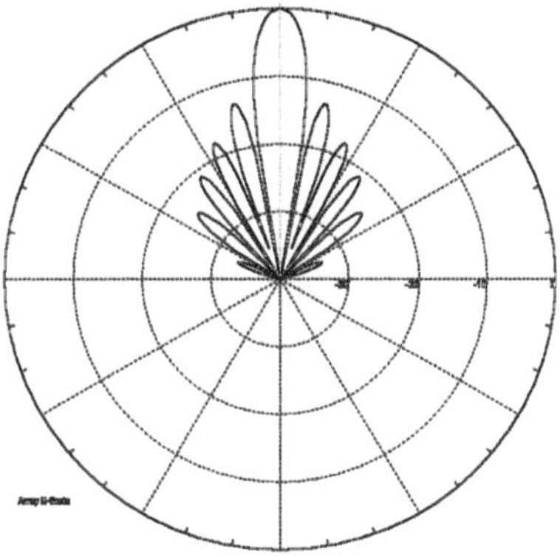

Special Considerations:

*Array of this type are typically large, in a particular example the vertical array was placed on a 500 foot tower, the horizontal array was placed on 100 foot towers. Low loss coax is required and a phasing mechanism is usually employed to steer the beam over various azimuth angles for scanning purposes and over the vertical to optimize propagation to a certain range.*

Testing:

*For mid sized and small apertures use far field range with transmitter/receiver or network analyzer. Near field and compact ranges work also. Use calibrated reference antenna. Use axial ratio measurements to confirm quality of required polarization. For large apertures, use radio star techniques, holography or satellites with known EIRP. Check Appendix 6 for details. For variations of performance due to propagation use beacons at various distances that emit id codes at specific times and frequencies. Use array and receiver to monitor beacons to achieve real time propagation values.*

Type: ***Sea States Studies Array***

Frequency of Operational: *5 Mhz to 50 MHz*
Gain Range: *0 to 15 dBi*
Bandwidth: *~ 5%*
Approximate Power Rating: *10 Kw*
Design Software: *IE3D, Ensemble, Moment Method*

Configuration:

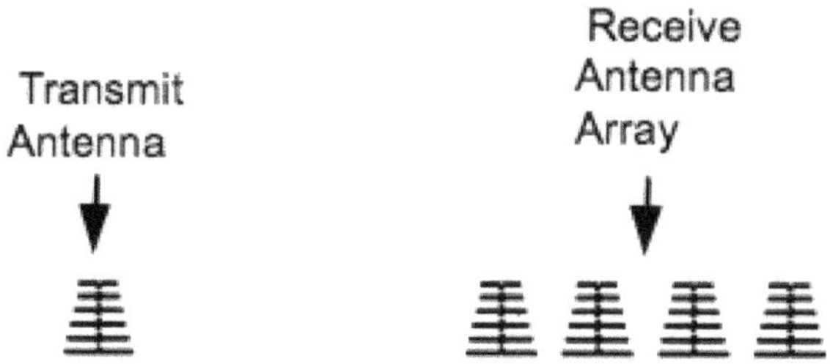


Design Approach:

*Spacing between elements = 0.5 to 0.8 Lambda*
*Number of elements = 2 to 16*

Construction and Theory of Operation:

*The is a bi-static radar arrangement where there is a separate transmit and receive antenna. These radars are situated close to a coast, on a beach or nearby hillside. The transmit antenna sends out a pulse that is received by the receive array. Using phase shifters or digitizing the individual receive element response to determine time of arrival allow the mapping of a wide swath of azimuth angle. Doppler information is detected to determine the direction and magnitude of the sea currents.*

*Azimuth Pattern*

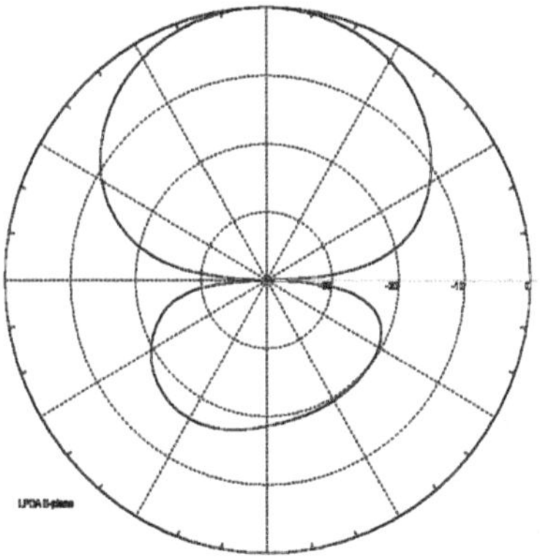

*Elevation Pattern*

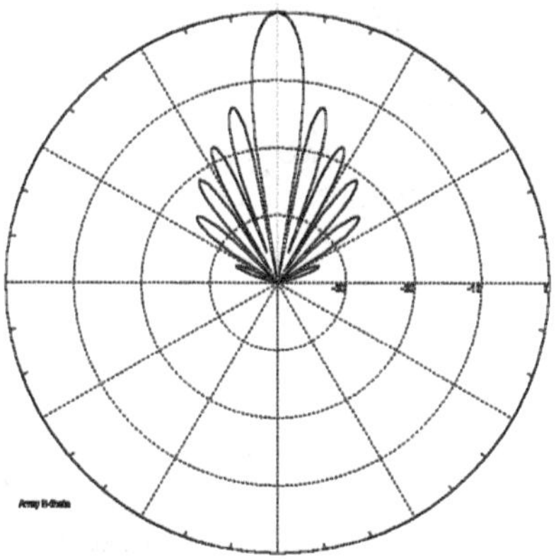

Special Considerations:

*Due to the significant ranges these radars are capable of, care must be taken to avoid range aliasing. This is due to echos from a greater distance than one the operators are interested in coming back and entering into the received data of a closer range gate.*

Testing:

*For mid sized and small apertures use far field range with transmitter/receiver or network analyzer. Near field and compact ranges work also. Use calibrated reference antenna. Use axial ratio measurements to confirm quality of required polarization. For large apertures, use radio star techniques, holography or satellites with known EIRP. Check Appendix 6 for details. For variations of performance due to prorogation use beacons at various distances that emit id codes at specific times and frequencies. Use array and receiver to monitor beacons to achieve real time propagation values.*

| | |
|---|---|
| Type: | ***Patch Array*** |
| Frequency of Operational: | *50 Mhz to 500 GHz* |
| Gain Range: | *0 to 42 dBi* |
| Bandwidth: | *~ 15%* |
| Approximate Power Rating: | *1 Kw* |
| Design Software: | *IE3D, Ensemble, Moment Method* |

Configuration:

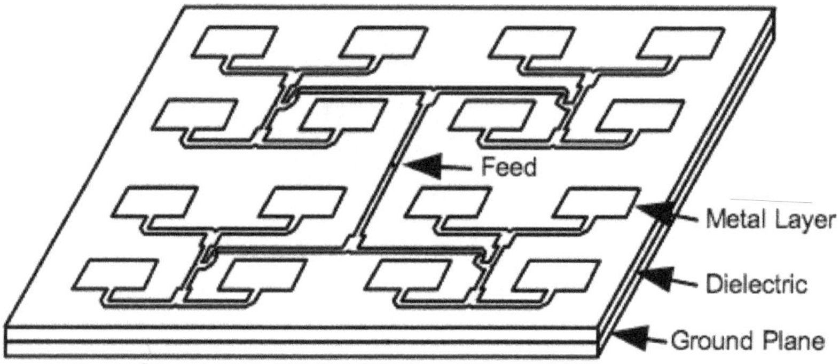


Design Approach:

*Spacing between elements = 0.5 to 0.8 Lambda*
*Number of elements = 2 to 256*

Construction and Theory of Operation:

*This is microstrip phased array, typically etched on copper clad dielectric material. The metal layer can also be etched on a thin material like G10 and placed on a dielectric layer like foam or 'hexcell' and thence on a conductive ground plane. Due to feed line losses, element numbers greater than 256 start to become inefficient. Array can be fed co planar or from a separate layer below.*

*Azimuth Pattern*

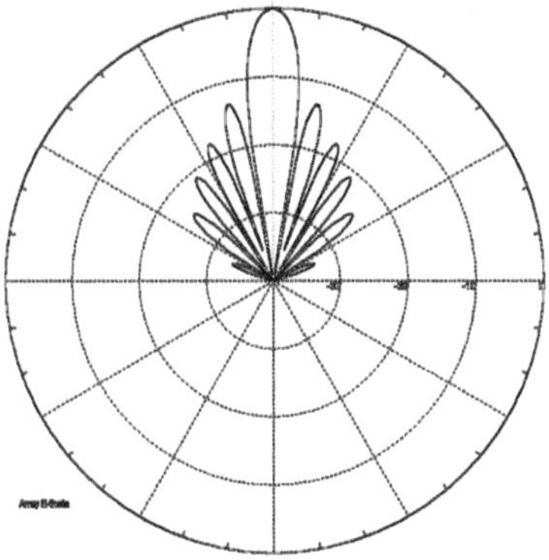

*Elevation Pattern*

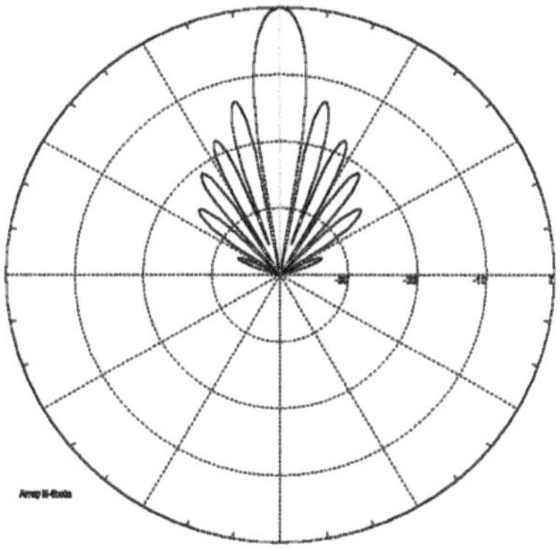

Special Considerations:

*Place radome over exposed elements to protect them from the weather if used outside. Use patch and line calculations in microstrip antenna section to design.*

Testing:

*For mid sized and small apertures use far field range with transmitter/receiver or network analyzer. Near field and compact ranges work also. Use calibrated reference antenna. Use axial ratio measurements to confirm quality of required polarization. For large apertures, use radio star techniques, holography or satellites with known EIRP. Check antenna testing appendix for details.*

| | |
|---|---|
| Type: | ***Helix Array*** |
| Frequency of Operational: | *5 Mhz to 50 GHz* |
| Gain Range: | *0 to 40 dBic* |
| Bandwidth: | *~ 25%* |
| Approximate Power Rating: | *10 Kw* |
| Design Software: | *IE3D, Moment Method* |

Configuration:

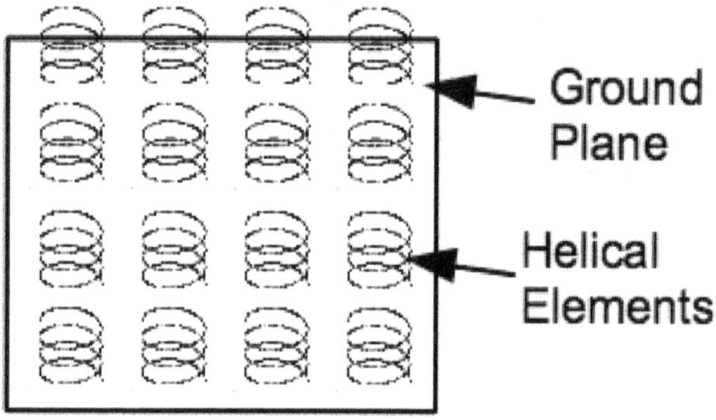


Design Approach:

*Spacing between elements = 0.5 to 0.8 Lambda*
*Number of elements = 2 to 256*
*Element design, use axial mode type*
*Right hand or Left hand circular performance is good, if linear is required, add right to left hand.*

Construction and Theory of Operation:

*An array of helices allows for wide band and steerable performance. In addition, circular polarization is easily achieved. Feed system is typically behind ground plane, using phase shifters and power combiner for steerable beam or just power combiner for broadside beam.*

*Azimuth Pattern* *Elevation Pattern*

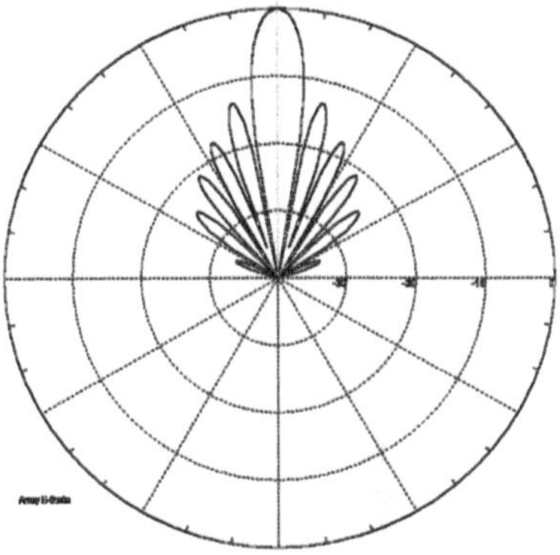

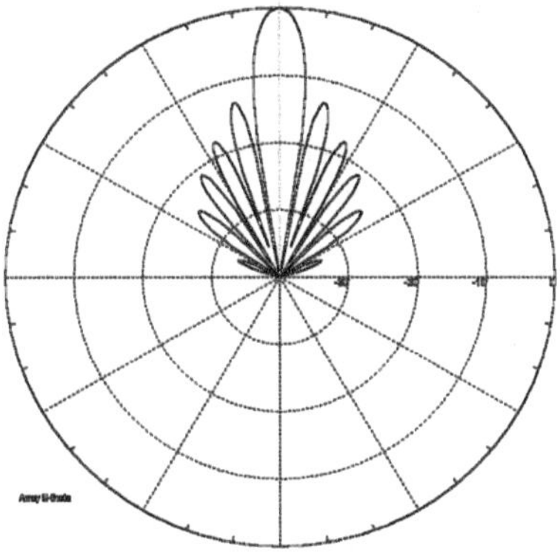

Special Considerations:

*Mutual coupling and scan blindness can affect performance. If both right and left hand circular polarizations are required, place left hand row next to right hand rows to minimize mutual coupling. Also, the use of cups and tapers can improve performance over bandwidths of better than 25%.*

Testing:

*For mid sized and small apertures use far field range with transmitter/receiver or network analyzer. Near field and compact ranges work also. Use calibrated reference antenna. Use axial ratio measurements to confirm quality of required polarization. For large apertures, use radio star techniques, holography or satellites with known EIRP. Check antenna testing appendix for details.*

Type: ***Drooped Dipole Array***

Frequency of Operational: *1 Mhz to 5 GHz*
Gain Range: *0 to 40 dBi*
Bandwidth: *~ 5%*
Approximate Power Rating: *10 Kw*
Design Software: *IE3D, Moment Method, NEC*

Configuration:

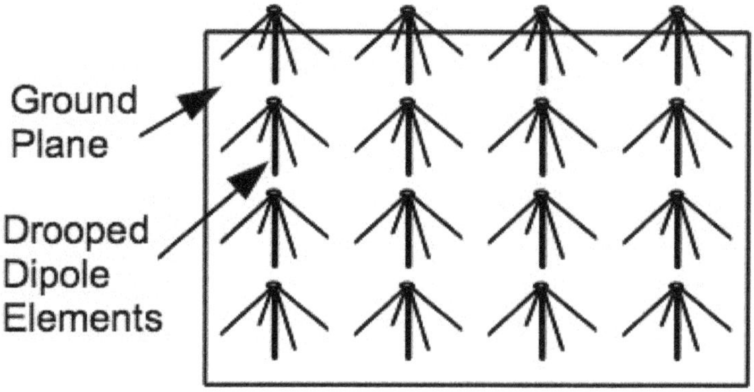


Design Approach:

*Spacing between elements = 0.5 to 0.8 Lambda*
*Number of elements = 2 to 256*
*Element design, use dual drooped dipole type, make sure dipole end is greater than .125 lambda above ground plane*
*Right hand or Left hand circular performance is possible with orthogonal dipoles, or vertical and horizontal.*

Construction and Theory of Operation:

*An array of drooped dipoles allows for reasonable bandwidth and steerable performance. In addition, circular polarization is easily achieved. Feed system is typically behind ground plane, using phase shifters and power combiner for steerable beam or just power combiner for broadside beam.*

*Azimuth Pattern*

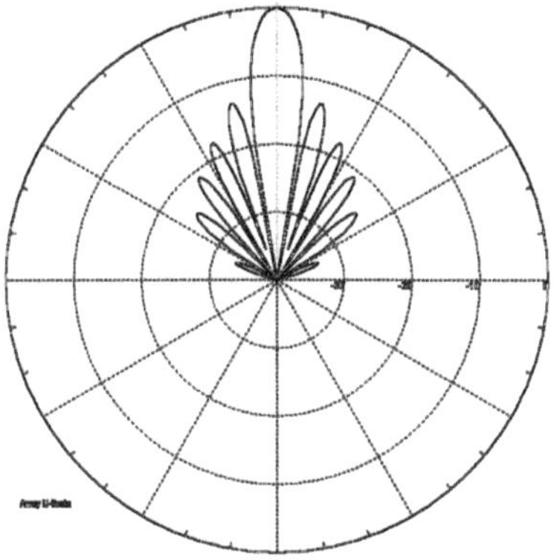

*Elevation Pattern*

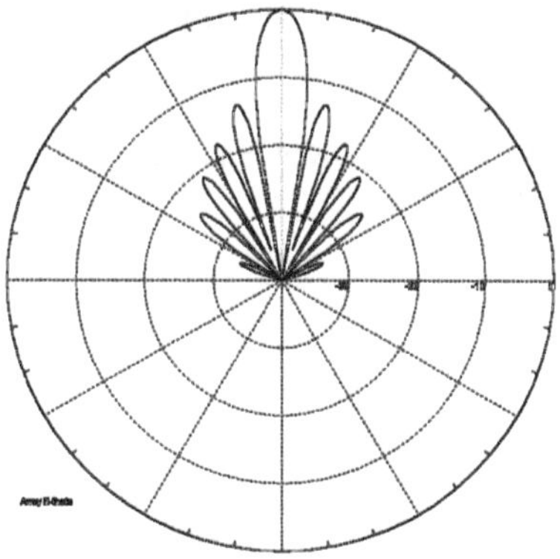

Special Considerations:

*Element to edge of ground plane spacing should be at least ½ lambda. Dipole ends should be at least 1/8 lambda from ground plane. If array is used for circular polarization, expect 1/3 bandwidth over decent axial ration.*

Testing:

*For mid sized and small apertures use far field range with transmitter/receiver or network analyzer. Near field and compact ranges work also. Use calibrated reference antenna. Use axial ratio measurements to confirm quality of required polarization. For large apertures, use radio star techniques, holography or satellites with known EIRP. Check antenna testing appendix for details.*

Type: ***Horn Array***

Frequency of Operational: *50 Mhz to 500 GHz*
Gain Range: *0 to 42 dBi*
Bandwidth: *~ 25%*
Approximate Power Rating: *1 Mw*
Design Software: *IE3D, GTD, Physical Optics, Geometric Optics*

Configuration:

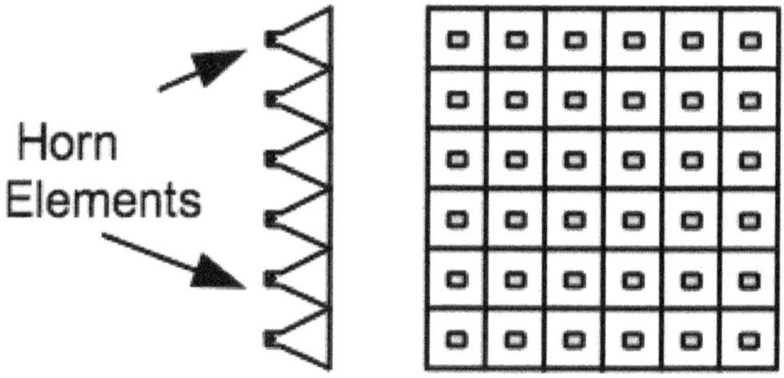


Design Approach:

*Spacing between elements = 0.5 to 0.8 Lambda*
*Number of elements = 2 to 256*
*Element design, use standard square or rectangular horn design*
*Right hand or Left hand circular performance is possible with orthogonal feed ports, or vertical and horizontal.*

Construction and Theory of Operation:

*An array of horns allows for reasonable bandwidth and steerable performance. In addition, circular polarization is easily achieved. Feed system is typically behind ground plane, using phase shifters and power combiner for steerable beam or just power combiner for broadside beam.*

*Azimuth Pattern*

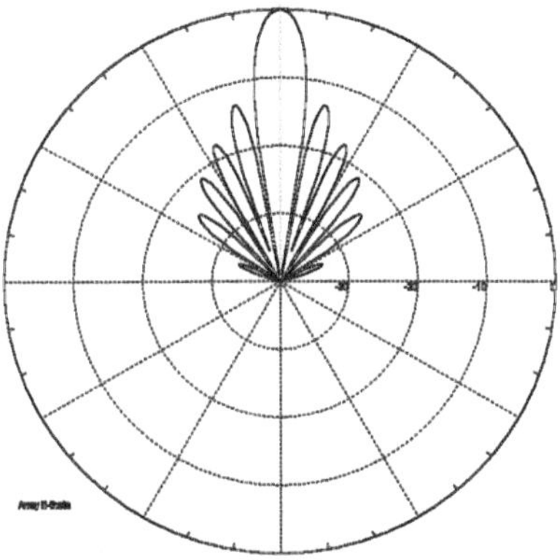

*Elevation Pattern*

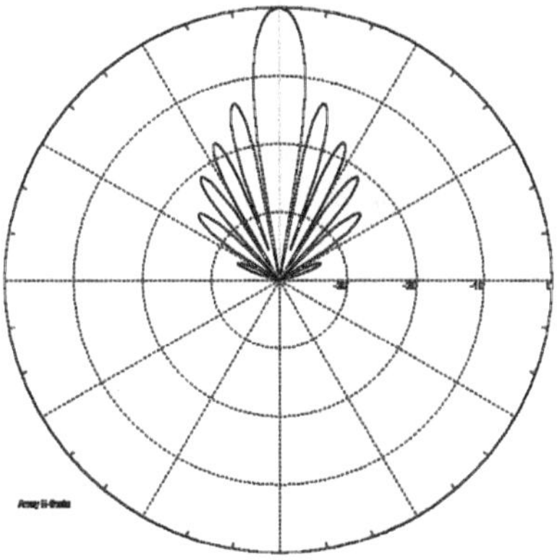

Special Considerations:

*Make sure horn edges touch. Add radome to keep horns clean. Mutual coupling can be an issue, check for scan blindness and expected operation at frequency of interest.*

Testing:

*For mid sized and small apertures use far field range with transmitter/receiver or network analyzer. Near field and compact ranges work also. Use calibrated reference antenna. Use axial ratio measurements to confirm quality of required polarization. For large apertures, use radio star techniques, holography or satellites with known EIRP. Check antenna testing appendix for details.*

Type: ***Slot Array***

Frequency of Operational: *5 Mhz to 50 MHz*
Gain Range: *0 to 2.12 dBi*
Bandwidth: *~ 5%*
Approximate Power Rating: *1 Kw*
Design Software: *IE3D, Ensemble, Moment Method*

Configuration:

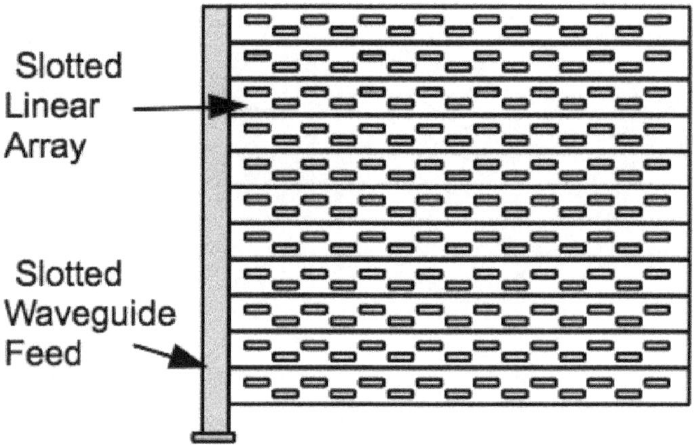


Design Approach:

*Spacing between slots = 0.5 guide Lambda*
*Number of slots = 2 to 256*
*Slot design,*
*For non-resonant design, frequency of operation other than center steers beam if slotted waveguide is used, beam steers diagonally across aperture*

Construction and Theory of Operation:

*An array of slots allows for reasonable bandwidth and steerable performance if designed for non-resonant mode. Feed system is typically behind ground plane for corporate type feed.*

*Azimuth Pattern*

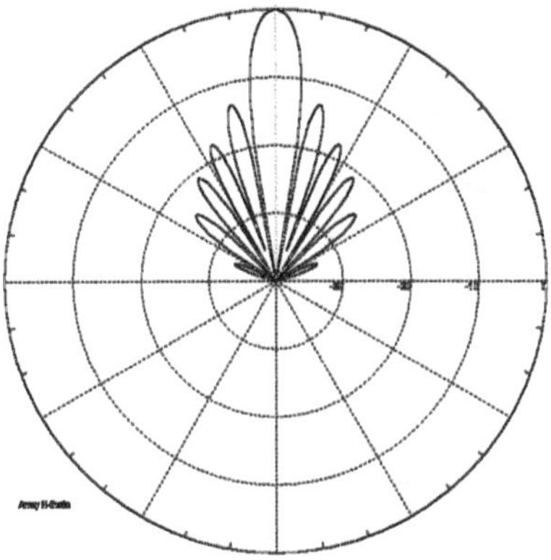

*Elevation Pattern*

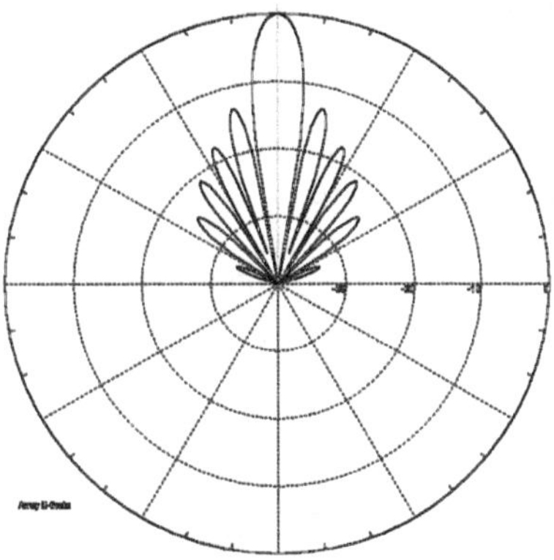

*Special Considerations:*

*Two types of arrays can be designed, resonant and non-resonant. Resonant have shorting plates at ends of waveguides, non-resonant have load. Non-resonant have greater bandwidth operation.*

Testing:

*For mid sized and small apertures use far field range with transmitter/receiver or network analyzer. Near field and compact ranges work also. Use calibrated reference antenna. Use axial ratio measurements to confirm quality of required polarization. For large apertures, use radio star techniques, holography or satellites with known EIRP. Check antenna testing appendix for details.*

Type: ***Short Backfire Array***

Frequency of Operational: *100 MHz to 50 GHz*
Gain Range: *0 to 35 dBi*
Bandwidth: *~ 15%*
Approximate Power Rating: *10 Kw*
Design Software: *IE3D, Ensemble, Moment Method*

Configuration:

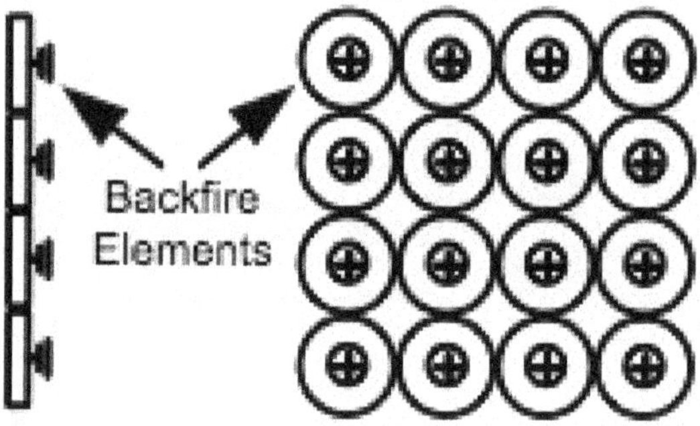


Design Approach:

*Spacing between elements = 2 lambda*
*Number of slots = 2 to 256*
*Use standard backfire antenna formulae*

Construction and Theory of Operation:

*An array of backfire elements allows for reasonable bandwidth and steerable performance. In addition, circular polarization is easily achieved. Feed system is typically behind ground plane, using phase shifters and power combiner for steerable beam or just power combiner for broadside beam.*

*Azimuth Pattern*

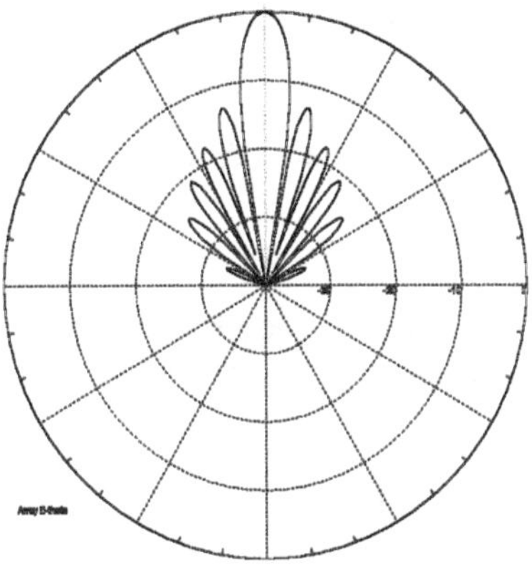

*Elevation Pattern*

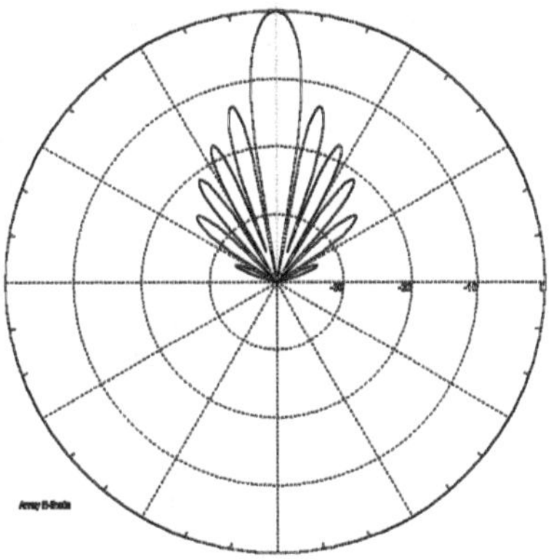

Special Considerations:

*Place elements next to each other. One ground plane can be used for complete array. Vertical, Linear, Right Hand, Left Hand polarizations are available. Feed network is behind ground plane.*

Testing:

*For mid sized and small apertures use far field range with transmitter/receiver or network analyzer. Near field and compact ranges work also. Use calibrated reference antenna. Use axial ratio measurements to confirm quality of required polarization. For large apertures, use radio star techniques, holography or satellites with known EIRP. Check antenna testing appendix for details.*

**Questions:**

1. Suggest an antenna array that has the following characteristics:

   a. Frequency of 200 to 250 MHz
   b. Gain of 35 dBi
   c. Linear Polarization

2. Your boss tells you price is a huge factor in the design of antenna array with the following characteristics:

   a. Frequency of 20 MHz
   b. Gain of 20 dBi
   c. Linear Polarization

   Can a wire array be an option?

3. Suggest an antenna array for the next generation rocket booster that is capable of going to the moon, with the following properties:

   a. Frequency of 2200 MHz
   b. Gain of 13 dBic
   c. Circular Polarization

4. Design a phased array with three independent moving elements. How can you keep the array in phase ?

5. Does a phased array have to made of the same elements ? Explain the consequences of using different ones.

6. What happens when the element to element distance is one wavelength, or multiples thereof ?

7. Do the same design rules apply for phased arrays in acoustic or visible light range?

# Chapter 15 – Direction Finding Antennas

Adcock Array from World War II (47)

A Doppler Type DF Array (35)

**Fundamentals**

Direction finding antennas, as the name implies are used to find the direction to the source of a radio transmission. This technology is very useful when looking for a ship in distress, aircraft in distress or other transmissions. There are several approaches to direction finding as defined below:

**Mechanically Steered**

This is the oldest and simplest technique for direction finding whereby an antenna or antenna array is moved in azimuth to find the source of transmissions. This can be done with a handheld antenna or on an azimuth scanning platform. First used in World War I, enemy transmission could be located on a map by using two or more directional antennas and getting a crossing bearing.

### Electrically Steered

Modern day direction finder antennas use electronic methods to steer the beam, which can be significantly faster than mechanical means. The allows for the quick determination of transmission source azimuth and / or elevation. In many cases a multitude of resonant elements are placed in a circle and turned on and off with PIN diode switches sequentially to create the effect of rotating a large directional antenna. This is the so called acoustic Doppler technique, where an artificial frequency change is spinning the array as mentioned above. The beam can be moved electronically very quickly, in milliseconds for instance.

### ADF

Automatic Direction Finder or ADF is principally found in aircraft and allows the pilots to locate the position of a fixed non directional beacon on the ground. This also allows the determination of wind and range. The ADF receiver uses two antennas, one normally on the belly of the aircraft which gives a vector to the transmitter on the ground; this antenna is typically called the "loop." In early planes the loop actually was a mechanically steered loop, in modern planes the mechanical component is replaced by two orthogonal loops and a component known as a "goniometer" which in effect steers the loop electronically. The loop antenna pattern is symmetrical however and cannot resolve the precise direction of the radio source due to a 180 directional ambiguity. In other words, a source aligned with the antenna from the North will have the same response from the South. The only way to correct this ambiguity is to have a second antenna, on aircraft this is known as the "sense" antenna and is usually a long wire spread between the fuselage and near the top of the vertical stabilizer. The phase of the sense will only add constructively in only one direction. The output of the receiver with the resolved direction controls a dial with a needle on a compass card. Sometimes the compass card is slaved to the magnetic heading the aircraft is flying. The pilot knows by looking at this instrument, the position of the radio station relative to the aircraft and can thus fly to it.

Using the ADF allow the pilot to gauge the wind direction. The technique is to fly directly to a non directional beacon, while holding the magnetic course, if the ADF needle starts to drift left or right, this is an indication of the rough position of the wind. The magnitude of the change indicates how strong the wind is.

While flying directly to the non directional beacon, if a pilot takes a 90 degree turn left or right and times how long it takes to deflect the needle 15 degrees more than the 90 degrees, the number of seconds divided by 10 is the rough distance from the aircraft to the beacon.

**Adcock**

An Adcock type of array is composed of at least 4 monopoles arranged (in the simplest form) ¼ wavelength apart in a a square formation, over a ground plane. In modern applications, each antenna element is turned on in rapid succession to simulate the physical movement of the antenna, thereby causing a doppler shift applied to the signal of interest. Using digital technology a correlation between the sequence of antenna activation (typically using a PIN switch) and the azimuth angle to the radiating source can be discerned. Acquisition times can be in the micro or milli seconds.

Applications for this type of array is in the nautical, military, telematics and amateur areas. In the early days of flight, these arrays were used in a transmission mode to produce a series of "beams" that pilots in the 1930s to 1960s followed for navigation. Listening on the frequency of transmission, the pilots would hear an "A" or dot/dash in Morse code, or a "N" or dash/dot. The letter would tell the pilots if they were left or right of course. A steady hum would indicated that they were precisely on course.

**Loop**

Loop antennas, either singly and rotated mechanically, or two crossed loops, rotated electronically, can be used to determine the azimuth to a transmitting source. This is useful for navigation and other directional finding operations.

These loops are greater than .1 wavelength in diameter and as a result are more sensitive to the electric field of a radio wave. The nulls are perpendicular to the plane of the antennas. Physically rotating the loop allows for the azimuthal determination of the transmitter by nulling out the signal. Because there are two nulls on this type of antenna, the ambiguity must be resolved somehow. On ships, the general direction to the source is typically known. In the case of aircraft, a second antenna, know as the sense antenna is included to resolve the ambiguity. This sense antenna is usually a wire strung between the fuselage and empennage. The sense antenna is more sensitive than the loop and when combined with the output of the loop antenna adds in phase in the true direction of the transmitter.

Modern design use more creative means to "move" the nulls. Using two loops set at orthogonal angles, high speed determination can be made of the angle to the transmitting source. This can be done with a device known as a goniometer or by digital means. The goniometer is a sphere with a loop of wire around one plane. It is set inside another sphere with two more loop of wire, connected to the antenna loops. By rotating the inside sphere the nulls of the crossed loops can be moved as well, thereby allowing an electronic circuit farther downstream to determine the angle to the source.

Tugboat with Crossed Loops for Direction Finder

**Butler**

Butler arrays are made up of a multitude of antenna elements connected to a series of hybrid circuits with 90 degree phase shifts.

Sampling the output of the antenna array from the various ports of the Butler matrix, one can "select" a multitude of beam positions, thereby allowing the position of the transmitting source to be determined by comparing the amplitudes of the of the output ports.

**Conical Scanning Feed**

Just after WWII, Bell Labs developed a radar system (the M-33) with an X-band Tracking antenna using a lens design. The feed of this antenna was made of a curved waveguide connected to a circular horn feed. This whole unit was physically rotated to create a conically scanning antenna beam. Once the general direction to the object tracked by the radar was determined, the system could be set to autotrack. As the feed was scanned, an error signal composed of +/- x and +/- y voltages was sent to the tracking electronics to determine which way to move the antenna.

**Monopulse**

Using four antenna in close proximity, typically horns touching each other, a pair of horns in the x direction and a pair of horns in the y direction are added in phase and differenced in phase to create error signal for both the x and y directions. The summed channels can be used for the signal of interest while the differenced channels are used for steering. These arrays are found in dish type installations and on military aircraft for threat detection. Performance improvement can be obtained using a Bayless Taper.

**Multipath Radar, Passive FM or TV Type**

A very interesting way of determining the direction to an aircraft, helicopter or airships is to passively monitor nearby FM or TV transmitters, looking for multipath effects. The position of the transmitters needs to be known and it is advantageous to have the transmitters spread around the area to allow a multitude of baselines. For a given transmitter, there is a direct and a reflected signal that can be detected, the time difference of arrival indicate the relative position along the baseline. Multiple baselines resolve the azimuth and elevation of the target. Multiple targets can be tracked using this approach. The electronics for this type of direction finder occupy a small volume using a simple antenna and a multiple channel receiver.

**Time of Flight**

Consider a transmitter at a distance from a series of randomly placed omnidirectional antennas. The time for the transmitted signal, in terms of phase, to reach each antenna (assuming different distances to the transmitter) is different. Therefore, if the phase is sampled and compared between the various antennas, the azimuth and elevation to the transmitter can be determined by simple trigonometry. This technique can be used with as little as three antennas. Using more of course, increases the accuracy of the measurements.

**Wullenweber**

This is a type of antenna known as a Circularly Disposed Antenna Array. It is a large circular antenna array used by the military to triangulate radio signal for radio navigation, intelligence gathering and search and rescue. The antenna consists of several concentric arrays of dipoles or monopoles backed by vertical ground planes. Each circle of antennas is larger in scope as it gets farther away from the center. Using a multitude of phase shifting beam formers, this array, usually very large in size, is used at HF frequencies (2 to 30 MHz) to observe the positions of radio sources. Invented in WWII by German Scientists, it has been used for many years to locate any shortwave station be it stationary, airborne or at sea. Using arrays placed at several positions across the earth, basic triangulation to a transmitting source in the shortwave bands can be determined quickly and accurately.

An example of a Wullenweber Antenna Array

Amelia Earhart's Lockheed 10, with ADF loop just above cockpit

Modern ADF indicator, wide needle points towards radio AM radio station or other non-directional beacon

**Multi-Armed Spiral**

Spiral antennas can be used for DF purposed by recording the phase and amplitude of emitter on each of a four armed spiral. The relationships of phase and time of arrival indicate the position of the source.

Multi spiral approaches, as in a phased array of spirals, have been used as well to take advantage of the same properties mentioned above, but with better resolution.

Adcock Array that works from 20 MHz to 3 GHz and can located emitters in milliseconds

**Phased Array direction finding**

Phased arrays can scan quickly to find a source of RF energy. They can also be used with a matrix for more instantaneous signal arrival angles. Additionally, if the outputs of all elements is digitized, later software analysis will reveal the position of emitters by virtue of their phase relationships. In fact, multiple emitters can be detected by this method. It works like a compound eye on an insect where all angles are detected at the same time.

An interferometer used in the same fashion offers significantly better resolution, but requires more post processing time.

**Design Examples**

The following pages contain examples of antennas discussed in this chapter. These examples represent a first order design approach. For a more complete and optimized antenna solution, use the example closest to the requirement needs and optimize by using good simulation software, creating prototypes, testing and finally modifying prototypes to optimize performance.

Type: ***Simple DF Loop***

Frequency of Operational: *10 Mhz to 50 GHz*
Gain Range: *2 to 6dBi*
Bandwidth: *~ 5 %*
Approximate Power Rating: *NA, Receive only*
*Design Software:* *IE3D, Moment Method, NEC*

Configuration:

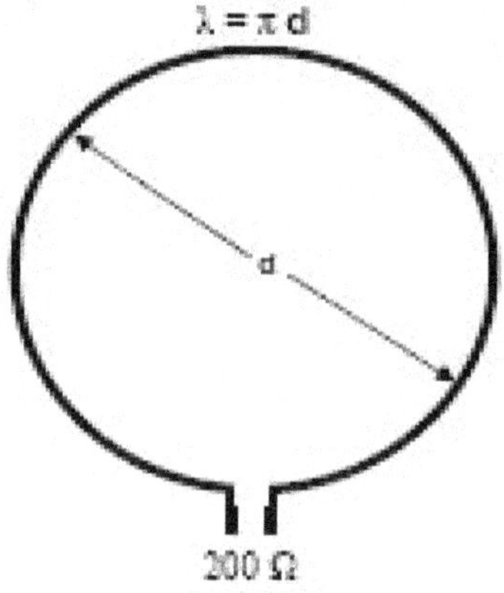


Design Approach:

*Diameter of Loop = 0.1 lambda*
*Thickness of Loop = 0.005 lambda*

Construction and Theory of Operation:

*This antenna is a basic loop in construction, but is moved mechanically around the vertical axis to find a null point in the direction of the emitter. In some design, the loop is moved with the aid of a motor in conjunction with a null position amplifier that moves the loop to a null and corrects a 1, 180 degree confusion with a separate "sense" antenna. In other designs, two loops can be constructed orthogonally and move electrically by use of a goniometer to archive the same results. Care must be taken to make the loop no larger than 0.1 lambda to avoid loss of deep nulls.*

*Azimuth Pattern*

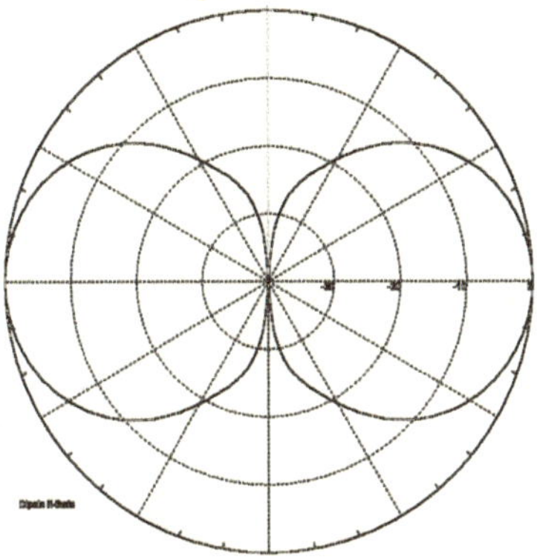

*Elevation Pattern*

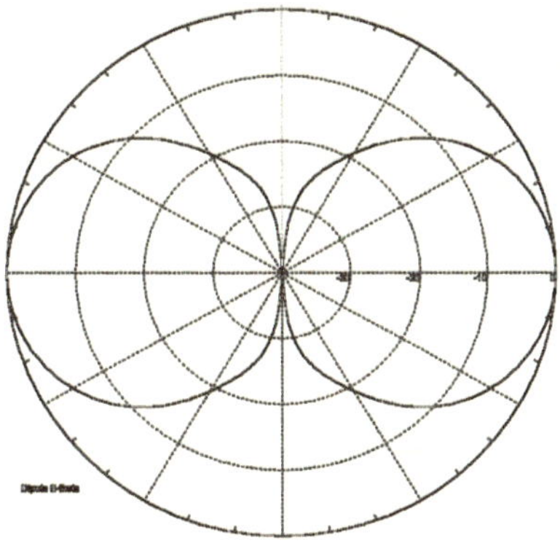

Special Considerations:

*Impedance match with a broadband device like a transformer or balun to optimize performance. Place on a rotator to use as a direction finding element, along with a "sense" antenna to resolve the 180 degree ambiguity from a two lobed pattern.*

Testing:

*Use far field range to measure pattern. Considering the small aperture relative to wavelength, the distance to the source antenna will not be great. Check Appendix 6 for more details. Examine the details of the null as this will allow direction finding optimization. The null should be deep and have a limited angle.*

Type: ***Doppler DF Array***

Frequency of Operational: *5 Mhz to 50 GHz*
Gain Range: *0 to 3 dBi*
Bandwidth: *~ 10%*
Approximate Power Rating: *NA, Receive Only*
Design Software: *IE3D, Moment Method, NEC*

Configuration:

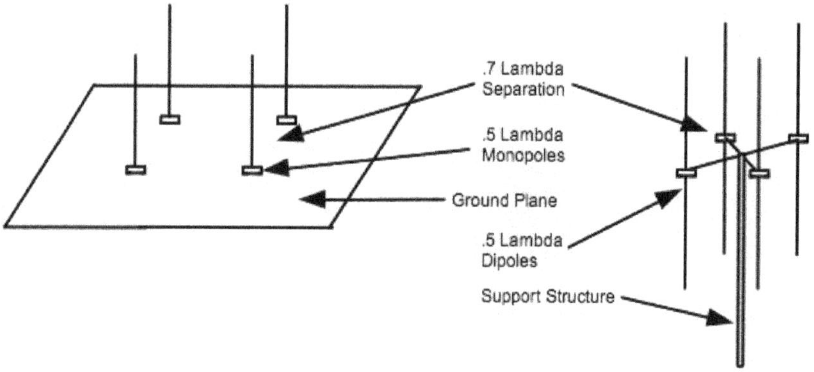


Design Approach:

*Length of Monopoles = 0.5 lambda*
*Separation between Monopoles = 0.7 lambda*

Construction and Theory of Operation:

*Essentially a monopole array whose elements are switch on and off in rapid succession to create a doppler shift in the received signal. The switching is done in a circular pattern with a phase detector in the receiver that resolves angle of arrival accurately. Modern systems scan in the millisecond time range and use digital electronics to determine bearing. Higher frequency systems can be surrounded by lower frequency systems to enhance bandwidth. Switching is done typically by PIN diodes driven by digital electronics.*

*Azimuth Pattern*

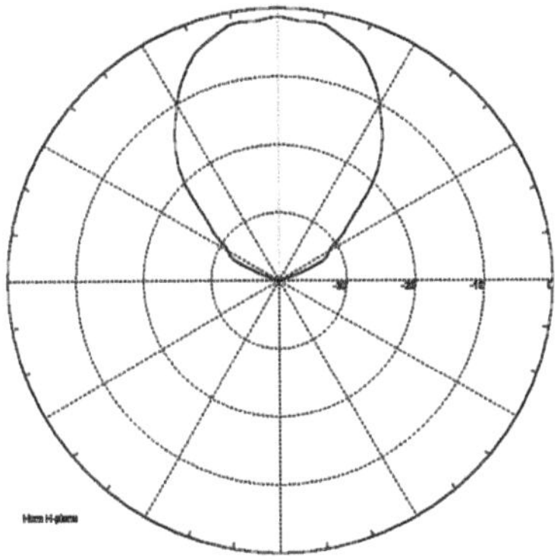

*Elevation Pattern*

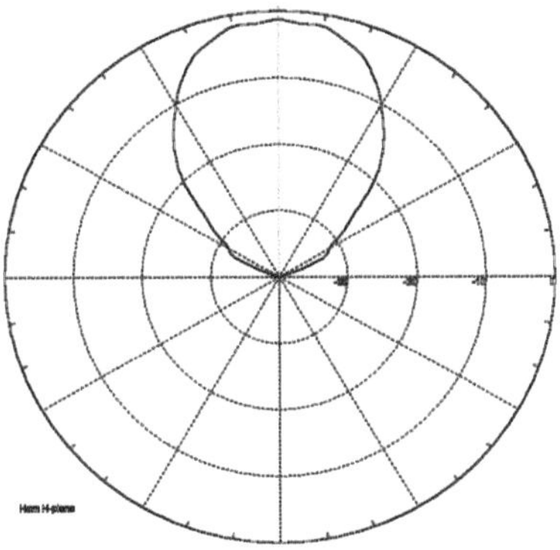

Special Considerations:

*Mare sure ground plane is large enough to create good patterns. Dipoles can substitute for monopoles if ground plane is not available. Place on cross beam support structure and extend support structure into the air.*

Testing:

*Use far field range to measure pattern. Considering the small aperture relative to wavelength, the distance to the source antenna will not be great. Check Appendix 6 for more details. Rotate array relative to source to verify the angle of arrival is accurate.*

Type: **Butler DF Array**

Frequency of Operational: *5 Mhz to 50 GHz*
Gain Range: *0 to 32 dBi*
Bandwidth: *~ 5%*
Approximate Power Rating: *1 Kw*
Design Software: *IE3D, Moment Method, NEC*

Configuration:

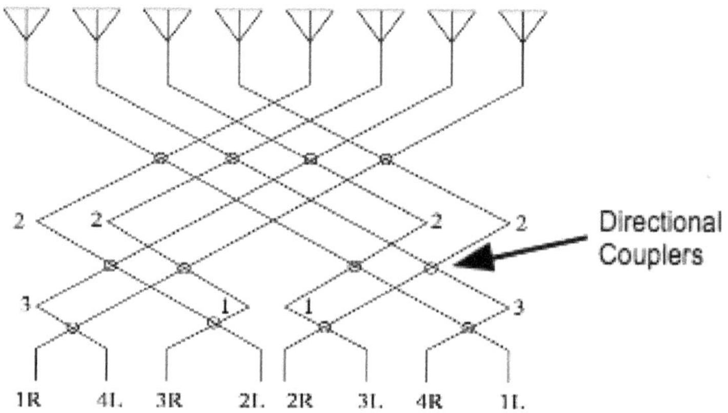


Design Approach:

*Types of elements = 2 to 64+, although matrix gets very complicated with large arrays*
*Separation between Monopoles = 0.7 lambda*

Construction and Theory of Operation:

*A series of elements made of either narrow or wide beam antennas can be arrayed to form a Butler DF array. 1D and 2D layouts are also permitted, giving 1D or 2D beam steering. Essentially, multiple receivers can be attached to the outputs or a single receiver with a switch matrix can be used. With multiple receivers, simultaneous observations can be done with the beams formed by the matrix. Circles on diagram are 90 degree hybrids.*

*Azimuth Pattern*

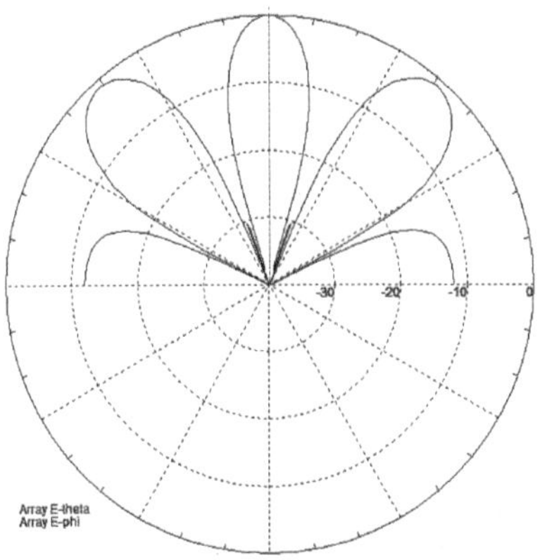


*Elevation Pattern*

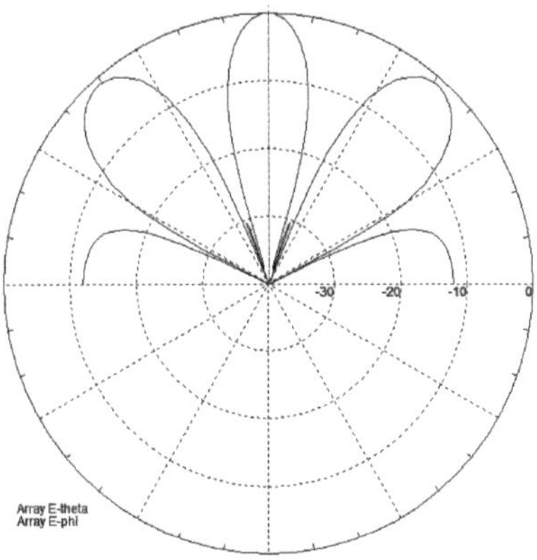


Special Considerations:

*For large arrays, matrix becomes complex. Digital techniques can simplify matrix to a degree.*

Testing:

*Use far field range to measure pattern. Considering the small aperture relative to wavelength, the distance to the source antenna will not be great. Check Appendix 6 for more details. Rotate array relative to source to verify the angle of arrival is accurate.*

Type: ***Adler Array***

Frequency of Operational: *5 Mhz to 50 MHz*
Gain Range: *0 to 2.12 dBi*
Bandwidth: *~ 15%*
Approximate Power Rating: *10 Mw*
Design Software: *IE3D, Ensemble, Moment Method, NEC*

Configuration:

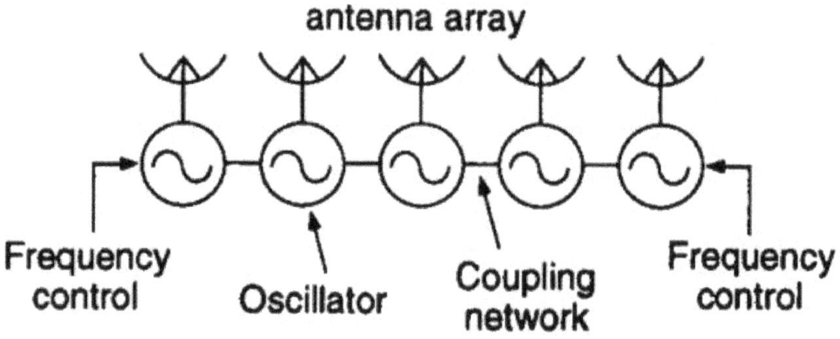


Design Approach:

*Element to Element separation = 0.5 to 0.7 lambda*
*Oscillator power output for transmit array = 1 mW to 1 Mw*
*Frequency control, multi bit digital*
*Coupling network = power combiner*

Construction and Theory of Operation:

*Wide beam or narrow elements can be used. Oscillators require fine control, use numerically controlled oscillators or phase locked loop types. Digital control should be employed to set oscillator frequencies. Beam position can be changes as quickly as frequency can be changed in the oscillators. Approach works with either transmit or receive arrays. Receive arrays require mixer to allow reception of intermediate frequency output.*

*Azimuth Pattern* *Elevation Pattern*

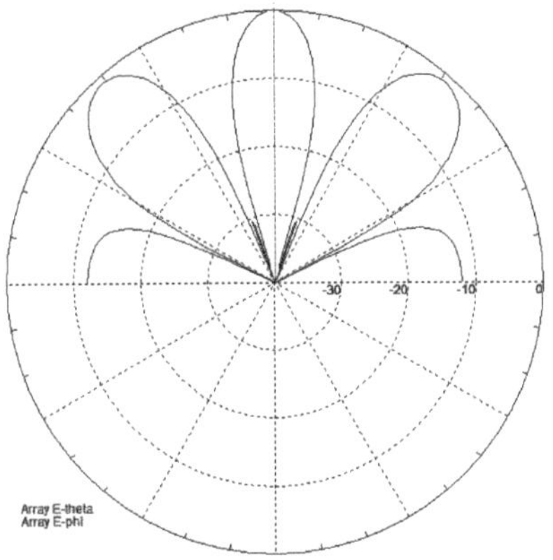


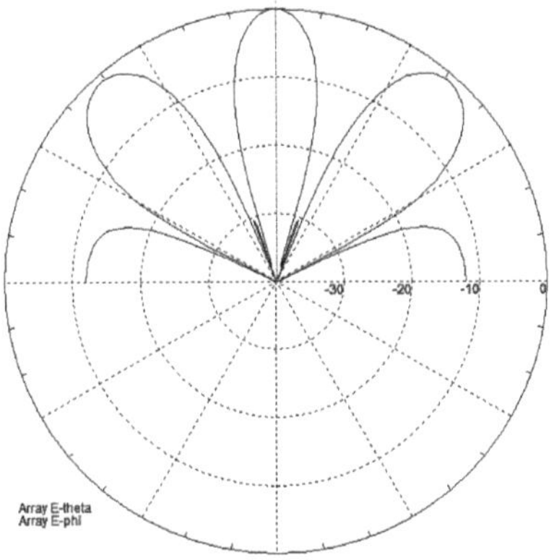


Special Considerations:

*1D or 2D layouts are possible, phasing network becomes complex with increase of number of elements. Phase stability on oscillators is very important,*

Testing:

*Use far field range to measure pattern. Considering the small aperture relative to wavelength, the distance to the source antenna will not be great. Check Appendix 6 for more details. Rotate array relative to source to verify the angle of arrival is accurate.*

Type: ***Conical Scan DF Antenna***

Frequency of Operational: *5 Mhz to 500 GHz*
Gain Range: *0 to 2.12 dBi*
Bandwidth: *~ 5%*
Approximate Power Rating: *1 Mw*
Design Software: *IE3D, Moment Method, Geometric Optics*

Configuration:

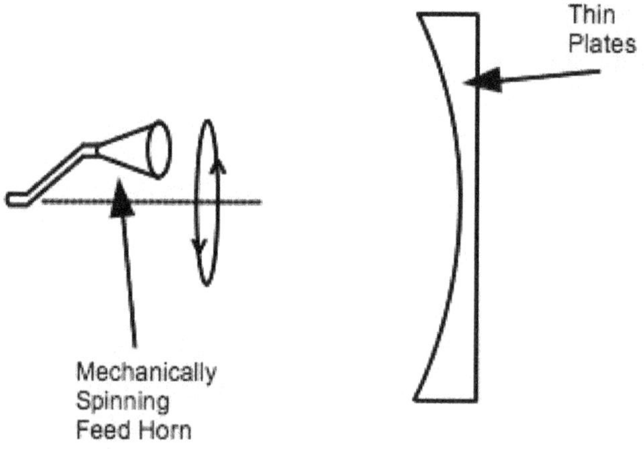


Design Approach:

*For example above, use lens design formula and circular horn design formula. Keep circle of motion within 10 lambda to avoid pattern distortions. Reflectors can also be used instead of lens*

Construction and Theory of Operation:

*Use an offset waveguide feed to a dish, spin it within approximately 10 wavelengths. Detect the position of the waveguide feed and determine +/- x offset and +/- y offset, feed into main antenna positioning electronics and determine optimal slew rate to track target of interest.*

*Azimuth Pattern*

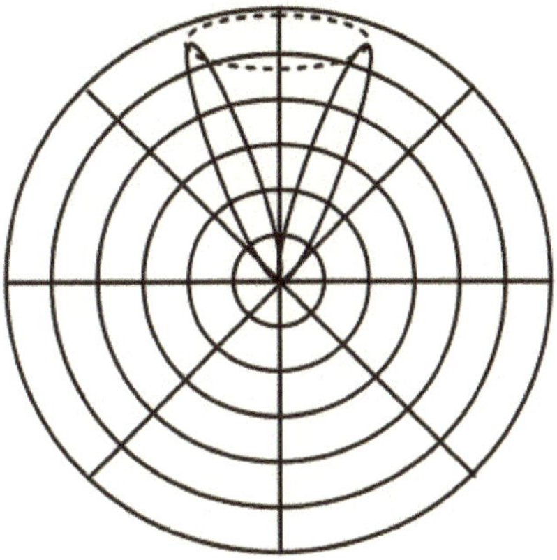

*Elevation Pattern*

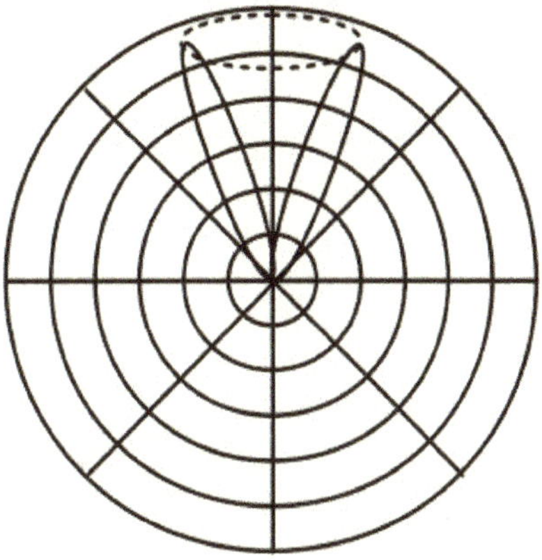

Special Considerations:

*For rotating horn assemblies, be aware that the polarization can spin as well if using rectangular waveguide. To avoid this problem, use rectangular to circular transition and use circular waveguide to feed horn. Keep spin circle diameter to within 10 lambda to avoid pattern distortions.*

*Take into account losses in rotating joint to determine accurate antenna gain.*

Testing:

*Use far field range to measure pattern. Considering the small aperture relative to wavelength, the distance to the source antenna will not be great. Check Appendix 6 for more details. Rotate array relative to source to verify the angle of arrival is accurate.*

Type: ***Conical phased array***

Frequency of Operational: *5 Mhz to 500 GHz*
Gain Range: *0 to 16 dBi*
Bandwidth: *~ 15%*
Approximate Power Rating: *10 Kw*
Design Software: *IE3D, Ensemble, Moment Method, NEC*

Configuration:

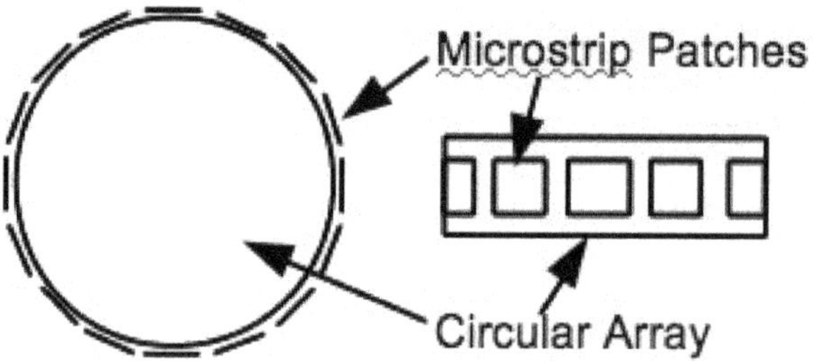


Design Approach:

*Use microstrip patch, slot or other suitable element design, set in circle and combine using Butler array or Doppler beam steering techniques.*

Construction and Theory of Operation:

*Normally this type of array is set on an aluminum ground plane such as a rocket body or aircraft fuselage. For aerospace applications, use a radome to protect conductive portions of antenna elements.*

*Azimuth Pattern* *Elevation Pattern*

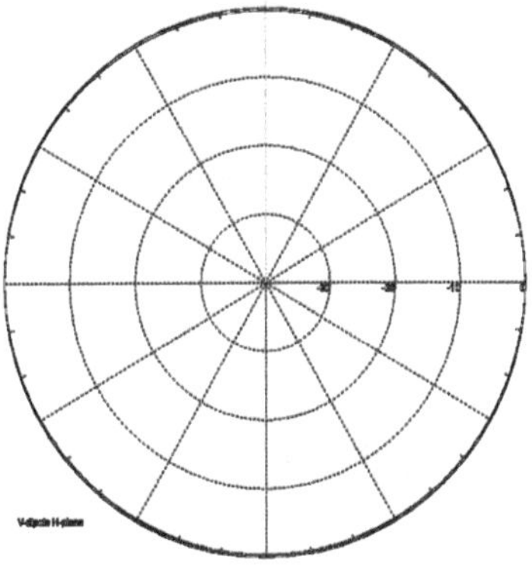

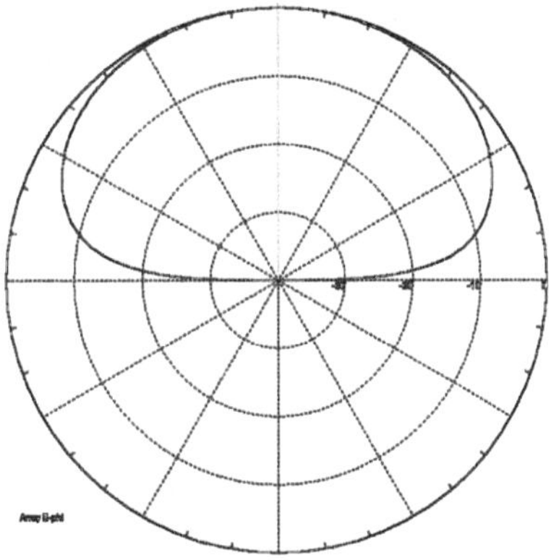

Special Considerations:

*Beamwidth is generally large in most applications, leading to course direction finding. For aerospace applications, test antenna array under temperature variations as well as shock and vibration that simulates flight.*

Testing:

*Use far field range to measure pattern. Considering the small aperture relative to wavelength, the distance to the source antenna will not be great. Check Appendix 6 for more details. Rotate array relative to source to verify the angle of arrival is accurate.*

Type: **_Monopulse DF Array_**

Frequency of Operational: *5 Mhz to 500 GHz*
Gain Range: *0 to 32 dBi*
Bandwidth: *~ 25%*
Approximate Power Rating: *10 Kw*
Design Software: *IE3D, Ensemble, Moment Method*

Configuration:

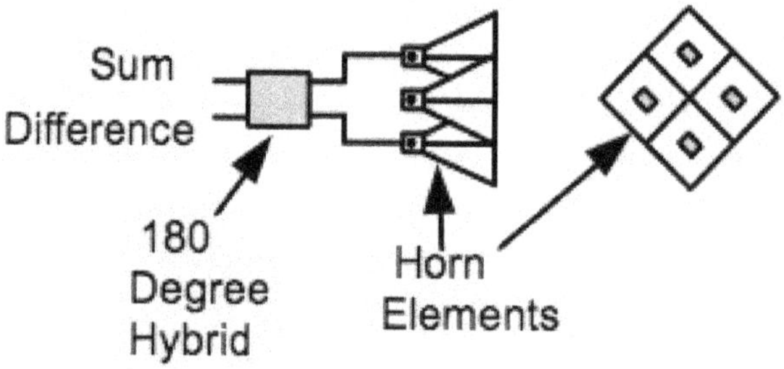


Design Approach:

*Typically this is a four element array with vertical elements and horizontal elements each attached to a hybrid which provides sum and difference outputs. The sum is used for course direction finding and the difference is used for fine position sensing.*

Construction and Theory of Operation:

*This antenna array is used commonly for maintaining an error signal for tracking radars or DF systems. Made of horns, microstrip or other suitable directional elements, its simple construction allow for high resolution target tracking.*

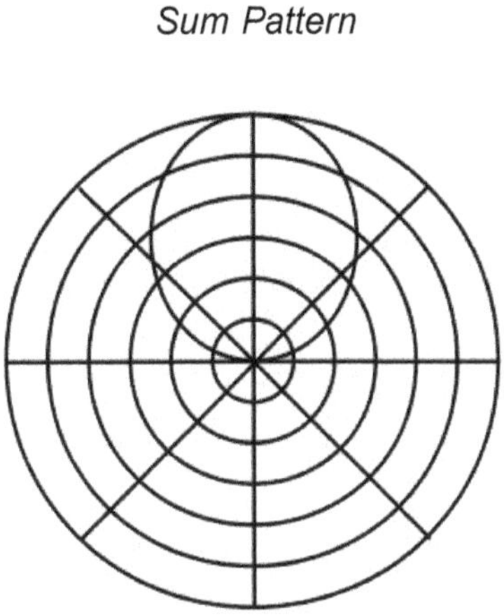


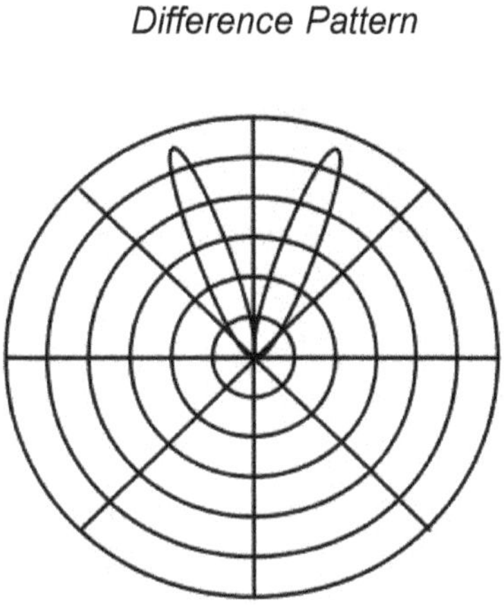


Special Considerations:

*Use radome for environmental protection if array is used in aerospace application.*

Testing:

*Use far field range to measure pattern. Considering the small aperture relative to wavelength, the distance to the source antenna will not be great. Check Appendix 6 for more details. Rotate array relative to source to verify the angle of arrival is accurate.*

| | |
|---|---|
| Type: | ***Time of Flight*** |
| Frequency of Operational: | 1 *Khz to 50 GHz* |
| Gain Range: | *0 to 32 dBi* |
| Bandwidth: | *~ 25%* |
| Approximate Power Rating: | *NA, Receive only* |
| Design Software: | *IE3D, Ensemble, Moment Method* |

Configuration:

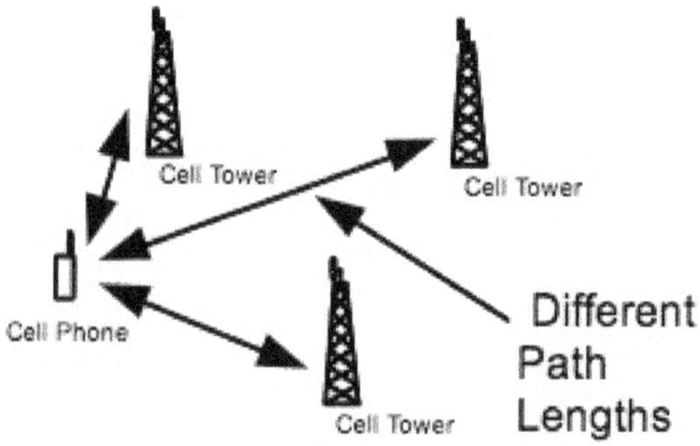


Design Approach:

*For the example shown above, several towers receive modulated energy from a transmitter, here a cell phone. The cell towers are frequency stabilized to GPS signals received by all towers. The time of arrival of the cell phone is compared by each tower to the reference time. Simple triangulation then takes place to determine the position of the cell phone.*

Construction and Theory of Operation:

*Existing cell towers or other towers with receive antennas can be used. Each tower must be frequency locked to a common standard. Correlation techniques are used on the received signals and compared trigonometrically to determine position.*

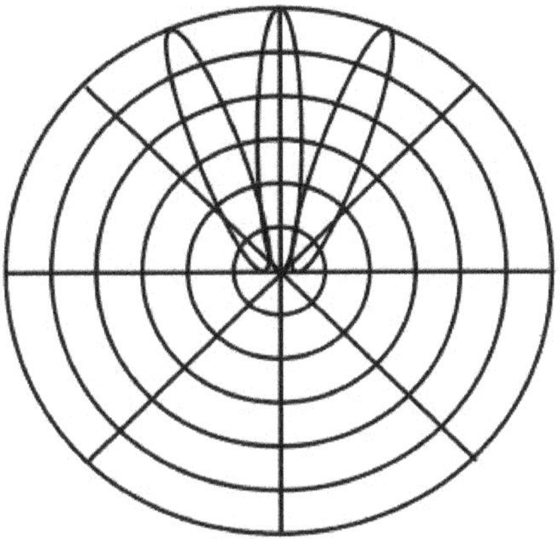


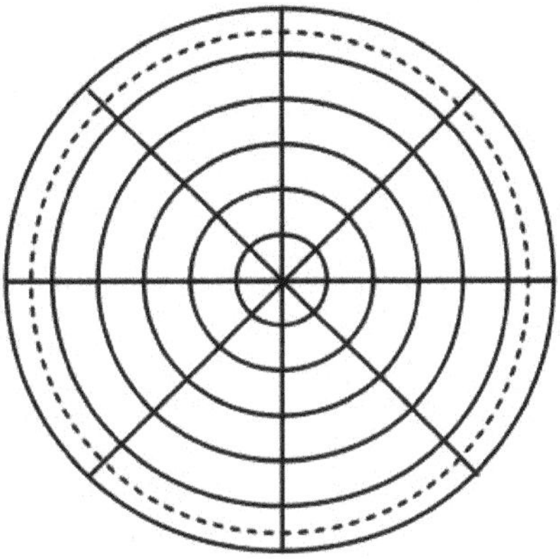


Special Considerations:

*Use high speed digital electronics to perform correlation math.*

*Line of sight is preferred to any multipath signal, as the reflected signals will give false positions as they are invariably longer.*

Testing:

*Place strong transmitter at unused frequency in area of interest. Test correlation function to verify position, then move transmitter to various positions with the field of view of the selected towers and map response. Use this map to correct real world transmissions if error is large.*

Type: ***Multipath DF Systems, Passive Radar***

Frequency of Operational: *5 Khz to 5 GHz*
Gain Range: *0 to 12 dBi*
Bandwidth: *~ 45%*
Approximate Power Rating: *NA, Receive only*
Design Software: *IE3D, Ensemble, Moment Method, NEC*

Configuration:

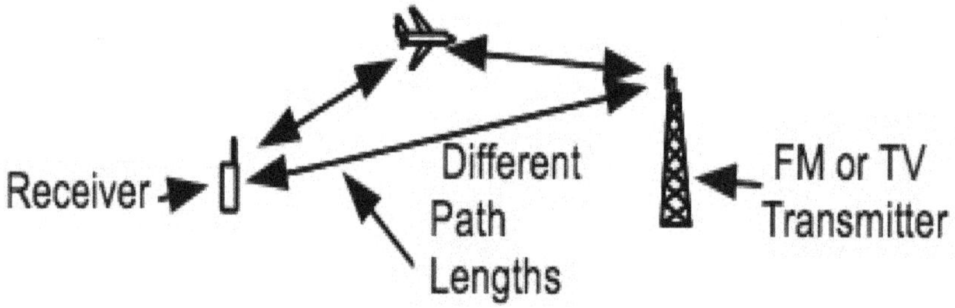


Design Approach:

*This is a receive only system that relies on the direct and reflected reception of local high power transmitter signals, like FM and TV channels. Generally, at least two transmitters are used to triangulate to the reflective object (shown as a plane above).*

*The signals are superimposed on one another and change phase relative to each other as the path lengths change. Applying trigonometric equations to the received signals allow for the three dimensional determination of position.*

Construction and Theory of Operation:

*Generally a receiver designed around the frequency of interest is used with the demodulated signal digitized. The two superimposed signals are discriminated in phase and the change in phase is observed over time. Using basic trigonometric equations the lengths of the baselines can be determined and then the angles from the transmitter and receiver to the object of interest can be displayed. Using multiple transmitters allows for the third dimension to be determined.*

*Azimuth Pattern*

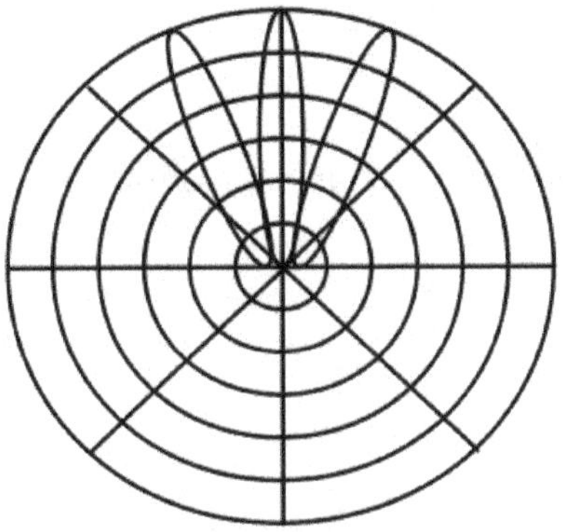

*Elevation Pattern*

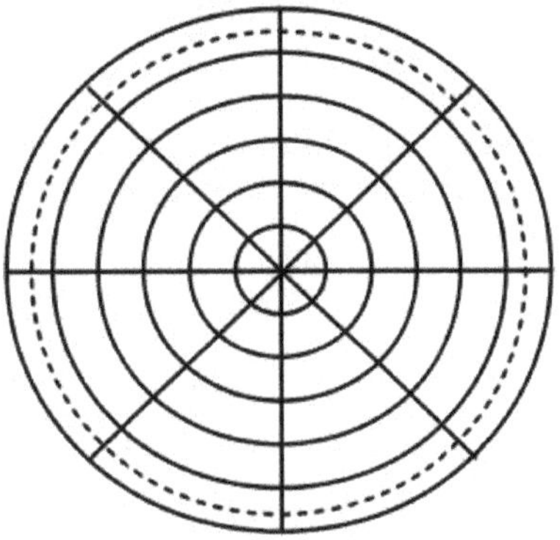

Special Considerations:

*Use strong transmitters for best results.*
*Observe phase stability of transmitters without reflective object to quantify accuracy.*
*Use phase stable receiver to avoid erroneous readings.*

Testing:

*Observe transmitter signal and verify accurate digitization of modulated waveform.*
*Use local radar observations for calibration of reflective object position.*

Type: ***Crossed Loop DF System***

Frequency of Operational: 1 *Mhz to 50 GHz*
Gain Range: *0 to 3 dBi*
Bandwidth: *~ 15%*
Approximate Power Rating: *NA, Receive Only*
Design Software: *Nec-Win Pro, Antennamax, Moment Method*

Configuration:

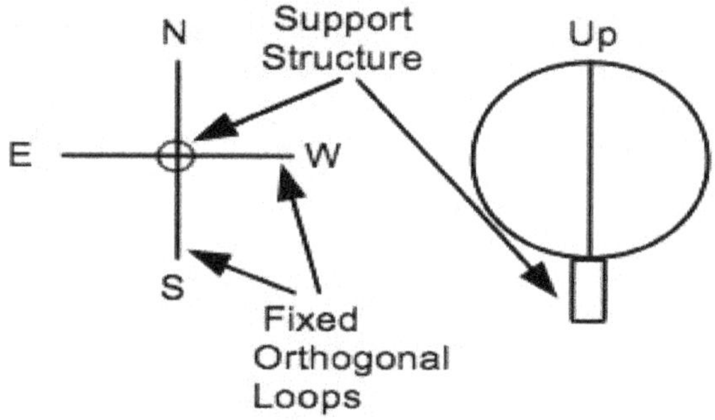


Design Approach:

*Antenna diameter is typically greater than 0.1 wavelength. Loops are fixed on support structure. Output of loops can be connected to goniometer or in modern systems, to receivers where phase can be observed. For shipboard systems 180 degree ambiguity is generally not a major issue, especially when observing land based signal emitters from the sea.*

Construction and Theory of Operation:

*These loops are greater than 0.1 wavelength in diameter and as a result are more sensitive to the electric field of a radio wave. The nulls are perpendicular to the plane of the antennas. Bandwidth is greater than a small loop so this antenna can be used as a direction finding component over many MHz of frequency. The advantage of using a crossed loop is that there is no need for mechanical scanning, azimuth to a signal of interest is obtained by connecting the loops to a "goniometer" or software defined radio with a fast analog to digital converter to measure phase and amplitude to calculate bearing.*

*Azimuth Pattern*

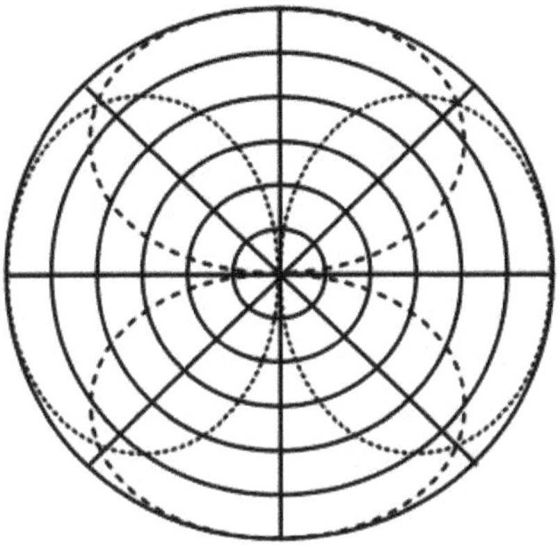

*Elevation Pattern*

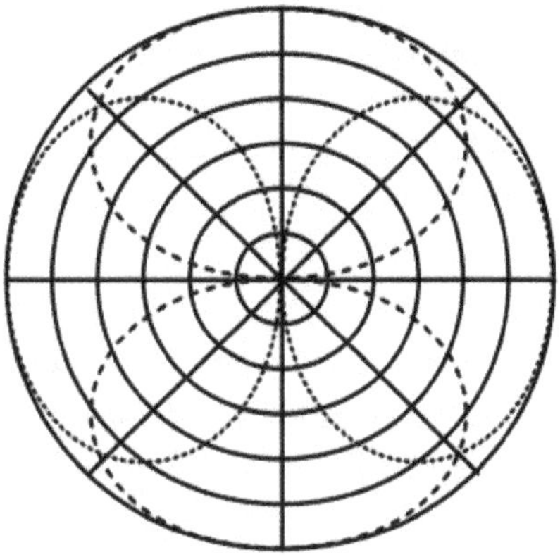

Special Considerations:

*Attach loops to electronics that discriminate phase and amplitude for high speed bearing determination.*

Testing:

*Use far field range to measure pattern. Considering the small aperture relative to wavelength, the distance to the source antenna will not be great. Check Appendix 6 for more details. Examine the details of the nulls as this will allow direction finding optimization. The nulls should be deep and have a limited angle.*

Type: ***Automatic Direction Finder (ADF)***

Frequency of Operational: *100 Khz to 50 MHz*
Gain Range: *0 to 3 dBi*
Bandwidth: *~ 35%*
Approximate Power Rating: *NA, Receive Only*
Design Software: *IE3D, Ensemble, Moment Method*

Configuration:

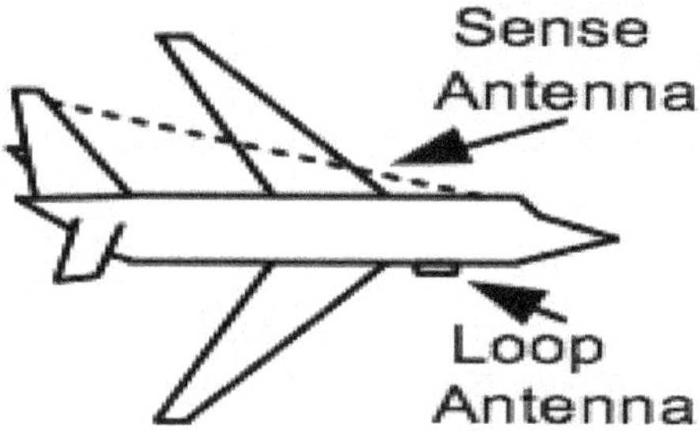


Design Approach:

*Antenna system requires two associated elements, a sense antenna, as long as possible in the case of an aircraft installation, and a loop antenna, either physically moved or dual loops. The loops determine the direction of arrival of the signal of interest and the sense corrects the 180 degree ambiguity as well as adds sensitivity.*

*Most systems for aircraft work between 200 Khz and 2 MHz. Other systems extend this range*

Construction and Theory of Operation:

*The sense antenna is usually a piece of wire that runs from the forward fuselage (in the case of an aircraft installation) to the top of the tail. The loop antenna is almost always on the bottom of the fuselage. The radio selects the sense (for greater sensitivity) or adds the loop to control a needle that points to the radio station of interest. The loop and sense antennas are designed to withstand the harsh aerospace environment. Some modern systems encase both the loop and sense in the same unit.*

*Azimuth Pattern*

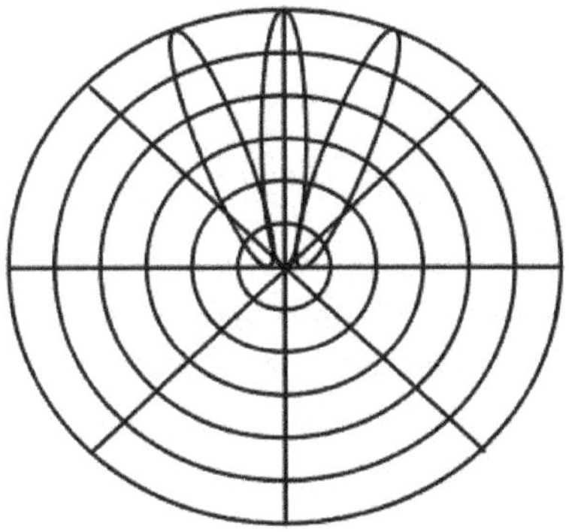

*Elevation Pattern*

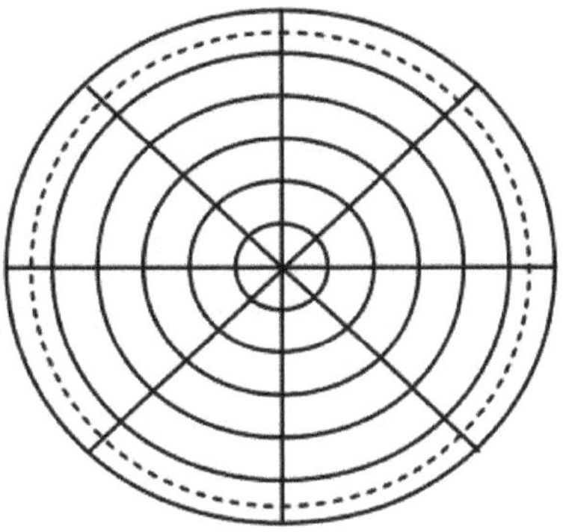

Special Considerations:

*Verify that the antenna system can withstand wide variations of temperature and altitude. Also, these elements can accumulate ice and can be compromised in terms of sensitivity.*

*In some cases, this type of antenna at the low frequencies of operation can detect the direction to thunderstorms. Special adaptations of this principle, notably "Stormscope," use loops and special algorithms to determine range and bearing to lightning strike, which are inherent in all large thunderstorms.*

Testing:

*Use known bearing to radio source to determine accuracy of installation. In aircraft, this is generally done by swinging the aircraft around in a circle on the ground or flying directly to the radio station and watching changes in direction. On a windless day, the path should be very straight.*

**Questions:**

1. Suggest an antenna type that has the following characteristics:

   a. Frequency of 200 to 400 MHz
   b. One degree azimuthal resolution
   c. 1 millisecond detection time

2. Your boss tells you price is a huge factor in the design of antenna with the following characteristics:

   a. Frequency of 20 MHz
   b. Gain of 20 dBi
   c. 1 arcsecond azimuthal resolution

   Can a yagi be used here?

3. Suggest an antenna design for the next generation rocket booster that is capable of going to the moon, with the following properties:

   a. Frequency of 2200 MHz
   b. 1 degree azimuthal resolution
   c. Tracks emitter while spinning

4. How could you use a phased array as a real time sky monitor ?

5. Is there a way to use radio telescopes as time of arrival measurement instruments and determine 3D positioning?

6. Can the doppler shift of a GPS satellite be used for positioning determination? In other words, without the demodulation of the transmitted signal?

7. If you retransmit a GPS signal received by a balloon package on a different frequency, then demodulate this signal, is there a way to do this very inexpensively?

# Chapter 16 – Radio Astronomy Antennas

The 100 meter Green Bank Telescope (GBT)

*'Thus the explorations of space end on a note of uncertainty. And necessarily so. We are, by definition, in the very center of the observable region. We know our immediate neighborhood rather intimately. With increasing distance, our knowledge fades, and fades rapidly. Eventually, we reach the dim boundary – the utmost limits of our telescopes. There, we measure shadows, and we search among ghostly errors of measurement for landmarks that are scarcely more substantial. The search will continue. Not until the empirical resources are exhausted need we pass on to the dreamy realms of speculation.'*

*Edwin Hubble*

**Fundamentals**

Some of the most interesting and innovative antenna designs have been accomplished by radio astronomers. Simply put, these antennas are designed for the highest gain, smallest beamwidth and lowest system temperature possible. These qualities are necessary for the further discovery of interesting astrophysical phenomena.

To date the discovery of noise storms on Jupiter, thermal radiation from the Milky Way, Pulsars, Quasars, emissions from ionized gas, masers, stellar jets and a host of other physical phenomena has been accomplished by these special antennas. In addition thermal radiation, synchrotron, spectral lines, absorption lines, Zeeman splitting, Doppler shifting are but some of the physical characteristics viewed by radio telescopes. The determination of the shape of the Milky Way, its core and the motion of the arms were discovered by these instruments as well.

Sensitivities of better than $10^{-26}$ watts per meter is typical of these antennas, with gains beyond 100 dBi and beamwidths (using interferometric techniques) in the micro arcsecond range. Antenna arrays spanning continents coupled with antennas in space have increased the resolution of these telescopes to many times better than the best optical instrument. New arrays are being built in the high deserts of Chile, operating at millimeter and submillimeter wavelengths. These instruments are rarely encumbered by the atmosphere and can be used day or night.

A radio telescope in Holmdel, New Jersey verified the existence of the 2.7 degree Kelvin afterglow of the creation of the universe. One in Puerto Rico is 1000' feet in diameter and has examined minute details of Pulsars, sent messages to our interstellar neighbors and starred in several movies.

As mentioned above, multitudes of telescopes can be connected together to synthetically create apertures exceeding the size of the Earth, which allow resolutions significantly better than any optical telescope, including the Hubble. These groups of telescopes are known as interferometers. Some are connected together like the Very Large Array in Socorro, New Mexico. Others, which have elements separated by hundreds or thousands of miles use extremely stable clocks that are used to reference the frequency of operation on all of the telescopes. One version of this type uses video tapes to record the individual outputs. These tapes are sent to

Socorro where they are correlated to create "baselines" across the United States and beyond. These baselines create the resolution equivalent to a telescope of the same physical size. This particular array is known as the Very Large Baseline Array. It consists of ten 90 foot diameter antennas with multiple operational bands. At the highest frequency, this array is capable of discerning details in the micro arcsecond range.

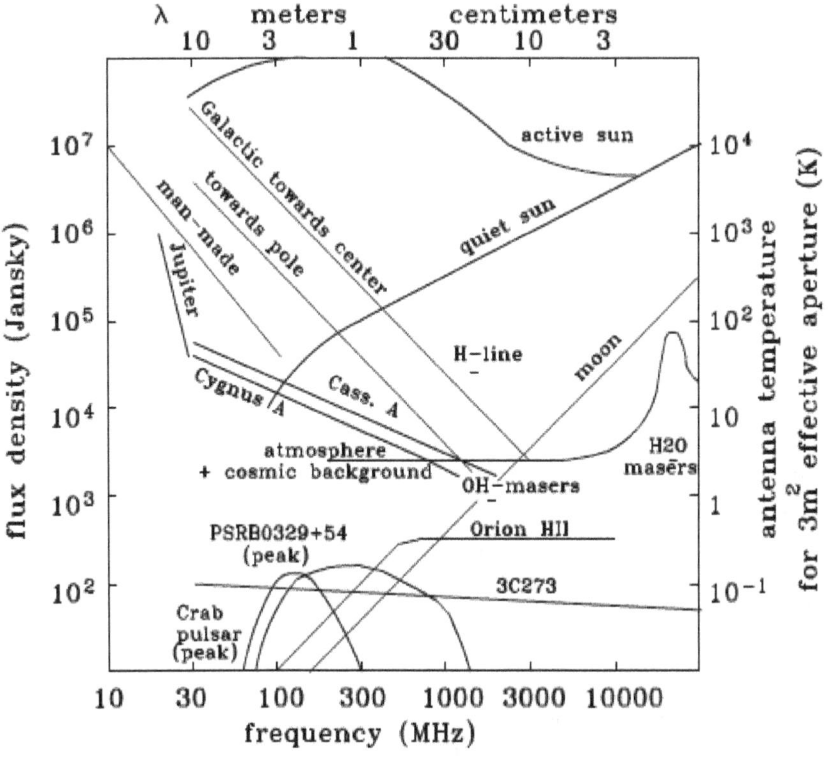


The Radio Astronomy Spectrum

**History**

In August 1931, and engineer working for Bell Telephone was tasked with finding out sources of interference for shortwave telephone traffic, then in its infancy. Jansky constructed a directional Bruce array, placed it on a wooden track using Model T tires. By moving the array around in azimuth, a measurement of the noise level from the sky and other terrestrial sources could be ascertained. With these words, the field of radio astronomy was born:

"The data give for the coordinates of the region from which the disturbance comes, a right ascension of 18 hours and declination of -10 degrees."

This position corresponds to the center of the Milky Way. The radio receiver was tuned to approximately 20.5 MHz, in the middle of today's shortwave band. Jansky noticed that the radio source followed a sidereal day as apposed to an "Earth" day and so was able to place the source in the celestial sphere. The receiver recorded a high noise level, and as the earth turned the peak of the galactic noise was significant and moving at a sidereal rate.

In 1937, an engineer in Illinois named Grote Reber, had read the results of Jansky's research and wanted to analyze the galactic noise further. He constructed a 30 foot prime focus parabolic dish in his back yard and placed within the focal point, several receivers, one at 160 MHz and another at 3 GHz. The antenna's mount was altitude over azimuth and considering it has a circular beam (as apposed to a fan beam like Jansky's apparatus), Reber was able to map the "radio" universe for the first time. Reber published several papers on his results in the *Astrophysical Journal* (by the way, the editor at that time, Otto Struve, concerned about the non standard language of Reber, took responsibility for the publications) and what resulted from Reber's efforts was the discovery of the remnants of a supernova remnant (Cassiopeia A), an extragalactic radio source (Cygnus A) as well as more definition of the radio emission from the galactic plane. Later, Reber was the first to publish a paper on radio emissions from our sun.

During WWII, radar operators in England, working at about 178 MHz noticed interference while scanning the skies at dawn and dusk. The emissions coincided with the position of the sun and as it rose during the day, and out of the antenna beams, the interference subsided. Scientists at the Cavendish radio astronomy group at Cambridge University were tasked with finding and hopefully nullifying the solar interference. The used search light reflectors and soon were able to pinpoint and track the solar radiation. It became apparent that the solar radio outbursts were not steady, solar "storms" were discovered and eventually the rise and fall of the number of storms over an 11 year period.

After the war, radar receivers and antennas were modified and used to make more accurate maps of the radio universe. In Australia, a group of researchers in Sydney started work on several telescopes, including a “Mill's Cross”, which discovered radio emissions from Jupiter.

Back in England, the discovery of the all important 21cm line, inaugurated spectral radio astronomy. With this discovery came the mapping of the motions of the arms of the Milky Way (this due to doppler shift), as well as the existence of emission from our galactic neighbor, the Andromeda Galaxy. In the 1950s and 1960s, several large telescopes were constructed including catenary types. Blackett, Lovell and others constructed a 218 foot diameter reflector, suspending wires on a ring of scaffolding poles and filling it in with 16 miles of wire. In addition, there was a 126 foot steel mast in the center holding the focal point antenna, this mast could be moved up to 15 degrees from the vertical to move the beam due to reflection, +/- 30 degrees. A significant number of discoveries and papers were obtained from this antenna. Later even more sophisticated antennas were constructed, like the one below:

The Jodrell Bank 250 foot Radio Telescope

Located in Jodrell Bank, England, this 250' radio telescope was build from battleship components in 1957. The elevation pinions are actually gun turret components salvaged from WWII battleships.

In 1967, Jocelyn Bell of Cambridge, after analyzing 100s of feet of chart paper, noticed anomalies in the records. The records were made from a new 81.5 MHz Mill's Cross telescope in England designed to examine quasars (or quasi stellar objects). Bell noticed that the recording rate of the chart paper was not fast enough to resolve all of the attributes of several of the sources recorded. Using oscilloscopes, faster chart speeds and audio techniques, it was discovered that these unique radio sources, now known as pulsars were in fact neutron stars rotating at very fast rates, some faster than 30 milliseconds per revolution. Confirmation was achieved through the gating of optical images at several optical observatories, including one at Flagstaff, Arizona. The gating was achieved by spinning a disk with a hole in it a rates slightly slower that the rate determined by Bell, the resultant image would slowly pulsate in intensity over a slower period.

In the U.S. The National Radio Astronomy Observatory was established and built several very large telescopes, including a 300 foot transit antenna, 140 foot high frequency antenna, an 85 foot telescope (used for one of the first extraterrestrial intelligence searches) and a large interferometer, used eventually to design the Very Large Array in Socorro, New Mexico.

Using interferometric techniques derived from English and American efforts, The Very Long Baseline Interferometer (VLBI) system was developed. As a telescope gets larger, the beamwidth gets smaller and finer details of astrophysical phenomena can be observed. From a practical point of view, a single telescope become prohibitively expensive over very large diameters. The solution is to separate one or more separate telescopes to improve the resolving power. These are know as interferometers and are usually just large enough to couple the outputs of the separate receivers together via long cables or microwave links. In the 1970s, the development of very stable oscillators and video tape recorders allowed the separation of telescopes to be expended to thousands of miles. Today there are orbiting elements of the VLBI network, allowing for resolutions in the micro arc-second range, vastly exceeding the resolution of optical telescopes.

Today, interferometers using millimeter and sub-millimeter

wavelengths have been built with more on the way. The advantage of these significantly higher frequency telescopes is the high resolution imaging of interstellar molecules, gas clouds, and other high energy phenomena.

One can imagine that in the future, arrays like the VLBI will be distributed in space with baselines of tens or hundreds of thousands of miles. This in combination with millimeter wavelengths will allow the detailed examination of many other solar systems.

The Atacama millimeter and sub-millimeter radio telescope at 16,000 feet elevation, Northern Chile. Note curved struts on antenna to the left, this is done to minimized scattering off of feed positioning structures.

**The Radio Astronomy Equation**

Generally speaking, the minimum flux sensitivity of a radio telescope can be described in the following:

$$\Delta S = \frac{(2 * k * Tsys)}{(Ae * \sqrt{(2 * \Delta T * \Delta V)})} \qquad [81]$$

k = Boltzmann's constant (1.38*10^-23)

Tsys = the system noise temperature (Antenna + Receiver)

Ae =the effective area of the telescopes

T = the observing time

V = the bandwidth of the received signal

Based on this formula, it should be obvious that the lower the system temperature, made up of predominately of the first low noise amplifier after the feed antenna, added to a combination of the cable losses and "back end" losses of the receiver, the better. Also the larger the aperture, longer the integration time, and wider the bandwidth, the better.

A consequence of a large aperture, then is lower beamwidth, or the ability to discern fine details of astrophysical phenomena.

**System Noise Temperature**

The system noise temperature heavily influences the ability to detect faint radio astronomical objects. It can be calculated by understanding the characteristics of the antennas, low noise amplifiers and the local environment. The following formula needs to be evaluated and entered into the radio astronomy equation to fully understand the detectability of radio sources:

$$Tsys = \frac{Ta}{Lf} + \left(1 - \frac{1}{Lf}\right) * Tout + Tr \qquad [82]$$

where:

Ta = Antenna Temperature at the "antenna port"

Lf = Front end loss (from "antenna port to preamp input"

Tout = reference temperature (290 degrees K for ambient)

Tr = Receiver noise temperature

**Beam Movement**

**Mechanical**

The most obvious way of moving a radio telescope beam around is with a moving structure, able to change elevation and azimuth as needed to measure source "temperature" or flux. This kind of structure enables tracking of a particular piece of sky to allow long term integration and thus improve sensitivity. Most modern radio telescopes employ this method, a few have equatorial mounts, whereby one axis points to the North (or South if below the Equator) pole. The other axis or declination, is set once and does not need to be moved further, enabling the equatorial axis to move with the rotation of the earth and thus track a source. This style is rare for radio telescopes but very popular with optical types. The largest example of an equatorial mounted radio telescope is the the 140 foot antenna at the National Radio Astronomy Observatory (NRAO) in Green Bank, West Virginia.

Another interesting method to mechanically move a telescope beam was by moving the focal point of a parabolic dish. In the 1950s this method was employed to perform a minimal scan around the area of the Andromeda galaxy, where the existence of radio emissions was proven. The area of coverage was two times the amount of deflection of the feed as per the optics of the arrangement. Today, there is a re-emergence of a moveable feed, albeit with a phased array. This allows significantly more sky coverage than with a single feed. See the following pictures of alt/az and equatorial type telescopes:

The Parkes, Australia 64 meter Radio Telescope, which brought the images of Neil Armstrong setting foot on the moon in 1969

The NRAO 140 foot 20 GHz Equatorial Mounted Radio Telescope

**Electrical**

Using a multitude of antennas separated by 0.5 to 0.75 lambda (for the simplest case) in a uniform pattern defines a phased array. In either one or two dimensional layout, the antenna will receive a signal well orthogonally from the plane of the aperture, if all of the antenna elements are phased the same relative to a summing feed point.

If, however, these element are phased sequentially, for instance by ten degree increments, the antenna beam will move off the orthogonal orientation and tilt towards the short end of the phase distribution. As a consequence, many beams can be constructed allowing a wide angle of view using a small beam size. This is the basic principal of many military radar systems and in radio astronomy has been used to great effectiveness.

These phase offsets were initially created by different lengths of coaxial lines and eventually, with electrically switched delay lines using relays, PIN diodes or other types of electronic switches.

In another manifestation, John Kraus discovered that if you place three helical antennas in a line, and add (in phase) their outputs, then physically spin the outer helixes, the resultant beam moves along the axis of the line. By adjusting the azimuth position of the outer elements to a particular angle, an stationary offset beam can be created.

**Earth Rotation**

**Transit**

If a radio telescope is set to a particular elevation and azimuth and set in place, the earth will of course continue moving and automatically scan the antenna beam across an arc coincident with a line of latitude. This techniques has been used often for allow a map of the radio universe to be built up by simply by setting the azimuth to North or South, then adjusting the elevation in single beamwidths every sidereal "day."

The following pictures depict two types of transit instruments, a Lloyd's mirror, using the sea as a reflector, and a catenary reflector, where the feed was moved to change the beam position. This telescope was able to detect the presence of radio emissions from the Andromeda Galaxy (M31).

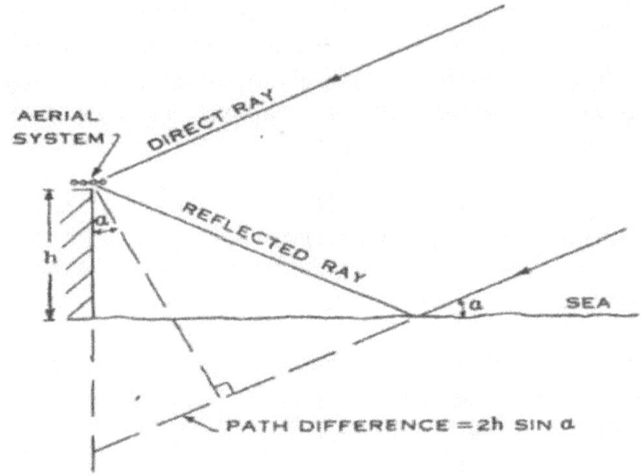


Lloyd's Mirror telescope, first used from the cliffs of Australia and New Zealand by Ruby Payne-Scott to measure the size of radio galaxies

Catenary Reflector dish with moveable feed, discovered Andromeda galaxy radio emissions

**Synthesis**

By the 1960s, radio astronomers realized that if they placed several antennas in an interferometric array (single antennas separated by many wavelengths) and tracked the astronomical source of interest, the movement of the earth would create a multitude of projected "baselines" and allow two dimensional imaging with great precision. These baselines change in length as the earth moves due to the fact that the relative orientation of the antenna array changes, looking from the radio source towards the array. This is the basic concept behind the VLA (Very Large Array, Socorro New Mexico) and the VLBA (Very Long Baseline Array, across the U.S. with elements in Europe and in space).

**Elements**

Radio Telescope elements are many time made up of many smaller antennas from microstrip patches to fully steerable parabolic dishes. The following examples show some of the more common types:

Example of a yagi array, capable of mapping the sun, Cassiopeia A and other strong radio sources

**Yagis**

These antennas have been used to form phased arrays, interferometers and other components of radio telescopes. Simple in construction and easy to reproduce, they are wideband and have reasonable gain.

**Horns**

Horn antenna have been instrumental in discovering major phenomena in our universe. The following horn discovered the Hydrogen emissions in our galaxy, in addition, this antenna found the doppler shift associated with moving clouds of Hydrogen gas, allowing astronomers to understand the movement of the Milky Way arms and associated components.

21 cm (1420 MHz) Horn that discovered the Hydrogen Line by Purcell, shown above (35)

This horn antenna located near Holmdel, New Jersey, discovered the 2.7 degree background radiation from the creation of the universe. Townes and Dicke, shown above, won the Nobel Prize for this discovery (35)

**Helixes**

Helix type antennas have been an important part of radio astronomy history. An inherent feature is that they are circularly polarized and can receive signals unaffected by the earth's ionosphere, at frequencies below 1 GHz. Linear type antennas are effected due to the Faraday rotation effect. Banks of helixes were designed and build by John Kraus and made some of the early detailed maps of the radio sky.

The 96 element helix array, designed by John Kraus and responsible for the early detailed mapping of the radio universe. (35)

**Parabolics**

The majority of radio telescopes are of the parabolic kind, including some interesting modifications of the generalized shape. Having shown several examples of large standard parabolas, in this section we shall show some of the variations on the theme that have contributed significantly to the study of the radio universe.

At Ohio State, the "Big Ear" was constructed, which is a combination of parabola as seen on the left side and a planar reflector as seen on the right side. This is a transit instrument and has a feed point near the right side centered in the little structure. When it was completed, they simply turned on the chart recorder and "started taking data". The amount of new radio sources recorded due the high sensitivity of this antenna allowed Ohio State to create a list of sources with several hundred entries. A similar design on a larger scale was constructed in Marseilles, France.

The Ohio State "Big Ear", designed by John Kraus and source of the Ohio State Radio Astronomy Survey. (35)

**Russian Ratan**

This design is a parabolic section, much like the rim of a standard parabola, which focuses the radio waves to a central point for reception. The advantage is significant resolution but with the inability to track sources and steer over large angles.

This telescope has very wide band capabilities and is being used for quasar and pulsar research. The feed for this antenna is at the very center of the circle of reflectors. Multiple feeds can in fact be used at different frequencies simultaneously is separate sections of the main reflector are used. To use the full telescope, an inverted cone reflector is used in the center with a feed horn placed beneath it.

The 600 meter diameter Ratan-600 (35)

## Arrays

### Mill's Cross

As shown in chapter 1, an un-filled aperture, made up of two lines of antenna elements, or Mill's cross has the resolution of a filled aperture of similar dimensions. Elements used in the construction of such arrays have been dipoles, smaller dishes, horns etc. with the predominant orientation of N-S for one array and E-W for the other array. Each array produces a fan beam which overlap in the sky at one particular spot. These arrays can be phased to cover a significant amount of sky with this common crossover point. By adding ½ wavelength to the summed arrays at a specified rate. This cross over point will change in and out of phase and thus modulate the summed signal from both arrays. Making a differential measurement of this modulation allows the observer to measure the radio intensity of that small spot on the sky.

Pulsars were discovered with this technique in 1967 and many papers have been written from observations with this style of telescope.

## Two element interferometer

A simple interferometer can be made up of two antennas, which can be single elements or arrays. The distance between the elements dictates the resolution of the array, the field of view of the interferometer is dictated by the elemental pattern. A series of fringes is observed with the summed outputs of both elements. Examining the fringe details reveals the size of the radio source. As the distance between elements increases (as in a synthesis array), the fringes become smaller. This attribute continues until the radio source is resolved, it is at this point that the observer can determine the extent of the source

**Multi element interferometers**

Extending this concept to more than two elements allows for the creation of multiple baselines. A two element interferometer has one baseline, a three element has three baselines, a four element has six, etc. The baselines are between any two of the multi element interferometer and create a "line" of measurement projected onto the sky. With multiple lines, a two dimensional picture of the radio source can be derived. The more elements, the better the clarity of the source.

The Very Large Array (or VLA) in Socorro, New Mexico is made up of 27 moveable 90 foot dishes. These dishes are typically placed in one of three configurations, giving low, medium and high resolution depending on the requirements of the science being conducted. Each dish has several low noise receivers set to about 330,660,1420 and 1666 MHz. Also, there are S band, C band and X band frequency capabilities for use with the NASA Deep Space Network and the Arecibo 1,000 foot dish. Each baseline between each possible pair of dishes is sampled, correlated, and run through a plotting algorithm to produce very detailed maps of radio sources.

The Very Long Baseline Array is made up of several 90 foot dishes, each with multiple feeds (10 bands) whose elements are placed from the Virgin Islands to California. Each element receives a signal, records it on wideband video tape, along with a reference signal derived from a very stable oscillator. These oscillators are typically Hydrogen Maser type or Cesium Beam with long term accuracies of one part in 10E-15 seconds. As such all the elements although not connected by cable or microwave link, are synchronized very accurately. All of the combined baselines are sampled as with the VLA, and maps are produced with resolutions in the micro-arcsecond range, significantly better than any optical instrument. These arrays have discovered such phenomena as stellar jets (which initially appeared to be moving several times faster than light) and amazing detail of the center of the Milky Way and the centers of other galaxies. A Japanese satellite with a large antenna, similar receivers and oscillators was recently flown to extend the baselines of several thousand miles to several tens of thousands of miles.

**Coelestat**

Invented by Russian radio astronomers in the 1950s, this antenna is made up of at least two elements back to back (for instance two dishes) aimed at two flat reflectors, the resolution, like the previous interferometers is dictated by the distance between the reflectors but the advantage is that a single receiver can be used without long cables and extra amplifiers.

**Ring Array CSIRO Solar Heliostat**

Located outside of Parkes, Australia, a ring of 15' diameter dishes were constructed in a circle of approximately 500' in diameter. This ring of antennas was fed in phase to a central laboratory where there was a phasing apparatus and a series of receivers set to VHF and UHF frequencies. The small dishes tracked the sun and the outputs where (in one manifestation) connected via receivers to transducers on one end of a barrel of water. These transducers were arranged in the same relative position and order as the outside array. Another set of transducers were arranged in a X-Y arrangement on the other end of the barrel and using the focusing properties of the water, were able to produce an image of the sun at several wavelengths. Using this apparatus, astronomers were able to watch ionized gases and highly magnetized loops move about the solar surface and above it. Eventually the water barrel was replaced by digital electronics capable of providing even higher resolution images. Several manifestations of this design have been build around the world specializing mainly in solar research.

**Lloyd's Mirror**

This type of radio telescope is always found near a large body of water, typically the ocean. A reasonably large antenna is pointed to the East (normally) and the elevation is set to the horizon. The combination of the direct rays and the reflected rays from the water's surface of the radio source create an inexpensive interferometer. The field of view is somewhat restricted with this approach, however many high resolution maps were made of the radio universe using this technique.

## Other Types

### Conical

A large array of conical spirals was assembled at China Lake in the 60's for the purpose of bouncing signals off of the Sun's corona. The conical spirals allowed for very broad bandwidths.

### Horn calibrators NRAO – Cass A

At Green Bank, West Virginia a horn antenna approximately 20 feet long was permanently placed at the same declination as the Cassiopeia supernova remnant. Data was taken with this transit instrument of the signal level on a daily basis for many years. In this way, the decay of power for this remnant was determined.

### VLF (Reber)

Grote Reber, the pioneering radio astronomer, in the 1970's moved to Tasmania to construct a very large series of arrays to explore the frequencies at and below 1 MHz. The research was difficult due to the fact that the ionosphere is typically opaque at these frequencies. On rare occasions, an "ionospheric lens" is created where by a significant amount of radio radiation can propagate through the atmosphere and at time focus and amplify the flux for easy detection on the ground. The lower the frequency of observation, the more predominant the effects of Earth's Ionosphere as mentioned above. The very lowest frequencies, from approximately 30 to 100 Khz, the effects of the magnetosphere can be observed. In fact there has been some research done trying to correlate the occurrence of a Gamma Ray Burster (GRB) to radio emissions at these very low frequencies.

### Decametric

Starting in the 1950's, signals from Jupiter were received using a Mill's Cross. Later arrays of wide band antennas, like log periodics, corner arrays with discone feeds and other wideband approaches were designed and built. The radio bursts were categorized into at least five types with the predominant emissions coming from three distinct Jovian latitudes. The closest moon, Io also has an impact on the emissions as it goes through a plasma torus surrounding the great planet. The wideband emissions have unique characteristics that in some way emulate similar outbursts from the sun. This led to the theory that Jupiter was actually a proto star and in fact produces more energy that is reflected from the sun. If the sun where to cease producing light for a moment, we would still be able to view Jupiter, as it glows due to its own energy output.

**Future Radio Telescopes**

The ability to observe detailed aspects of high energy physics, analyze molecular distributions in our galaxy and catalog some of the most fascinating attributes of the universe is the realm of radio astronomy. To that end, newer and more capable instruments are being designed and are coming on line soon. The requirements are better resolving power, more sensitivity, and broader bandwidth. No easy task for the antenna designer, but to better understand the dynamics of our universe, this challenge must be undertaken.

An example of a new telescope design with wide band phased arrays operating in the GHz range and a lower frequency array in the background.

# Appendix 1 – Historical Perspective

## Lightning, Magnets, Reflections, Shadows

Millions of years ago, as hominids were developing into bipedal savannah living beings, the wonder of fire and lightning created the first ideas about harnessing their power. As fear was certainly the first emotion, certain curious hominids ventured forth to examine the fires created by lightning and the craters made by meteors. Soon, these creatures learned to make fire and reshape still hot meteors into useful tools. Eventually, loadstones were discovered and static electricity could be created by rubbing certain materials together. In addition, reflections of light were observed from shiny bits of mica or quartz. Later, certain metals were able to be formed into mirrors. Lastly, shadows were observed around campfires or while walking in the sun. Although we take these phenomena for granted, many years ago they created wonderment. It is from these fundamental observations that electricity was discovered, compasses were made, reflectors were formed and principles of optics were discovered.

## Electricity

### Thales (624 – 546 B.C.)

In approximately 600 B.C. a Greek philosopher named Thales experimented with amber and silk to create sparks and attract small bits of organic materials like straw and feathers. He found that rubbing the silk and amber would create varying amounts of attraction or sparks. He also found that loadstone had varying degrees of magnetic attraction. The Greek word for amber is *elektron* and the word for magnetism is *magnesia*. From his observations and writings we now use terms like magnets and electronics.

Thales is known as the first philosopher, he theorized that all things are derived from water. He was a student of the Delphic Oracles, women who answer questions about life and the origin of the universe. Thales also made fundamental observations in mathematics and in fact, he was the first (or one of the first) astronomers to predict a solar eclipse. He spent time in Egypt to learn geometry and surveying, then examined the Babylonian records of the motion of the stars and sun; from these records he was able to correctly deduce the date of the eclipse. As a philosopher, he believed that philosophy was the trunk of knowledge with the branches being subjects like mathematics and electromagnetism. From these branches would grow more branches as major scientific subjects would be broken down into sub-categories. For instance, biology is now broken down into categories like microbiology and ecology. Philosophers who followed Thales, like Plato and Aristotle, furthered the notion of the tree of knowledge and contributed much to describing scientific phenomena.

Thales of Miletus

Miletus

**Gilbert (1544 – 1603)**

In the year 1600, William Gilbert experimented with magnets and charges, following the work of Thales. He invented the electroscope, an instrument that measures charges. The electroscope is constructed with a conductor at the top leading to a glass case that held a gold leaf suspended from the lead in. By applying a charge to the outside conductor, the gold leaf moves away from the inside conductor at an angle proportional to the charge. The glass case has graduations marked on it record the movement. The electroscope can examine the magnitude of either positive or negative charges. This instruments led to the creation of the volt meter and amp meter. It also led to the understanding of what effects different magnitudes of charges had on other phenomena. Gilbert also understood that the earth was a giant magnet and that loadstones could be fashioned into compasses and dip needles. His most famous book was called "De Magnete" and was full of valuable facts on electricity and magnetism.

Dr. William Gilbert

**Volta (1745 - 1827)**

Count Alessandro Volta was an Italian physicist and responsible for the determination of the electrical potential or volt. Typically known at that potential that places one amp across one ohm, this standard has become very important for the determination of power and the design of electrical circuits. Volts can be either direct current (DC) or alternating current (AC). Insofar as antennas work with AC, we will limit our discussions to this kind of potential. Volta invented the voltaic cell around 1800, he then connected several in series to create a battery. In 1820 Hans Christian Oersted discovered that a wire connected to a battery and placed near a compass causes the compass needle to swing. This was the first time that a relationship between electricity and magnetism was established.

Alessandro Guiseppe Antonio Anastasio Volta

**Ampere (1775 - 1836)**

Andre Marie Ampere, a French physicist, invented the solenoid and was able to characterize its operation. Ampere was a professor of physics and chemistry at Bourg in France at the early age of 26. He has been credited with the development of the subject of electrodynamics. In his honor, the amp has been defined as 1 Coulomb (6.3 x 10e18 electrons) per second. Finally, Watts were defined at Volts times Amps.

Andre-Marie Ampere

Ampere is also credited with the invention of the galvanometer, which measures the strength of magnetic fields.

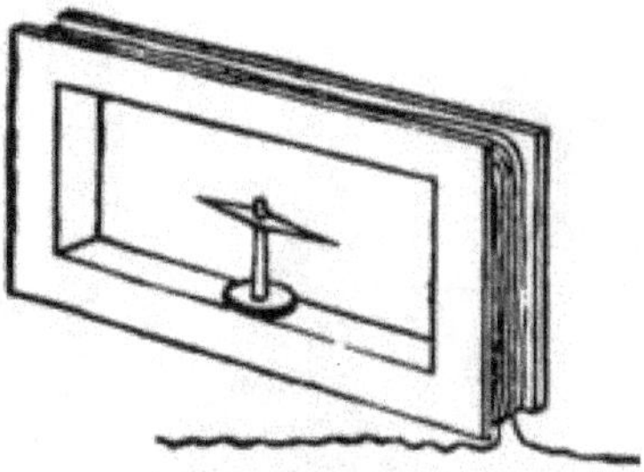

Simple Galvanometer

$$\Delta \times \boldsymbol{B} = \mu J + €\mu \, (\partial E / \partial t) \qquad [83]$$

Ampere-Maxwell Law

**Coulomb (1736 - 1806)**

Charles-Augustin de Coulomb introduced the inverse-square law for electrostatic forces, analogous to Newton's inverse-square law for gravity. Electric effects seem to involve action at a distance. This is known as Coulomb's law

A son of a well to do family, Coulomb studied at prestigious schools and worked for the British government designing coastal charts and military forts. He became interested in the strength of masonry.

He also studied elasticity and finally experimented extensively on the charges on surfaces, electrical and magnetic forces, producing mathematical theories to explain them. Ultimately he wrote a series of very important papers on this subject and laid the foundation for Maxwell's equations.

One coulomb is the amount of electric charge transported in one second by a steady current of on ampere. One coulomb is also the amount of charge stored by a capacitance of one farad to a potential of one volt (44)

Charles-Augustin de Coulomb

**Gauss (1777 - 1855)**

Carl Friedrich Gauss, German mathematician and scientist contributed to many aspects of science. A mathematician, astronomer and physicist, his contributions started early in his career when his first important works were published at age 21.

As a child prodigy he is credited with many feats of mathematical prowess including the ability to add a great many numbers in his head. After his death his brain was preserved and studied.

Gauss was able to quantify the actions of magnetic fields and derived formulas for the understanding of how moving charged particles generate magnetic fields.

Carl Friedrich Gauss

$$\vec{\nabla} \bullet \vec{E} = \frac{\rho}{\varepsilon_0}$$

Gauss's Law

**Faraday (1791 – 1867)**

Michael Faraday of London found that changing a magnetic field could produce electricity. This discovery of course let to AC and DC alternators and generators. Faraday's roots were in the bookbinding business, where he was able to read the latest scientific publications. An avid reader, he started to experiment with electromagnetism and along with discovery of electricity from varying magnetic fields, he found that a piece of bismuth was repelled by both ends of a magnet. This principle is known as diamagnetism. His discovery earned him several honorary degrees. He was also offered the presidency of the Royal Society but turned it down to further pursue his research.

Faraday introduced the concept of magnetic lines of force.

Michael Faraday

$$\Delta \times \mathbf{B} = -(\partial \mathbf{B} / \partial t) \qquad [95]$$

Faraday's Law

**Maxwell (1831 - 1879)**

James Clerk Maxwell published his mathematical theory of electromagnetism in 1860. This landmark theory initially caused much controversy amongst his fellow professors, primarily because he proposed many ideas that at that time could not be tested. In this publication, he proposed the possible existence of electromagnetic waves and at the same time postulated that if such waves could ever be produced, they would travel through free space. The speed of light itself, said Maxwell, is propagated as an electromagnetic wave, and should differ from electrically produced waves only in their wave length and frequency. Because Maxwell gave clues as to how such waves might be generated or detected, their real existence was not discovered until thirty two years later when Heinrich Hertz made his important discoveries. Maxwell is know today primarily for his four equations (actually derived from his famous book by Heaviside). These equations will be discussed later in some detail, but let it suffice to say that with these tools one can analyze almost all facets of electricity and magnetism. They relate the electric and magnetic field vectors, their sources, which are electric charges, currents, and changing fields. Maxwell's equations played as important a role in electromagnetics as Newton's laws did to classical mechanics. He died of cancer at age 48.

James Clerk Maxwell

**Heaviside (1850 – 1925)**

Oliver Heaviside was a self taught engineer from England, who spent several years studying Maxwell's book on Electromagnetism, *Treatise on Electricity and Magnetism*. Of it he said,

*"I saw that it was great, greater, and greatest, with prodigious possibilities in its power. I was determined to master the book...It took me several years before I could understand as much as I possibly could. Then I set Maxwell aside and followed my own course"*

As a result of his labors, Heaviside simplified Maxwell's 20 equations with 20 variables to four equations of two variables, which is what we use today.

Heaviside left school at age 16 to study at home in the subjects of telegraphy and electromagnetism. He continued to do so until age 18 when he took his first and only job as a telegraph operator. He published several papers during this period, then quit work at age 24 to return studying full time. He helped develop transmission line theory and develop the mathematics to describe various design methods. He also research the skin effect in transmission lines and in 1880 patented the coaxial cable, a ubiquitous device found in every home, office, car and aircraft. It was soon after this that he reformulated Maxwell's equations.

He predicted the existence of an conducting layer in the upper atmosphere, which is now known as the ionosphere (sometimes as the Heaviside layer), which allows high frequency communications over thousands of miles as it reflects off of the various ionized layers.

A recluse, he lived at home for most of his life, never married and was at time at odds with the established scientific establishment. His work however on Maxwell's equations greatly simplified them from a purely integral form to a more useful vector form and made them immensely useful. To this day we actually use the Heaviside forms to design antennas. In addition such terms as impedance, inductance and attenuation were coined by Heaviside. Other terms he coined include conductance, permeability and susceptance.

He also developed and advocated the use of vector methods to analyze electromagnetic circuitry.

Very prominent members of the Royal Society nominated him for inclusion based on the work he did with Maxwell's equations. His equations were known as the HH equations (Heaviside/Hertz) for many years as they allowed for the development of many new antenna types and concepts. In the 1930s, Albert Einstein recommended that they be renamed "Maxwell's" equations as that was the origin of the complete understanding of electromagnetism. We have used the latter term ever since.

During his explorations of Maxwell's equations, Heaviside introduced the concepts of dot products and cross products, still used widely in advanced mathematics.

Heaviside was the first recipient of the Faraday Medal and passed away in 1925.

Oliver Heaviside

Maxwell's Equations as simplified by Heaviside are as follows:

# Maxwell's Equations

$$\oiint \vec{E} \bullet \hat{\mathbf{n}}\, dS = \frac{q}{\varepsilon_o}$$

Gauss's Law

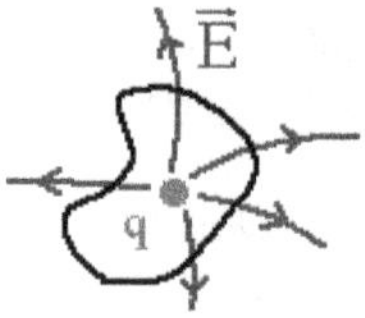


$$\oiint \vec{B} \bullet \hat{\mathbf{n}}\, dS = 0$$

(no monopoles)

$$\oint \vec{B} \bullet d\vec{l} = \mu_o\left(i + \varepsilon_o \frac{d}{dt}\Phi_E\right)$$

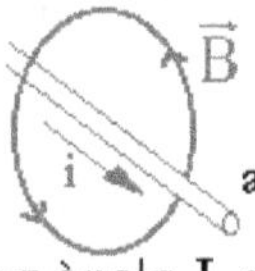


and

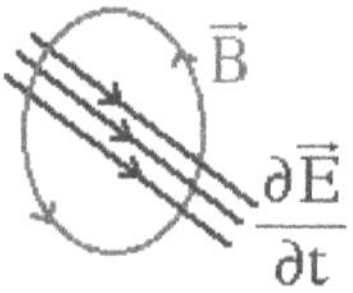


Ampère's Law

$$\oint \vec{E} \bullet d\vec{l} = -\frac{d}{dt}\Phi_B$$

Faraday's Law

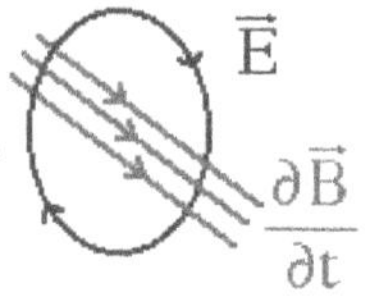


---

$$\vec{\nabla} \bullet \vec{E} = \frac{\rho}{\varepsilon_o} \qquad \vec{\nabla} \times \vec{B} = \mu_o\left(\vec{j} + \varepsilon_o \frac{\partial \vec{E}}{\partial t}\right)$$

$$\vec{\nabla} \bullet \vec{B} = 0 \qquad \vec{\nabla} \times \vec{E} = -\frac{\partial \vec{B}}{\partial t}$$

(Differential Forms)

where:

Equation one denotes the divergence of the electric field equals charge density divided by $\varepsilon_0$ . $\varepsilon_0$ is the permittivity of free space and turns out to be approximately 0.0000000000088541878176 .

Equation two shows that the divergence of the magnetic field is zero. In other words, magnetic fields do not have a beginning or end, they always exist in loops. Also, a magnetic monopole cannot exist for it implies that the magnetic field would emerge, then vanish into space.

Equation three indicates that the curl of the electric field is minus the rate of change of the magnetic field. The curl of the electric field is a measure of how "curly" it is, how much it twirls around in circles. The curl of the electric field is equal to... minus the rate of change of magnetic field. (The $\partial$ symbols and the *t* on the bottom are the mathematical symbols meaning "rate of change".) The rate of change means how fast the magnetic field is changing. If the magnetic field *isn't* changing, then the rate of change is zero, and the electric field is non-curly. If the magnetic field is changing, then the electric field goes curly. If the magnetic field is getting weaker, the rate of change is negative, and the electric field curliness is positive. If the magnetic field is getting stronger, the rate of change is positive, and the electric field curliness is negative - this just means that it rotates in the opposite direction.

*Curliness in the electric field pushes electric charges around in circles.*

*Electric charges going around in circles is an electric current.*

In other words: moving magnets around generate electric current. This is the basis for the invention of the electric generator.

Equation four shows that the curl of the magnetic field equals $\mu_0$ times the letter **J**, plus $\mu_0$ times $\varepsilon_0$ times the rate of change of electric field. So the curliness (or strength) of a magnetic field is equal to $\mu_0$ times current density. Current density is how much electric current is present in a certain area. So if you have an electric current, you will have a magnetic field curling around it. [44]

The $\mu_0$ (Greek letter mu and a subscript zero) term is just a constant number that makes the units of measurement come out right and is known as the permeability of free space. When the magnetic field is measured in *teslas* and current density in *amperes per square meter*, $\mu_o$ equals about 0.000001256637061.

Equation one says that the amount of electric field depends on the amount of electric charge. Equation four also shows that the amount of magnetic field depends on the amount of electric current .

When an electric field gets stronger or weaker, it generates a magnetic field.

If you assume the current density **J** is zero (i.e. there are no electric currents) evaluating equations three and four yields:

$$\nabla^2 \mathbf{E} = -\mu_0 \varepsilon_0 \frac{\partial^2}{\partial t^2} \mathbf{E}$$

$$\nabla^2 \mathbf{B} = -\mu_0 \varepsilon_0 \frac{\partial^2}{\partial t^2} \mathbf{B}$$

It should be obvious that these two equations are identical, except that one refers to the electric field, while the other refers to the magnetic field. In essence, what these two equations say is that an electric field that changes in time (on the right) generates an electric field that changes in space (on the left) in a precisely determined way, and likewise for magnetic field. In fact, these equations describe the motion of ripples of electric and magnetic field, which travel through space like waves. (44)

Solving these differential equations, one can determine the speed at which these waves travel. The speed works out to depend on those numbers $\varepsilon_o$ and $\mu_o$, thus:

$$Velocity = \frac{1}{(\sqrt{(\varepsilon o * \mu o)})} \qquad [88]$$

James Clerk Maxwell, in 1865, formulated this theory of electricity and magnetism, and wrote down the equations that describe their actions . He wrote down the terms of these equations based on experiments with electric batteries, bits of wire, and magnets. He also realized that he needed to define some constants ($\varepsilon_o$ and $\mu_o$) to make the numbers come out right. He measured the values of those constants using batteries and wires and magnets. The numbers didn't make any sense to him at the time and appeared to be some constants that nature required.

Applying some algebra to these equations, one can generate an equation that only refers to the electric field, and an almost identical equation that refers to the magnetic field. Solving these equations describe the motions of waves of electric and magnetic fields. With further algebraic calculation the speed of the waves are realized depend on those values measured for $\varepsilon_o$ and $\mu_o$.

If one takes the values measured for $\varepsilon_o$ and $\mu_o$, multiplies them together, takes the square root, and then take the reciprocal. The answer is a speed, so it has units of speed, in this case meters per second. And the answer is very close to 300,000,000 meters per second. Converted into miles, that's a little over 186,000 miles per second. James Clerk Maxwell, a brilliant physicist, immediately recognized what this number was... the speed of light.

In addition, the impedance of free space can be defined as:

$$Zo = \sqrt{\frac{\mu o}{\varepsilon o}}$$

Solving for *Zo* obtains a value of 377 ohms.

## Early RF

### Hertz (1857 – 1894)

Heinrich Hertz was a German physicist and did his most famous work at the Karlsruhe Polytechnic and University of Bonn. He studied under Helmholtz in Berlin and it was he was introduced to Maxwell's electromagnetic theories. He was able to produce and detect electromagnetic waves through various apparatus. In addition, he was able to demonstrate their properties of reflection, refraction and interference. Hertz used a tunable dipole with movable capacitive spheres activated by a spark gap as a transmitter. He also used a loop of wire spark gap in the center as a receiver. By tuning the dipole (by moving the spheres along the wire), Hertz was able to resonate the system. He was able to receive the charges up to hundred feet away. In addition, Hertz used large metal plates to demonstrate reflection and a prism of paraffin to demonstrate refraction, and a lens of pitch to demonstrate focusing of radio waves. Later, Hertz was able to design and produce a cylindrical reflector with feed. Unfortunately, Heinrich Hertz died prematurely and as a result science lost one of its most promising disciples.

Heinrich Hertz

**Tesla (1856 – 1943)**

Nicolai Tesla is probably best described as an inventor, showman, and scientist. An immigrant from Yugoslavia he came to New York in 1884. His first experiments were designed to create large amount of voltage using AC current, transformers and various filter combinations. He produced several large voltage sources that were capable of illuminating gas contained in glass spheres and long tubes without direct connection. The New York Worlds fair of 1890 showcased some of his inventions in a spectacular light show. Later in his career, he built a laboratory in Colorado Springs, Colorado. This lab was designed primarily to generate millions of volts using the famous "Tesla Coil" principle. The current consumption became so great during one night he literally caused all of the power to go out in the city. One of the principles he discovered and that found practical purpose was the remote reception of energy through the air using a transmitter / receiver pair. Tesla also theorized that energy could be transmitted with sufficient quantities to supply cities from power plants located across the Atlantic Ocean. Scientists did not take Tesla seriously at the time, primarily due to his tendency to sensationalize discoveries. Recent examination of his work, however has called into question who really invented the radio. In fact, the patent Marconi was awarded for the invention of radio at the early part of the century was overturned in favor of Tesla in 1939.

Tesla used the principle of resonant circuits, spark gaps, and transformers to get sufficient energy to radiate an electromagnetic field through a top loaded monopole. The loading was accomplished using a capacitive sphere. Modulating the spark gap allowed transmission of intelligible signals. To receive, Tesla used resonant circuits (at the same frequency as the transmitter, of course) connected to a coherer and finally to a loud speaker or headphones. The coherer was made up of iron filings contained within a tube and had a rectifying diode effect. Demodulation took place in the coherer and was then used to drive the headphones.

Nicola Tesla

Tesla experimenting with wireless transmissions from his laboratory in Colorado Springs. Photo is a composite.

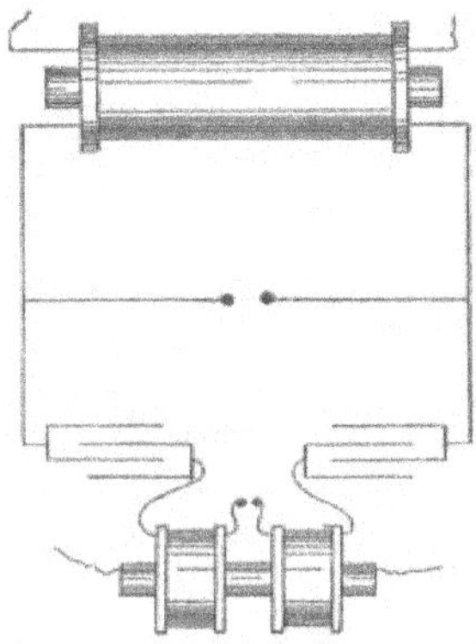

Tesla's Spark Gap Transmitter

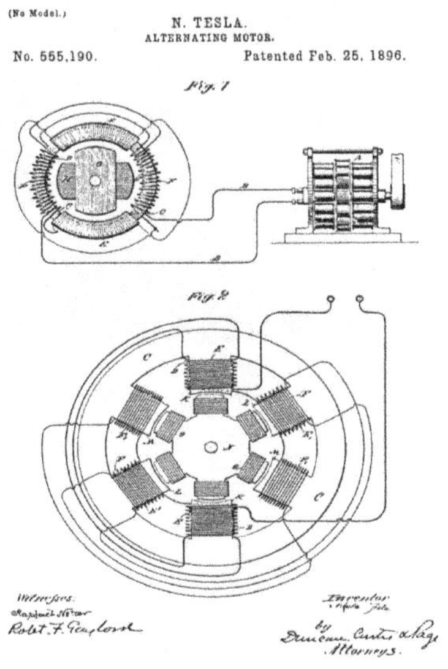


Tesla’s AC Motor Patent (the first Oscillator)

**Marconi (1873 – 1939)**

Guglielmo Marconi was an Italian inventor and had an intense interest in the design of Heinrich Hertz's antennas and radio system of 1886. He discovered that by improving power of the transmitter by stretching a wire high in the air and improving Hertz's detection techniques, he could transmit information at great distances. His experiments worked for 12 miles in 1898 and extended to thousands of miles by 1903. Radios of his design we on ships and coastlines to enable communications with vessels at sea. The contributions of radios systems were dramatized in the USS Republic and HMS Titanic disasters. Before these events, ships were completely isolated when at sea. The antennas he used were typically constructed of inverted cones suspended from 100-200 meter towers.

Guglielmo Marconi

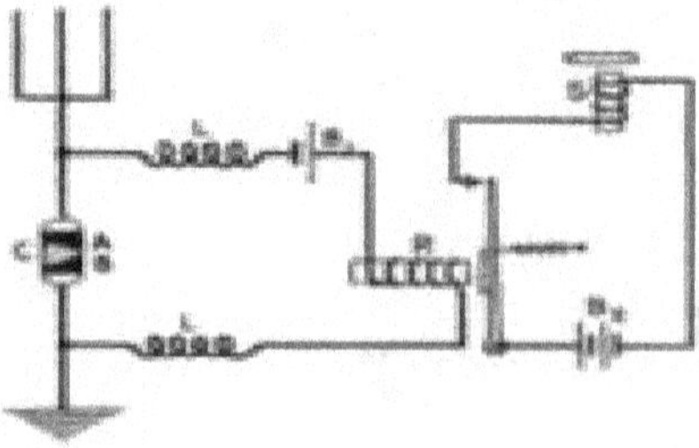

Example of Coherer Receiver

**DeForest (1873 – 1952)**

Lee DeForest was a professor at Yale and is most famous for the invention of the vacuum tube. He also designed the first triode tube that was capable of improving the performance or transmitter and receiver sets. As a consequence, DeForest designed and installed the first 5 radio stations for the Navy. Later in his career, he worked extensively on the development of talking motion pictures. He received many awards for his inventions during his career including the Gold medal at the St. Louis exposition in 1904. The antennas he used consisted of long wires and inductive loading.

Lee DeForest

**Fleming ( 1849 – 1945)**

Sir John Fleming was the inventor of the "Fleming" valve. This essentially was tube and acted as a diode. This meant that it was capable of rectification and was used extensively in radios. Edison had been able to evacuate the air out of a tube for the purposes of making a bulb. After much experimentation, Fleming was able to activate a "filament" on one side and connect it to a cathode. On the other side of the bulb was a anode that "received" electrons emitted by the cathode. Because the electrons are negatively charged, they move in only one direction inside the diode. As a consequence, attaching a transformer to create voltage on one side of the filament plus cathode and the other side to the anode allow only the positive portions of the AC waveform to pass to a load attached from the another transformer. The resulting positive waveforms can be smoothed by a capacitor and in turn source DC current. Diodes like these can be used for the rectification of signals after amplification inside a radio receiver. As a result of his work, Fleming was able to design and build a radio receiver in 1904.

Fleming worked for a while with Maxwell, Edison and Marconi.

Sir John Ambrose Fleming

**Armstrong (1880 – 1954)**

Major Edwin H. Armstrong as the inventor of the super-heterodyne receiver. In 1921, he and several other radio amateurs set up transmitters and receivers transmitting signals at 200 meters across the Atlantic Ocean. Using a beverage antenna, the transmission of information was successful in the communications attempt. As a result, long distance communications were then maintained at this and lower wavelengths. Armstrong is also credited with the invention of FM radio and in later life spent much of his time defending his patent.

Edwin H. Armstrong

**Karl Jansky (1905 - 1950)**

As a result of Armstrong's ( and others) work, it soon became obvious that long distance communications at short and long wavelengths were subject to changes in the atmosphere, in particular, changes in the ionosphere caused major changes in the ability of communicate occur at certain wavelengths. In general, it was found that nighttime communications were better than daytime. Depending on the time of day or night, there was a maximum usable frequency that if exceeded, caused blackouts. Much research was being performed in this area in the laboratories to try to understand this phenomena. Karl Guthe Jansky was working at Bell Telephone laboratories during this period and was assigned the task of analyzing the variability of long distance communications. To do this he designed and built a directional antenna array known as a Bruce curtain. This consisted of a serpentine driven element in front of a reflector of similar dimensions. At 14 meters frequency, (approximately 21.4 MHz) this antenna was quite large and was supported by automobile tires on a concrete circular track. This antenna was movable in azimuth so that Jansky could analyze the positions and intensities of atmospheric noise. After several months of work, Jansky noticed that a faint hiss could consistently be heard in the direction of the Milky Way. As a consequence of his discovery, Jansky founded the field of Radio Astronomy and now has the honor of having his name used as the primary unit of measurement (flux density).

Karl Guthe Jansky

**Reber (1911 – 2002)**

Grote Reber was the only person to follow up on Jansky serendipitous discovered after World War II. He built the first prototype parabolic dish in his back yard in the 1930s. With this dish (9 meters in diameter) he was able to make the first maps of the radio universe. This antenna was a prime focus reflector with an altitude over azimuth mount. He used several feed systems, 145 MHz, 400 MHz, and 900 MHz.

Later, Grote moved to Tasmania to build low frequency phased arrays at around 2 MHz to explore the radio Universe. Some of his equipment still remains to this day.

Grote Reber

## Early Antennas

### Tesla Top Loaded Monopole

Although it is unclear if Tesla initially intended his various high voltage apparatus to transmit radio waves. It is obvious now that these structure radiate large amounts of radio energy.

Tesla soon discovered this and constructed large high voltage transmitters and detectors similar to Hertz's. As a consequence, Tesla is now credited with the construction of the first transmitter / receiver pair. He was able to understand the importance of greater power as applied to transoceanic distances.

In fact, Tesla at one point was constructing huge 300 foot wooden towers on Long Island for the purpose of communications and electric power distribution around the world but was unable to complete the transmitter. However, communication techniques flourished with his many inventions. In any event, his antennas took on the form of inductively coupled monopoles with capacitive spheres placed on top. Many were built and he understood how to resonate structures with power sources to make efficient transmitters.

His multitude of inventions included transmitters and receivers with components like antennas, transformers, inductors, capacitors, and resistors. In addition, Tesla considered ship detection and collision avoidance in 1898 a precursor to radar systems. Tesla also wrote about guidance and remote control. In fact, Tesla was able to design a remote control "torpedo." for use by the Navy. All of these innovations are used today in modern communications equipment.

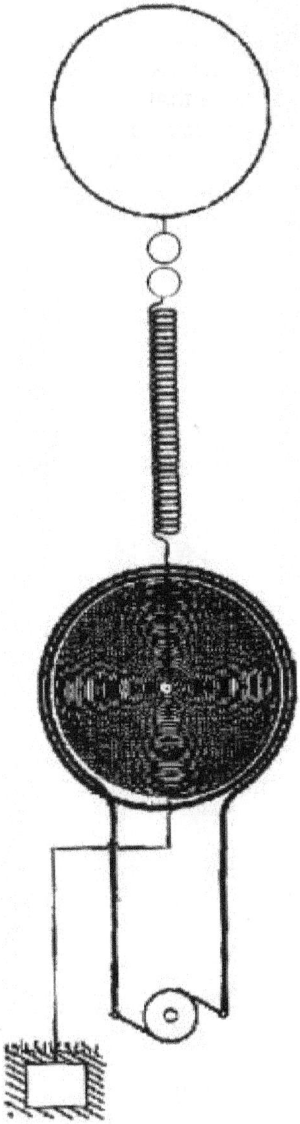

Tesla's Top loaded monopole, or “magnifier”, built in Colorado Springs, 1899

**Marconi "Cone"**

Marconi's early antennas took the form of a large (200' tall) conical structures. These systems were typically made with four towers supporting a wire ring that then supported wire that came to a point at ground level. By driving this point against ground, Marconi was able to radiate signals in an omni-directional pattern at short wave frequencies. These structures are well known today and are used for broadband transmission and reception. Variations on this technology include discones and biconical radiators.

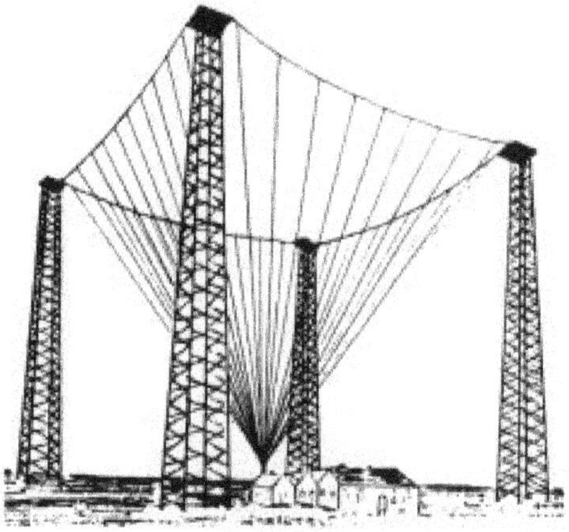

Marconi Cone Antenna, c. 1901

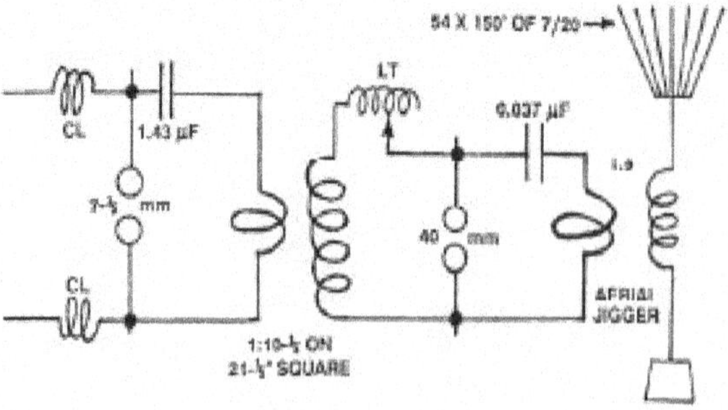


Marconi Cone Antenna Transmitter

Marconi at his radio

This cone antenna and transmitter were paired with a similar system on the Irish coast to produce the first trans-atlantic messaging system. Before this, underwater cables were used, which were very expensive.

These radio systems operated at very low frequencies compared to today's operations. Signal were sent at 500 KHz, 1 and 2 MHz as this was the limit of the technology available at the time. These frequencies are used today as the "AM" band and one can imagine the difficulty in listening to signals thousands of miles away using these frequencies. Nightime operations of course were the best and even today, if you listen in the evening on the AM band, you might here stations at great distances. The Titanic for instance had the two lower bands in operation during their voyage. Propagation at these low frequencies is problematic as there is a significant amount of terrestrial noise. As the frequency of operation moved up the spectrum, signal clarity increased until the use of the ionosphere was not longer possible, around 30 MHz.

Although Marconi was credited with invention of the first radio system, this title was rescinded recently as it was proved that Tesla was sending signals before Marconi.

Marconi curtain Antenna, 1901

**Hertz Dipole and Reflector**

Clearly, Hertz was well ahead of other physicists with his development of dipoles, loaded dipoles and reflectors. He constructed several variations of each and was able to propagate frequencies between 30 meters and 30 centimeters (10 MHz and 1 GHz). Other experimenters were able to go even higher in frequency with horns an waveguides, it is interesting that work at higher frequencies like these was progressing near the turn of the century, however after this point activity was idle until World War II when radar systems where being developed. Hertz was able to use his dipoles and reflectors to prove Maxwell's theorems and show optical properties of radio waves in the form of reflection, refraction, polarization and focusing. Had he not died prematurely, he would have advanced the art of antenna design to even greater heights. Today dipoles and reflectors are used extensively in the communications industry. Few designs of any sort have had the usefulness of at least 100 years.

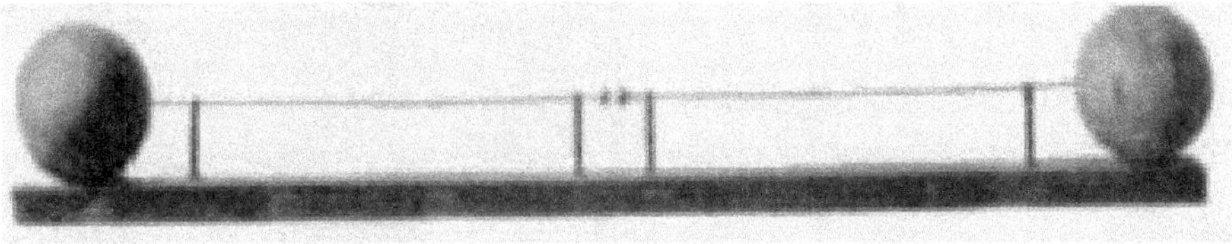

Hertz Dipole, 1898

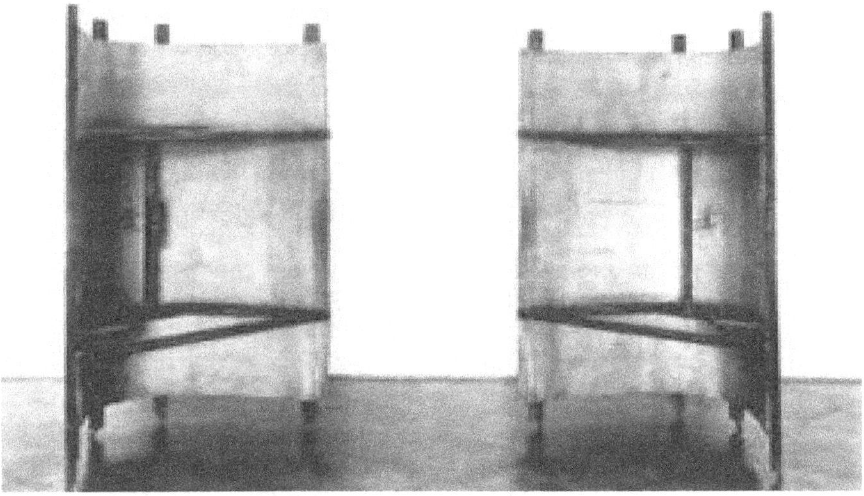

Hertz Transmit and Receive Reflectors

Hertz Polarizing Grid and pitch covered Prism

**Amateur Radio Antennas**

Because of the work of Marconi and Tesla, several visionaries realized the potential of communications and antennas design. These designers were both professional and non professional. A great deal of transmitters, receivers and antennas were designed for communication and great strides in this field were made by these visionaries. As a result government allowed the licensing of radio amateurs in starting in 1912. As their ranks grew, the American Radio Relay League came into being and was made up of many talented designers working toward a common goal. Today at least 150,000 radio amateurs are in existence worldwide, designing all kinds of advanced communication technology and antenna systems. Initially, amateurs used spark gap sets and telegraphy to communicate. Soon, they had a multitude of stations on the air and were able to apply their hobby to the professional world. Many antennas types were designed by these experimenters and articles were published on their innovations and performance. Today, some of the leading antenna designers are radio amateurs and enjoy designing antennas for such applications as low frequency short wave to moon-bounce to shuttle communications. Contributions from this group have been enormous and too numerous to discuss in this work.

Early Amateur Radio Antenna

Early Amateur Radio Station, showing loop and long wire antennas, transmitter is a spark gap type, receiver is crystal type

**Early Radio Astronomy Antennas**

As mentioned before, Carl Jansky built the first antenna for use in characteristizing shortwave propagation. In essence this antenna (known as a Bruce aerial) acted like a phased array with directional qualities. As a consequently, Jansky was able to point the antenna to large "spots" the sky and make measurements. Using rudimentary receiving equipment, he was able record atmospheric and exo-atmospheric radiation. The style of the Bruce aerial is that of a broadside serpentine with a reflector of slightly larger dimensions behind it. It was fed in the center and a feed cable led to a building where the receiving equipment was housed. The antenna was built to operate at a frequency of about 20 MHz. This antenna was built in 1927 and is still intact and viewable at the National Astronomy Observatory in Greenbank, West Virginia. After the work of Jansky and the amateur radio pioneers came WWII where most radio amateurs were called into duty. During the war, the government strongly considered getting rid of the amateurs in favor of a radio operators solely for military use. This mandate was changed based mainly on the effort of the first president of the ARRL (Amateur Radio Relay League). As a result, many new innovations from the amateur community including antenna designs were put into use.

Karl Jansky and the first Radio Telescope, a Bruce Aerial

# Appendix 2 – Radar Antennas

Typical Shipboard Radars

## General Description

A basic radar systems consists of transmitter, receiver, data reduction system. The transmitter typically is capable of high power, especially in a pulses mode. Although there are several different types of of radar systems, pulsed types are most prevalent for use in tracking aircraft and examining weather. A pulsed radar creates two major considerations for antenna designers. The first is average power and the second is peak power. In a typical air traffic control radar, the transmitter can put out at least 20 KW in a peak pulse, which can lead to arcing. Because the pulses typically are repeated at a rate of between 1 Khz and 15 Khz, the average power must be calculated. Average power can lead to heating of the antenna components.

Another main consideration for a radar antenna design is the shape of the beam, including the magnitudes and positions of the sidelobes. For aircraft radar systems a cosecant squared pattern is best for ground coverage. In the case of air traffic control radars, low sidelobes near the ground are important to mitigate ground clutter.

Radar systems are typically monostatic, or designed with one antenna that is used for both transmission and reception. Bistatic systems use two antennas, one for transmission and the other for reception which minimizes the extra complexity of a transmit / receive switch but has the problem of isolation between the transmit pulse and the receiver sensitivity.

To determine the gain of the antenna for a particular radar system, one should understand the basic calculations for a radar system, especially the basic radar equation and the radar cross section equation.

Assuming a monostatic radar system, the power density illuminating a target at range R is:

$$S = \frac{(P * G(\theta, \phi))}{(4 * \pi * R^2)} \qquad [94]$$

where:

P = Power

G = Gain over area $\theta, \phi$

Gain is determined by:

$$G(dBi) = 10 * \log\left(\frac{(4 * \pi * Ae)}{\lambda^2}\right) \qquad [95]$$

where:

A = Area of antenna

lambda = wavelength in same units as Ae

The magnitude of the reflected energy from the target is known as radar cross section and is defined by:

$$RCS = \frac{(Reflected\ Power)}{(Incident\ Power\ Density)}$$

[96]

or:

$$RCS = \left( \frac{(Ps(\theta r, \phi r, \theta i, \phi i))}{((Pt * G)/(4 * \pi * Rt^2))} \right) \quad [97]$$

where:

Ps = Power over target area or EIRP

Pt = Power of transmitter

G = Gain of antenna

Rt = Range

Once we have the radar cross section, we can now calculate the power received at the radar antenna due to the reflected energy from the target:

$$\frac{Pr}{Pt} = \left( \frac{((G^2 * \lambda^2) * RCS)}{((4 * \pi)^3 * R^4)} \right) \quad [98]$$

where:

Pr = Power received by the radar antenna

Pt = Power of transmitter

G = Gain

R = Range

**The Radar Equation**

The most useful way to understand the antenna requirements for a radar systems is to evaluate the radar equation. This is done to determine what size antenna is required to produce the desired signal to noise level and ultimately produce the most detailed image or range measurement possible.

A designer typically balances the output power of the transmitter against the gain of the antenna. The gain of course is directly related to the beamwidth and as a consequence the expected target dimensions can drive the antenna size as well. Consider the classic radar equation:

$$Pr = \frac{(Pt * G * Ae * RCS * F^4)}{((4\pi)^2 * Rt^2 * Rr^2)} \quad [99]$$

where:

Pt = Transmitter Power

G = Gain of the Transmitting Antenna

Ae = Effective Aperture area of the Receiving Antenna

RCS = Radar Cross Section, or scattering coefficient of the target

F = Pattern Propagation factor

Rt = Distance from the transmitter to the target

Rr = Distance from the target to the Receiver

In most radar systems, the transmit and receive antennas are the same, this formula can therefore be simplified to:

$$Pr = \frac{(Pt * G * Ae * RCS * F^4)}{((4\pi)^2 * R^4)} \quad [100]$$

This shows that the received signal strength declines as the forth power of the range. The pattern propagation factor F = 1, is a simplification for a vacuum without interference. The propagation factor accounts for the effects of multipath and depends on the characteristics of the environment.

## Radar Antenna Types

### Boundary Layer or Wind Profilers

In the case of boundary layer radars, the inertial subrange is determined. This is done by evaluating the meteorological phenomena a designer wants to examine vs. the available radar bands and their concomitant bandwidths. The size of a bubble of air will need to be ½ wavelength to reflect the radar signal, this fact determines such things as maximum operational height and the range resolution. These radars typically are built with phased arrays positioned horizontally to look up into the air and determine wind velocity and direction at several altitudes. A typical boundary layer radar antenna has a minimum of 5 "beams", including one vertical and 4 others at (e.g.) 15 degree offsets from vertical and pointed North, East, South and West. Using these beam positions, Doppler motion of the air is determined and thus speed and direction. The vertical beam determines vertical velocity and with acoustic sounder, can measure temperature at various altitudes as well.

### Air Traffic Control

In the case of air traffic radars, the frequency of operation is typically between 400 MHz and 5 GHz to best illuminate the aircraft and minimize reflections due to weather. These radar antennas can be scanned (like the radars one typically sees at airports), tracking (like an altitude/azimuth positioner used to follow the positions of aircraft for the military) or phased arrays (like the Cobra Dane or other flat radar antennas used to track multiple targets and minimize jamming).

### Weather

In the case of weather radars, rain is best illuminated by radar working from approximately 8.5 GHz and above. Insofar as there are several frequency bands where there is significant water absorption, radars around 20, 40 and 60 GHz are rare. One exception to this rule is a radar system used on the Space Shuttle, which operates at roughly 14 GHz and is used to locate space debris, help navigate the Shuttle to satellites in need of repair and establish approaches to the International Space Station.

### Millimeter

Above 60 GHz, there exist a class of radars known as the millimeter radars, which offer high imaging resolution and can be used to guide aircraft through fog and haze during landing and takeoffs. Antenna configurations for this type include offset parabolas, lenses and quasi – optical designs.

### Lidar

Finally, a class of radars operating at infrared and optical frequencies use classical antenna design formulas to determine reflector size and / or lens specifications. Lidars are a typical example of this type of antenna. These are used to determine air movement and composition.

## Special Considerations for Radar Antennas

The presence of high power drives most antenna designs used for radars. When using a reflector type design, it is common to use waveguides in the construction of the feed assembly. Waveguides can accommodate very high power levels, the highest of which might require the use of pressurization with dry air or neutral gas to minimize the generation of arcing. Powers of up to many megawatts has been used using waveguide. In normal radar systems the power is pulsed at a certain rate, known as the pulse repetition frequency. The level of the pulses is determined by the transmitter design. The pulse level has a direct effect on the arcing potential and the average power, determined by multiplying the peak power (or pulse level) times the amount of time the pulse is active (or pulse width). The average power will influence the heat generated on the antenna and thermal considerations like heat sinking are examined here.

The size and shape of the radar antenna beam is also important. For instance, air traffic control radars are usually rectangular or elliptical in shape, making a beam shape that allows the greatest coverage of height and the narrowest coverage of azimuth.

Integrated Cross Pole Ratio (ICPR) is also a factor for polarized Doppler radar systems. There is a need to have balanced cross pole response in both polarizations to minimize distortion in the measured backscatter. This is typically an important requirement for weather radar systems used today.

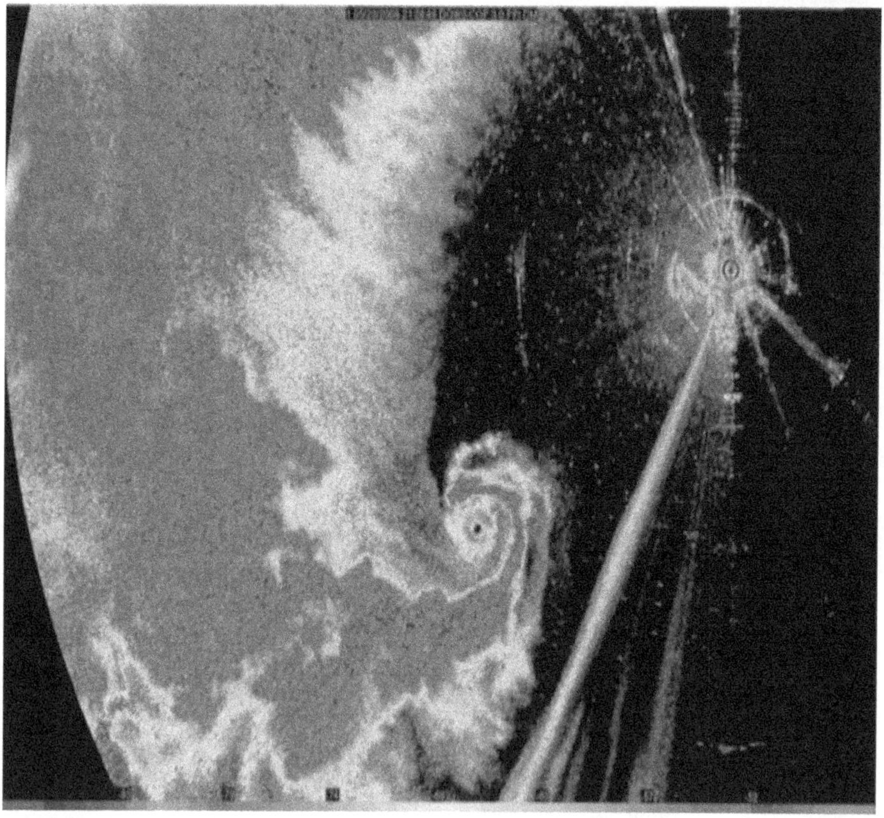

A Plan Position Indicator or PPI display, showing a tornado and outlying storms

Arecibo 300 Meter Radio Telescope and Planetary Radar
Puerto Rico

Nestled in a natural bowl formation in the jungles of Puerto Rico, this radar antenna, with a spherical surface and Gregorian feed, has mapped the surfaces of Venus, Mars and many asteroids with a multi Mega watt transmitter and low frequency ability. In addition, signals were sent out from this antenna with SETI greetings to nearby star systems. The site has been used for several feature films including "Contact".

Overhead View of HAARP Array, Alaska

Used over several low frequencies, including 7.4 and 9.4 MHz, this radar has enough power to warm the ionosphere. It has been used for a multitude of scientific experiments and consists of a multitude of rhombic type phased array elements.

The HAARP project directs a 3.6 MW signal, in the 2.8–10 MHz region of the HF (high-frequency) band, into the ionosphere. The signal may be pulsed or continuous. Then, effects of the transmission and any recovery period can be examined using associated instrumentation, including VHF and UHF radars, HF receivers, and optical cameras. According to the HAARP team, this will advance the study of basic natural processes that occur in the ionosphere under the natural but much stronger influence of solar interaction, and how the natural ionosphere affects radio signals. (45)

Wind Profiler Array and Aerostat Radars (55)

**Wind Profiler**

The wind profiler array shown is constructed from a series of yagis with imbedded phase shifters. The array is sparsed, with more element density in the center and less at the edges to affect a Taylor like amplitude distribution, lowering sidelobes to around -30 relative to the main beam. The frequency of operation is 449 MHz and this system is used to keep the Aerostat in the background from being launched in poor wind conditions. Wind velocity and direction can be determined up to 18,000 feet above the surface with this radar. The building below the antenna array houses a 2 Kw transmitter and other electronics.

The Aerostat shown is much like those found along the southern U.S. Border, the radome underneath the balloon houses a sophisticated military grade radar that can locate moving airborne targets as well as map ground vehicle movement. These radars are suspended on tethers up to 15,000 feet and can monitor several thousand square miles of ground and airspace. To give a sense of scale, the model aerostat shown above is actually larger than a Boeing 747. Typical radars used in aerostats are from F-15s or other sophisticated acquisition, targeting and tracking aircraft. Communications repeaters are also housed in the lower radome that allow for significant increases in ground based transceiver ranges.

Air Traffic Control Radar (ASR-9) showing two feeds for elevation determination and a top mounted Identify Friend or Foe (IFF) or Transponder Antenna. This is a Cosecant Squared Reflector system.

Found at most large airports, this radar is used to map aircraft in flight using a rotating assembly, as shown. The two feeds allow for some elevation discrimination, the main reflector is shaped to provide an elliptical beam with the smaller dimension placed vertically. The flat array on the top of the main antenna is for activating and receiving transponder signals. The top array works around 1.05 GHz and the main antenna 1.3 GHz. In practice the upper array shows the identification, altitude and vertical speed of an aircraft on the main air traffic control displays. The lower antenna is for "painting" the aircraft or finding "primary" targets, such as small planes without transponders and other objects.

These radar systems are connected together over a nationwide grid and can send sector scans of most any airspace to government and military installations as needed.

Cobra Dane Military Aircraft Radar System

Situated near several U.S. Coasts, this radar is an electrically steered active phased array capable of mapping a multitude of targets at great range. The elements, of which there are thousands, all have integrated phase shifters allowing adaptive radar techniques to be employed. This includes multiple beams, steerable nulls (to minimize interference) and very quick target acquisition.

Smaller versions of this radar can be found on many U.S. Navy ships allowing for electronic beam stabilization (as the ship rolls) and 3-D tracking.

Example of Mobile Doppler Weather Radar (56)

Weather radars come in a variety of types, some are designed for aircraft and typically work at 9.3 GHz. Others like the one shown above, are set on alt-az mounts to allow pointing and mapping of thunderstorms and fronts. Placed on a flatbed truck as shown allows researchers to drive to close proximity of a tornado and map the speeds and distribution of water particles. Some weather radars are made of two antennas closely spaced and can also be used for determining doppler velocity of the cloud droplets. One antenna is used for transmitting a constant or CW signal, these transmitters also will move this CW over frequency, using frequency or FM modulation. The other antenna in the pair is for reception, with a high degree of isolation from the first antenna.

Example of a Planetary Radar Antenna, Australia

Used to map planets, asteroids and the moon, these large antennas along with powerful transmitters are used to watch near earth objects and track satellites. The used of digitally encoded signals from radars such as these allow for very fine resolution to be realized when looking at planets in our solar system. Also, the location and details of asteroids and even space debris around our Earth can be examined.

Example of Shipboard Radar

The radar shown above is typical of the shipboard radars, however software associated with its output can detect oil spills. The highly rectangular design creates a beam that is highly elliptical, causing a significantly large vertical profile and a narrow azimuth footprint. This compensates for the rolling of the ship. Many of these radars work in the 14.5 GHz range employing various inter pulse periods (for different ranges) and power in the several Kw range. The newer systems use digital techniques for display and signal recovery.

This type of radar is also used for the detection of birds around airports. The wide elevation beam allows for quick detection and can delay the departure or landing of commercial aircraft to increase safety margins. Birds have caused a multitude of serious aviation accidents, including the destruction of jet engines and the downing of aircraft. The miraculous landing of a US Air A320 twin engine commercial flight out of La Guardia in 2009 was a prime example of the risks of flying in the proximity of large birds.

Example of Electrically Steered Phased Array Radar

This is an example of an X band targeting radar that can track multiple targets simultaneously. Made up of several thousand individual elements which are phase and amplitude controlled, this radar can create "beams" in microseconds. Coupled to the attitude reference system in the aircraft, the output display is thus stabilized for simplified target acquisition and selection.

Agile and adaptive phased arrays used in radar have the technological edge in modern military aircraft. In addition, modern radar systems can send display information via satellite link to a command bases located far from the operational theater to allow commanders to select and prioritize targets.

This type of radar also has the capability of tracking multiple targets and minimizing the effects of jamming by use of adaptive phase array algorithms.

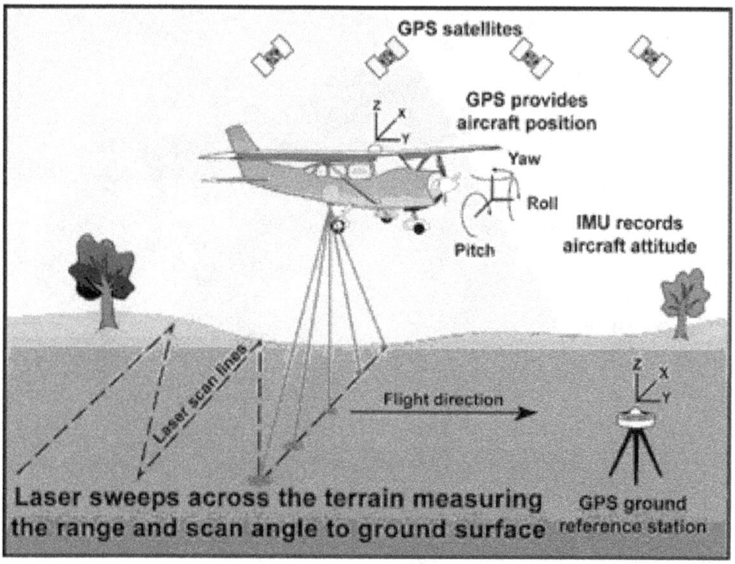


Example of Lidar System

Lidars (or Laser Radars) employ similar architectures as radars at lower frequencies. Composed of a transmitter (Laser), antenna (Telescope) and receiver (Photodetector), these systems can map ground terrain, clouds, aerosols, dust and crops to name a few applications. Based on the backscatter of a laser beam, significant information can be obtained with these instruments.

NOAA operates several Lidar systems for use in viewing weather phenomenon. These systems use high power laser diodes, fiber optics and 12" mirrors to scan a beam horizontally from atop a large van or truck. From the data, scientist can tell wind speeds and direction, humidity and temperature profiles.

**Passive Radar, Bi-Static**

There are no dedicated transmitters in a passive radar system. A special receiver monitors the phase and amplitude of local TV and radio stations to measure the difference of arrival between a direct and a reflected signal. This allows the bistatic range of the object to be determined. In addition to bistatic range, a passive radar will typically also measure the bistatic Doppler shift of the echo and also its direction of arrival. This allows the location, heading and speed of the object to be calculated. In some cases, multiple transmitters and/or receivers can be employed to make several independent measurements of bistatic range, Doppler and bearing and hence significantly improve the final track accuracy.

TV signals, FM signals GSM cellular bas stations, HDTV are a few of the transmission sources used by passive radar systems.

Fundamentally, a passive radar must do the following to track a target:

Reception of the direct signal from the transmitter(s) and from the surveillance region on dedicated low noise, linear digital receivers

Digital beam forming to determine the direction of arrival of signals and spatial rejection of strong in-band interference

Adaptive filtering to cancel any unwanted direct signal returns in the surveillance channel(s)

Transmitter specific signal conditioning

Cross correlation of the reference channel with the surveillance channel to determine object bistatic range and Doppler

Detection using constant false alarm rate (CFAR) scheme

Association and tracking of object returns in range/doppler space known as line tracking

Association and fusion of line tracks from each transmitter to form the final estimate of an objects location heading and speed. (77)

Many aircraft can be tracked using this system with accuracies of 100's of meter vertically and 10s of meters horizontally. Several commercial systems are in use today, they have the advantage of low cost, no moving parts, easy deployment and good accuracies. They do however, require transmission sources and are not as accurate as standard air traffic radar systems.

Several government agencies have demonstrated passive radar systems capable of making a synthetic aperture image of an aircraft target. Using multiple transmitters at different frequencies and locations, a large data set in Fourier space can be built for a given target. Reconstructing the image of the target can be accomplished through an inverse fast Fourier transform. Target classification and Radar Cross Section can be determined as well.

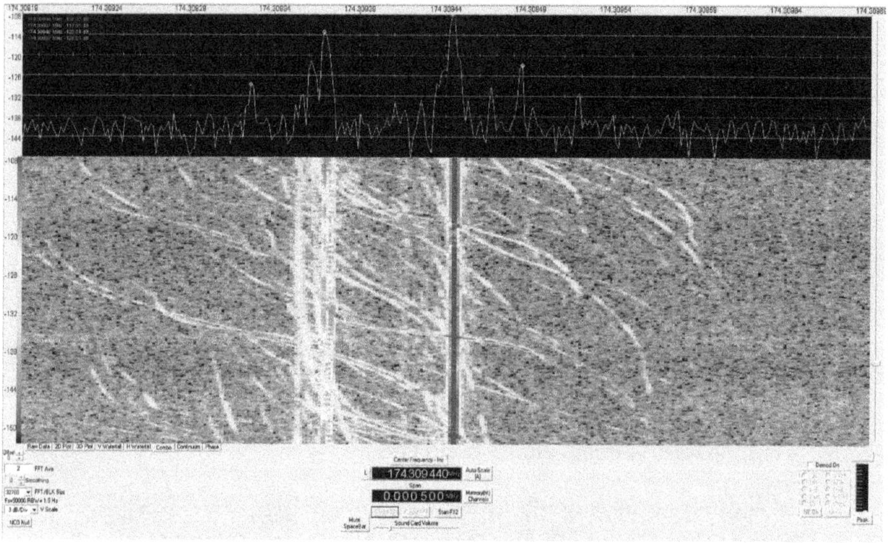

Passive Radar display showing aircraft trails

**Advanced use of Radar Data**

The output of radars can now be post processed to determine small details of thunderstorms and the creation of tornados. As seen below, details of the extent and movement of rain cells can be determined and the amount of rainfall from these cells measured. In addition, with the use of polarimetric radar data (using both vertical and horizontal polarizations) the types of particles within thunderstorms can be determined. This includes the position and magnitude of ice, hail, snow and rain.

TITAN is a software program developed by NCAR for the advanced observation of weather information, mostly radar, although other measurements can be integrated.

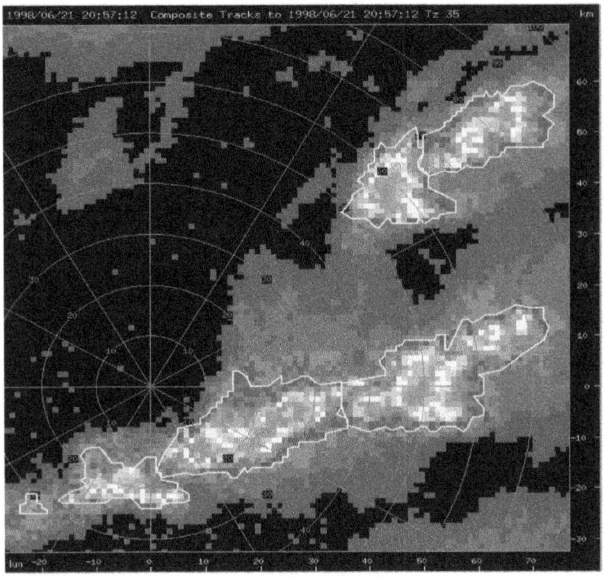

TITAN display of radar returns determining the extent and intensity of thunderstorms

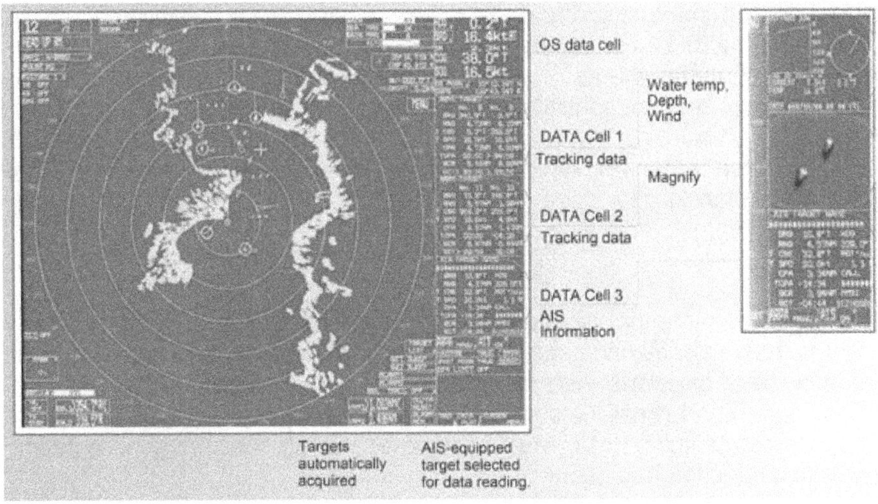


Marine Radar display showing coastlines and boats

# Appendix 3 – Small Antennas

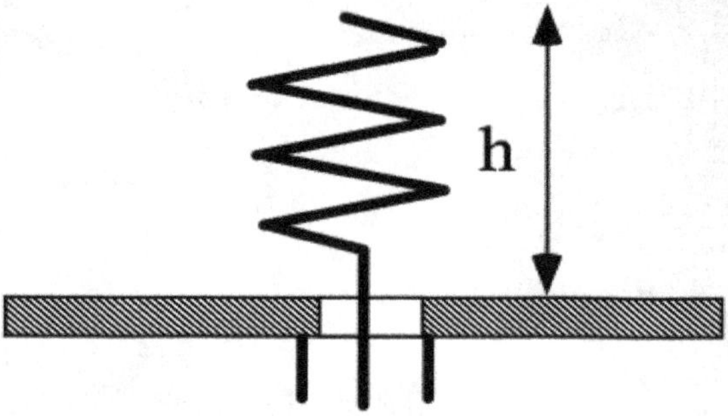


$$h < \frac{\lambda_0}{4}$$

Small antennas are a result of pressure to make antennas less obtrusive and immune to breakage. Cell phones are a classic example of small antenna implementations. In these devices, stub antennas as shown above dominate the market. Internal antennas do not have as much gain as external types and are sensitive to hand and head proximity.

These antennas typically have higher Q (lower bandwidth) and lower gain than ½ and ¼ wave monopoles. Due to improved coverage by cell phone towers and higher operational frequencies, these small stubs work adequately.

Small antenna designs are accomplished by providing matching circuits between the transceiver and the radiator to present a proper impedance to the transceiver and provide a "virtual" match to the antenna. In addition, high Er materials also allow the creation of low profile antennas. Creative combinations of these two approaches is usually how an antenna like this goes into production.

For more information, start with papers written by Wheeler, then follow up with references in Appendix 9.

# *Practical Antenna Design*

## Fundamentals

Start the design by understanding the required bandwidths and evaluate the Friis formulas (path loss) using a typical receiver sensitivity of -100 dBm and transmit power of .6 watts. Careful analysis will reveal why cell tower are spaced so closely and with relatively high gain antennas and power levels.

Small antenna options include monopoles less than a ¼ wavelength in size, dielectrically loaded patch antennas, serpentine layouts on a printed circuit board, and small "F" type antennas with loading circuitry, to name a few. Based on the space available, choose the best option keeping in mind that the more external the antenna the better the performance. After the initial design is accomplished through a good electromagnetic simulation program, or on paper if need be, prepare a representative piece of hardware that will emulate the final form of electronics that the antenna is to be connected to.

Use this hardware to match the antenna as well as evaluate patterns and gain. The goal is to provide the best match to the electronics while maintaining good gain bandwidth and uniform pattern.

## Gain

The smaller the antenna relative to a standard ¼ wave monopole on a ground plane, the lower the gain. Internal antennas suffer the most due to losses incurred by the surrounding materials and proximity to conductive components. A designer can be fooled easily by designing an antenna with a nice match but very poor radiating qualities. Both attributes must be optimized for best performance.

Consider the orientation of the antenna in normal use, this might allow the designer to optimize (for instance) horizontal gain vs. vertical gain. Using the internal component effects can be done for this purpose. Again the use of a good electromagnetic software simulation package is recommended to evaluate the antenna layout. Genetic optimizers can be used here to design the best solution.

**Polarization**

Polarization diversity suffers significantly as the antenna gets smaller. The ability to design circularly polarized antennas for instance become even more difficult. Expect energy to be distributed between vertical and horizontal polarizations. If in fact the device the antenna is designed for will be connected to a mobile device such as cell phone or telematics device, carefully observe the patterns in both polarizations to best understand how to design the "base" station, if applicable. If the design is for a mobile to mobile application, then try to minimize the cross pole nulls to enhance operation. If the designer has control over the base station design, consider base stations facing each other with orthogonally polarized antennas. For instance, vertical to horizontal or left hand to right hand. Consider also slant linear designs where both polarizations are used but without a phase shift between them. In these cases the optimum coverage can be realized. Finally, use a good RF coverage software suite that takes into account terrain, foliage and distance.

**Patterns**

Understandably, the pattern uniformity suffers significantly with small antennas, especially those in internal settings. The interaction with surrounding material both conductive and non-conductive can be debilitating. Again, as mentioned above, understand the application, evaluate the problem using good software tools and pick the best solution. In the case of patterns, uniformity and lack of deep nulls should drive the design approach. Obviously if one wants a more directional antenna for non mobile use, this can be accomplished as well.

Evaluate patterns using a good antenna range (either near or far field) and pay attention to cross-pole patterns as well as co-pole.

Finally, take patterns at band edges as well as centered operational frequencies. In most applications a compromise will be in order while evaluating different design approaches.

**Applications**

The cell phone is an obvious example of a small antenna application. Keep in mind that at PCS frequencies (~ 1700 to 2100 MHz) the antenna on a common cell phone is about ¼ long and will operated adequately. For frequencies below the PCS applications (viz 820 to 880 MHz and some GSM operations at 940 MHz) the same size antenna is by definition small and will perform less reliably for a given transmitter power and receiver sensitivity. As such, the cell tower placements become very important for good coverage.

Besides the cell phone there are several interesting applications for small antennas, one of which is the inventory scanner, where hand held devices in stores are used to scan the UPC codes on merchandise and send the codes and quantities to a resident store placed receiver. These devices use packet technology with automatic error correction to maximize high quality data transfer in an electromagnetically complicated environment. In other words, the transmission of a data packet can be done several time until good data is correctly transmitted. In many cases a check sum is added to the data packet which is the sum of the ASCII character decimal equivalents added together and modulo 256 converted to derive a unique value for the data string in the packet. The receiving computer's software observes its own check sum calculation and compares it to the check sum transmitted by the inventory scanner. If it does not match the store computer asks for a retransmission of the packet. Tacking place at reasonably high data rates, the retransmission is hardly noticed. Use of small antennas in this case eliminates broken monopoles and allows for an aesthetic design.

Other types of small antennas also include the "RF Tag" designs where for very low frequencies, loops are implemented and using a standard LC circuit, resonate at the desired operational frequency. In higher frequency applications dipole and loaded dipole designs dominate. These applications rely on close proximity to a scanner to work well. Some of these devices employ a diode that upon reception of the scanner frequency, emits a harmonic that is detected and signals the proximity of the device. Further advances include the use of CMOS circuits that upon scanning, can emit a harmonic with some data as well. The driver here is in very low cost and the industry is maturing to the point that someday most goods in a store will have a rf tag of some sort.

Another popular application of a small antenna is with telematic devices, or those devices which are designed to receive positioning information (normally from GPS) and upon query, send this information to a cell tower or other receiver. In the case of LoJack, the transceiver receives an activation code from a local transmitting tower and then beacons its own identification code to direction finding receivers, typically in police vehicles. Relative direction and range is displayed leading to the stolen car. Many more telematic designs employ the GPS approach whereby information is transmitted containing latitude, longitude, speed and direction of travel. For fleet operations this has become a necessity. Also, for applications like over the road freight hauling, this solution works well. In most of the telematic applications it can be highly desirable to have a hidden small antenna, to avoid detection. This is achieved by placing resonant antenna structures in less than desirable places in a car, truck or piece of construction equipment. These small antenna applications require designs which are impervious to the proximity of conductive and non conductive parts. In some cases, an active antenna tuner is employed to optimize operation. The best solution is to restrict the installation positions and test for every conceivable application.

**Transimpedance**

If a small antenna is receive only, there are several options available to optimize performance. Instead of using a matching circuit, one should consider a transimpedance amplifier as shown below. This can take the form of an op amp for low frequencies or a FET for higher. The advantages are low noise and wideband operations. Several fine designs are available for receivers at low frequency through UHF using these techniques.

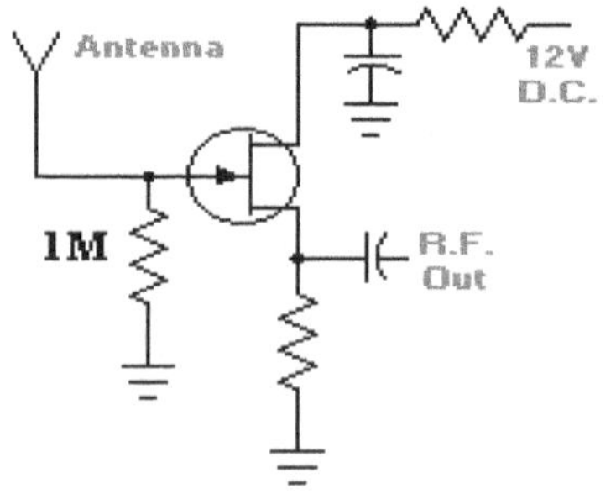


Transimpedance amplifier with antenna

**Materials**

**High Er**

Use of high dielectric constant materials such as alumina and barium titanate allow the designer the flexibility of designing small antennas which operate relatively efficiently over reasonable bandwidths. The key is to use a material with a low loss tangent. Several favorite materials are used in this antenna style, refer to Appendix 7 for more details.

**Plastics**

Low loss plastics are hard to find, with the exception of some foam plastics. A designer should minimize the thickness and expect lower radiation efficiency in the presence of large amounts of plastic. These materials however are used effectively for radomes and support structures, just not as substrate material. Refer to Appendix 7 for more details.

## Popular Small Antenna Types

Several antenna designs lend themselves well to use in small antenna applications. Resonant shapes of various configurations include the following:

1. "F"
2. Dipole
3. Loop
4. Ceramic loaded monopole
5. Structures imbedded in PC board
6. Slot

The factors that drive antenna selection include:

1. Amount of available space
2. Type of materials in close proximity to radiating element
3. Pattern requirements
4. Gain requirements
5. Bandwidth

Given sufficient room, always get the radiating element away from materials that will absorb radiation. If possible, use outside apertures and antenna placement.

Place pattern nulls in areas of little use, for instance cell phone s can have nulls facing the ground.

# Appendix 4 – Unusual / Interesting Antennas

## Introduction

There have been some very interesting antenna requirements that have allowed the creative designer to think "out of the box", using the basic Maxwell's equations and standard testing techniques. These requirements include the need to transmit signals in unique environments, like underwater or under ground.

Creative antenna designers have solved difficult problems required of antenna, like operations in harsh environment, very low frequency operation and conformal requirements.

The following pages show some of these interesting designs, although far from complete, they hopefully will indicate to the reader the amount of latitude an antenna designer should use to solve challenging problems.

Decal type antenna that is placed on automobile windows and can be used at VHF, UHF and Cellular frequencies

**Airborne VLF Antennas**

Antennas have been unreeled from high flying aircraft, submarines, spacecraft as well as high velocity projectiles. In terms of propagated through the waver, very low frequencies work the best, necessitating long antennas for efficient operation. For spacecraft, the need for compacting the shape for launch in a restricting shroud mandates release and deployment mechanisms. In this case reflectors in excess of 100 feet in diameter have been flown using compressed wire mesh, inflatable shapes and flat metal bands that extend much like a tape measure used by carpenters.

Picture of E-4B Aircraft, with trailing wire antenna (cone like object at end of fuselage) which deploys up to 28,000 feet from aircraft and hangs vertically during flight when aircraft orbits.

**US Patent # 1,315,862**

**Radio Signaling System**

**(9 Sept. 1919)**

**James H Rogers**

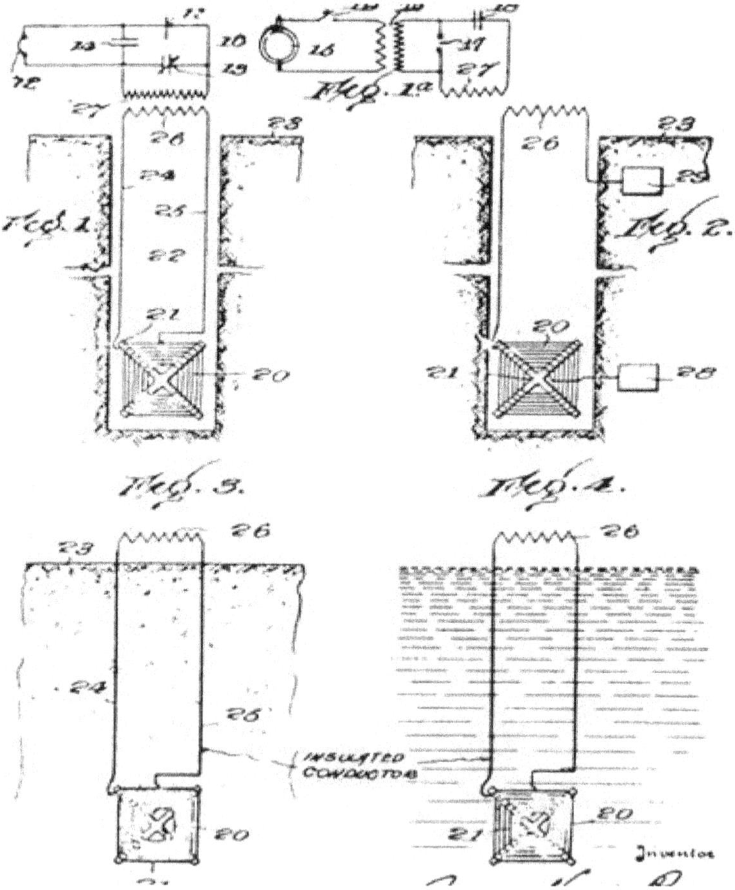


Example of an underground antenna patent

**Ground Penetrating**

For ground penetrating applications, VHF and UHF antennas are required to be small, employing dielectric loading techniques and serpentine approaches. In this field there are also a class of antennas which are buried underground to allow communications during nuclear and conventional attacks. The placement of antenna in concrete and buried well in the ground has been achieved and perfected.

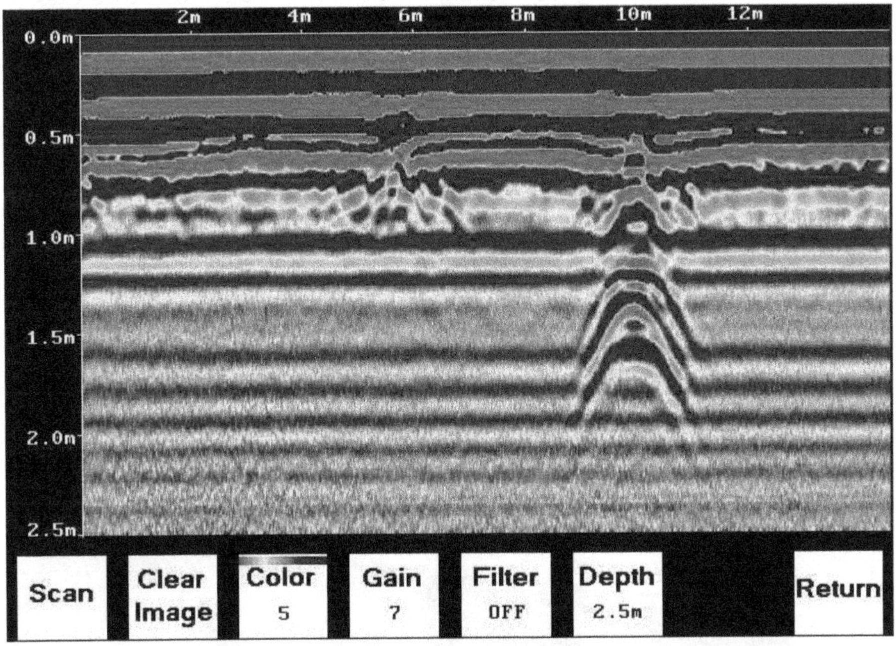


Image from a ground penetrating radar using small antennas

**Human Body Antennas**

Antennas designed to be worn on a human body have special considerations, none the least of which is variable dielectric loading requiring special matching techniques. In this case the selection of frequency is very important, as the higher ones tend to be absorbed. In the future, members of the military will require radio contact both digital and analog to transmit images and receive instructions.

**Stealth**

There is also a class of antennas that must operate in a stealth mode, or hidden to be effective. Many types of these antennas have been manufactured to be placed inside of cars and other vehicles (e.g. LoJack) or the new aesthetic requirement to have the standard car antenna placed in a window to remove any opportunity to have the standard whip removed or damaged. Aircraft can move through the air more efficiently if the communications and navigation antennas are conformal, this is also true of rockets and the Space Shuttle. In the case of stealth aircraft, all antennas are specially designed to be conformal and non obvious. Finally, due to the great amount of cell phone tower placement, alternatives to a metal structure have introduced flag pole "towers," fake buildings made of plastic and fake trees holding antennas.

Shuttle Landing at Edwards Air Force Base, with conformal antennas on top surface, just behind windshields, necessitating a roll maneuver

As far as antenna design is concerned, the progression historically has been from slide rules to calculators to computers to sophisticated modeling programs capable of iterative design cycles. In other words, these programs can take several parameters and incrementally change them, calculate the performance, and track the results. One more interesting approach has been to use the Genetic algorithms, which allow a wide open design philosophy to be applied to a particular antenna requirement. The design examples from this kind of software sometimes are very unexpected. In the future, expect approaches like this to quickly suggest optimal antenna design for a given application.

An example of an antenna designed with genetic algorithms[44]

**Balloon Reflector for Ground Penetrating Radar**

Another interesting antenna design was part of a proposal to NASA for a ground penetrating radar system to be used in a ballon floating above the Martian surface. The radar used the spherical surface of the upper part of the balloon, which was metalized with a thin conductive coating. The surface formed a spherical reflector that, in conjunction with a proper field created a single beam or in the case of a phased array feed system, multiple beams that could penetrate the surface up to several meters. The balloon was designed to float at 72,000 feet above the surface and operate at a frequency of 300 MHz.

Balloon in Flight showing spherical shape

**Russian Vostok Antennas**

In the 1960s, the Russian Space Agency developed a series of communications and telemetry antennas used in their Vostok series spacecraft. These antennas appeared to be folded loops and used at HF, VHF and UHF frequencies. The antennas were paired, on opposite side of the spacecraft to produce an essentially omni pattern. Although the capsules are today essentially the same as the ones developed in the 1960s, the antennas have been replaced with conformal types, either microstrip or cavity.

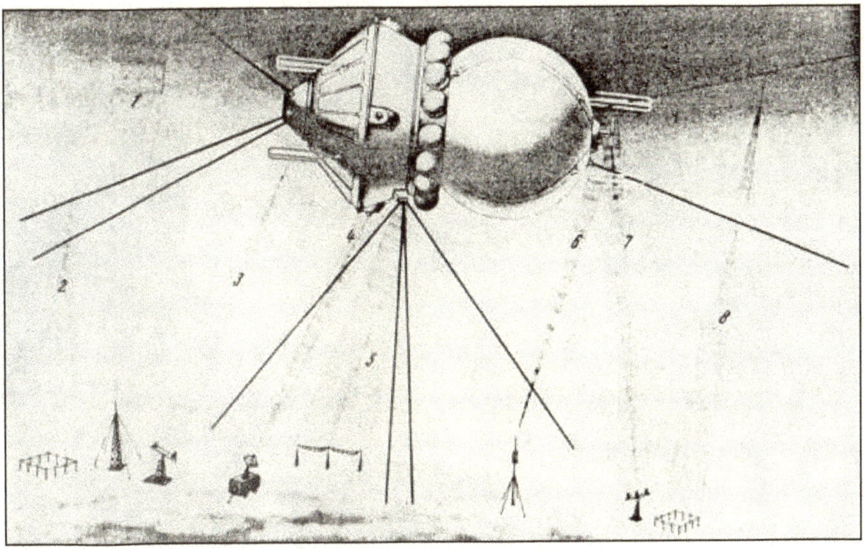


Vostok Antennas and Ground Facilities

**Apollo Command Module S Band Array**

In the 1960s, NASA developed and built a series of space capsules that went to the moon and returned. These capsules included all life support, telemetry and communications equipment for three astronauts for up to a week in space. Communications and telemetry were relayed through an array of S band dishes that were deployed outside of the capsule. TWTs were used as the power amplifiers and the antenna array was fully steerable. Instead of one large dish, NASA engineers took the safe approach of using four smaller dishes, for redundancy. These arrays worked to the moon and back (~235,000 Miles) sending data from the lunar surface as well as TV pictures of the astronauts' activities on the surface. There were also redundant receivers and transmitters used in this system. Finally, omni antennas were attached to the space capsule for shorter range communications, or in the case of a complete failure in the S band array, lower data rates could be sent back to earth.

The Apollo Command Module showing the S band array

**Echo**

In the 1960s, NASA designed and flew two spherical balloons into orbit to be used as reflectors. The balloons (two of which flew, a third did not make it into orbit) were made of .0127 mm thick mylar polyester film. Using a set of beacons at 107.9 MHz for telemetry, the balloon was used as a reflector for 960 and 2390 MHz transmissions. Echo 1 and 1A were 30.5 meters in diameter, Echo 2 was 41.1 meters in diameter. Ultimately the first space based TV and voice transmissions were successfully conducted over transcontinental and intercontinental distances.

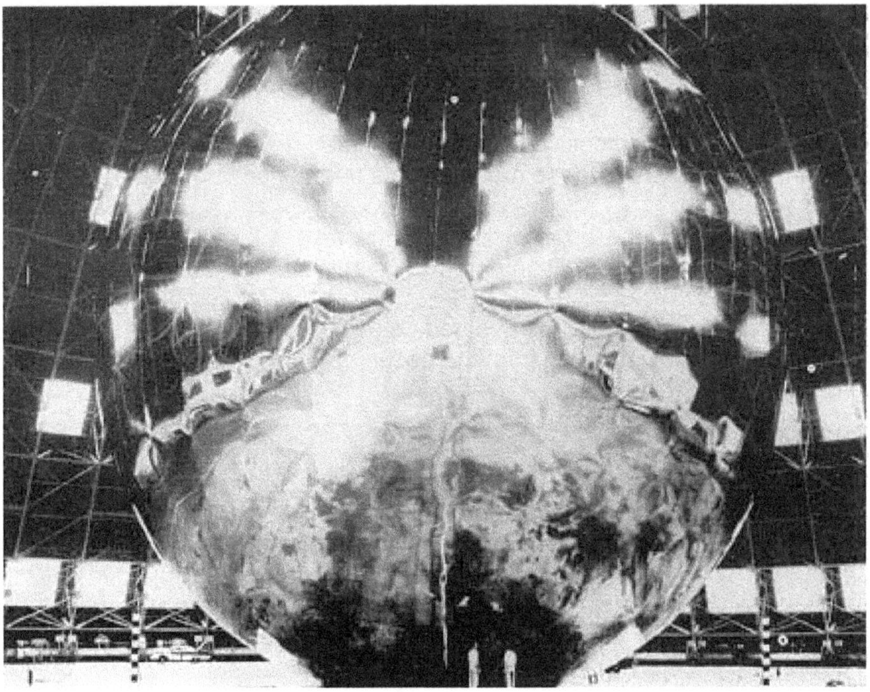

NASA's ECHO Balloon Reflector in Airship Hangar, note men in foreground. Capsule for balloon was about the size of a suitcase.

**Metamaterials**

This term was coined in 1999 by Rodger M. Walser who defined the special materials as:

'Macroscopic composites having a manmade, three-dimensional, periodic cellular architecture designed to produce an optimized combination, not available in nature, or *two or more responses* to specific excitation.'

Also known in electromagnetics as a material which exhibits negative refraction. These materials are important in optics and photonics as they show promise in the design of microwave and optical components like beam steerers, modulators, band-pass filters, lenses, microwave couplers and antenna radomes. The materials are typically constructed of periodic structures like wires or "C" shaped devices which have specific capacitive and inductive qualities.

An important feature of these material is the ability to exhibit negative refractive indices, not found in nature. In addition, several other interesting properties of these materials has been observed, like:

Snell's law still applies however the rays will be refracted on the same side of normal on entering the material.

The Doppler shift is reversed, e.i., the frequency of a radiation source moving away from the observer *increases.*

Cherenkov radiation points the other way.

The time-averaged Poynting vector is antiparallel to phase velocity. This makes this "left" handed material create wave fronts moving in the opposite direction to the flow of energy.

The "right hand rule" commonly known in electromagnetics, is reversed. The electric field, magnetic field and wave vector follow opposite movements.

Some interesting examples of this unique antenna material have been achieved, like:

A *Superlens*, where a negative refractive index provided resolution three times the diffraction limit and was demonstrated at microwave frequencies.

A *Cloaking device* was demonstrated using this material and made an object invisible at microwave frequencies. It is a narrowband phenomena however.

**Theoretical Basis**

Metamaterials allow an electromagnetic wave to convey energy (or have a group velocity) in the opposite direction to its phase velocity. J. B. Pendry (the first to theorize a practical way to make this material) had the initial idea to make metallic wires aligned along propagation directions that could provide the negative permittivity. He demonstrated that on open ring

('C' shape) with an axis along propagation direction could provide the negative permeability as well as a periodic array of wires. The 'C' shapes allow for the actions of inductance (the conductive path of the 'C') and capacitance (the gap).

This phenomenon is sure to have applications in the future, even now agile arrays are being constructed as are lenses in the Tera Hertz regime.

**ESTAR Multi-Beam Array**

In addition to unusual shapes, there are antenna requirements that demand multiple beams, as in the case of receiving several satellite signals at once or tracking several radar targets at once. The use of Rottman and Luneburg lens are examples of but a few solutions to multi-beam antennas. Adaptive phased arrays also are capable of producing multiple beams.

E-Star Array, with Simultaneous Beams

During the 1990s, an array was developed to view several Ku band satellites simultaneously. The construction was unique, using several state of the art technologies to achieve the challenging performance required.

The array elements were helices configured in a rectangular fashion, set in reflector cups and tapered to maximize performance across a

500 MHz bandwidth. The elements were set in 16 rows by 32 columns, columns 1,3,5...31 where composed of right hand circular helices, columns 2,4,6....32 where left hand circular helices. This approach minimized mutual coupling effects and allowed the array to be compact. The feed of each helix was inserted (at one wavelength intervals) into a waveguide feeding each column. The feed of the waveguide terminated in a low noise amplifier. Each of the 32 low noise amplifiers feed an input to a Rottman lens combiner using a special dielectric loading to minimize the size. The output of the lens opened into an arc of ports where low noise block down converters could be placed at positions analogous to the position of the geosynchronous satellites in orbit, thereby allowing reception of several satellite signals simultaneously using the complete gain of the array.

Another E-star innovation was the combination of a reflector and a Rottman lens to again view several geosynchronous satellites simultaneously.

**Shaped Reflector**

For the purposes of creating beams of very specific dimensions, shaped parabolas have been designed and used extensively, particularly in space. Several major aerospace companies employ such designs. Derived from physical and geometric optics, these reflectors look more like potato chips than dishes. As a consequence of their special shape, the antenna beam can take the shape of specific regions on earth, like countries, continents or islands. The drop off of signal strength can be surprisingly steep if the antenna is designed properly.

Commercial satellites use these antennas to cover certain well defined geographic areas for the purposes of broadcasting commercial TV and data.

**MRAS**

In the 1980s, an antenna array was designed to be used as a radio telescope, in particular as an element of the Very Long Baseline Array. The intent was to be able to deploy a high gain antenna system to optimize the U,V coverage of a particular radio source. This array was know as the Mobile Radio Astronomy System (or MRAS) and consisted of at least 90 three Meter dishes with a 10 band feed system. The outputs of the individual elements were sent down fiber optic cables to a central receiving and control station. This station also housed a hydrogen maser for stable frequency control and several wide band magnetic tape recorders to store the output data. The elements were placed in a sparsed arrangement to take advantage of the lower sidelobe performance inherent in the layout.

The individual elements tracked in azimuth and elevation, controlled by signals coming up the fiber optic cable. The feed of the elements consisted mostly of stacked microstrip patches, where for instance a C band sub array was placed over an L band patch, the patch became the ground plane for the sub array. There were four panels of stacked patches arranged on a mechanically rotating feed mechanism. Sidelobes for the elements were minimized by the microstrip layout, in addition very high efficiencies were achieved due to the patch arrangements.

Inside the feed mechanism were several low noise amplifiers along with the associated hardware to switch feeds and convert RF signals to optical.

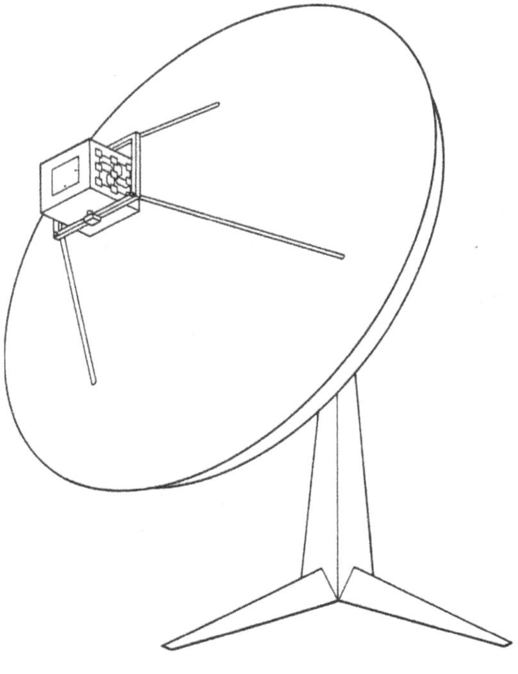

MRAS Element

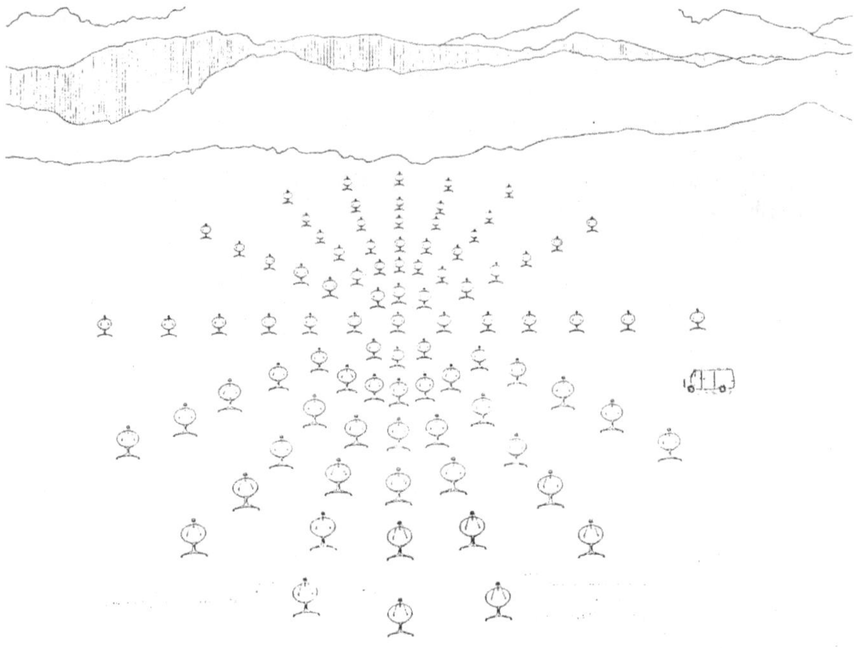

MRAS Array

**RDOW (Rapid Doppler On Wheels), Radar Antenna**

Another interesting antenna employees a two dimensional beam steering concept using waveguide arrays. The subarrays are composed of waveguides with slots cut into the narrow wall. As frequency of operation changes, the delay of phase between apertures allows the beam to steer.

The subarrays are fed by another waveguide with slots cut into the narrow wall. As the phase changes in the feed waveguide, so too does the phase distribution in the sub arrays. As a consequence, between 6 and 10 frequencies are used simultaneously to affect a multi beam radar system that can quickly slice through thunderstorms, tornadoes, hurricanes and other interesting meteorological phenomenon.

This system, in essence can quickly produce a 3D image similar to a medical CT scan or tomographic image.

The RDOW Array, using multiple beams on Tornado

**Nanoantennas** (79)

Invented in 2007 by then graduate student Kenneth Jenson ( a colleague of Alex Zettl), a nano-antenna consists of a nanotube of very small dimensions, about 500 by 10 nanometers (millions of an inch). The antenna works by detecting the motions of electrons in an electric field and physically moving in synchrony. Gravity plays almost no effect, as the scale of the structure is so small. Small electric fields therefore play a dominant role in the operation of this device. As a result of the excellent conductive qualities of a nanotube, electric currents flow from one end to the other, usually to where a conductive pad is placed and further amplification can occur.

Nanotubes are generally thought to have been discovered in 1991, by a Japanese physicist, Sumio Iijima, from the tips of graphite electrodes that emitted arcs. These structures are very strong and conduct electricity better than copper, silver or gold.

If an electric voltage is applied to the nano-antenna, amplification occurs, in fact a small radio of nano-dimensions has been built with surprisingly good results. The entire radio consists of a small antenna connected to a conductive electrode, with another electrode near the other, rounded end. With voltage applied, the antenna physically resonates, amplifies the signal and drives a small speaker.

The frequency of resonance can be adjusted by changing the amount of current on the electric field surrounding the antenna.

**FFT Defined Antenna Beams**

There has been significant advancements in the field of Synthetic Aperture Antennas and Radar systems in the last few years. Initially Synthetic Aperture Radars (SAR) where optically based where the phase and amplitude of the received signals of a generally linear array where sampled as the array was moved over the ground or in space. The outputs were sampled by moving photographic film whose rate of motions was synchronized to the speed of the aircraft or spacecraft. Later this film was processed using Fourier optics into high resolution images. The "Synthetic" part of this technology come from the distance the array moved in space or over the ground, as the radar output data was sampled.

Further development allowed the electronic sampling of phase and amplitude using digital techniques as apposed to photographic film. Modern surveillance radars systems used by the military and other agencies use this approach extensively.

Next, the phase and amplitude were sampled using the In-phase and Quadrature (I and Q) data available from the output of the radar receivers. Fast Fourier Transform (FFT) algorithms can now be applied to this I and Q data to sample the spacial frequencies in multiple dimensions allowing for several interesting applications, including:

Location of multipath source positions, much like passive radar using GPS signals.

Elimination of false echoes in radar data.

Elimination of sidelobe contamination.

Additionally, multiple feeds within a cylindrical reflector can be used in conjunction with FFT beamforming techniques to effectively create a multitude of beams across the long dimension of the cylinder.

**High Power**

High power antennas require special design requirements, to mitigate voltage breakdowns (or arcing). These types of antennas can been seen with TV and Radio transmission antennas as well as high power radar systems. Even satellites must carefully consider the transmission of several thousand watts (e.g. 150 watts per channel times 24 channels) through waveguide, diplexers and feed horn assemblies. In the case of radars, some of these devices operate with millions of watts necessitating very careful design including pressurized waveguides and heat dissipation.

High power antennas are in a class of themselves, where the high currents involved can cause arcing, multipaction and other disastrous effects. It is important for the designer to learn from successful designs and test properly. The highest power antennas, in the Megawatt range have been radar systems, typically using pressurized waveguide. Next in line of power is broadcast stations, starting with TV, followed by FM and AM stations. In many cases, short wave broadcast stations can have significant amounts of power.

Other than waveguide and usually at lower frequencies than radar, coaxial feed lines are employed to transfer the power. These special coaxial lines are usually rigid, pressurized and large. They can be made of brass or aluminum.

Antenna radiators used for high power applications are reflectors for radar systems and dipoles, slots and bowtie types for TV and Radio.

High Power TV antenna

**Low Frequency**

Low frequency communications have been around since the inception of radio and operate from between 3 Khz to 550 Khz (or the lower AM band). Navigation signals like LORAN are present here, as well as communications to submarines. During the formative periods of radio, transoceanic telegraphy was the best way to send messages between continents. Marconi and others designed and build several stations for the purpose of communicating with ships and foreign shores.

Between the years 1918 and 1924, a multitude of low frequency sites were build at frequencies between 10 and 50 Khz. The antennas necessarily had to be large as the wavelengths at these frequencies are around 10,000 to 20,000 meters. As dipoles and yagis are not practical at these wavelengths, top loaded monopoles, loops and loaded long wires were used. For transmitters, it was common to use an alternators spinning at the frequency of resonance. Power levels could be very high and stability was reasonable if proper mechanical care was taken. The mode of operations was code or CW.

21 Khz radio station in Sweden build in the 1920s, for sending messages to the U.S. Note antenna system, consisting of six 127 meter tall top loaded monopoles. This station is on the air a few time per year. [44]

Another important use for low frequency is the communications to and from submarines in the ocean. Propagation depths can be quite deep. The presence of noise from the Earth's magnetosphere in the form of charged particles needs to be dealt with in these systems but with creative digital techniques, very high quality messages can be sent.

In the United States, many power plants employ low frequency radios and antennas to transfer status information.

Also, these frequencies are used to transfer time information from the National Institute of Standards and Technology. Many clocks and watches receive this data to set their time precisely.

Alexanderson Alternator, used for VLF station in Sweden, rotated at 21 Khz, supplying high currents to antenna depicted in previous picture. [44]

Today, low frequency (LF) and very low frequency (VLF) operations are still in use, for long range communications to submarines, the sending of time signals by WWV and for navigation purposes (LORAN). Amateur experimentation is also taking place. Interesting monitoring of the Earth's magnetosphere can be done at these frequencies. Periodically, “whistlers” can be heard, which are charged particles traveling between magnetic poles.

The actions of the solar wind and extra solar radio emissions are being studied at these frequencies as well.

**Russian Antenna Array**

This Radar design uses a multitude of wide band apertures arrayed together to increase gain.

"Shipant" Radar System

**Chinese FAST Antenna**

As of 2016, the Chinese are building the largest aperture radio telescope ever. With a diameter of 500 meters, this telescope will have unprecedented gain over a very wide band of frequencies.

Chinese FAST Antenna

**Spherical Antenna Arrays**

These antenna arrays are design for complete spherical coverage, very useful in space applications where the vehicle must maintain a radio link in any orientation.

**Plasma Antennas**

Significant work has been performed on these unique antennas. They are composed of an enclosure full of ionizable gas, like neon. Once activated, the gas is highly conductive, more so than most metals. As a result, if the enclosure is shaped like a phased array, ground plane or single element; it becomes active or invisible depending on if its activated. When off, the gas is not conductive at all and invisible at radio wavelengths. The concept of using ionized gas is not new, over 100 years ago, an engineer proposed using this technique by blowing gas into the air from two different angles that crossed at altitude. Once this occurred, an antenna was formed.

A Plasma FM antenna

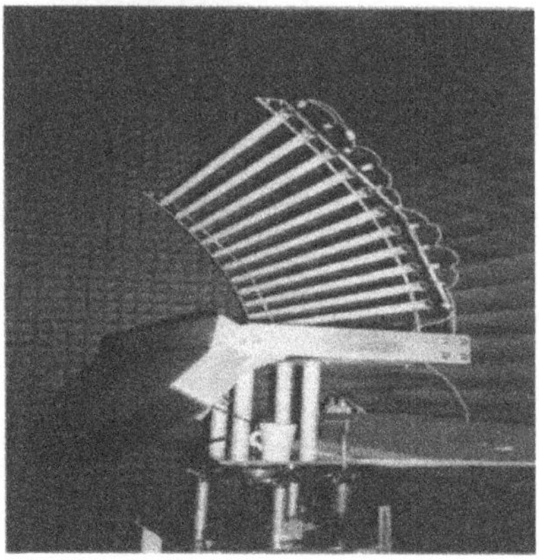

Plasma Reflector

# Appendix 5 – Propagation

**Fundamentals**

A comprehensive description of propagation phenomena will not be attempted here, only the basic concepts will be visited. Refer to the bibliography for further reading. Propagation is the science of understanding how radio waves move from transmitter to receiver. This science is very dependent on frequency, transmission media, meteorologic phenomena (including space conditions) and antenna placement. Although there are more variables, these few can five us a rough idea as to what will happen under varying conditions.

At low, medium and high frequencies (1 KHz to 30 MHz), the earth's ionosphere and magnetosphere have a profound impact on the ability of radio waves to propagate. Typically, the maximum usable frequency (MUF) can be calculated and forecast to allow short wave broadcasters, Ham radio operators and military communications specialists to best select an operating frequency that will be optimum There are diurnal changes due to the several ionospheric layers moving around the earth. There are also solar variations due to the number of sunspots and other space related weather that influence communications at these frequency as well.

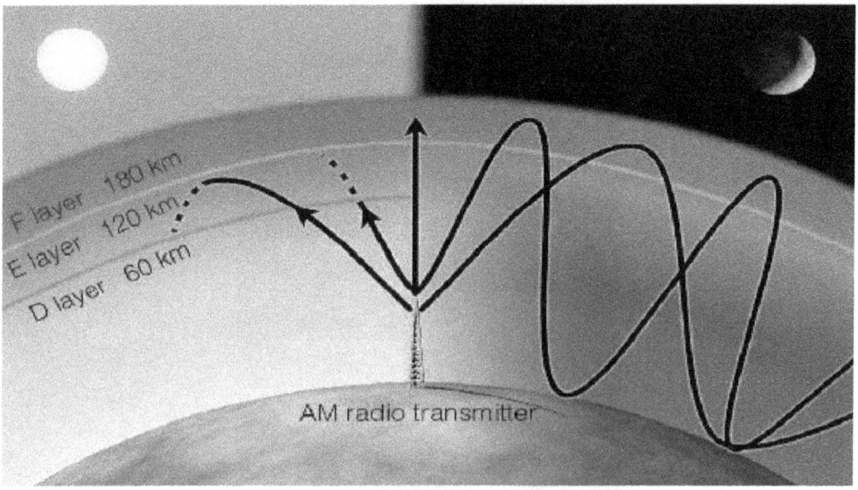


The Ionosphere at low frequencies

For our purposes we define operating frequencies as those below about 100 MHz, as this is where the ionosphere has the most effect. During the day, the characteristics of ionospheric reflection change due to the charging of various layers of particles in our atmosphere. Three main layers have been defined, the D, E, and F layers, the E layer at some point can split in two, changing propagation qualities.

When the ionosphere is active during the day we can present a simple model for the expected long distance communications capabilities. Short wave stations and Ham radio operator follow these rules and use Internet resources from several different laboratories to predict the quality of transmissions. Normally, the morning hours allow good communications for frequencies from 15 to 50 MHz, later near noon, the range changes to approximately 10 to 20 MHz, during the afternoon, 7 to 15 MHz works well, early evening, 5 to 12 MHz, late evening 3 to 10 MHz. This is very simplistic but usable for most casual listeners and radio operators. One big factor not mentioned yet is the role of the solar cycle on propagation. During a period of about 11 years, the sun goes through a cycle of minimum to maximum sunspot production and general activity that sends varying amounts of charged particles towards the rest of the solar system. These charged particles include x-rays and neutrons and can effect the characteristics of the ionosphere to a great degree. During the height of the activity, long distance communications over short wave frequencies can be highly enhanced. Auroras also peak during this level of activity.

The best way to predict how communications will be at these frequencies is to review the online resources and look for the Maximum Usable Frequency (MUF) predictions, these are generally very accurate and take into account solar activity.

**Meteor Burst**

One very interesting way of propagating signals at low frequencies over great distances is to use the ionized trails of billions of micrometeorites that hit the earths atmosphere every day. The best time to use this technique is around 4 to 7 am as this is the time when the Eastern horizon is facing directly in the orbital plane of the Earth.

The optimum frequencies for this purpose is around 30 to 50 MHz with 40 MHz being the best and most used by meteor burst communication system manufacturers. Ranges of several thousand miles can effectively be used for this technique. The trick is to understand the the ionized trails for the meteors only are active for periods less than about a second. Normally, a beacon from one end of the communications link is monitored by the other end of the link. When sufficient signal strength is detected, a quick burst of information in the form of a packet is transmitted. The period of transmission might only be 1/2 of a second, but is sufficient to for instance, send a position and status of an over the road truck.

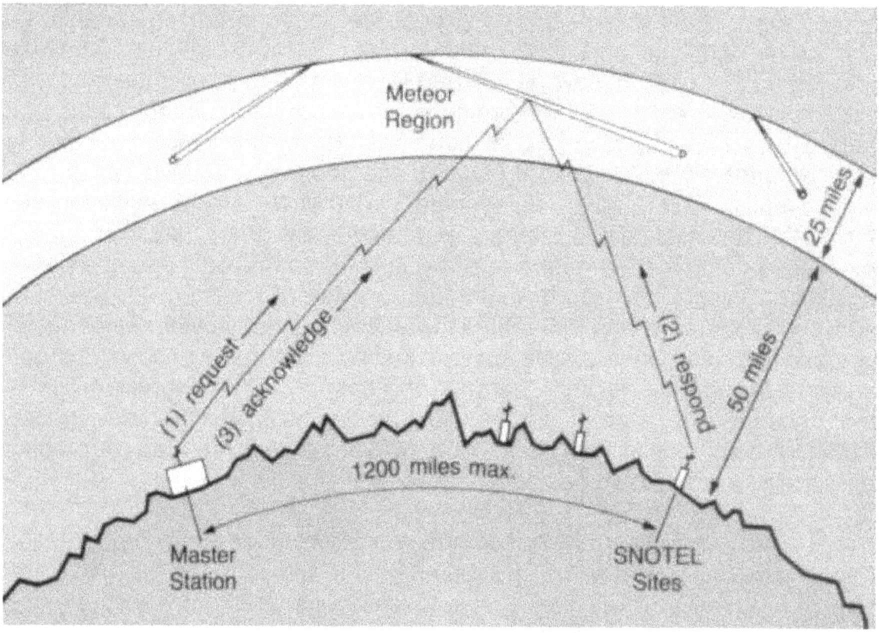


Meteor Scatter Operations

### Radio Horizon

For frequencies above about 50 MHz the distance a radio transmission can be sent is dictated by the radio horizon. This includes the heights of the transmitting and receiving antennas. Generally this horizon can be defined by the following formula:

$$Distance(Miles)=\sqrt{(1.5*Altitude(Feet))} \qquad [102]$$

The value 1.5 can be changed to 1.6 for smooth open surfaces or to 1.4 for highly forested, hilly areas. There is also a 4/3 rule about radio propagation over the horizon, but this is only an estimate and frequency dependent. For more detailed analysis, for instance of a cell phone or FM radio coverage area, there are several simulation programs that are very useful. These programs include the effects of rough terrain, foliage, transmitter power and receiver sensitivity. The receiver sensitivity includes such factors as conversion gain, which is derived from efficient modulation techniques.

**Atmospheric Ducting**

At VHF and UHF frequencies, a phenomena where radio waves are trapped in the tropospheric layers can if used properly propagate signals effectively. Much research was performed in this field and several military systems are in use today that take advantage of this phenomena. One of the advantages is the low intercept probability for receivers between the two points of communication. This is sometimes referred to as the Fata-Morgana effect.

**Urban Propagation Considerations**

As anyone with a car radio or cell phone will attest, communications are not always reliable when in the presence of skyscrapers and other large structure. It is important for cell phone operators for instance to utilize good propagation simulation software to fill in the gaps and minimize dropped calls. FM radio stations try to handle the problem by using high power levels and lower frequencies of operation. The urban environment is highly frequency selective with lower being better. Line of site operations like satellite links have to be carefully chosen whereas AM radio signals are effected to a much lesser degree.

**RFI Considerations**

Radio Frequency Interference (RFI) challenges the quality of radio links, especially in urban settings. RFI is made up of minimally filtered transmitters that are sending out their primary frequency as well as harmonics. In addition, RFI can be created by modulation "skirts" around the fundamental frequency of transmission. The Federal Communications Commission (FCC) sets limits on these emissions, but at even low levels, reception can be compromised. Different modulation techniques can cause various problems as well with digital typically causing interference at odd harmonics of the primary frequency. In very dense environment like large cities, the preponderance of RFI causes the actual noise floor of receivers to rise, which should be taken into account when evaluating link performance. An analysis of the local environment that is to be used for a link is prudent. Using a good spectrum analyzer and checking over the period of a few days, recording the results for analysis later, is the best approach.

It is of course very important, and mandatory in most places to have the radio system verified for compliance to FCC regulations (or the appropriate country's regulations).

The antenna is a very important component of the RFI profile. First, it acts like a filter over the passband of interest. To mitigate interference (both transmitted and received), an addition passive filter might be in order. Second, the antenna can be a source of interference not related to the transmitter by a process called Passive Intermod or PIM. PIM is caused by dissimilar metals in the construction of the antenna, junctions such as flanges and coaxial connectors. PIM can also be caused by components such as filter, diplexer, duplexers, transmit/receive switches, etc.

The worst conditions for PIM production are under high heat and high power. PIM requires at least two transmitted signals to be created, such as cell tower antennas where there are multiple carriers. Satellite antennas also are a source of PIM products, especially in their associated feed structures where the power levels are high. To calculate potential PIM products, use the same techniques as RF designers use to calculate Intermodulation Distortion Products by evaluating harmonics, then twice one frequency plus the first, then twice one plus twice the second. A matrix should be build with all combinations of first, second and third frequencies (etc.) to see what products fall in a receive band or other protected band. See attached software for further details. PIM mitigation includes designing with similar metals (all aluminum for instance), clean/polish flanges and coaxial connections and shielding if necessary.

**In Building Considerations**

Propagation inside buildings can be a complex issue due to the fact that the wall, ceiling, doors etc. are all made of various materials. Some of these materials reflect radio waves, some absorb and some diffract. Also, frequency selection is very important to balance construction with communication needs. Generally speaking, high frequency satellite links are not possible inside of a building, lower bands such as L might work better. These lower bands include GPS, Inmarsat and the PCS type cell bands. Even lower works better, for instance VHF and UHF bands are used very often for inside building communications. Typical of these bands is Orbcomm, Ham radio, and personal radio applications. FM and AM radio typically work well due to their lower frequency of operation. There has been a significant amount of research done in this area due to the ubiquitous amount of Wi-Fi, intranet and hospital applications. Refer to I.E.E.E. Journals and other publications before designing a system for in building use. Also, there have been several good software programs developed to evaluate performance under specific building morphologies.

**In Vehicle Considerations**

Yet another challenge is how to effectively design an antenna for use *inside* a vehicle. LoJack, various law enforcement agencies and other covert type operations require antennas well hidden inside any type of car, truck, construction or farm vehicle.

The rules are the same in some ways as in building applications, where lower frequencies tend to work best. GPS for instance requires a view of the sky restricting the placement of the antenna to just under the dashboard or just under the hat shelf in the rear of a standard car. Orbcomm operations in the VHF band have a bit more flexibility but considering the low powers involved in the link, placement near a window is recommended. VHF and UHF radios with stealth like antennas usually work best when the antennas are placed in one of a few basic areas.

To summarize:

| Application: | Frequency: | Placement Options: |
|---|---|---|
| AM Radio | .5 – 1.7 Mhz | Under dash, in window (laminated) in door posts |
| Short Wave | 2 – 30 MHz | Door posts, or windows (laminated) |
| HF and VHF | 30 – 300 MHz | Door posts, windows (laminated), roof liners, plastic bumpers, tail lamps |

| | | |
|---|---|---|
| UHF | 300 – 1 GHz | Door posts, windows (laminated), roof liners plastic bumpers, tail lamps just under dash or hat rack |
| L band | 1 – 2 GHz | Just under dash or had rack tail lamps assemblies,inside third stop light module |
| S band | 2 – 4 GHz | Tail lamp assemblies, third stop light module, just under dash or had rack |
| C band + | 4 + GHz | Tail lamp assemblies, third stop light module |

These suggested antenna placements are very general, there have been several interesting antenna designs that can for instance use S band within the roof liner, but these can get expensive.

Keep in mind that in all of the above cases, detuning and narrow banding of the “open air” antenna will occur. To counter this, tune antenna to specific installation positions if possible, add tuning circuitry for broader install applications, consider magnetic type designs (AM radio ferrite loops, for instance) and certainly design for broadband as the proximity of people or cargo can further detune an antenna inside of a vehicle. After a design is implemented, measure the antenna pattern by (for instance) using a spectrum analyzer and antenna around a circle of closely place equidistant points relative to the vehicle. Typically, the lower the frequency, the more uniform the pattern, meaning that at frequencies above say 400 MHz, nulls and pattern squint can occur, sometimes to a significant degree. Also, be aware that the pattern has a vertical component as well. The same dependance on frequency will be evident in this vertical beam.

Finally, if a design is realized for a large volume application, be sure to educate the installation technicians about the hazards of improper antenna placement. Create a matrix and clearly denote any special considerations, like stretching out an antenna if necessary or not placing radiating elements near or on metal surfaces. Perform link tests if possible after each installation.

# Appendix 6 – Antenna Testing

All antennas need to be tested, not just for contract compliance for instance but for the minimizing of unwanted attributes. The law of unexpected consequences applies here, were the best intentions of an antenna designer is often met with antenna qualities that are non compliant with the original requirements. In addition, such problems as passive intermod, or in the case of a phased array, mutual coupling can easily occur causing unwanted performance characteristics.

It is very common to have a design go to a testing facility several times during the design cycle, to enable the complete understanding of the antenna's attributes and the minimization or elimination of unwanted effects.

Antenna testing includes VSWR tests and pattern tests but can also include thermal cycle, thermal vacuum, passive intermod and high power tests. In the case of satellite antennas, there could be additional tests for shock, vibration and other mechanically related tests before an antenna can fly.

The best way to approach an antennas test philosophy is to properly document required performance in a specification document, then write a test plan, followed by test procedures that clearly spell out testing technique and specific test limits. In addition, statements of work are sometime written to allow an outside vendor to clearly understand the requirements of the antenna design. There can be several layers of this kind of documentation including for instance in the case of a satellite antenna, feed horn documentation, feed assembly documentation, feed network documentation (the feed and reflector in this case) and finally the performance documentation when the antenna network is attached to the spacecraft. Design Reviews (preliminary and critical) are typically held during the design cycle of space aircraft antenna applications. This is not a bad idea even if the designer is working on another type as it allow other engineers to evaluate and help with the process. Sometimes, test readiness reviews are held to prepare for a test sequence along with non compliance meetings in the case of a failure during a test.

## VSWR Testing

The initial test during an antenna design cycle is how well an antenna is matched to its feed. This feed might be coax cable or waveguide. In either case, it is imperative that the optimal match be obtained to best transfer the electromagnetic energy to and from the antenna. There are several methods to make this measurement. In simplistic terms, they are scalar and vector analyzers.

### Scalar

Scalar refers to measurements made without phase, just amplitude. One method is to measure the amount of energy that is reflected back from a frequency source to ant antenna. The simplest way to do this is with a directional coupler which is a device that is sensitive to the direction of movement of electromagnetic energy. Usually a three port device, there is a port for incoming energy, one for outgoing energy and the third for the reflected energy. The input to output port is usually very low loss, and set to the characteristic impedance that needs to be measured. The third or coupled port is typically 10 to 40 dB attenuated from the "through" port, so as not to influence the measurement to a great degree. A typical setup includes a generator set to the frequency of interest on port one, the antenna on port two, and a spectrum analyzer or receiver on port three. Logically, a designer wants the minimal amount of energy to appear on port three, indicating that the highest degree of energy is going from the generator to the antenna. When the generator is swept in frequency, the designer will be able to determine the match bandwidth of the antenna an the optimum frequency of operation. Modifying element lengths or feed point positions on the antennas will influence the output of port three.

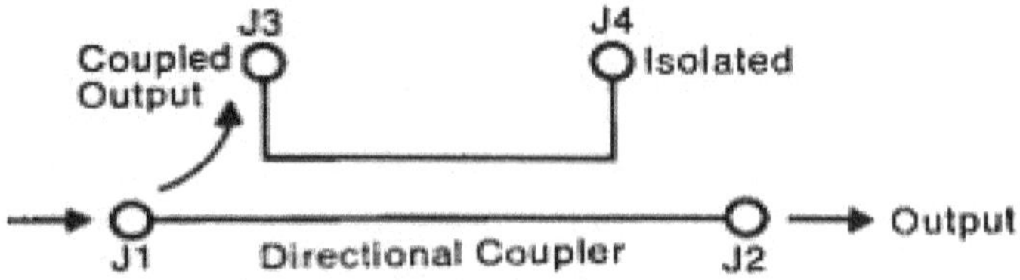


Diagram of a directional coupler, terminate J4 with load, place antenna under test on J2, generator on J1, observe return loss on J3

The diagram above is of a dual directional coupler where return loss can be measured as shown and by placing a load on J3, incident power can be observed on J4. This is convenient for monitoring output power levels.

The same effect can be realized by using a Wheatstone bridge circuit as diagramed below.

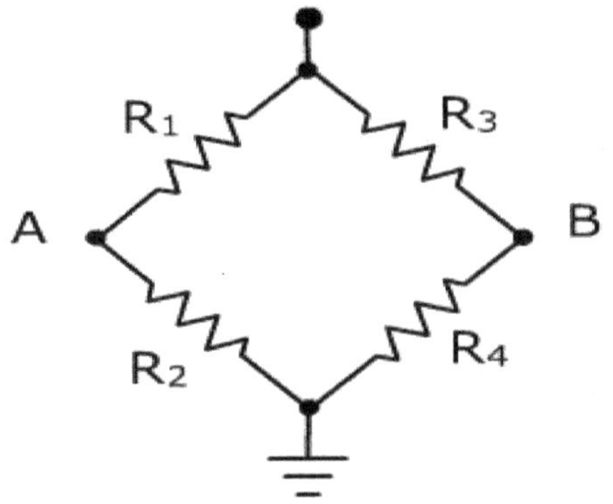


A diode between A and B rectifies the mismatch. All values are 50 ohm, R1 is the antenna under test.

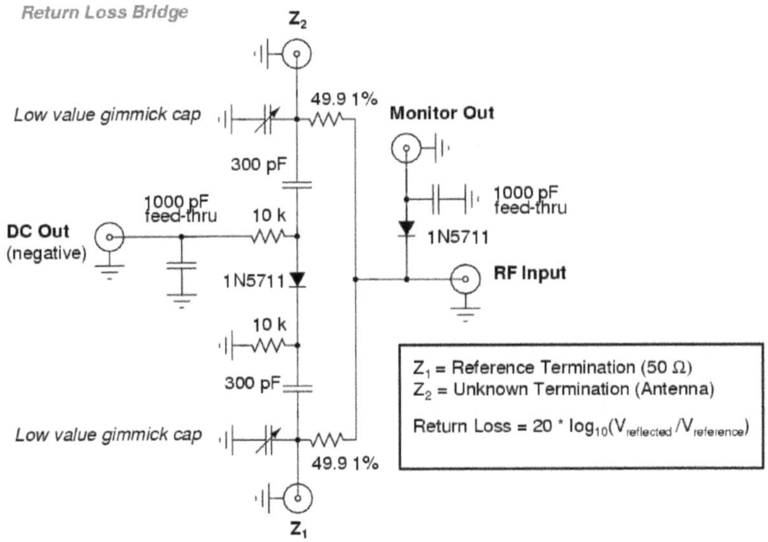


An example of a bridge detector (32)

In this case the reflected energy causes the diode to rectify the returning AC waves, causing a DC voltage to be present between the cathode and ground. This is a good wideband, sensitive approach to making a VSWR measurement. Use a diode capable of rectification at the frequency of interest. Also use low inductance resistors, like carbon chip types and make a "tight" layout to minimize inductive influences of connecting lines in the case of an etched circuit. Using microstrip designs, make sure that the lines are the correct (e.g. 50 ohms) impedance. Verify operation with good reference loads and shorts.

**Vector**

A sliding line is one of the most fundamental methods of accurately measuring phase and amplitude, the basic requirement for vector analysis. With these two variable one can plot antenna responses on a Smith chart and get a complete picture of the complex characteristics of an antenna performance. The sliding line is made up of a low loss waveguide section with a thin slot cut into the top. A probe extends down approximately into the center of the cavity and is attached to a trolley device that can move the probe up and down the cavity over at least a wavelength of the frequency of interest. The probe is attached to a diode detector which is then attached to an amplifier and meter assembly. The generator is connected to one end of the cavity, the antenna to the other end and the trolley is moved across the slot to indicate the amplitude vs. phase of the antenna response. The top of the sliding line has fiducial markings. The operator can move minimum to minimum in the amplitude readout to determine ½ wavelength positions and phase. These values are plotted on a Smith chart with the amplitude or magnitude of the reflected wave being plotted radially from the center point, which is the perfect match to the impedance of interest. The phase is plotted as "azimuth" on the Smith chart with 180 degrees corresponding to a full rotation on the perimeter. Smith charts are marked on the perimeter in terms of phase to and from a load (or antenna in this case). To calibrate, a short is placed on the port that will go to the antenna and the position is plotted on the far left of the Smith chart, an open port (or calibrated open) places the measurement on the far right of the chart and a calibrated 50 ohm load (or characteristic impedance of the slotted line and antenna) places the measurement in the very center of the chart by moving the trolley to achieve the minimum amplitude. The AUT or Antenna Under Test is now attached, the change in amplitude is observed and the return loss (convertible to VSWR) is observed and recorded. Changing the frequency of the generator and plotting reveals the frequency response of the AUT. By changing the physical parameters of the antenna, the optimum design can be obtained.

In lieu of the sliding line, vector network analyzers are used today. They in effect make the same measurement as the slotted line however without all of the painstaking work. They usually come with a variety of display options including Smith chart, VSWR, Return Loss, Insertion Loss, Phase, Polar and Group Delay. They are basically calibrated in the same way as the sliding line but in a much more automatic way. Usually the frequency of interest is programmed in or applied to the appropriate port. Then the calibrations take place over the entire bandwidth. A new manifestation of the vector network analyzer is the PNA or Programmable Network Analyzer, which offers even more features and a significant increase in sensitivity.

Vector network analyzers are also used for measurements other than basic impedance match determination. For instance, losses of cables, filter, diplexer, etc. are measured with an Insertion loss setting. Characteristics of filters and amplifiers can be measured with the group delay feature. The insertion loss type measurement is often used as the basis of the antenna gain measurement in far and near field range operations.

**Using a Network Analyzer to measure distance to a Reflection**

An interesting measurable quantity if a designer has a Scalar or Vector Network Analyzer is the distance from an antenna to a reflective surface. In essence the Analyzer can be used as a radar using a well matched antenna and an open area. To observe the phenomena, use the following procedure:

1. Warm up the Network Analyzer for a prolonged period (30 minutes) to stabilize the electronics.

2. Calibrate the Analyzer

3. Attach a good antenna, preferably with a gain > 19 dBi and a VSWR better than 1.1 at the center frequency.

4. Point the antenna into a large open area or good absorbent material.

5. Verify that the display is a classic smoothed "V" shape in the VSWR or Return Loss Mode using a rectangular display.

6. Place the display in the Memory mode and display the memory as well as the real time display.

7. Point the antenna up approximately 10 Degrees and mount in a solid fashion, observe any changes in display.

8. If the display is within 5% of the memory display, re-save the plot in memory and again display both traces.

9. If the Analyzer has a "Normalize" feature use this to flatten the display.

10. At approximately 10 feet away from the boresite of the antenna, place an aluminum or other reflective plate perpendicular to the antenna beam.

11. Notice on the display a superimposed sine wave, place in memory.

12. Measure the change in frequency between the peaks or troughs of the sine waves. If the analyzer has the capability of transferring data via GPIB, IEEE or other suitable format, send to a computer for higher detail analysis.

13. Use the following formula to calculate the distance to the plate:

$$Distance = \frac{C}{(2 * \Delta F)} \qquad [105]$$

where:

$C$ = Speed of Light

$F$ = Frequency between peaks

If the data can be sent to a computer, take the Fourier Transform of the data (normalized or direct), use a peak detector subroutine to find the frequency of the superimposed sine wave. Use integration techniques to increase sensitivity, either on the raw data or the transformed data. This technique has been used to find the direction, turbidity and other parameters of the wind at low altitudes by pointing the antenna into the sky.

This technique has also been used to detect very fine changes in the reflective qualities of a room, such as detecting very small positional changes of the room's contents.

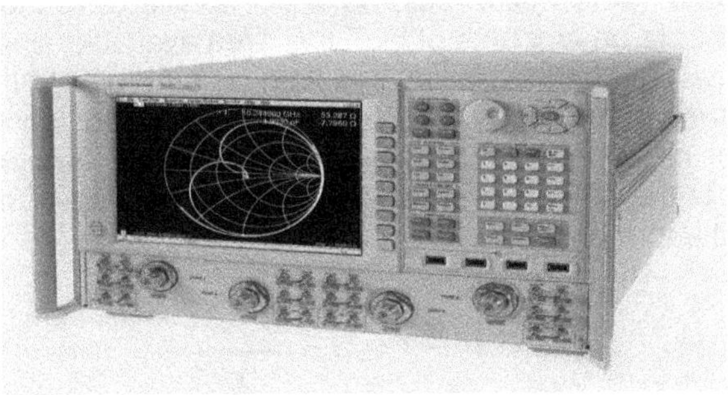

A Modern Network Analyzer

**Antenna Gain Measurements**

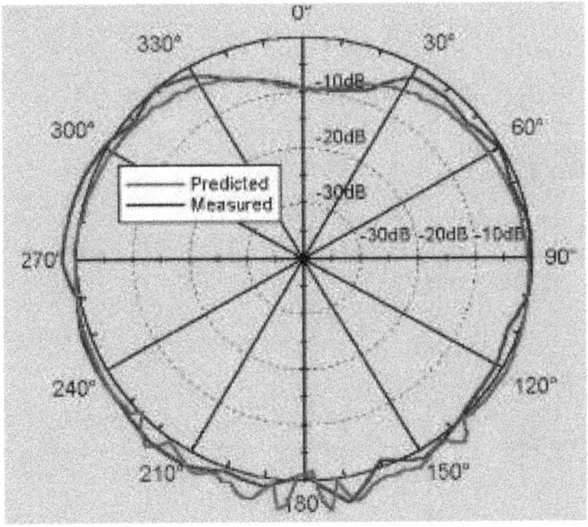


Polar Pattern showing predicted vs measured performance

**Picking the right range**

One of the first major decisions that has to be made regarding the measurement of the gain of antenna is the type of range that will be used. There are several types to choose from including, far field, near field (several types), compact, holographic and mode stirring.

Several factors must be examined carefully before making a choice in ranges. One major factor is space available, if there is a lot out outside space or large amounts of high bay area available, one can consider a far field range. But before money and time is committed to this type, make sure that the range capabilities (distance between source and test antennas) will be in the far field for the foreseeable frequency and gain requirements of the antennas to be developed at this site. To calculate the far field, use the formula:

$$Far\ Field = \frac{(2D^2)}{\lambda} \quad [106]$$

Measurements inside this distance will create inaccuracies in the measurements, including oblation of the main beam, increase in sidelobe levels and inaccurate gain determination.

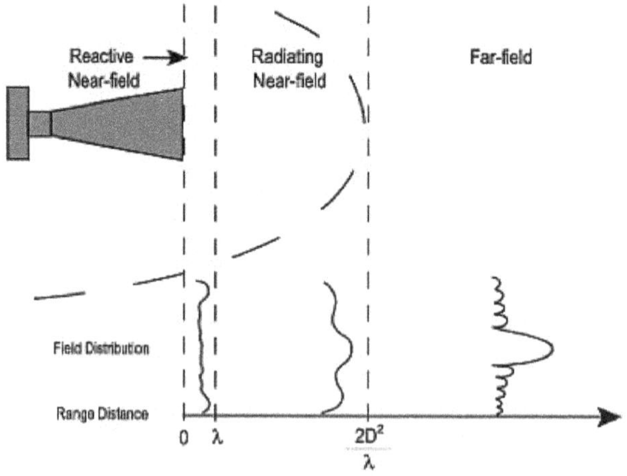


Far field vs near field

Far field ranges can be inexpensive however, if this is a decision driver. An outdoor range can do a nice job with little outlay of cash. There are a few issues which have to dealt with, this will be addresses below.

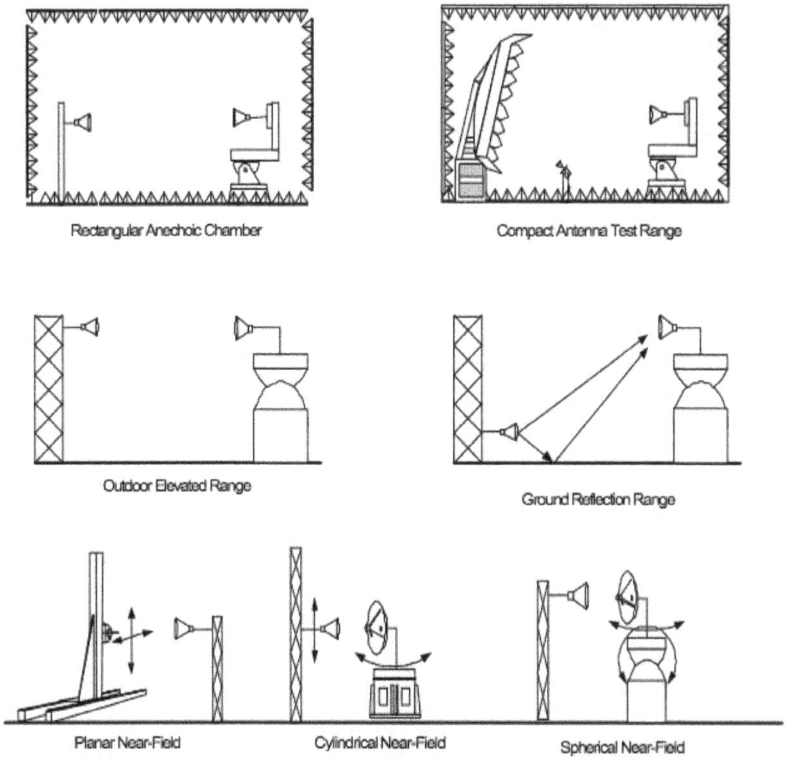


Examples of several types of antenna test ranges (44)

Second in terms of cost is a far field indoor skeleton range can be considered. This is a range in an open, preferably high bay area, where panels of absorbent material are arranged typically in a rectangle, with the source antenna at one end and the test antenna at the other. Absorber should be placed on the floor and hung above the range if possible.

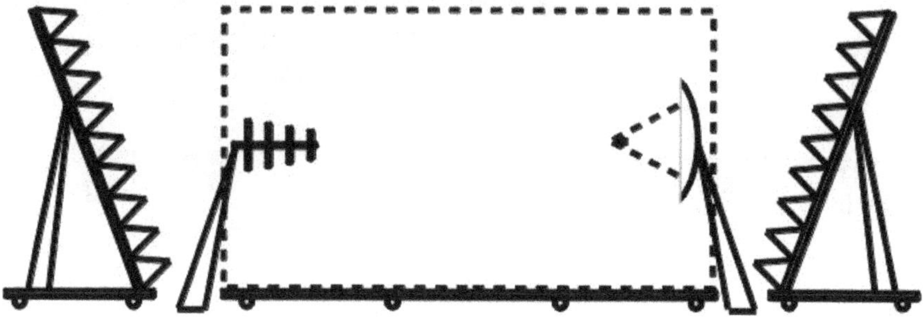

Diagram of Skeleton Range

Next a compact range can be considered, if inside measurements are to be made, but a longer far field distance is needed. Compact ranges are made typically of a source antenna, a corrugated reflector to bend or fold the transmitted signal from the source antenna in such a way as to compress the range to available space considerations. These ranges have an effective operating frequency range that should be taken into consideration. Of course there is an added cost for the use of the reflector and alignment of the range is critical.

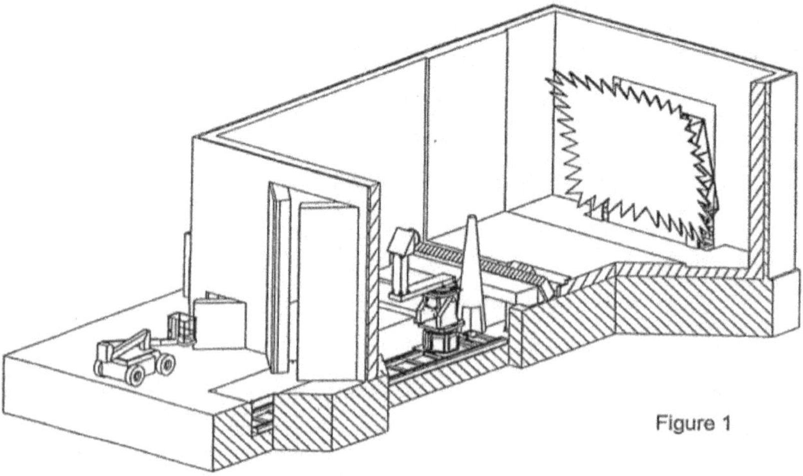

Figure 1

Details of a Modern Compact Range (44)

Today, serious consideration should be given to a near field range. This type does not require the far field distance limitation to be taken into consideration. As a consequence, large apertures at high frequencies can be measured accurately in a relatively small area. A near field range comes in three basic types, spherical, rectangular and cylindrical depending on the way the probe is moved across the antenna aperture. The rectangular type is made of an x-y table whereby the measuring probe is moved in a grid like pattern across the antenna under test.

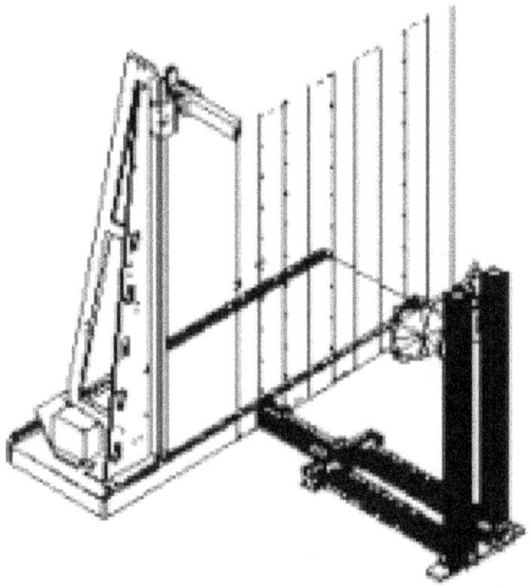

An example of a rectangular near field antenna test apparatus

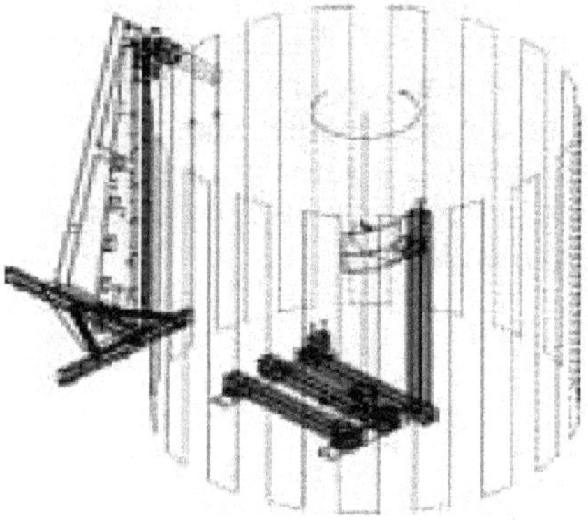

A cylindrical type has the antenna moving about a (typically) vertical axis while the probe is scanned vertically.

An example of a cylindrical near field antenna test apparatus

A spherical near field range has the probe moving equidistant in a sphere around a test antenna.

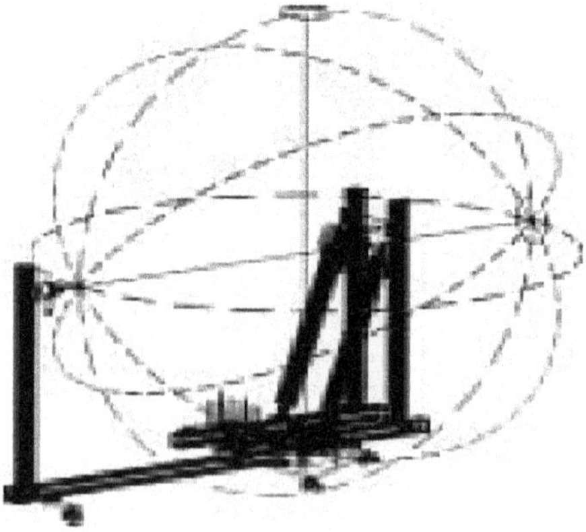

An example of a spherical near field antenna test apparatus

Taking a rectangular near field range as an example, there are several factors which all of these types of ranges have in common. First the fundamental measurement is in both phase and amplitude. For instance, a good vector network analyzer will work in the S12 or S21 mode. One port of the analyzer is connected to the antenna under test and the other to the sampling probe. These probes can be made by building a balun type element and surrounding it (with the exception of the very end) with absorber material. A second type of probe is a piece of waveguide with the walls tapered at the opening. These waveguide probes come in single polarization or dual.

Normally, a sample of phase and amplitude is taken at ½ wave increments across the antenna aperture. The values are placed in a software array which is then Fourier transformed. This process converts the individual measurements to a combined far field pattern and can be calibrated by either scanning a standard gain horn and using it as a comparison or calculating directivity from the initial transformed data. Several companies have perfected this type of range and should be contacted for further information. Calibration of a near field range is done best with an eighteen term measurement technique pioneered by the National Institute of Standards and Technology. See references at end of book. These terms include probe compensation, room scattering effects, cable effects, mechanical tolerances to name a few. It is advised to examine this type of calibration before declaring gain and pattern measurements accurate with confidence.

Another type of range that is getting increasing interest is a mode stirring chamber. This configuration includes a test probe placed into a fully reflective chamber, which included a mode stirring "paddle" inside to allow a full set of reflected waves to be present and measured. The probe first tests a standard gain horn, followed by a test antenna, the phases and amplitudes are measured in both, a reflection chamber algorithm is applied and the difference between the two antennas is then calculated. This is done over wide bandwidths if needed and again typically uses a good vector network analyzer for measurement in the S12 or S21 modes of operation.

**Spherical Wave Expansion**

Waves arc outwards towards the antenna under test in most chambers. To calculate the effects use:

$$Expansion(degrees) = \frac{(2 * \pi)}{\lambda} * L \qquad [105]$$

where L = Distance between points on the antenna under test

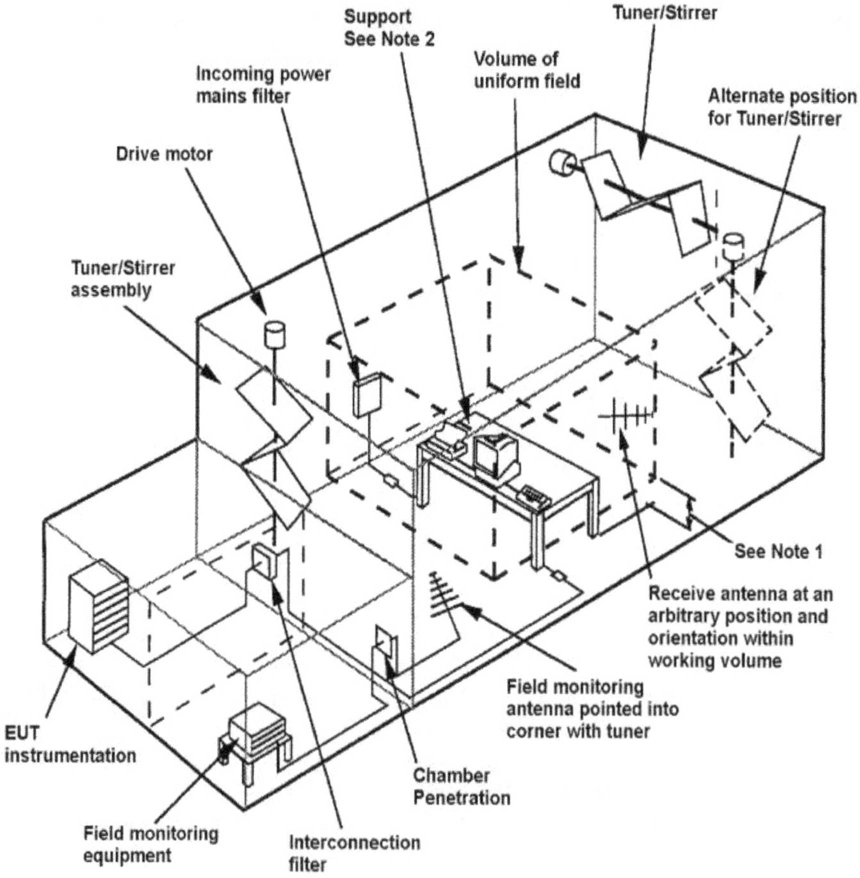


Mode Stirring Chamber (55)

Note 1: Equipment under test support needs to be non conductive and non-absorptive.

Note 2: Test volume should be at least $\lambda$ /4 at the lowest usable frequency from any chamber surface, field generating antenna or tuner assembly.

| **Parameter:** | NF Planar | NF Cylindrical | NF Spherical | FF Outdoor | FF Anechoic | FF Compact |
|---|---|---|---|---|---|---|
| **High Gain** | Excellent | Good | Good | Adequate | Adequate | Excellent |
| **Low Gain** | Poor | Good | Good | Adequate | Good | Excellent |
| **High Freq.** | Excellent | Excellent | Excellent | Good | Poor | Excellent |
| **Low Freq.** | Poor | Poor | Good | Good | Fair | Poor |
| **Gain** | Excellent | Good | Good | Excellent | Good | Excellent |
| **Sidelobes** | Excellent | Excellent | Excellent | Good | Poor | Good |
| **Axial Ratio** | Excellent | Excellent | Excellent | Good | Poor | Good |
| **Multipath** | Good | Good | Good | Adequate | Adequate | Good |
| **Complexity** | Moderate | Moderate | High | Moderate | Low | High |

Antenna Test Decision Matrix, NF = Near Field, FF = Far Field

**Robotic Range**

Robotic ranges are testing facilities that use a large robotic arm to perform any type of near field probing to produce antenna patterns. Essentially the robot can place the near field probe anywhere within a spherical space around an antenna or antenna system. This give maximum flexibility to the engineer for measurement approach. These test facilities were developed to a great extent at the National Institute of Standards and Technologies (NIST) in Boulder, Colorado. This facility uses state of the art network analyzers to measure phase and amplitude response from antennas to frequencies above 600 GHz.

It is anticipated that these types of ranges might dominate the near field measurement approaches as the equipment (robotic arm) is not very expensive and custom installations are not required. These techniques can also be applied to in situ measurements for large aperture reflectors and phased arrays.

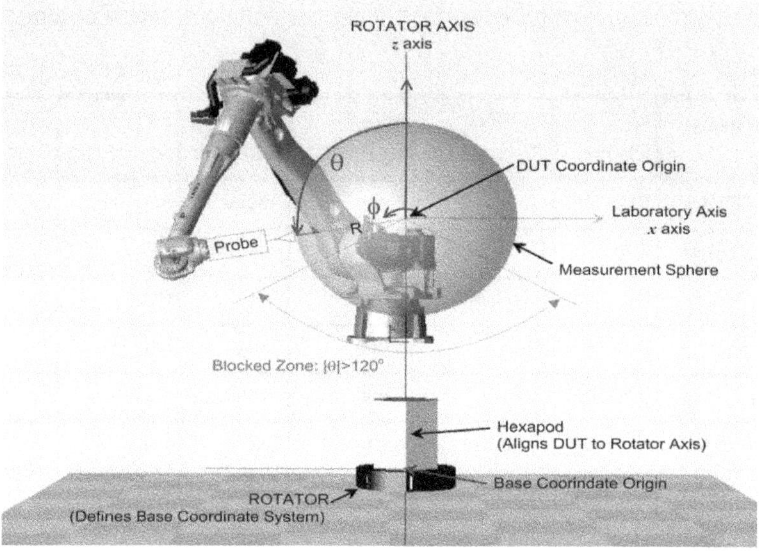


Robotic Arm Geometries

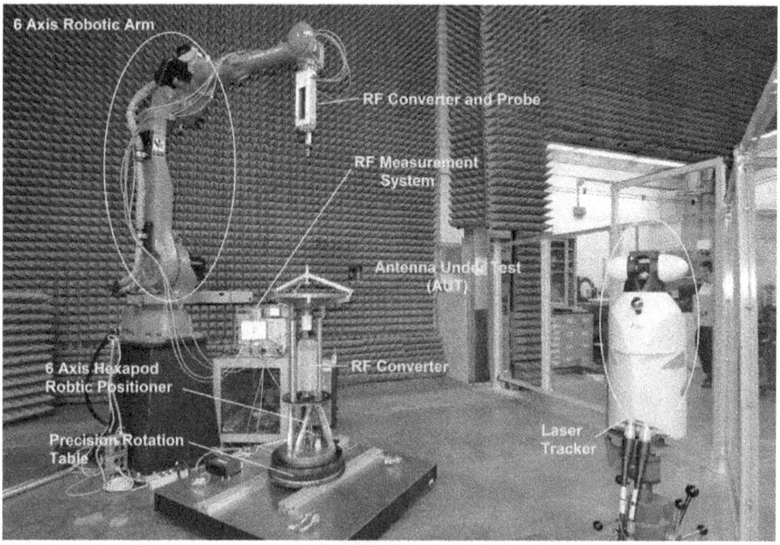


Robotic Range Details

## Details of Far Field Antenna Measurements

### Substitution Method

Historically, the most common type of antenna measurement has been using a far field range (indoor or outdoor), placing an antenna on a rotating platform with an orthogonally rotating head (see diagram), using a source antenna at the far end of the chamber and finally, comparing the measurements of the test antenna with a standard gain horn.

The antenna under test is rotated about the vertical axis (or Theta) while a pattern is taken, then the antenna can be rotated 90 degrees on the head (or Phi) to observe cross pole or orthogonal pole performance. A standard gain horn is placed at the exact same position as the test antenna and the same measurements over the same frequencies are made. Standard gain antennas come with calibration curves that have been carefully derived from NIST. Antennas which become standards and are referenced to a standard gain antenna are known as a secondary standard. These secondary standards are not as accurate but can serve the purpose of facilitating high volume antenna production or initial antenna design implementation.

### Three Antenna Method

Another popular method of determining gain is to use three antenna, each with unknown gain. Fundamentally, antenna one is measured against antenna two, then one versus two, then two versus three. Applying the following formula obtains the gain from antenna one:

$$G1=\sqrt{\left(\frac{(Pr1 * Pr2 * L3)}{(Pt * Pr3 * L1 * L2)}\right)} \qquad [106]$$

where:

*G1* = Gain of Antenna 1

*Pr1* = Power for 1 and 2

*Pr2* = Power for 1 and 3

*Pr3* = Power for 2 and 3

*L1* = Losses for *Pr1*

*L2* = Losses for *Pr2*

*L3* = Losses for *Pr3*

*Pt* = Transmit Power

**Deriving Gain through Directivity Measurements**

There are several near field antenna measurement facilities that test antennas or antennas assemblies by first measuring directivity then applying corrections based on input power vs. full pattern energy integration. For instance, the measurement of satellite antenna, or antennas take a significant testing range in terms of size and complexity. Although standard gain horns can be used as reference, the variations in temperature and time required for full probing of the patterns cause changes in phase and amplitude, causing inaccuracies in the final antenna performance calculations. In many facilities, the standard gain horn is measured before and after an extended test session (sometime taking days for complete satellite antenna testing), however during the probing, the test carriage is placed over the center of the antenna under test and previous measured data is compared, this is to compensate for phase and amplitude drift.

After the antenna under test is probed, the corrections are applied to determine gain, using:

$$Gain(dBi) = 10 * \log\left(\frac{(Directional\ Power)}{(Total\ Power)}\right) \quad [107]$$

**Rectangular vs. Polar Plots**

While rotating an antenna in a far field setting, patterns are taken by measuring azimuth (or Theta) angle vs. amplitude. If the pattern is circular it is known as a polar plot, which is most representative of how the antenna will perform in the real world. Sometimes however, it is convenient to plot azimuth angle on the x axis and amplitude on the y axis producing a rectangular plot. In this case it is easier to compare sidelobe symmetry and other factors. For marketing purposes, the polar pattern is most often used.

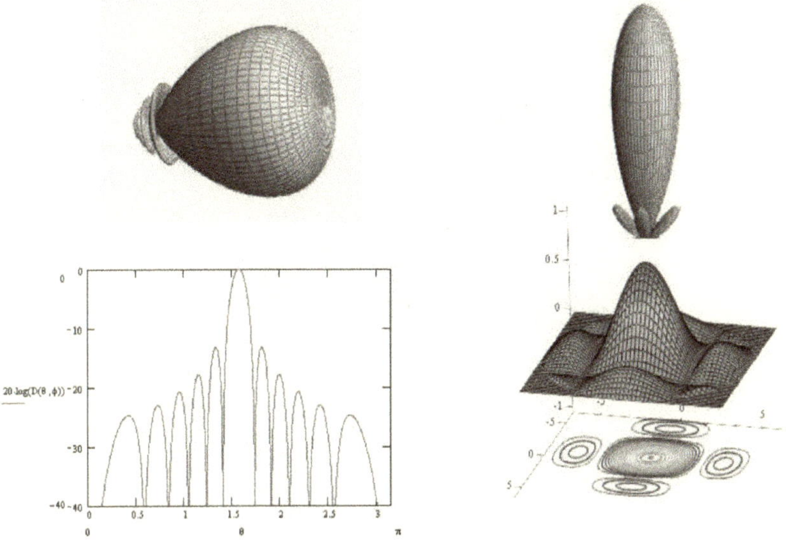


Samples of Antenna Patterns, showing Rectangular (L) and RDP (R)

**E plane vs H plane patterns**

To most completely understand the fundamental quality of an antenna pattern, several "cuts" should be made. These cuts refer to the orientation of the source antenna relative to the test antenna.

When the test antenna is vertically polarized with the polarization axis on the vertical plane, then rotated and measured, this is known an an H-plane cut. The source antenna is oriented in the same way for a co-pole measurement, and orthogonally for a cross-pole measurement.

When the test antenna is vertically polarized with the polarization axis on the horizontal plane, then rotated and measured, this is known an an E-plane cut. The source antenna is oriented in the same way for a co-pole measurement, and orthogonally for a cross-pole measurement.

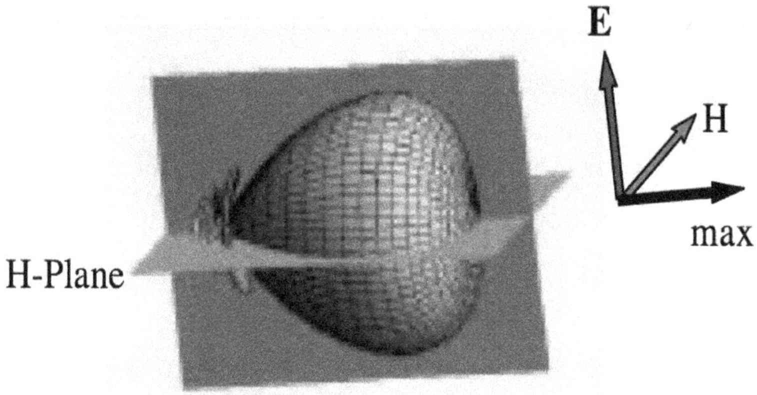


E and H pattern definition

**The Radiation Distribution Plot (RDP)**

When the phi (head) axis is turned in small increments, say every 2 degrees and the azimuth or theta axis is rotated 360 while measurements are made, a complete spherical measurement of an antenna performance can be ascertained. When plotted on a rectangular graph of theta vs phi, it is known as a Radiation Distribution Plot or RDP. The RDP is the most comprehensive measurement of an antennas performance and should be performed in the most critical design environments.

These RDPs can be displayed and/or interpreted in a three dimensional approach to best understand main lobe and sidelobe structure. Using a decent 3D modeling software yields good insights into the complete antenna performance.

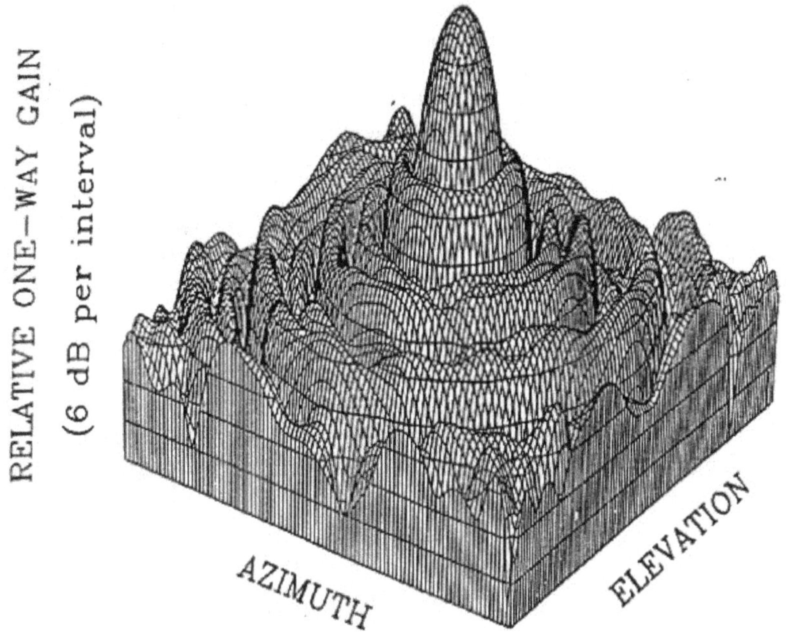


Example of Radiation Distribution Plot

**Testing Circularly Polarized Antennas**

There are several ways to determine the quality of a circularly polarized antenna. The simplest way is to monitor amplitude while spinning the source antenna. The antennas are set to boresight first, then the source antenna is spun, changing the angle (Theta) will indicate the extent of *axial ratio* as the antenna is moved from boresight. Any variation in the pattern shows asymmetrical behavior. In consumer antenna design, variations less than 3 dB are considered reasonable. However in the commercial and aerospace industry, significantly more stringent specifications are required, with axial ratios of much less than 1 dB.

A second way to determine axial ratio is to observe the phase and amplitude of a linear vs circular antenna. In other words, if the test antenna is circular and the source antenna is linear, turning the source antenna by 90 degrees should cause a 90 degree phase shift in the test antennas output. If the source antenna happens to be dual polarized, then the phase shift between ports should stay 90 degrees apart during rotation. Whether the vertical vs horizontal port is ahead by 90 or behind by 90 degrees dictates whether or not the test antenna is right hand or left hand polarized. For a more concise definition, see appendix 6.

Near field ranges perform the same measurement in orthogonal polarizations to determine the quality of the circularly polarized antenna. They do so by measuring both amplitude and phase, comparing the phase to determine the quality of the circularly polarized far field response.

As mentioned before, axial ratio is the amount of gain in orthogonal linear polarizations when looking at a circularly polarized antenna. When a left hand antenna is set to look boresight against a right hand circular antenna, this measurement is known as cross pole. Cross pole isolations or measurements are reasonable above 30 dB, higher for aerospace applications.

To calculate Axial Ratio from measured right and left circular responses, use:

$$AR = 10\log\left(\frac{(Er + El)}{(Er - El)}\right)^2 \qquad [112]$$

where, Er and El are measured Right and Left responses.

In addition, the Jones equations can be calculated based on antenna measurements to determine axial ratio of circularly polarized antennas. Jones determined that you could use 4 complex numbers to define any polarization state. This approach is used for instantaneous measurements and therefore is narrow band.

Finally, Time Gating using high end network analyzers can remove interference in antenna measurements by only recording phase and amplitude details in a certain space around the antenna under test. This has the effect of minimizing multi-path effects and other interference sources.

**Time Domain Reflectrometry**

One interesting type of measurement to determine feed line impedance is the time domain method. This type of measurement can be used for other components but they must be passive. The technique used is predicated on the use of a pulse generator which imparts energy into the test cable, then observes the time domain (not frequency domain) response, much like an oscilloscope. Deviations from the characteristic impedance, usually 50 ohms is shown by variation from a horizontal line charted on the display. This is a great method to determine how junctions are working. For instance, a coaxial cable transitioning to a waveguide, if there are any deviations from 50 ohms, a bump will occur at the time, convertible to distance, from the pulse generator.

This technique is very useful for long coax runs, especially with adapters or other transitions. The distance a typical reflectometer can measure can be up to many kilometers.

**Testing Large Aperture Antennas**

Large aperture antennas require long far field distances. There comes a point where this approach is not an option, near field techniques can measure aperture up to the size of the near field measuring apparatus, but this pieces of equipment can get very expensive above a certain size.

Generally, for apertures above 10 meters, it is best to use other approaches to determine the quality of the antenna under test. Options include the use of radio stars, holographic techniques and direct surface measurement.

**Radio Stars**

Above about 10 meters in size and assuming operational frequencies of UHF and above, antennas have the capability of using "bright" radio sources as a far field source to examine gain and sidelobe quality. These radio sources tend not to be very polarized and as a result, cross pole measurements using this technique is not advised. These sources however can be very small in angular extent and very stable in output power.

The general approach to using this method is to either scan across a known point source radio star or to allow the earth's rotation to do the scanning automatically. The antenna design engineer should calculate the expected beam width and gain, then observe the radio star drift through the antenna beam, looking for beamwidth and maximum gain. From the peak gain the antenna beam should be allowed to drift well beyond the radio star to allow the analysis of sidelobe structure. Be a advised however that sidelobe beyond the second or third set will be hard to discern.

If the expected beamwidth of the antenna is above one ½ degree, then the sun is by far the best radio star to use, as it is by far the brightest radio source in the sky. The suns radio output is very broadband, with energy available across the entire spectrum, however the radio emissions are variable and follow an 11 year cycle, which is associated with the number of sunspots on the solar surface. When using this technique, always contact an agency which monitors the solar constant at various frequencies for calibration. One such agency is the National Oceanic and Atmospheric Administration, another agency is the space weather data center at NIST.

The calibration temperature of the sun or other radio stars is usually expressed in flux units, or Janskys. A Jansky (named after Carl Jansky) is equivalent to 10E-26 Watts per meter.

To use a radio star or the sun, the power received by the antenna is determined by several factors, including gain, power of the radio source etc.

The power received from an antenna over a solid angle Ω, received by the effective area A and delivered to a receiver is:

$$Pr = .5 * \left( \frac{(2 * k * Tsys * \Delta v)}{\lambda^2} \right) * \Omega * Ae \qquad [108]$$

where:

$k = 1.38 * 10^{-23} W / K^{-1} / Hz^{-1}$ (Boltzmann's Constant)

$\Delta v$ = Bandwidth in Receiver

*Tsys* = System Temperature

*Ω* = Angle of received signal

*Ae* = Effective Area of Antenna surface

Several qualities can be determined by using this measurement approach. First is beamwidth, where the relationship of gain to beamwidth can be examined. The focus quality of the antenna (assuming a parabolic dish) can be determined. Another quality is the sidelobe performance, where the expected levels relative to the main beam can be observed. The expected angular positions should also be compared as well.

**The Holographic Technique**

A more modern antenna test technique for large aperture antennas is to use a radio star, the test antenna and another co-located smaller antenna. The smaller antenna looks at the same radio source as the larger and measures phase and amplitude to enable the accurate measurement of the large antennas surface geometry. Several companies have been created to do this type of test and can be found on the Internet and at I.E.E.E conferences. Surface distortions will affect gain and sidelobe performance, using this approach allows for the readjustment of the surface to a predicted shape. In addition, the movement of the surface over elevation angle can be determined. See reference section, Appendix 9 for more details.

Because the far field pattern of an antenna is related to the Fourier Transform of the radiating surface, careful and complete measurement of the far field can yield, by using the inverse Fourier Transform, the phase distributions across the radiating or receiving antenna surface. Using radio stars, satellites or even terrestrial sources, a designer can determine with great accuracy, the positions of the surface plates or other reflecting surfaces. This technique is used on many large antennas such as those used for radio astronomy and NASA applications.

**Physical Measurements**

Techniques to measure the surface of a large dish using photogrammetry or direct survey using a theodolite yield good positional information that can be used by a good physical or geometric optics modeling program. Determination of focal point, RMS surface deviations and influences of feed mounting hardware can be assessed. Photogrammetry techniques include the photographing of a multitude of targets placed on the reflector surface. The photographs, usually taken with a high resolution digital camera, are digitized and using standard trigonometric approaches can determine the three dimensional positions of the targets. The points are then translated into a file, used by one of several modeling programs. Direct survey includes the use of a good theodolite on a stable platform, measuring the position of many targets placed on the surface of the antenna. Again, the three dimensional positions of the targets is determined and appropriate files generated. For a complete understanding of the antenna performance it is best to do the aforementioned measurements at several elevation angles, to quantify reflector deformations due to gravity.

**Photogrammetry**

This technique is especially useful for large aperture reflectors. Reflective targets are uniformly placed on a reflectors surface. A special camera is placed at the focal point of the antenna and measurements are taken of the position and the distance to each reflector. After all of the measurements are completed, software takes the vales measured and creates a topographical map of the surface and how it deviates from a perfect parabola (for instance). Main reflectors as well as sub-reflectors can be measured in this fashion, obtaining very accurate results when the values are placed into antenna design simulation software packages where gain and antennas patterns can be determined.

One of the advantages of this approach is to examine the changes in surface characteristics when a large parabola changes elevation angle. Gravity tends to pull the surface out of perfection, especially at low elevation angles.

Modifications on the surface can be applied once the photogrammetric measurements are applied and analyzed. In some cases, active surface plates can move in relation to elevation angle to keep the surface as close to perfect as practical. This allows for the use of these types of apertures at much higher frequencies. The cameras used come in two varieties, stereoscopic and monoscopic. The imaging area inside the camera is typically a high density CCD surface. The camera is calibrated at the factory before use in the field.

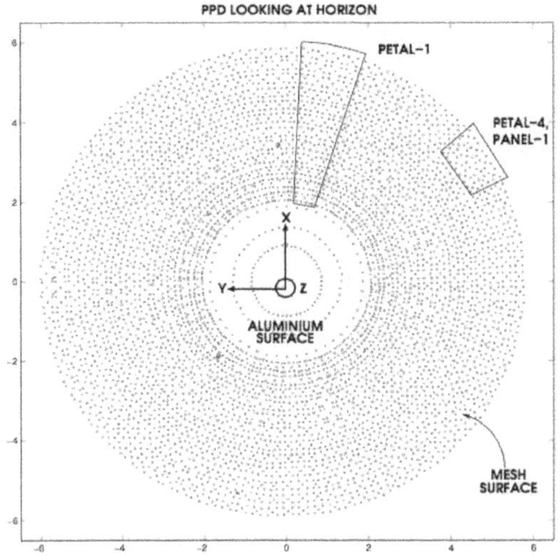


Antenna surface with a multitude of measuring targets

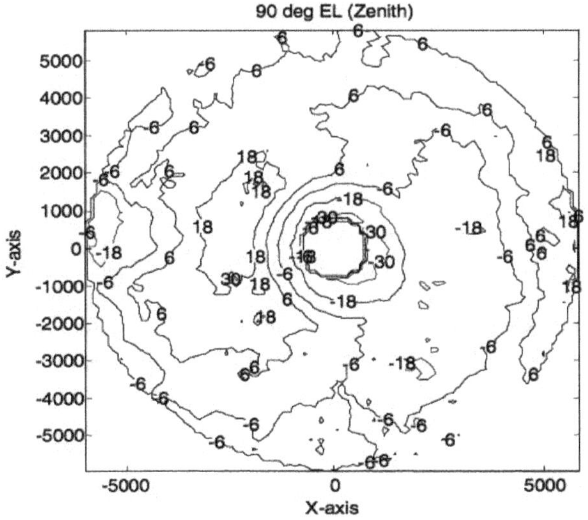


Resultant measurements and topographical plot from surface in above diagram

**Satellites**

If for instance the large aperture is designed for use as common satellite transmission frequencies, these geostationary and non-stationary satellites can be used successfully for the determination of antenna quality. Most of these satellites have highly polarized and highly stable radio outputs.

Most is not all geostationary satellites have beacons on board that radiate stable emissions near the normal outputs bands of the satellites.

Mid level satellites such as GPS have steady outputs radiating over at least a 10 MHz bandwidth, the output levels are low however and using these satellites as sources should only be attempted with large aperture, greater than 10 meters at L band.

Low level satellites, or LEOs often have output transmission in the VHF and UHF bands. For instance Orbcomm, Amsat and various weather satellites all transmit at these lower frequencies. Check with the Amateur Radio Relay League (ARRL) for more details on Amsat, Orbcomm.com for their satellite system properties and NOAA for details on their weather "birds."

**Aircraft**

One final method yet risky method to evaluate performance of a large aperture is to use a small transmitter in an aircraft that flies a grid pattern over the antenna under test. Because the aircraft's position is hard to determine and movement from turbulence can modify the incident power on the ground, this approach should be used with the understanding that there could be large sources of error inherent in its use. On many occasions, this kind of test, without a transmitter, has been used to evaluate a low power radar system. Range determination can be observed as well as signal level quality.

**Mode Stirring Chambers**

One interesting method for determining gain on an antenna is to place is inside a completely reflective chamber with a source antenna. A large mode stirring paddle is also placed in the chamber. As the paddle is moved continuously, the reflections are stirred and as a result, every possible phase and amplitude of reflection can be obtained from the source antenna to a test antenna. Comparing the results of a standard gain horn to a test antenna can yield the gain over a given frequency band. Beamwidth can be estimated from the gain, however the sidelobe structure is not easily observed. Using this method is convenient for testing low gain antennas from omni to standard gain types. The equipment required to make a measurement in this type of chamber is usually a network analyzer with port one attached to a source antenna and port two to the antenna under test. The output of the network analyzer is sent to a computer for response integration and final gain determination. As the paddle moves inside the chamber, ripples and undulations occur on the analyzer display, a multitude of these must be observed with a computer to get a complete picture of the antenna under test performance.

**Multi-Path Calculations**

A good estimate of field strength in a complex environment can be found by using the following equation:

$$E = 2 * Ed * \sin(2 * \pi) * \left( \frac{\eta}{(2 * \lambda)} \right) \qquad [110]$$

where,

E = Resultant Field Strength from direct and reflected path

Ed = Direct ray Field Strength

$\eta$ = Length difference between direct and reflected path

or,

$$\eta = 2 * Ht * \left( \frac{Hr}{d} \right) \qquad [111]$$

where,

Ht = Height of Transmitter

Hr = Height of Receiver

d = Direct distance length

**Special Considerations when Measuring Antenna Performance**

There are several point to remember when testing antennas, this section elucidates a few of these points.

**Determining Gain Standard**

Calibrating an antenna at NIST can be an expensive proposition with costs in excess of $2,000 a day. Although the reliability and quality of the measurement is superb, there seldom is a need to know the gain of a test antenna to within .01 dBi. As a consequence, many manufacturers make standard gain horns for use by other facilities. These standards are NIST traceable, meaning that there is a paper trail from the antenna purchased from one of these vendors to a test on an identical antenna at NIST. This latter antenna is used as the standard at the facilities building the gain standards. The reliability of these antenna (from the most reputable companies) is to within 1 dB across the frequency of operation. Example vendors of good standard gain antennas are Scientific Atlanta and Waveline. Each antenna comes with a calibration sheet that show gain vs. frequency. These types of standards are known as secondary standards. Making standards from these and so on dilutes the confidence of a good antenna gain number and should be avoided.

Gold standards are useful in the manufacturing process, as they have been carefully constructed and measured against a known standard gain horn. These standards are rarely used for any other purpose other than comparison to an identical antenna in mass production. To produce antennas in large quantities requires comparison of each new antenna to the gold standard. At some point confidence is obtained as the quality of the production antennas and only periodic comparisons to the gold standard will be necessary. It is highly advised however that all antennas be checked for VSWR and random checks against a gold standard be made on a regular basis. For very high volume production, manufacturing techniques must be documented and strictly adhered to as production personnel will not have the understanding and expertise to know what will effect an antennas performance. When any new processes or components are introduced it is imperative that they be tested throughly and a comparison to the gold standard is made and documented. A phenomenon known as "manufacturing creep" can occur when several small changes are made to the production process that conspire together to cause the antenna to loose performance. The push for lowering cost is the main reason for the introduction of reduced performance parts and the resultant degradation in performance.

### Calibrating Equipment

In all cases of antenna design and manufacturing, the use of high quality, calibrated equipment is necessary. For network and spectrum analyzers, a calibration cycle should be maintained to send the equipment to a certified calibration facility on a periodic basis. The periods are typically 6 months to a year depending on the recommendations of the equipment manufacturers.

There are several problems that occurs is the slow changing of measurement values due to extended use; for instance range and linearity of a Db scale. It is difficult to maintain a below dB measurement value over a large dynamic range. In other words, .1 dB over 100 dB of measurement range will lead to inaccuracies if the equipment is not calibrated properly and periodically. The same holds true for phase measurements.

### Gain vs Directivity

Gain is best described as the measurement of the overall performance of an antenna (directivity) minus the losses due to cables, sidelobes and other factors. Usually, we talk about the efficiency of an antenna which encompasses all of these losses. Some antenna, like offset parabolas are highly efficient, near 70%. For elements themselves, microstrip patches by themselves can achieve efficiencies of better than 90%.

The way an antenna measurement can quantify an efficiency is to compare the expected gain to a measured gain. Typically, this is done by simply comparing a standard gain antenna vs the test antenna. The difference in dB is converted to percentage, for instance if a test antenna is 3 dB below the standard gain, it is considered to be 50 % aperture efficient.

Another way to determine gain from an antenna is to measure as much of the complete sphere of radiation and integrate the pattern. This is a preferred method for satellite manufacturers where commercial geo-synchronous antennas are not designed to have a single focused beam but an oblate shape that covers the intended market to the greatest extent. For instance, satellite beams are shaped like the United States or countries within Europe.

**G/T**

A good figure of merit for antenna quality typically used in satellite or radio astronomy antennas is the ratio of Gain over Temperature. The temperature here refers to the total system noise temperature which includes contributions from the low noise amplifier, losses in the feed system (cables, waveguides, etc.) and noise picked up by the antenna sidelobes viewing the ground at ambient temperatures.

The system temperature (T) is related to the system noise factor by:

$$Noise\ Temperature(K) = 290 * 10^{(NF^{10})} - 1 \qquad [108]$$

or where F is the noise figure expressed in dB:

$$F = \log(-1) * \left(\frac{NF}{10}\right) \qquad [109]$$

Where NF is expressed in degrees K

To measure G/T using the sun first measure the ratio of the sun to cold sky, a region of minimal radio flux:

$$Y = \frac{Sun}{(Cold\ Sky)} \qquad [110]$$

Then use the following equation to calculate the G/T value:

$$\frac{G}{T} = (Y-1) * 8 * \pi * k * \left(\frac{L}{lambda^2}\right) \qquad [111]$$

where:

Y = Sun noise rise expressed as a ratio (not dB)

k = Boltzmann's constant 1.38 * 10^-23 joules/degree K

L = Beamsize correction factor (discussed below)

$\lambda$ = Wavelength in meters

F = Solar Flux Density in watts / meter^2 / Hz (discussed below)

Beamsize Correction (L)

This is dependent upon the antenna beamwidth, and approaches unity for small dishes with beamwidths larger than a few degrees. If the dish (or other type of antenna) has a beamwidth larger than 2 or 3 degrees, set L=1, otherwise

$$L = 1 + .38 * \left(\frac{Ws}{Wa}\right)^2 \qquad [112]$$

where:

Ws = Diameter of the radio sun in degrees at the frequency of interest

Wa = Antenna 3 dB beamwidth

The diameter of the radio sun (Ws) is frequency dependent. Typically it is .5 degrees for frequencies above 3 GHz, 0.6 degrees for 1420 MHz and 0.7 degrees for 400 MHz.

Solar Flux Density (F)

The United States Air Force Space Command operates a worldwide solar radio monitoring network with stations in Massachusetts,

Hawaii, Australia and Italy. These stations monitor solar flux density at 245, 410, 610, 1415, 2695, 4995, 8800, and 15,400 MHz. If the antenna under test is near these frequencies the measured values from the USAF can be used directly, otherwise you can interpolate between them.

The solar flux density measured by the USAF must be multiplied by $10^{-22}$ in order to get the units to be used by the G/T equation.

The Solar Flux can be obtained by going to:

gopher://solar.sec.noaa.gov:70/00/latest/curind

http://www.drao.nrc.ca/icarus/www/current/current.flx

http://www.ips.oz.au/Main.phpCatID= 5&SecID=3&SecName=Learmonth%

20Observatory&SubSecID=4&SubSecName=Radio%20Flux&

LinkName=Quiet%20Solar

http://www.sec.noaa.gov/ftpdir/list/radio/45day_rad.txt

**Scaling**

For physically large antennas, especially arrays, it is sometimes advantageous to scale the antenna up in frequency and therefore down in size. For instance, a large 10 MHz phased array used for direction finding can be modeled at 1 GHz and at 1% the size.

In general this works well in an anechoic chamber and can reliably indicate gain and beamwidth parameters. Care should be taken when scaling large antenna systems to take into account other "real world" variables like ground permittivity and local conductive structures.

# Appendix 7 – Terminology and Definitions

## Active VSWR Measurement

This is an impedance measurement of each element in an array for a given set of phase and amplitude weights. This includes effects of mutual coupling between elements and changes for every array beam position. This is very important due to the fact that mutual coupling at some beam angles can conspire to produce poor element VSWR and as a result, cause scan blindness in certain directions. With good electromagnetic simulation tools, these mismatches can be revealed and with prudent element design and placement, optimized.

## Adaptors

Coax to waveguide

Typically metal devices with waveguide flanges (with several types of hole patterns) with shorted waveguide ends. About ¼ wavelength from short is a probe about 1/8 wavelength inside cavity connected directly to the center pin of a coax connector. Tuned to 50 ohms and as wideband as the waveguide itself.

Between series

There is always a need for between series adaptors, typical are N to SMA, BNC to N, F to BNC, and about every other combination. These types of adapters are found in most every microwave or antenna laboratory.

## Admittance

The inverse of impedance and is measured in Siemens. This is the measure of how easily a circuit or device will allow a current to flow. Admittance is not only a measure of the ease with which a steady current can flow (conductance, the inverse of resistance), but also takes into account the dynamic effects of susceptance (the inverse of reactance)

## Antenna Factor

This is defined as the ratio of the incident electromagnetic field strength (V/m or uV/m) to the voltage on the line connection of an antenna:

$$AF = \left(\frac{E}{V}\right) \quad [112]$$

in a 50 ohm system, the antenna factor is is related to the antenna gain G at wavelength $\lambda$ thus:

$$AF = \frac{9.73}{(\lambda * \sqrt{(G)})} \quad [113]$$

## Array Factor

This is a quantification of the results of adding several antenna elements into an array. There are impacts on directivity and pattern. This factor quantifies the effect of combining radiation elements in an array without the element specific radiation pattern taken into account. The overall pattern of the array is determined when both the elemental patterns and array pattern is combined. For instance the array pattern is truncated off axis when using radiating elements with low beamwidths.

## Array Thinning

Much the same beamwidth and sidelobe performance can be obtained in large phased arrays if the number of elements in a specific manner to minimize cost and complexity. For instance in a linear array of dipoles, assuming the total length remains the same, a taper can be applied which removes several selected elements in a symmetrical pattern which will lower the array gain by a minimal amount and retain the same beamwidth. In two dimensional arrays the same approach can yield significant reductions in cost and complexity. Although there are several approaches that have been used in the past, minimum redundancy and statistical sparsing algorithms have been most popular.

Minimum redundancy: In the case of a linear array, a binary

aperture distribution would in the sample case of 8 elements would follow the spacing pattern 1,1,2,4,4,2,1,1 or 1,3,6,8,6,3,1

Sparsing: "Random number generator [0,1] times aperture distribution [0,1] time scale factor k, which delineates level of thinning required. When the random number is higher than the product of k time the aperture distribution, the element is removed and therefore a lower k produces greater thinning. Thinning reduces gain by the ratio of the remaining elements to the initial total, but array size determines beamwidth. The number of elements determines the average sidelobe level at 10 log(N) dB." (44)

## Aperture Blockage

Any structure that is in the field of view of an aperture is considered a blocking component. Typically, these components include feeds, sub-reflectors and associated support structures. Offset antennas eliminate these blockages and thus are more efficient.

## Attenuation

Generally the losses (often described in dB per unit length) due to components in a circuit or the space between transmitting and receiving antennas. For dielectrics the attenuation of the propagated waves is:

$$Attenuation(dB) = 9.1 * 10^{-8} * F * \tan(\delta) * \sqrt{(k)} \qquad [114]$$

where:

F = Frequency

$\tan\ \delta$ = Loss Tangent

k = Permitivity

## Axial Ratio

This is the ratio of the orthogonal components of an E-field. A circularly polarized antenna contains two such fields 90 degrees out of phase. If they are equal in magnitude then the axial ratio is 1 (or 0 dB). Measurements of this quantity are typically made on axis, however the quality of the antenna performance is best realized when calculating axial ratio over a certain beamwidth, for instance +/- 30 degrees from boresight. This is due to the fact that axial ratio tends to degrade away from the main beam position. In addition, cross pol measurements tend to degrade when axial ratios are poor.

## Babinet's principle

The theorem concerning diffraction that states that the diffraction pattern from a solid body is identical to that from a hole of the same size and shape except for the overall forward beam intensity. This principle allows for the design of slot antenna where a rectangular slot of ½ wavelength in a ground plane has similar properties as a dipole.

## Bandwidth

Gain Bandwidth

Defined as where the gain of an antenna is above 3 dB from the peak gain. In other words, 3 dB below peak, through peak, to 3 dB above peak.

Impedance Bandwidth

Defined as where an antenna VSWR is below a certain value, typically 2:1, depends on specification requirements.

Mathematically, bandwidth can be defined as:

$$BW = \frac{(VSWR-1)}{(Q*\sqrt{(VSWR)})} \quad [115]$$

where,

$$Q = \frac{(Energy\ Stored)}{(Power\ Lost)} \quad [116]$$

## Bazooka

This antenna can be made of a ½ wavelength of coax with

the shield opened at the center. A balanced feed line feeds the two shields and the presence of the center conductor (floating) reacts with the radiating shields to create greater bandwidth. This antenna has high efficiency and can be easily constructed. See ARRL references and publications.

## Beam Feed

For many earth station designs, losses of any kind need to be minimized. Although waveguides are low loss compared to coaxial feeds, even better performance can be obtained be a beam feed layout, where a Cassegrain style dish antenna has the energy from the secondary reflector project the energy through a focal point and thence to a series of flat and parabolic reflectors (making the beam more or less parallel in nature) finally to a feed horn in the base of the antenna which transmits or receives the energy. The result of not using waveguides from the feed point to the receiver minimized any losses and as a consequence, the efficiency of the antenna is improved.

## Beamwidth

Defined as the 3 dB points on either side of the main beam peak gain. This value can be approximated by the formula:

$$Beamwidth(Degrees)=\left(\frac{\lambda}{D}\right)*57.3 \qquad [117]$$

## Boresight

Refers to the line directly from the center of the antenna or mechanically where the antenna is pointed.

## Boresight Power Level

The amplitude difference between the E and H planes, on boresight.

## Bridge or VSWR Bridge

A typically three port device that can measure return loss by means of directing reflected signals to an output port. The other ports are an input from a signal generator and a test port. The output port can be fed to a spectrum analyzer or to a detector / oscilloscope combination. This device can be calibrated with a short and good 50 ohm load.

## Calibration

Holographic

This type of calibration technique is typically used on a reflector and employs a second, smaller reflector near the first that is used in concert to examine the phase and amplitude distribution across the face of the larger reflector using holographic transform algorithms.

Dicke Receiver types

This method is used to calibrate the signal intensity as received by an antenna. Basically it is constructed by switching between the main feed and a calibrated signal source or temperature controlled load. The switch is controlled by an electronic circuit that also controls a synchronous detector that in effect derives the difference between the calibrated source and the measured source, taking out the effects of (for instance) temperature drift and noise figure drift in a receiver. A consequence of using a system like this is that the total power received by the unknown signal of interest is reduced by the amount of time the calibration signal is being measured. For instance, if the calibration signal is sampled at a 50% rate then the signal of interest is also sampled at 50% reducing the integration period by a factor of 2. Black plate interrupting beam (ala NRAO millimeter wave observatory at Kitt Peak) This calibration technique follows the basic philosophy of the Dicke receiver type whereby a load, in this case a black plate placed periodically in front of the feed horn of a dish antenna, introduces a signal (a black body radiator) into the feed horn. This radiator has a known temperature which is measured by the receiver system and used as a calibration point. As in the case of the Dicke type calibration, the unknown signals measured by the receiver are referenced to the known black body radiator.

Injected signal

Typically using a directional coupler, this type of calibrator adds a known signal magnitude into the receiver chain. The signal is at a known frequency and known amplitude. This technique is best used by a receiver with a wide bandwidth where the known injected signal is only a part of the total received band and other signal within the band can be compared to the known signal.

Differential

The simultaneous comparison of known calibration signals to the unknown is sometimes used to constantly calibrate the antenna/receiver chain.

## Capacitance

Is the measurement of the amount of electric charge stored for a given electric potential. The energy in joules stored in a capacitor is equal to the work done to charge it.

## Choke Ring

A conductive device usually surrounding a single antenna element that is made of concentric circles ¼ wavelength apart and with walls that are ¼ wavelength tall. This device is used to minimize multipath effect and proved a uniform phase and amplitude distribution around the antenna. Applications include GPS survey antennas and satellite feeds for reflector type antennas.

## Circulator

This is a device that has typically 3 ports, waveguide or coax connected that contains a ferrite toroid or similar device. Port 1 is low insertion loss to Port 2, Port 2 is low insertion loss to Port 3 and Port 3 is low loss to Port 1. If for instance a transmitter or amplifier output is connected to Port 1, then the output will be on Port 2. If the output of Port 2 in this case presents a poor VSWR, the reflected energy is sent to Port 3. This is useful in the case of radars or TWTs, where damage can occur on poor output port VSWR. Port 3 in this case is terminated with a load, capable of handling any expected poor VSWR reflections. In radar systems, the transmitter can be connected to Port 1, antenna to Port 2 and receiver to Port 3 where the receiver is protected by a limiter, which does not allow large reflected power to enter the receiver.

There are two fundamental types of circulators, one with a doughnut shaped ferrite and one integrated into a waveguide coupler. The latter is used for high power applications. See also, Isolator.

## Coax

A type of feedline with concentric components. Typically composed of a center core of solid or stranded wire made of copper or aluminum. Around the center is a dielectric from air (with spacers) to polyethylene or teflon etc. The dielectric is made of low loss material (low loss tangent) to minimize propagation losses. Surrounding the dielectric is a ground shield, solid or braided. The ratio of the diameters of the center conductor to the shield dictates the impedance of the coax. This type of feedline is considered un-balanced which dictates what type of matching circuit will be required.

## Corner Reflector

Designed initially by Hertz, this simple antenna works well for many applications. Made up of two or three flat reflectors connected at 90 degrees. The feed antenna, typically a dipole, is located ¼ wavelength from the connecting apex.

## Current Sheet

This is a class of array antennas formulated by Monk of the Ohio State Electrosciences lab. Wheeler and others contributed to this design as well. It generally consists of dipoles connected end to end via interdigitated capacitors. The array has very good broadband performance and each dipole has a feed point that can be used in phase with the other elements for a single boresight beam or phased with the other elements to steer the beam as desired.

**CW**

Continuous Wave, where the amplitude of the signal is pulsed in the case of morse code transmissions or steady in the case of signal generators or specialized radar systems where the transmissions are steady and the reflected signal are detected by virtue of their doppler shifts.

**Dielectric**

A non-conducting material designed to be used as an insulator for particular antenna designs such as microstrip. The two important quantities to define the quality of a dielectric is the dielectric constant (Er) and loss tangent (tan). Er can be viewed as a refractive index and tan can be viewed as a loss. High Er dielectrics (or substrates) make microstrip antennas small for a given wavelength, they also lower the gain and lower the bandwidth. High loss tangents cause a lowering of gain for a given application. In antenna design, materials are chosen with low loss tangents to optimize performance.

**Diffraction**

The physical phenomena that allows radio waves to “bend” over flat plates or around the corners of buildings. It is caused by interference between the radiated energy and the induced currents in the object.

**Diffraction Fence**

A metal fence, typically with a saw tooth top that is used to minimize ground reflections in antenna test ranges. The size of the “tooth” dictates the cutoff frequency of the fence itself.

**Diplexer**

A device that combines two signals, through bandpass filters, to a common port. The two ports are highly isolated from each other but have very low insertion loss to the common port. Commonly used in commercial broadcast satellites.

## Directional Coupler

A 3 or 4 port device that allows the sampling of rf energy going in a particular direction. A 4 port device typically has an input port, output port, forward and reflected ports. The forward and reflected ports sample the rf signal with some level of attenuation, usually above 10 dB. A common application is to have the input port connected to a transmitter, the output port connected to an antenna, the forward port connected to a sampling device to measure power output and the reflected port connected to a sampling device to measure VSWR.

Lange

A type of coupler made of parallel lines usually made in microstrip. A variety of Lange couplers uses four lines with line 1 shorted to line 3 and line 2 shorted to line 4 on the opposite end. The coupling between the 4 lines allows for the coupling action between both ends of line one and both ends of line 4. The length, width and separation between the lines is critical.

Cross guide

A coupler used on waveguide to sample forward and reflected energy. The sampling guide is placed orthogonally over the main waveguide, a slot is cut into both waveguides whose orientation allows for the forward energy to go one end of the sampling waveguide and the reverse energy to go to the other end. Both ends have a waveguide to coax adapters for the sampling ports.

Waveguide

A waveguide couple is composed of 2 waveguides attached on the long wall with slot cut into both to allow the forward or reverse energy to propagate from the main waveguide to the sampling waveguide.

Coaxial

This type of coupler has 3 or 4 ports, an input, an output, a forward and in the case of 4 ports a reflected. A 3 port device can have either forward or reflected outputs. These devices have internal circuitry to make the desired coupling direction at the desired level.

## Directivity

This is a measurement of the performance of antenna without the inefficiencies added. It is the ratio of the achieved radiation intensity in a particular direction to that of an isotropic antenna. Or, in other words, the measure of the concentration of radiation in the direction of the maximum. Directivity and gain differ only by the efficiency.

For a directional beam:

$$Directivity(dB) = \frac{(4*\pi)}{((Half\ Power \text{ in } E\ plane)*(Half\ Power \text{ in } H\ plane))} \quad [114]$$

in radians

or

$$Directivity(dB) = \frac{41{,}253}{((Half\ Power \text{ in } E\ plane)*(Half\ Power \text{ in } H\ plane))} \quad [115]$$

in degrees

## Dispersion

The transverse modes of waves confined laterally within a waveguide generally have different speeds and field patterns depending upon their frequency, compared to the size of the waveguide. An associated phenomenon is group delay with waveguides.

## Diversity

This is a technique that uses (normally) 2 antenna separated by ½ wavelength. Because of complex multipath propagation, one or the other antenna will have a higher signal strength (either to or from the other end of the link) and can be selected to provide a superior signal to noise ratio.

Amplitude

For amplitude diversity, the signal strength is measured on each antenna at millisecond intervals, the determination is made as to which antenna is best suited to provide a better communications link and used thereafter. PIN diodes are typically used to select which antenna is being measured. The detected output or RSSI (Received Signal Strength Intensity) is measured and a microprocessor makes a decision as to which antenna to use.

Polarization

In the same fashion as amplitude, two or more antennas are compared in terms of amplitude. In this case the antennas are placed at different polarizations to detect which one offers the best RSSI. In complex environments like urban jungles and in building venues,

Time

Another type of diversity is to use accurate time standards at two or more ends of a communications link and carefully choreograph the transmissions. In this way significant increases in efficiency can be realized.

Beam

Beam diversity requires an antenna with multiple beams, a phased array or Butler type matrix for example. In this way the best signal path can be selected and in fact an increase in traffic can be accommodated by using the same frequency but different antenna pattern.

## Doppler Shift

This is the change in frequency due to the movement of the distance between transmitter and receiver. This is characterized with the following formula:

$$\Delta f = \frac{Fo}{c} * \sin(\theta) \qquad [116]$$

where,

Fo = Emission frequency

c = Speed of Light

$\theta$ = angle of approach

## Ducting

This is an atmospheric phenomenon where signals get trapped between several atmospheric layers. In some cases low attenuation of the original signal has been measured. The main drawback is the ability to predict when ducting will occur. More often an existing link has lower signal strength while a link many miles away at the same operating frequency observes another signal.

## Edge Effect

Essentially, this effect is due mainly to diffraction across edges of metallic components and has the ability to bend the radio waves around the components. The phenomena is observable in optics as for instance a laser is directed at a sharp edge, like a razor blade. The light from the laser is observable at angles in the shadowed area on the other side of the edge. In the radio regime, signal can be detected on the shaded area of similar edges.

## EIRP

Standing for Effective Isotropically Radiated Power, it is the addition of the antenna gain (in dB) to the power of the transmitter (also in dB). This is an important value when determining the power per square meter at the receiver end. Commonly used in satellite link calculations, it is for instance the defining value for geostationary communications satellites broadcasting programming received by backyard dishes.

Mathematically:

$$\text{EIRP(dBm)} = \text{Transmitter Power (dBm)} + \text{Antenna Gain (dBi)} \quad [116]$$

and:

$$\text{EIRP(dBW)} = \text{Transmitter Power (dBW)} + \text{Antenna Gain (dBi)} \quad [117]$$

## Electrically Short Antenna

Dipole with a length less than an electrical half wavelength, or vertical monopole with a length less than an electrical quarter wavelength.

## Evanescent Mode

A waveguide propagation mode at a frequency below the cut-off wavelength. In this mode, the amplitude of the wave diminishes rapidly along the waveguide, but the phase does not change; applied in certain special waveguide filter designs.

## Far Field

(or) the Franhoufer zone
Defined as that point where the far field pattern is fully formed. In mathematical notation:

$$Far\ Field = \frac{(2 * D^2)}{\lambda} \qquad [117]$$

Far field measurements are only accurate at and beyond this distance. At distances less that the far field, beam widths are oblated and sidelobes are larger than what will be observed in the true operation of the antenna. Distances less than the Franhoufer zone are referred to as the Fresnel zone (or Near Field).

## Feed

That component in a reflector antenna system that broadcasts the radio energy to the reflector. The feed design must take into account the amount of power to be transmitted and how the reflector is to be illuminated. Normally, the 6 to 9 dB levels of the feed pattern are placed on the edges of the reflector to minimize antenna noise temperature. Also the feed establishes the polarization that will be transmitted or received. Feeds can be made of horns (very common) dipoles, spirals, microstrip elements, log periodics as well as other suitable single element antenna.

## Field Probe

This device is used to examine the quality of an antenna test range. It typically is a small antenna or horn that is scanned in a flat plane across the far field area of a chamber or outdoor area where the antenna under test is to be examined. The probe measures both phase and amplitude and thus establishes the planarity of the test zone. Normally values less than a dB and only a very few degrees are acceptable over the test area. Values in excess of these numbers degrade the measurement quality.

## Field Strength

Essentially, the amount of radio energy in a given area, generally described in watts per meter. The devices that measure this are composed of wide band detectors that integrate the signals and present the output on a meter or other similar device. Mathematically:

$$Power\ Density\,(W/m^2) = \frac{P}{(4*\pi*r^2)} \qquad [118]$$

where,

P = Transmit Power radiated isotropically.

or,

$$E\,(v/m^2) = \frac{\sqrt{(30*P)}}{r} \qquad [119]$$

And for a a directional antenna,

$$Pd\,(W/m^2) = \frac{P*G}{(4*\pi*r^2)} \qquad [120]$$

where,

G = Gain of antenna

## Free Space Impedance

The impedance of free space, $Z_0$, is a physical constant relating the magnitudes of the electric and magnetic fields of electromagnetic radiation traveling through free space. That is, $Z_0 = |E|/|H|$, where |E| is the electric field strength and |H| magnetic field strength. It has an irrational value, given approximately as 376.7 ohms. This is good for air as well as the vacuum in outer space.

## Friis Transmission Formula

This formula is used to calculate the power received from a given set of antennas, power levels and range.

$$Pr = \frac{(Pt * \mathrm{Gt} * Gr * \lambda^2)}{(4 * \pi * R)^2}$$

[121]

where,

Pt = Power of transmitter

Gt = Gain of transmitter antenna

Gr = Gain of receiver antenna

R = Range

## Flux Density

This term is used widely in radio astronomy and relates to the amount of energy at a given frequency hitting the earth from a distant radio source. The unit of measure is a Jansky (named after the father of radio astronomy) and is calibrated at 10^27 watts / meter / Hz

## Focal Length

For reflector antennas, this is the distance between the center of a dish (assuming prime focus style) and the position where the feed is to be placed. It can be estimated by:

$$Focal\ Length = D^2/16 * x \qquad [122]$$

$D$ = Antenna Diameter in inches
$x$ = Antenna Depth in inches

or

$$FL = \frac{Radius^2}{(4 * Dish\ Depth)} \qquad [123]$$

## Forward Power

Power leaving a source and traveling toward a load. In contrast to reflected power, which returns from the load if mismatched.

**FM**

Frequency Modulation, when a signal is varied in frequency as apposed to amplitude. This is the basis of FM radio in cars and homes. FM has advantages in terms of noise immunity but is slightly more complex to modulate and demodulate.

## Frequency Selective Surface

Any conductive surface that resonates at a frequency of interest. For instance, in a reflector system a “dichroic” subreflector will reflect any energy hitting it if it has conductive elements one half wavelength in size. Any other radiation will proceed through the reflector to another feed. Frequency Selective Surfaces (FSS) are used in reflectennas and reflect arrays whereby the resonant surfaces can reflect energy in a pattern much like a phased array.

## Front – to – Back ratio

This is the measurement of the energy in front of an antenna (e.i. The direction it is pointed in) to the back of the antenna opposite the boresite. Ratios for reflector antennas are generally high, over 30 dB in most cases and lower for antennas like yagis where the ratio might be below 10 dB. Poor front-to-back ratios cause an increase in antenna noise temperature and can allow interference to be detected easily.

## G/T

A figure of merit for (typically) large satellite antennas. It is the ratio of the gain of the antenna to the system noise temperature. See the antenna testing section for more details. The higher the number the better with typical reflector systems have numbers about 20 dB / degree K. This measurement is useful to accurately measure the expected performance vs. the real performance. Large aperture antennas can use the sun as a source to calculate this value.

$$\mathrm{G/T} = AntennaGain - (10 * \log(Tsys)) \qquad [124]$$

where:

G/T is in dB per degrees K

Antenna Gain in in dBi

Tsys is system temperature in degrees K

## Gain

This is the ratio of how sensitive an antennas is relative to detectable sphere around it. For aperture based antennas, it is calculated with the following formula:

$$Gain(dB) = 10 * \log\left(\frac{(4 * \pi * Ae)}{\lambda^2}\right) * N \qquad [125]$$

where:

$\pi$ = 3.1415926

*Ae* = Aperture size in same units as lambda

$\lambda$ = wavelength

N = Efficiency of the antenna

## Grating Lobe

In the case of phased arrays, where the inter-element spacing is greater than one half wavelength, an aliasing effect conspires to create sidelobes that can approach the magnitude of the main lobe. At one wavelength spacing an end fire condition can occur that sends substantial energy in a direction orthogonal to the main beam.

## Ground Losses

Portion of transmitter power intended to be delivered to an antenna system that instead is lost due to various resistances in the antenna ground system. This is often particularly significant in short monopoles because the ground connection handles the full antenna circuit.

## Group Delay

This is a measure of the transit time of a signal through a device under test (such as a filter) versus frequency. This is important in communications systems where components with large group delay numbers can distort signal propagating through the devices.

## Hound's Tooth Polarizer

A series of pins, varying in length in a filter like fashion, placed in a waveguide that allows for the formation of circular polarization in a feed system.

## Hybrid Coupler

A passive device where the input power is equally divided between two output ports. In many cases the output ports are phase offset by 90 or 180 degrees. This device is useful in the design of high power transmitters and receivers using I/Q (In phase, Quadrature) conversion.

## Impedance

This is the measure of opposition to time-varying electric current in an electric circuit. This value is used extensively in antenna design theory to denote the number of ohms measured or required for a match. More accurately, it is measured in a complex manner with a real component and an imaginary component. For instance:

$$Z = 48 + j10$$

This impedance indicates that the real part is 48 ohms and the reactive part is 10 ohms and inductive. Using a Smith chart, the complex conjugate of 48 + j10 can be placed on the graph, and if the chart is normalized to 50 + j0 ohms, a significant amount of information can be gleaned. For instance, what is required to move the value closer to 50 +j0, or a perfect match in a 50 ohm system. Also, the phase and amplitude can be measured as well as the VSWR.

## Index of Refraction

For a dielectric material, the index of refraction is:

$$n \simeq \sqrt{(k)} \quad [126]$$

where:

n = Index of Refraction

k = Dielectric Constant

## Inductance

A measure of the ratio of the magnetic flux to the current, measured in Henries. This is a useful quantity in the analysis of antenna systems as it, in addition to capacitance give a good measure of the resonant frequency of the device.

## Integrated Cross Polar Ratio

Integrated Cross Polar Ratio (ICPR) is defined as the integrated cross-pol response weighted by the co-pol response of the entire pattern, but typically only the main beam contributes to ICPR for high gain antennas. For instance, if the cross-pol sidelobe within the -10 dB point of the beam are < -32 dB, then ICPR will also be < -32.
If you have a -38 dB cross-pol sidelobe that dominates the cross-pol response within the main lobe, then ICPR will be very low.

## Isolation

Port to Port

Measured in dB, this is value denoted the amount of separation of for instance a transmitted signal to a received signal, when they both flow through a single device, like a diplexer. In bi-static radar antenna design this is the measurement of the separation between the transmit and receive antenna.

Polarization

This is the measurement of separation of powers between orthogonal planes of radiation. Usually, it means comparing vertical to horizontal in a linear system or right hand to left hand in a circular system. It can also mean ratios of elliptical or ratios between orthogonal slant linear systems. It is measured in dB with good values in excess of 30 dB.

## Isolator

A ferromagnetic device typically made of 3 ports with port 1 having low insertion loss to port 2, port 2 having low insertion loss to port 3 and port 3 having low insertion loss to port 1. High insertion loss is to found in the opposite direction, hence the isolation. Assuming good VSWR on all ports, this device can be used to act as a separation device for a radar system, where the transmitter is connected to port 1, the antenna to port 2 and the receiver to port 3. Care must be taken however to limit the amount of power that goes into the receiver port due to the fact that if the antenna VSWR becomes poor, the reflected energy from the transmitter will find its way into the receiver.

## Magic T

A waveguide device that combines several signals and gives (depending on the input phase relationships), sum or difference signals on the outputs. The device is made up of a waveguide "T" with an orthogonal port connected to the center extending vertically. This device can be used for monopulse antennas where sum and difference channels allow for the near instantaneous determination of signal location.

## Magnetism

This is the phenomena whereby certain material attract or repulse each other. In antenna theory, is is the field orthogonal to the electric field. Maxwell's equations explain the actions of a magnetic field and its applications in antenna design. These fields can be created by magnets or electromagnets, where coils of wire are energized by electric current. The first discussions about magnetic properties were formulated by Thales.

## Magnetron

A magnetron is a high power microwave oscillator in which the potential energy of an electron cloud near the cathode is converted into r.f. energy in a series of cavity resonators similar to the one shown on Page 21. As depicted, the rear wall of the structure may be considered the inductive portion, and the vane tip region the capacitor portion of the equivalent resonant circuit. The resonant frequency of a microwave cavity is thereby determined by the physical dimension of the resonator together with the reactive effects of any perturbations to the inductive or capacitive portion of the equivalent circuit.

## Match

A device that couples at least two different electromagnetic components to each other in the most efficient way. In antenna theory 50 ohm impedances are very common, to connect a 50 ohm antenna to a 75 ohm feed line (for instance) requires a matching device with a 1.5:1 impedance ratio to best transfer energy. In addition, impedances are best described in complex terms, whereby there is a real and an imaginary component. Matching devices which take this into account are the most efficient. There are several types of matches, a few which are described below.

Balun

A device which matches balanced to unbalanced components. An example of an unbalanced component is a coaxial feed line, an example of a balanced component is a dipole or ladder type feed line. This device can be made of a transformer with specific taps and ground routings that connect the two dissimilar components.

Transformer

A device which can match two dissimilar impedances by setting the ratio of winding about a core or in air, to the same ratio as the impedances that need to be matched.

Delta

Typically used in yagi designs, this device takes the output of an unbalanced feed, such as coax as distributes the energy to both sides of a dipole acting as the driven element. Sliding components along the extent of the dipole are manipulated to optimize the match.

Gamma

Another yagi matching device, this design uses only one side of the of the driven element dipole, which is connected to the center conductor of the feed coax.

## Mode

In electromagnetic theory, this is the phenomena where currents set themselves up in convenient positions to propagate energy. These patterns of vibration set themselves up in confined reflective areas. In coax this is known as TE01 mode, in waveguides, there are numerous modes which can be developed inside the confining walls of the guide. In antennas, modes appear in the feed horns of dish designs. Care must be taken to use only the modes which are intended for use in a particular design. Unwanted modes generated from geometries in the waveguides and/or feeds, can cause changes in the impedance and frequency response of the antennas. Using a good electromagnetic emulator can mitigate most of the undesired effects.

## Mode Stirrer

In chambers with metallic walls, modes will appear that vary the amplitude and phase distribution inside the chamber. To analyze the chamber or antennas within the chamber in a more accurate way, a paddle, or mode stirrer is placed inside to vary the amplitude and phase of the modes to affect an integratable measurement, which more accurately depict the qualities of the antennas inside.

## Multipaction

In high power systems such as communication satellites and radar systems a breakdown of high power waveguide channels can occur, leading to significant loss of power and possible physical damage. Multipaction calculators are available from the European Space Agency and include such input parameters as power and coatings on the insides of waveguides. This calculation must always be performed when designing a high power system. The choice of waveguide material and frequency of usage can mitigate the harmful effects of the electron breakdown that can occur. Multipaction is usually only a concern in a high vacuum environment like space.

## Mutual Coupling

This is a phenomena which occurs with the close spacing of antenna elements in a phased array or similar device. In the case of a phased array, the elements are tuned to the same frequency and as a result are resonant to each other, thereby detuning each other, normally this is due to inductive or capacitive loading and as a result the resonant frequency of the array move lower in frequency. In the case of antenna elements in close proximity but not operating at the same frequency, mutual coupling effects are less pronounced, but can still detune the individual antenna or distort the patterns. To mitigate this problem, careful selection of radiating elements is recommended with low sidelobes in the direction of the next element. Also, if possible, stagger the polarization attributes of adjacent elements, for instance, a right hand circular next to a left hand circular. Use a good 3D electromagnetic modeling simulator to examine the effects. Designing the elements to a higher resonant frequency for instance, might minimize the array effects.

## Near Field

This is area where energy from antenna has not yet formed into a coherent pattern. For instance, in a phased array, this is an area where the radiating elements have a measurable amplitude and phase profile that will, in the far field, combine to create the final antenna pattern. The near field far field boundary is defined as:

$$Far\ Field = \frac{(2 * D^2)}{\lambda} \qquad [127]$$

where:

D is the largest dimension of the antenna aperture,

$\lambda$ is the wavelength of the energy being propagated

The near field is the same for either a transmitting or receiving antenna. See 'Theory of Reciprocity' for more details. For antenna testing, the near field can be measured in both phase and amplitude, then a Fourier transform is performed to generate the resultant far field pattern. Due to the fact that large apertures can result in large far field distances, near field probing can be the best test method to determine antenna quality. Care must be taken to compensate for the effects of the probe and chamber these measurements are taken in. Refer to the 18 step near field calibration procedure, pioneered at the National Bureau of Standards.

Fresnel

This is the name for the near field zone, as apposed to the Franhoufer zone or far field. It best describes the actions of the electromagnetic waves close the radiating aperture.

## Noise Figure

Every antenna has a noise figure, which is an important component for discerning the sensitivity of the device. Noise temperature is dependent on the design, local environment and the attributes of the field of view. Designs which have poor sidelobe and poor front to back performance will generally have a higher noise temperature due to the fact that they "see" the ground around them. This ground is at an ambient temperature of the earth or roughly 293 degrees Kelvin. In space, this is not the case so designs of this type are still useful. If the local environment is noisy, contributions to the sidelobe and front to back increase the antenna noise temperature. For instance, an antenna on the top of a mountain generally is less noisy than one used in a dense urban environment. In terms of the field of view, is the antenna is point vertically to the sky, the noise temperature is generally very low, sometimes in the 10s of degrees depending on frequency of operation. See the radio astronomy section to best understand this environment. If the antenna is pointed horizontally, the earth temperature effects the performance as well as any closely located electromagnetic radiators in the field of view. Interestingly, an antenna located in space where the "outside" temperature is low can have a high noise temperature if it is pointed towards earth (again 293 degrees Kelvin).

LNA

This noise figure dominates the total noise figure of a receiver and is given in either dB or degrees K. It is defined as the noise that the amplifier contributes to the input signal. It can be dependent on the temperature of the amplifier and/or the amount of random fluctuations of the electrons in the amplification mechanism (e.i. Tube or transistor, etc.)

## Null

This is defined as a dip in the antenna pattern where phase

from various parts of an antenna aperture conspire to minimize the gain at certain angles relative to the antenna boresite.

Steered

Using both amplitude and phase control over a phased array, the position of the main beam, sidelobes *and* nulls can be manipulated to place these attributes in selected positions. See adaptive array.

## Octave Bandwidth

Twice the initial frequency. For instance, 2 to 32 MHz represents 4 octaves; 2-4, 4-8, 8-16 and 16-32.

## Ohm's Law

This law states that Ohms = Volts / Amps. This is one of the most fundamental law in electrical theory and is meaningful for both DC and AC circuits.

## OMUX

Abbreviation for Output Multiplexer, this device is typically found in satellites and combines the outputs of several amplifiers, like TWTs or Solid State High Power Amplifiers into one waveguide or coax to be connected to the feed of an antenna. OMUX units can be designed with filters and combiners to minimize the interaction between output channels.

## ONET

Abbreviation for Output Network, this device is typically found in satellites and is composed of a series of filters, corresponding to each input channel of the satellite, being fed by a single waveguide. The output of the ONET will be multiple feeds for the individual TWT (Traveling Wave Tube) or SSPA (Solid State Power Amplifier) whose signals will eventually be recombined for transmission.

## Optics

Geometric

A design approach for reflectors that uses much of the same formulas for visual and infrared lens and reflector design.

Concepts such as F/D, reflector shape, subreflector shape, polarization and gain are described.

Ray tracing

This is a method for calculating the path of waves through a system with regions of varying propagation velocity, absorption characteristics, and reflecting surfaces. When used in antenna design (like a reflector), ray tracing often relies on approximate solutions to Maxwell's equations that are valid as long as the waves propagate through and around objects whose dimensions are much greater than the operating wavelength.

Holographic

This is a technique that processes a significant amount of direct far field measurements, then using the Fourier transformation to construct the complex aperture illumination. The inverse transform is used to show the individual components, be they elements of a phased array or panels on a dish, and their phase and amplitude response. In this way, adjustments can be made to optimize the phases and amplitudes of the elements or panels for best overall performance.

Physical

This is the branch of optics that evaluates interference, diffraction, polarization and other phenomena for which ray approximation from geometric optics is not valid. Geometric optics ignores wave effects, physical optics uses electromagnetic theory to evaluate antenna performance.

## Organ Pipe

This is a type of resonant cavity used in the construction of high power amplifiers such as Klystrons. Working much in same way as a classic organ pipe, the length of the cavity has resonant properties with energy being applied at the base. In addition, waveguide assemblies have been designed as filtering components using this principle.

## Orthomode transducer

This is a microwave duct component of the class of microwave circulators, sometimes known as a polarization duplexer. Generally, it is a three port device that has a feed port, vertical port and horizontal port. The ports work over the same frequency bands as the waveguide used to make the device.

## Pattern

This is the response of the antenna, generally depicted in the far field. It shows the direction, gain, sidelobes and front to back ratio of each measured polarization. From these patterns, the quality of the antenna can be determined.

### Offset

An offset pattern is one where the main beam is not at boresite. This is desirable for instance with antennas placed on mountains or tall towers where a down tilt is necessary for uniform coverage. An offset pattern can be achieved by incrementally changing the phase on radiation elements of a phased array. Another way to offset the pattern is to move the focus of a dish, opposite the direction of the desired beam tilt.

### Generation

Pattern generation refers to the mechanism required to produce a pattern. Normally this means an antenna positioner, either altitude over azimuth, or phi over theta (where theta is azimuth and phi is rotation parallel to the horizon).

### Grating Lobes

This are attributes of a phased array where the element to element spacing is close to one lambda where the array with end fire and place energy orthogonal to the main beam. This of course lowers overall antenna gain, potentially raises noise temperature and allow interference to enter the pattern. For sensitive radar systems, such as boundary layer radars, grating lobe fences are sometimes constructed to minimize any local interference to the radar receiver. The ideal element to element spacing is .5 lambda, but the size of the elements and/or the mutual coupling make this spacing impractical. Standard practice is to space elements about .7 lambda. Any farther apart allows the generation of grating lobe degradation.

### Sidelobes

These are pattern attributes that are lower in sensitivity than the main lobe but if not carefully considered can allow unwanted energy to enter the pattern or unwanted energy to escape. Generally, sidelobes are the next most sensitive component of an antenna pattern. There are several methods to mitigate the magnitude of sidelobes, such as Taylor, raised cosine, Chebychev tapering or other amplitude tapering techniques. In sophisticated adaptive arrays, amplitude control is used along with phase control to move the beams and minimize sidelobe magnitudes.

### Permittivity

This is the physical quantity that describes how an electric field affects and is affected by a dielectric medium, and is determined by the ability of a material to polarize in response to the field, and thereby reduce the total electric field inside the material.

### Permeability

This the the degree of magnetization of a material that responds linearly to an applied magnetic field. Permeability can vary with the position in the medium, the frequency of the field applied, humidity, temperature, and other parameters.

### Phase Center

The phase center is the point from which it appears that ant antenna radiates spherical waves. This is important for antennas like GPS, where the phase center variation over elevation (assuming an antenna pointed vertically) dictates the accuracy of the positional information transmitted from the satellites.

### Phase Shifter

A device that changes the phase from the input to the output in smooth or incremental parts. This device is used extensively in phased array antennas. The design challenges are to make this device with very low loss and/or to make the device handle significant power.

PIN

This is diode type switch that is often used for phase shifters. This device switches in and out various lengths to create the phase shifts. The device is turned on (switched on) by forward current, typically at low (~1,2 volts) and 10s of milliamps of current. The device is turned off by reverse current, with a high voltage and a very few milliamps of current. The higher the reverse voltage, up to the breakdown voltage of the device, the greater the isolation, especially if high power is being used, as in a radar system. Normally phase shift increments for this type are in terms of "bits", which for a standard 4 bit phase shifter include 180, 90, 45 and 22.5 degrees of phase delay. A designer should evaluate how many bits are necessary to economically create the beams of interest. In some cases, 3 bits (180, 90, 45 degrees) are sufficient to create a workable antenna. Digital electronics is normally used to hold the various voltage static while the antenna is in operation. Hot switching is not recommended in transmitting arrays.

Ferrite

This device is capable of a smooth variation of phase by applying a voltage across this device, which interacts with the permeability of the ferrite core to delay the phase of the input rf.

Trombone

This is an older device which has an advantage over the two previous devices described above: it is very broadband, usable in very wide band phased arrays. It is mechanical in nature, looking much like a trombone with an electric motor that moved the "slide" back and forth. In many cases there is also a position feedback device to verify the placement of the slide. Also, this type of phase shifter is by far the lowest loss type. The drawbacks are that is cannot be changed quickly like the two previous examples.

## Passive Intermodulation (PIM)

When two or more signals of different frequencies are mixed together, they form additional signals that are typically harmonics of either input frequency. This is due to the non linear effects of the junctions of dis-similar metals. A diode effect occurs causing the intermodulation products to occur. This is a real concern for high power systems such as radars and communication satellites and must be mitigated by the proper choice of materials and thorough testing. PIM testing is typically done in a chamber using two signal generators as sources fed into the system under test and a spectrum analyzer with either a probe or direct connection to the system. The analyzer is used to examine the intermodulation products in terms of frequency and magnitude to determine the impact on the system under test. This test should be carried out under the temperature range of the proposed system installation. In addition, if the system under test is to be used in humid climates, the relative humidity should be varied as well.

## Polarization

This is a property of waves that describes the orientation of their oscillations. For transverse waves, like radio waves, it describes the orientation of the oscillations in the plane perpendicular to the wave's direction of travel. A more detailed explanation used in industry is Ludwig's 3rd definition. Polarization is usually described in either linear or circular terms, however elliptical polarization also exists but is less useful than the other types. Linear describes radio waves that are highly oriented in one plane of propagation. Wire antenna for instance are lineally polarized and will work transferring information only with another antenna oriented in the same way, in other words vertically or horizontally. These are typical orientations on an earth based systems but are not exclusive. In other words, a transmitting and receiving antenna can be in any orientation as long as they are parallel. Applying rf power to two antennas at orthogonal orientations, but in phase creates a slant linear condition, offsetting the incident energy by 90 degrees creates circular polarization and any two phases other than 90 create elliptical polarization.

Isolation

This is the degree, usually factored in dB of difference between signals transmitted or received between orthogonal polarizations. This is normally vertical vs. horizontal or left hand vs. right hand polarizations. Values in excess of 30 dB are considered good.

Axial Ratio

In a circularly polarized antenna, this is the amount of deviation, usually described in dB, when comparing a circularly polarized antenna and a spinning linear antenna. If the variation is less than 3 dB, the CP antenna is considered reasonable, but variations less than 1 dB are much more desirable.

Diversity

In a antenna system using two polarizations, this is the concept of taking advantage of the isolation between orthogonal polarizations for the purpose of sending independent streams of communications via each.

Orthogonal

This is the opposite polarization, in other words, vertical vs. horizontal or left hand vs. right hand circularly polarized radiation.

Circular or CP

Generally defined as one linear polarization delayed relative to the other orthogonal polarization by 90 degrees of phase.
Left Hand – from the viewpoint of the transmitter, these waves rotate to the left and away from the antenna.
Right Hand – form the viewpoint of the transmitter, these waves rotate to the right and away from the antenna. A simple example is placing a standard bolt on a table, the thread orientation is that of a right hand CP antenna.

Elliptical

This is a CP antenna without equal power shared by the orthogonal linear polarizations or with a phase difference other than 90 degrees.

## Polarizer

Iris

This is a type of waveguide device which consists of a circular waveguide equipped with a number of elliptical irises inside arranged at regular intervals, resting on parallel planes and all oriented in the same way, i.e. their longer axes all belonging to the same axial plane.

Pin

In circular waveguide, where vertical and horizontal polarizations have been inputed in phase, a series of pins, sometimes in the form of adjustable screws, are used in on plane to delay on polarization by 90 degrees and allow a circularly polarized output.

Septum

In a circular or square waveguide, this is a conducting plate that is attached to the center of the waveguide in a longitudinal manner. The device is typically stepped down from full extent of the waveguide section to the floor of one of the walls. This step sequence is designed to allow one of the input signals to be delayed 90 degrees relative to the other, allowing for the formation of circularly polarized waves.

OrthoMode Transducer (OMT)

This waveguide device allows for combining of two signals with an output of one signal delayed relative to the other by 90 degrees, forming a circularly polarized output. Essentially, a signal propagating down the broadwall of the waveguide is combined with another signal fed orthogonally along the narrow wall of the original waveguide.

Hybrid

This is a coaxial or microstrip device that looks like a square loop with the "left" and "right" sides of the square thicker than the upper and lower portions. There are four ports, corresponding to two input and two outputs. If one input is terminated with a 50 ohm load and the other input connected to a signal source, one output port will have a 0 degree phase output and the other port will have a 90 degree phase output. These two outputs, if fed into a circular waveguide or into orthogonal inputs to a dual polarized antenna, will create a circularly polarized signal. Most hybrids are designed to have 0 and 90 degree output, but some have 0 and 180 for special applications. Multiple hybrids such as these can be used to form matrices which allow multiple beams from an array of elements, check Butler and other similar matrices for more information.

## Poynting vector

This is the energy flux in W/m^2 of an electromagnetic field. Co discovered by John Henry Poynting and Oliver Heaviside. This also can be described by:

$$S = E \times H \qquad [122]$$

where:

S = the Poynting vector

E = Electric field

H = Magnetic field

At distances greater than about four wavelengths, the Electric and Magnetic fields become proportional to each other.

## Probe

A device which can sample a RF field while causing a minimum of perturbation. These devices are used in near field antennas ranges and EMC chambers. They can be made of waveguides, cut off and tapered at the edges or with coax.

### Loop

A type of balun where a gap is placed in a coaxial cable folded back onto itself; useful for frequencies between 1 and 20 GHz.

### Sniffer

This can be a device made of ferrite rods placed at 90 degree orientation. Loops around the rods send current to detecting diodes, which rectify the signals and send a DC current to an amplifier and meter for the purpose of measuring an electric field. This device is best used at low frequencies.

## Propagation

The movement of a radio signal through a medium or venue. This includes space or terrestrial or indoor locations. Software simulators have been created to take into account the many variables when evaluating propagation. This includes the ionosphere for lower frequencies, foliage and vegetation, buildings, earth curvature, terrain, and many other factors. Understanding this phenomenon is key to understanding the quality of a communications link.

## Q

The degree of resonance in an electrical circuit. A

narrowband antenna or filter is said to have high Q if the impedances are very well matched. Also, a quality factor of a reactive element. Equal to the ratio of the inductive or capacitive reactance to the total series loss resistance of the tuned circuit. For instance, an inductor could be evaluated with:

$$Q = 2 * \pi * \left(\frac{L}{R}\right) \qquad [123]$$

or,

$$Q = \frac{1}{(f * R * C)} \qquad [124]$$

or,

$$Q = \frac{(Energy\ stored)}{(Energy\ Lost)} \qquad [125]$$

where,

L = Inductance in Henries

R = Resistance in Ohms

f = Frequency in Hertz

## Quite Zone

On an antenna test range, where the antenna under test is to be placed, there needs to be an area of minimal phase and amplitude variation. Any changes in amplitude and phase cause inaccuracies in the test results of the antennas. As a general rule, the quite zone should have less than 1dB of amplitude change and less than $\lambda/8$ in phase changes. The extent of the quite zone in an anechoic chamber or outside range should be at least 1/3 more than the largest dimensions of the antenna under test or at least two wavelengths larger, which ever measurement is greater. To examine the extent of the quite zone, use a non conductive test rig that enables a horn or probe to be used at the frequency of interest. Scan the probe in an X-Y fashion over an area larger than the antenna under test (1/3 more) or use a polar probe fixture to accomplish the same goal. Map the area and plot at the test frequencies to be used. In a taper chamber, the quiet zone will be acceptable down to lower frequencies than that of a rectangular chamber. To estimate the distance from the source antenna to an acceptable quite zone, use the following formula:

$$Range = \frac{(2 * D^2)}{\lambda} \qquad [126]$$

where:

D = largest dimension of the quiet zone

$\lambda$ = wavelength in same dimensions as D

## Radiation Resistance

That part of an antenna's feed point resistance that is caused by the radiation of electromagnetic waves from the antenna. This is caused by the radiation reaction of the conduction electrons in the antenna. Generally, the higher the radiation resistance, the higher the efficiency will be.

## Radio Refractive Index

This is defined as the ratio of the velocity of a wave phenomenon such as radio waves in a reference medium to the phase velocity of the medium itself. Said another way, It is a measure of how much the speed of light (same as radio waves) is reduced inside the medium.

Or,

$$N = \frac{C}{Vp} \qquad [127]$$

where N= Refractive Index

C = Speed of Light (300,000,000 meters per second)

Vp = Phase Velocity

Another way to define this is:

$$N = \sqrt{(\epsilon r * \mu r)} \quad [128]$$

where $\epsilon r$ = material's relative permitivity (or Dielectric Constant)

$\mu r$ = material's relative permeability (or Loss Tangent)

This can further be defined for radio waves in the range of 1 MHz to 1 GHz in the Earth's atmosphere as:

$$N = 1 + \frac{(0.373 * e)}{(T^2)} + \frac{(77.6 * 10^{-6} * p)}{T} - \frac{40.3 * (Ne)}{f^2} \quad [129]$$

where:

e = partial pressure of water vapor (hPa)
T = absolute temperature (K)
p = atmospheric pressure (hPa)
Ne = number density of free electrons ( $m^{-3}$ )
f = radio frequency (MHz)

## Reciprocity

In antenna theory, an antenna that transmits well will receive well at the same frequency. Care should be taken with this definition as a receive (low power) antenna will not perform well as a transmit (very high power) antenna does.

## Refraction

This is the change in direction of a wave due to a change in speed. This phenomenon is normally observed when a wave passes through one medium into another. Refraction is described by Snell's law, which states that the angle of incidence is related to the angle of refraction by:

$$\frac{(\sin(\theta 1))}{(\sin(\theta 2))}=\frac{V1}{V2}=\frac{N2}{N1} \quad [130]$$

or:

$$N1*\sin(\theta 1)=N2*\sin(\theta 2) \quad [131]$$

where:

V1 and V2 are the wave velocities through the respective media $\theta 1 \text{ and } \theta 2$ are the angles between the normal plane and the incident waves respectively N1 and N2 are the refractive indices. The principles of refraction are important in the design of lens antennas as well as understanding the actions of radio waves through different materials including astronomical phenomena.

## Resistance

This is the ratio of the degree to which an object opposes an electrical current through it, measured in ohms. With a uniform current density, an object's electrical resistance is a function of both its physical geometry and the resistivity of the material it is made from, thus:

$$R=\frac{(\iota*\rho)}{A} \quad [132]$$

where:

R = resistance in ohms

$\iota$ = length in meters

$\rho$ = resistivity in ohms/meter of the material

A = cross sectional area in meters^2

## Resonance

The frequency at which a circuit of a resistor, capacitor and inductor has an impedance which is only resistive. The inductive and capacitive reactances are equal and opposite.

## Retroreflectors

These are (typically) pyramidal devices composed of triangular pieces of conductive material (when applied to the radio spectrum) that allow an incident wave to reflect directly back to the source. Used in maritime and aviation applications, these devices allow an on board radar to receive a strong reflected signal. These devices are used on floating buoys, on sailboats and on the ground near the approach end of a airport runway.

Frequency Selective

For certain applications it is necessary to have a retroreflector that works at only a narrow range of frequencies. Using microstrip technology, these retroreflectors can be designed to receive radiation and reflect it back over a narrow range of input angles as well. A typical application for such a device is for the tracking of a missile from a launching point, where the tracking radar's signal is enhanced but radars at angles other than up range have a much attenuated reflection.

## Return Loss

This is defined as the reflection of the signal power resulting from the insertion of a device in a transmission line (or optical fiber). It is usually expressed as a ratio in dB relative to the transmitted signal power. Mathematically, this is described by the following equation:

$$RL = 20 * \log\left(\frac{Pr}{Pt}\right) \quad [133]$$

where:

RL is in terms of dB

Pr = Power Received

Pt = Power Transmitted

Also, in terms of Impedance, this can be described with the following equation:

$$RL(dB) = -20 * \log\left|\frac{(Zo - Zl)}{(Zo + Zl)}\right| \quad [134]$$

where:

Zo is the impedance toward the source

Zl is the impedances toward the load

## RF Voltage and Current

The voltage on an antenna is based on Root Mean Square power and can be calculated using:

$$V(rms) = \sqrt{(Power * Z)} \quad [135]$$

where Power is in Watts and Z is the characteristic impedance of the antenna. In similar fashion, current can be calculated using:

$$I(rms) = \sqrt{\left(\frac{Power}{Z}\right)} \quad [136]$$

Lastly, the impedance can be calculated by measuring the RMS voltage and RMS current and using:

$$Z=\frac{V}{I}$$

[134]

## RMS

Root Mean Square voltages can be measured by using:

$$V(rms)=\frac{V}{\sqrt{(2)}} \quad [135]$$

Peak voltages at those measured with suitable detector that rectify AC waveforms to DC values.

## RMS Errors

These errors are sometimes referred to as "RMS Losses" as they are the deviations from a perfect surface (be it paraboloidal or otherwise) and have an average RMS vaue across the surface. With this average value, one can calculate the gain loss thus:

$$Loss(dB)=685*(\frac{RMS}{\lambda})^2 \quad [136]$$

## Satellite Terms

Ring Around

This is the leakage of energy from the transmitted band to the receive band on a commercial satellite. Due to the high energy (many thousands of watts in many cases), the close proximity of components and the high sensitivity of the receiver systems, care must be taken to avoid the result of ring around, namely oscillation, intermodulation products in the receive band, PIM and other harmful effects. The mitigation techniques include careful filtering of both the transmitted and received bands, careful placement of waveguides and associated components and high isolation between the transmit and receive portions of the antenna system.

Footprint

The "designed" pattern of a commercial satellite antenna system. This is done by shaped reflectors (non parabolic) and/or distributed feed systems.

Orbital Slot

The position of a satellite in the geo-synchronous orbit, at 23,400 miles above the earth. The slot is designated by the longitude it is directly over. Normally, similar frequency operations are separated by 2 degrees on this plane. For instance overlapping Ku band transmitters will be separated by this amount. Also, the sidelobe performance of earth terminals have to fall into the formula:

$$\text{Envelope} = 29 - 25(\log(\text{Theta})) \qquad [137]$$

where Theta is the angle from the boresite of the antenna.

Link Budget

The most important attribute of a commercial (or any other) satellite. It includes transmit power, losses in the transmission system, antenna gain, receiver sensitivity, losses in the receiver systems, path loss, loss due to the atmosphere and many other important factors. This calculation is carefully examined before the final design of any satellite is performed.

## Scan Loss

For an array, this is the loss incurred as the effective area of antenna changes as the beam progresses away from the perpendicular. In other words, a circular array viewed from a 45 degree offset appears to be an ellipse. The beamwidth properties changes as a result as does the gain. Scan loss can also be attributed to the elemental losses over incident angle, that, convolved with the array factor can drop the gain of the antenna.

Blindness

This is a result of the interaction between element in an array or mutual coupling. Described in complex terms like, 50 -j0, showing real and imaginary (in this case reactive) impedances. The mutual coupling between antennas as viewed from different incident angles, can conspire to lower the gain of the array in that direction. The higher the coupling the higher the potential of significant scanning loss.

## Scattering

This is the general physical process whereby radiation, such as RF and light is forced to deviate from a straight trajectory by one or more non-uniformities in the medium through which they pass. This also includes reflection both diffuse and non diffuse. There are numerous types of scatterers, including particles, fluids bubbles and a host of others.

## Secondary Reflector

In a reflector system this is a second reflector that takes the energy from the main reflector and forms a beam then sends it to a horn or other suitable feed mechanism. These reflectors can be hyperbolic, elliptical or partially transmissive, in the case of dichroic reflector designs. These latter elements allow certain frequencies (generally the lower bands) through to a focal point behind the reflector and send the higher frequencies onto a suitable feed mechanism generally located on the surface of the main dish.

## Skin Depth

This is defined as the depth below the surface of the conductor at which the current density decays to 1/e (or about .37) of the current density at the surface, using the following formula:

$$Skin\,Depth = \sqrt{\left(\frac{(2*p)}{(2*\pi*f*\epsilon*\mu)}\right)} \qquad [138]$$

where:

$p$ = resistivity in ohms per meter (see Materials Properties)

$\pi$ = 3.1415926

f = Frequency in Hertz

$\epsilon$ = Material Resistivity in Ohms

$\mu$ = Henries / m or $4 * \pi * 10^{-7}$

## Skin Effect

This is the tendency of an alternating electric current to distribute itself within a conductor so that the current density near the surface of the conductor is greater than that at its core. The skin effect causes the effective resistance of the conductor to increase with the frequency of the current.

## Sidelobes

These are areas of sensitivity in an antenna beam that are of less magnitude than the main beam. The are predictable in position and size. In many cases, sidelobes can create problems as they allow interference or false echoes (in the case of radar) to be introduced to the rest of the RF system. Sidelobes can be minimized by using taper techniques and careful design.

### First

The first sidelobes are those closest to the main beam. For a boresite antenna beam, these sidelobes are typically symmetrical and can be predicted with any one of many software tools. These sidelobes are the ones with the most gain and are the most capable of receiving interference or in the case of a transmission system, most capable of spreading radiation in unintended areas.

$$29 - 25 \log ( \theta ) \text{ envelope}$$

Using this equation, the profile of an antenna beam,

including sidelobes can be graphed. This profile cannot be exceeded for any antenna transmitting or receiving to a satellite in geo-synchronous orbit. The satellites in this orbit are separated by two degrees with in the same frequency band. This envelope minimizes interference between nearby satellites.

Knee

First sidelobes that do not form a null between their peak and the main lobe form bumps or sometimes called knees around the main antenna beam, having the effect of widening the area of sensitivity around the main beam.

Vestigial lobe

This lobe occurs when the first sidelobe becomes joined to the main beam and forms a shoulder. Essentially the same as the knee definition above. Again, this has the effect of adding sensitivity in space beyond the dimensions of the main beam.

## Spectral Index

In radio astronomy, this is the spectral distribution of energy, leading to a greater understanding of the mechanism creating the energy. A standard shape of a radio source shaped on black body radiation mechanisms can be altered by other physical processes such as Synchrotron, Zeeman effect, Balmer series and several other types of radiation. The spectral shape shows the distribution of energy from these various mechanisms and tells much about the makeup of the radio source.

## Spectrum

This is the range of frequencies over which certain types of radiation propagate. There is the overall spectrum that includes all types of radiation called the electromagnetic spectrum. There are also sub band within this spectrum that include gamma rays, x-rays, ultraviolet, light, infrared and radio waves. All energy produced in the universe of a wave nature is observable in the electromagnetic spectrum.

## Spillover

In a reflector type of antenna, the energy that is not captured by the reflector is considered lost or "spilled over." The consequences of a large amount of spillover is loss of gain and higher antenna noise temperatures.

## Squint

This is the unintended movement of an antenna beam due to problems in focus or general assembly of the antenna.

## Taper

In an antenna system such as an array or reflector, this is the purposeful adjustment of the amplitude across the aperture to create lower sidelobes. There are several profiles of amplitude adjustment, as listed below.

Taylor

Using coefficients derived by Taylor, this is a very popular approach to tapering. The magnitude of the taper affects the amount of sidelobe suppression in an antenna system, such as an array. Taylor coefficients can be found in the CRC handbook of mathematics and other electromagnetic textbooks.

Amplitude

This is the most common form of tapering where by the distribution of energy to the various element of an array is altered in such a way as to improve sidelobe performance. The consequences of tapering in this way is to lower the main beam slightly and broaden it out as well. The elements at the perimeter of an array are attenuated the most, tapering to zero attenuation at the center of the array.

Chebycheff

This is an alternate amplitude tapering methodology as compared to a Taylor distribution. Compare this type of tapering with a good software tool to see if the antenna beam shape best fits the application. Based on Chebycheff equations typically used for filter design.

Cosine Pedestal, Raised

This is an alternate amplitude tapering methodology as compared to a Taylor distribution. Compare this type of tapering with a good software tool to see if the antenna beam shape best fits the application. Based on adding an integer to a standard cosine curve.

Cosine Squared

This is an alternate amplitude tapering methodology as compared to a Taylor distribution. Compare this type of tapering with a good software tool to see if the antenna beam shape best fits the application. Based on a cosine squared distribution.

## Twin lead

This is a form of feedline which has balanced sides for easy distribution of power to an antenna, like a dipole, bowtie, folded dipole or even yagi. This type of feedline comes in various standard impedances, including 300 ohms (for old TV type applications) to 600 ohms and various other values.

## Traveling Wave Tube

A transmitting power amplification device consisting of a helix like conduction embedded in a vacuum tube. High voltages are applied to acceleration grids and other components of the tube to create powers up to several thousand watts. These devices are used extensively in satellites due to their compact sizes, wide bandwidths and reliability.

## VSWR

Voltage Standing Wave Ratio, defined as the ratio of the amplitude of a partial standing wave at an antinode (maximum) to the amplitude at an adjacent node (minimum), in an electrical transmission line.

## Waveguide

This is a structure which guides waves, such as electromagnetic waves, light or sound waves. In the RF regime, these waveguides are typically rectangular in nature, dimensioned in several standard shapes. Square or circularly shaped waveguides are also used to accommodate multiple polarizations. These components are low loss and capable of handling significant amounts of power. In addition, they have very low leakage, useful for Tempest type applications.

**Wavenumber**

This is a wave property inversely proportional to wavelength. Wavenumber is the spatial analog of frequency or the number of wavelengths per unit distance, sometimes 2*pi the distance or the number of radians of phase per unit distance.

# Appendix 8 – Charts and Nomographs

## Impedance / Admittance Chart (Smith)

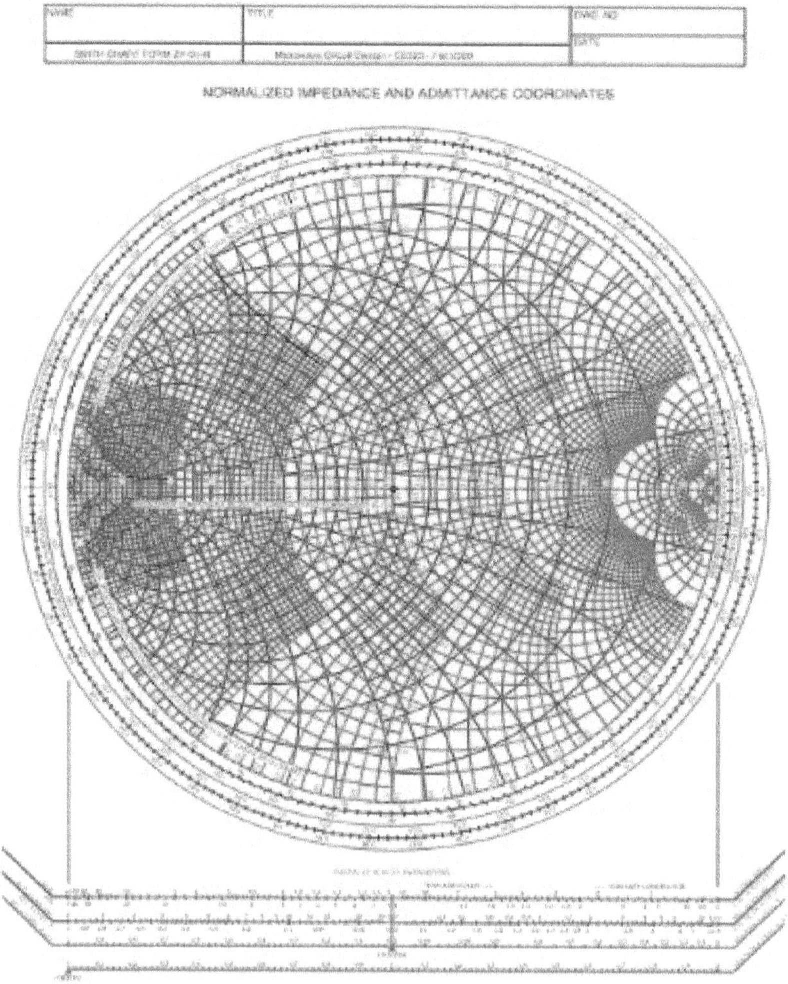

## Frequency vs. Wavelength

| Frequency (MHz) | Wavelength (Meters) |
|---|---|
| 0.1 | 3,000.0000 |
| 0.25 | 1,200.0000 |
| 0.5 | 600.0000 |
| 1 | 300.0000 |
| 2 | 150.0000 |
| 5 | 60.0000 |
| 10 | 30.0000 |
| 20 | 15.0000 |
| 50 | 6.0000 |
| 100 | 3.0000 |
| 250 | 1.2000 |
| 500 | .6000 |
| 1000 | .3000 |
| 2500 | .1200 |
| 5000 | .0600 |
| 10000 | .0300 |
| 25000 | .0120 |
| 50000 | .0060 |
| 100000 | .0030 |
| 250000 | .0012 |
| 500000 | .0006 |
| 1000000 | .0003 |

## Frequency vs. Wavelength (Graph)

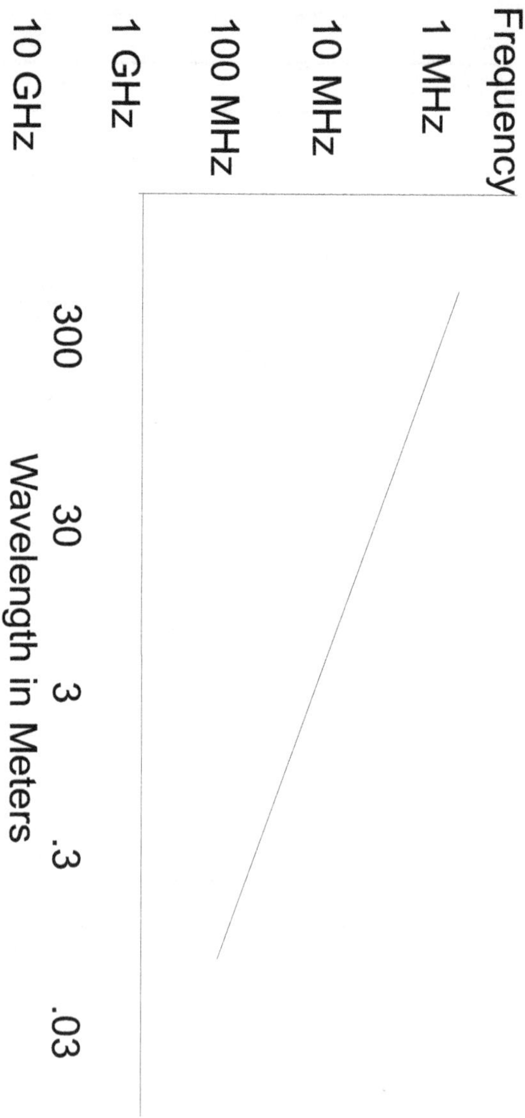

## Frequency vs. Directivity (dBi) for NRAO VLT

| Frequency in MHz | Directivity for 100 Meter Aperture (dBi) |
|---|---|
| 10 | 20.4 |
| 20 | 26.42 |
| 50 | 34.38 |
| 100 | 40.4 |
| 250 | 48.36 |
| 500 | 54.38 |
| 1000 | 60.4 |
| 2500 | 68.36 |
| 5000 | 74.38 |
| 10000 | 80.4 |
| 20000 | 86.42 |
| 50000 | 94.38 |
| 100000 | 100.4 |

## Aperture vs. Directivity at 1 GHz

| Aperture in Meters | Directivity in dBi for 1 GHz |
|---|---|
| 1 | 20.4 |
| 2 | 26.42 |
| 5 | 34.38 |
| 10 | 40.4 |
| 20 | 46.42 |
| 30 | 49.94 |
| 40 | 52.44 |
| 50 | 54.38 |
| 60 | 55.96 |
| 70 | 57.3 |
| 80 | 58.46 |
| 90 | 59.48 |
| 100 | 60.4 |
| 200 | 66.42 |
| 300 | 69.94 |

## Aperture vs. Directivity at 10 GHz

| Aperture in Meters | Directivity in dBi for 1 GHz |
|---|---|
| 0.25 | 28.36 |
| 0.3 | 29.9 |
| 0.5 | 34.38 |
| 1 | 40.4 |
| 2 | 46.42 |
| 5 | 54.38 |
| 10 | 60.4 |
| 20 | 66.42 |
| 30 | 69.94 |
| 40 | 72.44 |
| 50 | 74.38 |
| 60 | 75.96 |
| 70 | 77.3 |
| 80 | 78.46 |
| 90 | 79.48 |
| 100 | 80.4 |
| 200 | 86.42 |
| 300 | 89.94 |

## Power in Watts vs. dBm

| Power in W | dBm | dBW |
|---|---|---|
| 0.001 | 0 | |
| 0.002 | 3 | |
| 0.01 | 10 | |
| 0.02 | 13 | |
| 0.1 | 20 | |
| 0.2 | 23 | |
| 1 | 30 | 0 |
| 2 | 33 | 3 |
| 5 | 37 | 7 |
| 10 | 40 | 10 |
| 20 | 43 | 13 |
| 50 | 47 | 17 |
| 100 | 50 | 20 |
| 200 | 53 | 23 |
| 500 | 57 | 27 |
| 1000 | 60 | 30 |
| 2000 | 63 | 33 |
| 5000 | 67 | 37 |
| 10000 | 70 | 40 |
| 20000 | 73 | 43 |
| 50000 | 77 | 47 |
| 100000 | 80 | 50 |
| 200000 | 83 | 53 |
| 500000 | 87 | 57 |
| 1000000 | 90 | 60 |

$$dBm = 10 * \log(Power \text{ in } Milliwatts) \quad [139]$$

$$dBW = 10 * \log(Power \text{ in } Watts) \quad [140]$$

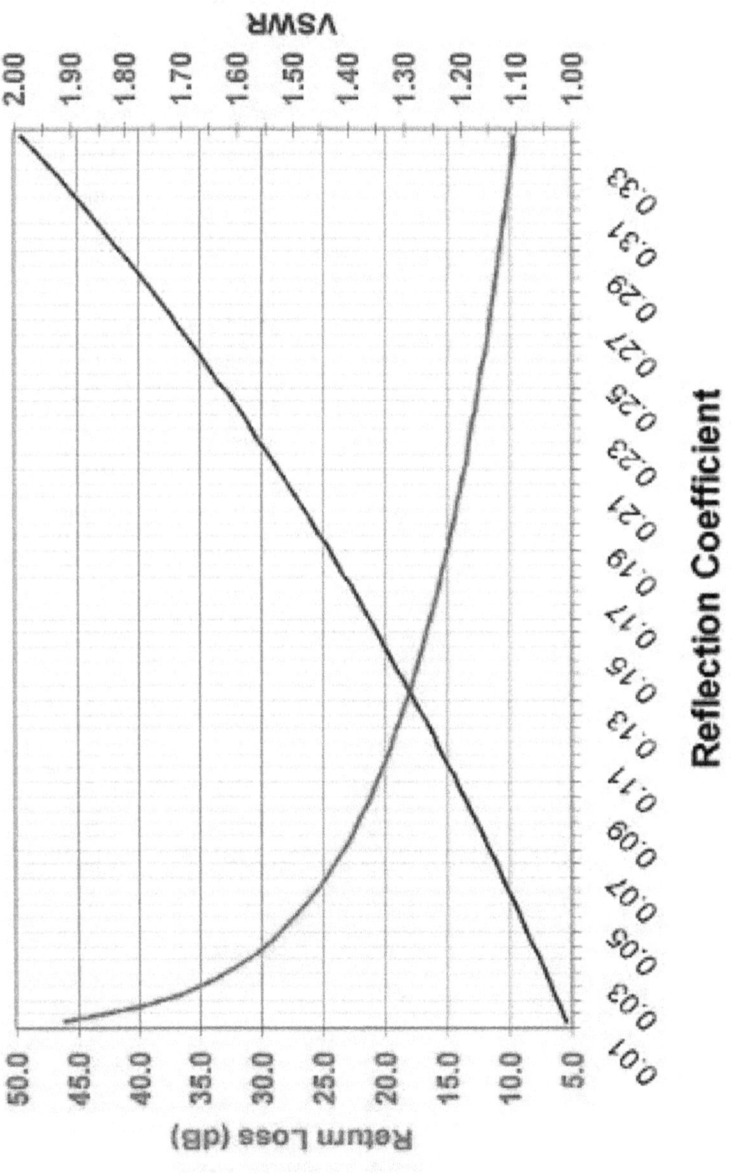
VSWR
2.00
1.90
1.80
1.70
1.60
1.50
1.40
1.30
1.20
1.10
1.00
Return Loss (dB)
50.0
45.0
40.0
35.0
30.0
25.0
20.0
15.0
10.0
5.0
Reflection Coefficient
0.01
0.03
0.05
0.07
0.09
0.11
0.13
0.15
0.17
0.19
0.21
0.23
0.25
0.27
0.29
0.31
0.33

| VSWR | Return Loss (-dB) | Reflection Coefficient | Reflected Power (%) | Mismatch Loss (dB) |
|---|---|---|---|---|
| 1.00 | 250.00 | 0.00 | 0.00 | 0.00 |
| 1.05 | 32.26 | 0.02 | 0.06 | 0.00 |
| 1.10 | 26.44 | 0.05 | 0.23 | 0.01 |
| 1.15 | 23.13 | 0.07 | 0.49 | 0.02 |
| 1.20 | 20.83 | 0.09 | 0.82 | 0.04 |
| 1.25 | 19.08 | 0.11 | 1.23 | 0.05 |
| 1.30 | 17.69 | 0.13 | 1.71 | 0.07 |
| 1.35 | 16.54 | 0.15 | 2.23 | 0.10 |
| 1.40 | 15.56 | 0.17 | 2.78 | 0.12 |
| 1.45 | 14.72 | 0.18 | 3.38 | 0.15 |
| 1.50 | 13.98 | 0.20 | 4.00 | 0.18 |
| 1.55 | 13.32 | 0.22 | 4.80 | 0.21 |
| 1.60 | 12.74 | 0.23 | 5.50 | 0.24 |
| 1.65 | 12.21 | 0.25 | 6.20 | 0.27 |
| 1.70 | 11.73 | 0.26 | 6.80 | 0.30 |
| 1.75 | 11.29 | 0.27 | 7.40 | 0.34 |
| 1.80 | 10.88 | 0.29 | 8.20 | 0.37 |
| 1.85 | 10.51 | 0.30 | 8.90 | 0.40 |
| 1.90 | 10.16 | 0.31 | 9.60 | 0.44 |
| 1.95 | 9.84 | 0.32 | 10.20 | 0.48 |
| 2.00 | 9.54 | 0.33 | 11.00 | 0.51 |
| 2.50 | 7.36 | 0.43 | 18.00 | 0.88 |
| 3.00 | 6.02 | 0.50 | 24.90 | 1.25 |
| 3.50 | 5.11 | 0.56 | 31.00 | 1.60 |
| 4.00 | 4.44 | 0.60 | 36.00 | 1.94 |
| 4.50 | 3.93 | 0.64 | 40.60 | 2.25 |
| 5.00 | 3.52 | 0.67 | 44.40 | 2.55 |
| 10.00 | 1.74 | 0.82 | 68.20 | 4.81 |
| 20.00 | 0.87 | 0.90 | 86.90 | 7.41 |

# Noise Figure Calculations

| NF (dB) | T (K) | NF (dB) | T (K) | NF (dB) | T (K) | NF (dB) | T (K) |
|---|---|---|---|---|---|---|---|
| 0.1 | 7 | 1.1 | 84 | 2.1 | 180 | 3.1 | 302 |
| 0.2 | 14 | 1.2 | 92 | 2.2 | 191 | 3.2 | 316 |
| 0.3 | 21 | 1.3 | 101 | 2.3 | 202 | 3.3 | 330 |
| 0.4 | 28 | 1.4 | 110 | 2.4 | 214 | 3.4 | 344 |
| 0.5 | 35 | 1.5 | 120 | 2.5 | 226 | 3.5 | 359 |
| 0.6 | 43 | 1.6 | 129 | 2.6 | 238 | 3.6 | 374 |
| 0.7 | 51 | 1.7 | 139 | 2.7 | 250 | 3.7 | 390 |
| 0.8 | 59 | 1.8 | 149 | 2.8 | 263 | 3.8 | 406 |
| 0.9 | 67 | 1.9 | 159 | 2.9 | 275 | 3.9 | 422 |
| 1 | 75 | 2 | 170 | 3 | 289 | 4 | 438 |

$$Noise\ Temperature\,(T) = 290 * (10^{(Noise\ Figure/10)} - 1) \qquad [140]$$

$$Noise\ Figure\,(NF) = 10 * \log\,(Noise\ Factor)\ dB \qquad [141]$$

# Noise Temp, Noise Figure and Noise Factor Notes:

0 K = -273 deg C

273 K = 0 deg C

290 K = 17 deg C (ambient temperature)

System Noise Temperature is referred to the input of the LNA

Antenna Noise Temperature is referenced to the flange or connector specified by the manufacturer.

The Noise Temperature of the LNA refers to the input of the LNA

The Noise Temperature of the cable after the LNA refers to the input of the cable

The Noise Temperature of the receiver refers to the input of the receiver

System Temperature or Tsys = Antenna Noise Temperature +

290 * (1-waveguide gain) +

LNA Noise Temperature +

Cable Noise Temperature +

Receiver Noise Temperature

The sensitivity of a receiver is defined as:

$$Psens(dBm) = -174 + 10\log(B) + TSys(dB)$$

where:

B = Bandwidth in Hz

Example:

Antenna noise temperature = 35 K (mainly ground pick up noise)
Waveguide feeder gain = -0.25 dB (0.944), temperature = 290K
LNA gain = 50 dB (100000), input noise temperature = 75 K
Cable loss or attenuation = 20 dB or cable gain = -20 dB (0.01)
Cable noise temp= 290 K
Indoor receiver noise figure = 9 dB
Indoor receiver input noise temperature = 290 * (10^(9/10)-1) =2013.5519 K

Tsystem = 35 * 0.944 = 33 Noise contribution of the antenna
+ 290 ( 1 - 0.944) = 16 Noise contribution of the waveguide

+ 75 Noise contribution of the LNA
+ 290/100000 = 0.0029 Noise contribution of the cable
+ 2013.5519/(100000 * 0.01) = 2.0135519 Noise contribution of the indoor receiver
= 126.0164519 K

# SPIRAL CHART RELATES NOISE FIGURE AND NOISE TEMPERATURE

Here's a 40-inch-long conversion chart for receiver and amplifier noise that's neatly rolled into a spiral to combine accuracy and convenience. The chart is arranged for maximum accuracy at the low-noise end of the scale.

*A. C. Hudson, Radio and Electrical Engineering Division, National Research Council, Ottawa, Canada*

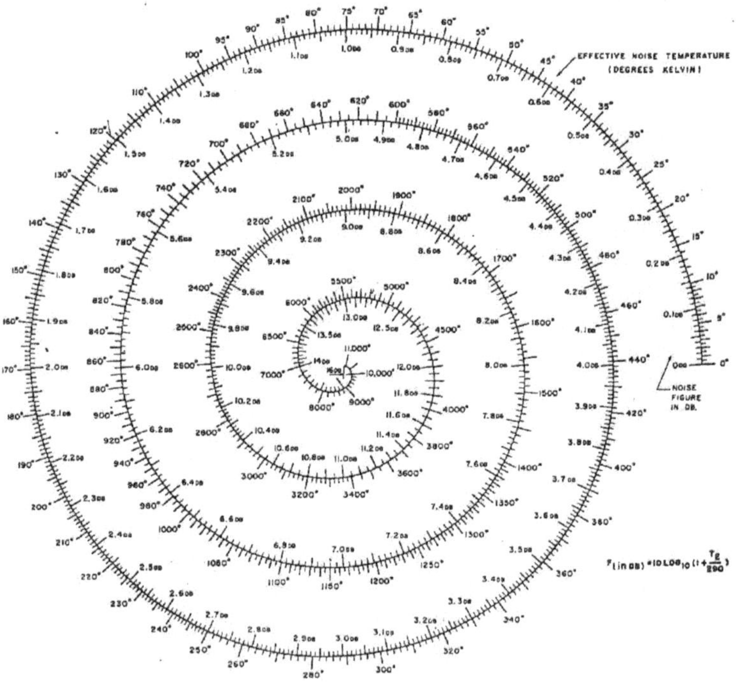

# Conversion of dB to Power Ratio

| dB | -V Ratio | -P Ratio | +V Ratio | +P Ratio |
|---|---|---|---|---|
| 0 | 1.0000 | 1.0000 | 1.0000 | 1.0000 |
| 1 | 0.8913 | 0.7943 | 1.2200 | 1.2590 |
| 2 | 0.7943 | 0.6310 | 1.2590 | 1.5850 |
| 3 | 0.7079 | 0.5012 | 1.4130 | 1.9950 |
| 4 | 0.6310 | 0.3981 | 1.5850 | 2.5120 |
| 5 | 0.5623 | 0.3162 | 1.7780 | 3.1620 |
| 6 | 0.5012 | 0.2512 | 1.9950 | 3.9810 |
| 7 | 0.4467 | 0.1995 | 2.2390 | 5.0120 |
| 8 | 0.3981 | 0.1585 | 2.5120 | 6.3100 |
| 9 | 0.3548 | 0.1259 | 2.8180 | 7.9430 |
| 10 | 0.3162 | 0.1000 | 3.1620 | 10.0000 |
| 11 | 0.2818 | 0.0794 | 3.5480 | 12.5900 |
| 12 | 0.2512 | 0.0631 | 3.9810 | 15.8500 |
| 13 | 0.2239 | 0.0501 | 4.4670 | 19.9500 |
| 14 | 0.1995 | 0.0398 | 5.0120 | 25.1200 |
| 15 | 0.1778 | 0.0316 | 5.6230 | 31.6200 |
| 16 | 0.1585 | 0.0251 | 6.3100 | 39.8100 |
| 17 | 0.1413 | 0.0200 | 7.0790 | 50.1200 |
| 18 | 0.1259 | 0.0159 | 7.9430 | 63.1000 |
| 19 | 0.1122 | 0.0126 | 8.9130 | 79.4300 |
| 20 | 0.1000 | 0.0100 | 10.0000 | 100.0000 |

# Path Loss in Db for Various Distances (Miles)

| Miles | 100 MHz | 1 GHz | 10 GHz | 100 GHz |
|---|---|---|---|---|
| 1 | 76.58 | 96.58 | 116.58 | 136.58 |
| 2 | 82.6 | 102.6 | 122.6 | 142.6 |
| 5 | 90.56 | 110.56 | 130.56 | 150.56 |
| 10 | 96.58 | 116.58 | 136.58 | *156.58* |
| 25 | 104.54 | 124.54 | 144.54 | 164.54 |
| 50 | 110.56 | 130.56 | 150.56 | 170.56 |
| 100 | 116.58 | 136.58 | *156.58* | 176.58 |
| 250 | 124.54 | 144.54 | 164.54 | 184.54 |
| 500 | 130.56 | 150.56 | 170.56 | 190.56 |
| 1000 | 136.58 | *156.58* | 176.58 | 196.58 |
| 2500 | 144.54 | 164.54 | 184.54 | 204.54 |
| 5000 | 150.56 | 170.56 | 190.56 | 210.56 |
| 10000 | *156.58* | 176.58 | 196.58 | 216.58 |
| 25000 | 164.54 | 184.54 | 204.54 | 224.54 |
| 50000 | 170.56 | 190.56 | 210.56 | 230.56 |
| 100000 | 176.58 | 196.58 | 216.58 | 236.58 |
| 250000 | 184.54 | 204.54 | 224.54 | 244.54 |
| 500000 | 190.56 | 210.56 | 230.56 | 250.56 |
| 1000000 | 196.58 | 216.58 | 236.58 | 256.58 |
| 2500000 | 204.54 | 224.54 | 244.54 | 262.6 |

## Path Loss in Db for Various Distances (Kilometers)

| Kilometers | 100 MHz | 1 GHz | 10 GHz | 100 GHz |
|---|---|---|---|---|
| 1 | 72.45 | 92.45 | 112.45 | 132.45 |
| 2 | 78.47 | 100.41 | 120.41 | 140.41 |
| 5 | 86.43 | 106.43 | 126.43 | 146.43 |
| 10 | 92.45 | 112.45 | 132.45 | 152.45 |
| 25 | 100.41 | 120.41 | 140.41 | 160.41 |
| 50 | 106.43 | 126.43 | 146.43 | 166.43 |
| 100 | 112.45 | 132.45 | 152.45 | 172.45 |
| 250 | 120.41 | 140.41 | 160.41 | 180.41 |
| 500 | 126.43 | 146.43 | 166.43 | 186.43 |
| 1000 | 132.45 | 152.45 | 172.45 | 192.45 |
| 2500 | 140.41 | 160.41 | 180.41 | 200.41 |
| 5000 | 146.43 | 166.43 | 186.43 | 206.43 |
| 10000 | 152.45 | 172.45 | 192.45 | 212.45 |
| 25000 | 160.41 | 180.41 | 200.41 | 220.41 |
| 50000 | 166.43 | 186.43 | 206.43 | 226.43 |
| 100000 | 172.45 | 192.45 | 212.45 | 232.45 |
| 250000 | 180.41 | 200.41 | 220.41 | 240.41 |
| 500000 | 186.43 | 206.43 | 226.43 | 246.43 |
| 1000000 | 192.45 | 212.45 | 232.45 | 252.45 |
| 2500000 | 200.41 | 220.41 | 240.41 | 259.41 |
| 5000000 | 206.43 | 226.43 | 246.43 | 266.43 |
| 10000000 | 212.45 | 232.45 | 252.45 | 273.45 |

## Path Loss in Db for Various Distances (Meters)

| Meters | 100 MHz | 1 GHz | 10 GHz | 100 GHz |
|---|---|---|---|---|
| 1 | 12.45 | 32.45 | 52.45 | 72.45 |
| 2 | 18.47 | 40.41 | 60.41 | 80.41 |
| 5 | 26.43 | 46.43 | 66.43 | 86.43 |
| 10 | 32.45 | 52.45 | 72.45 | 92.45 |
| 25 | 40.41 | 60.41 | 80.41 | 100.41 |
| 50 | 46.43 | 66.43 | 86.43 | 106.43 |
| 100 | 52.45 | 72.45 | 92.45 | 112.45 |
| 250 | 60.41 | 80.41 | 100.41 | 120.41 |
| 500 | 66.43 | 86.43 | 106.43 | 126.43 |
| 1000 | 72.45 | 92.45 | 112.45 | 132.45 |
| 2500 | 80.41 | 100.41 | 120.41 | 140.41 |
| 5000 | 86.43 | 106.43 | 126.43 | 146.43 |
| 10000 | 92.45 | 112.45 | 132.45 | 152.45 |
| 25000 | 100.41 | 120.41 | 140.41 | 160.41 |
| 50000 | 106.43 | 126.43 | 146.43 | 166.43 |
| 100000 | 112.45 | 132.45 | 152.45 | 172.45 |
| 250000 | 120.41 | 140.41 | 160.41 | 180.41 |
| 500000 | 126.43 | 146.43 | 166.43 | 186.43 |
| 1000000 | 132.45 | 152.45 | 172.45 | 192.45 |
| 2500000 | 140.41 | 160.41 | 180.41 | 199.41 |
| 5000000 | 146.43 | 166.43 | 186.43 | 206.43 |
| 10000000 | 152.45 | 172.45 | 192.45 | 213.45 |

# Path Loss in Db for Various Distances (Feet)

| Feet | 100 MHz | 1 GHz | 10 GHz | 100 GHz |
|---|---|---|---|---|
| 1 | 2.13 | 22.13 | 42.13 | 62.13 |
| 2 | 8.15 | 30.06 | 50.09 | 70.09 |
| 5 | 16.11 | 36.11 | 56.11 | 76.11 |
| 10 | 22.13 | 42.13 | 62.13 | 82.13 |
| 25 | 30.06 | 50.09 | 70.09 | 90.09 |
| 50 | 36.11 | 56.11 | 76.11 | 96.11 |
| 100 | 42.13 | 62.13 | 82.13 | 102.13 |
| 250 | 50.09 | 70.09 | 90.09 | 110.09 |
| 500 | 56.11 | 76.11 | 96.11 | 116.11 |
| 1000 | 62.13 | 82.13 | 102.13 | 122.13 |
| 2500 | 70.09 | 90.09 | 110.09 | 130.09 |
| 5000 | 76.11 | 96.11 | 116.11 | 138.11 |
| 10000 | 82.13 | 102.13 | 122.13 | 146.13 |
| 25000 | 90.09 | 110.09 | 130.09 | 154.09 |
| 50000 | 96.11 | 116.11 | 138.11 | 162.11 |
| 100000 | 102.13 | 122.13 | 146.13 | 170.13 |
| 250000 | 110.09 | 130.09 | 154.09 | 178.09 |
| 500000 | 116.11 | 138.11 | 162.11 | 186.11 |
| 1000000 | 122.13 | 146.13 | 170.13 | 194.13 |
| 2500000 | 130.09 | 154.09 | 178.09 | 204.09 |
| 5000000 | 138.11 | 162.11 | 186.11 | 212.11 |
| 10000000 | 146.13 | 170.13 | 194.13 | 220.13 |

# Path Loss Nomograph

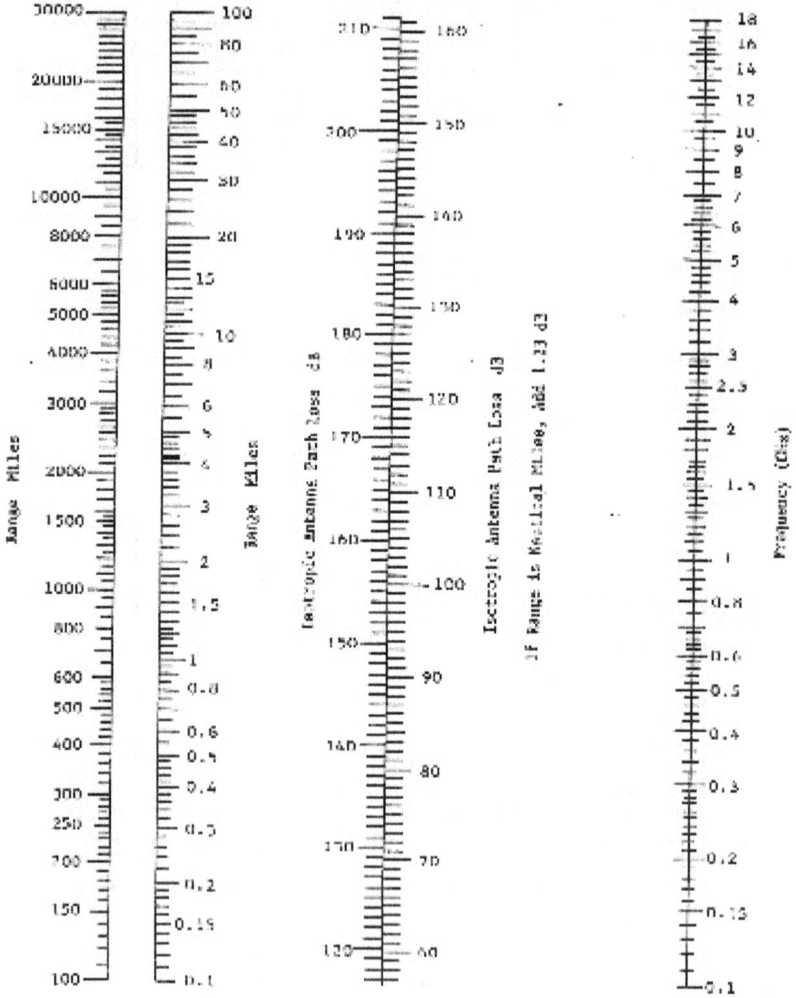

## FCC Power Density Calculations, Direct

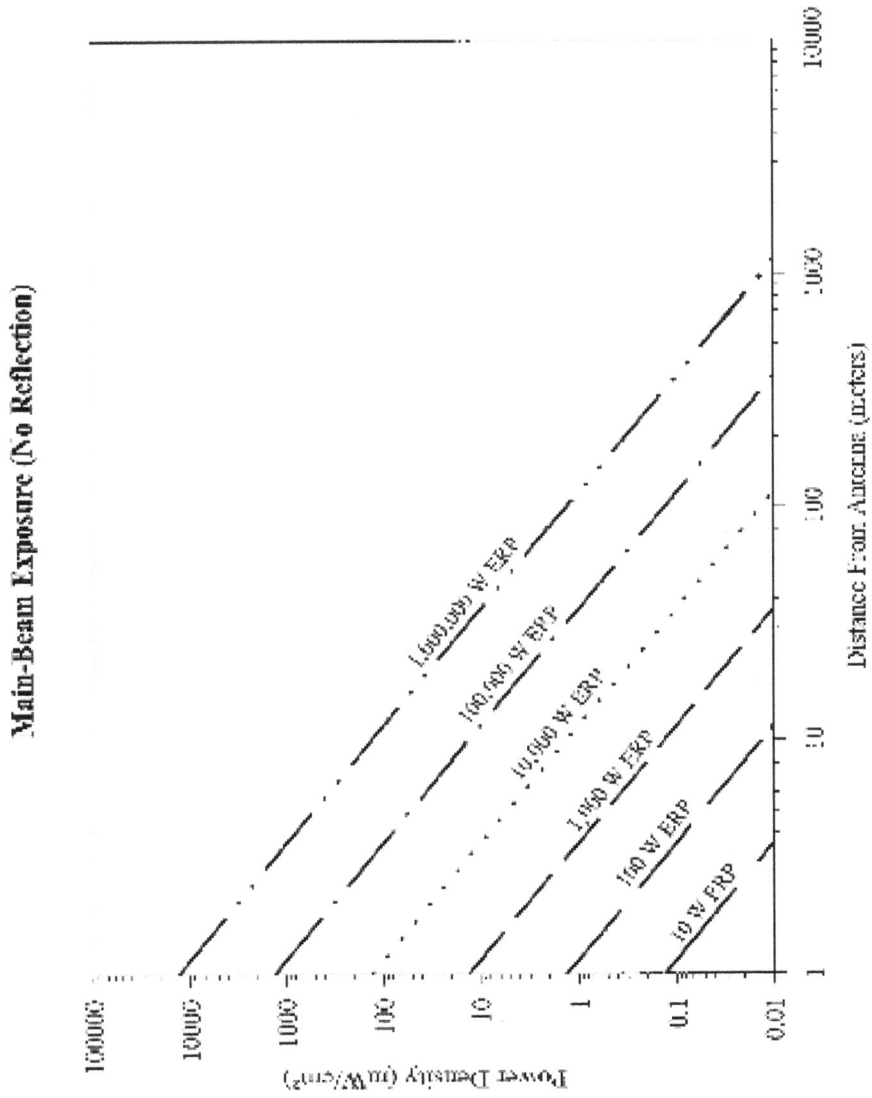

# FCC Power Density Calculations, Reflected

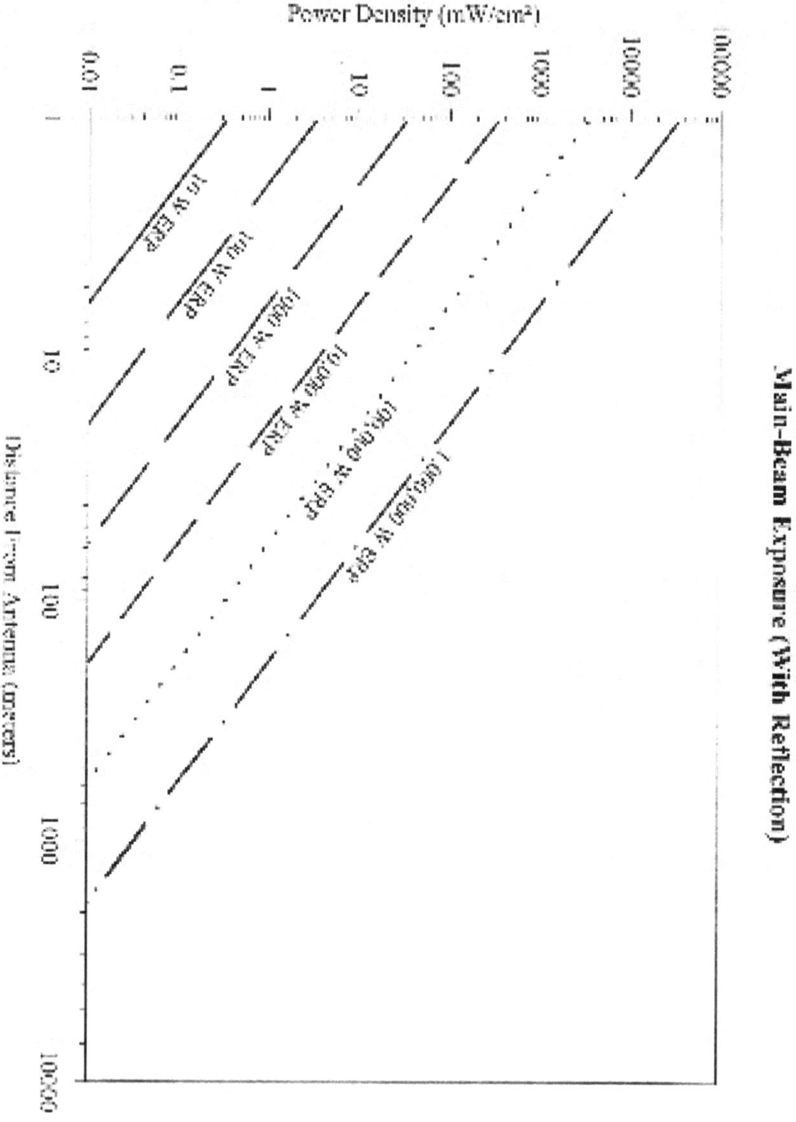

## FCC Limits for RF Exposure

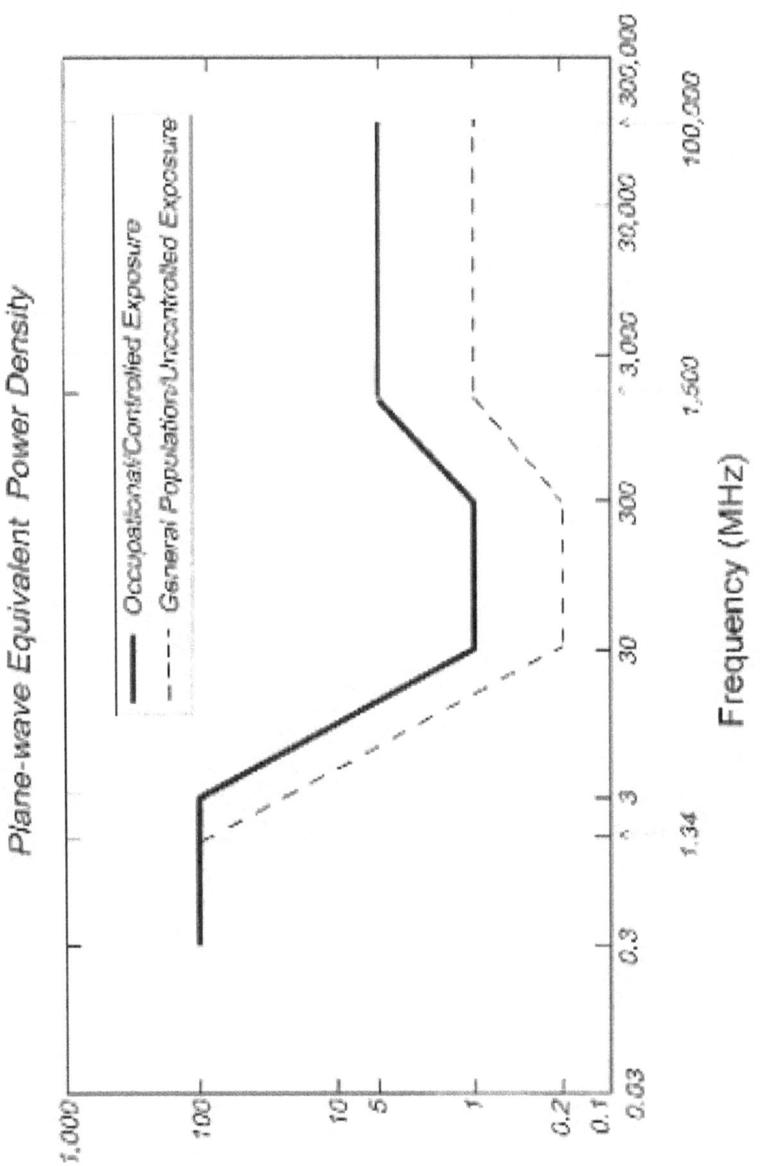

*2nd Edition*

# Polarization Mismatch Calculations

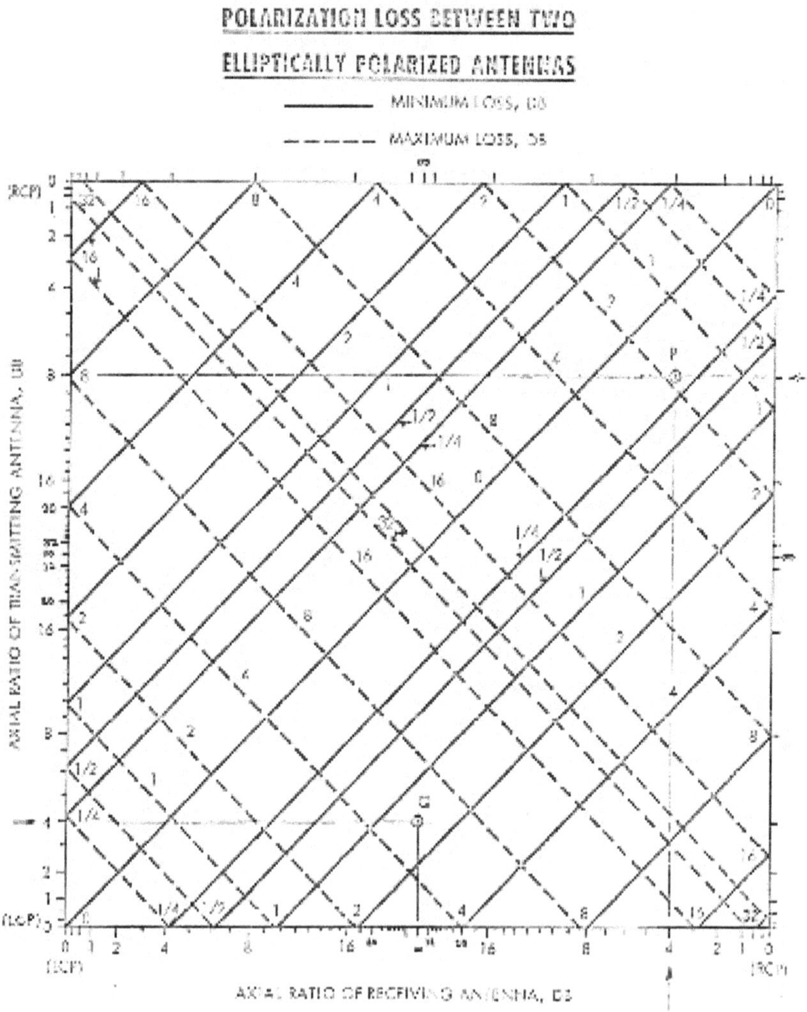

## Wind Loading, PSF vs. Velocity

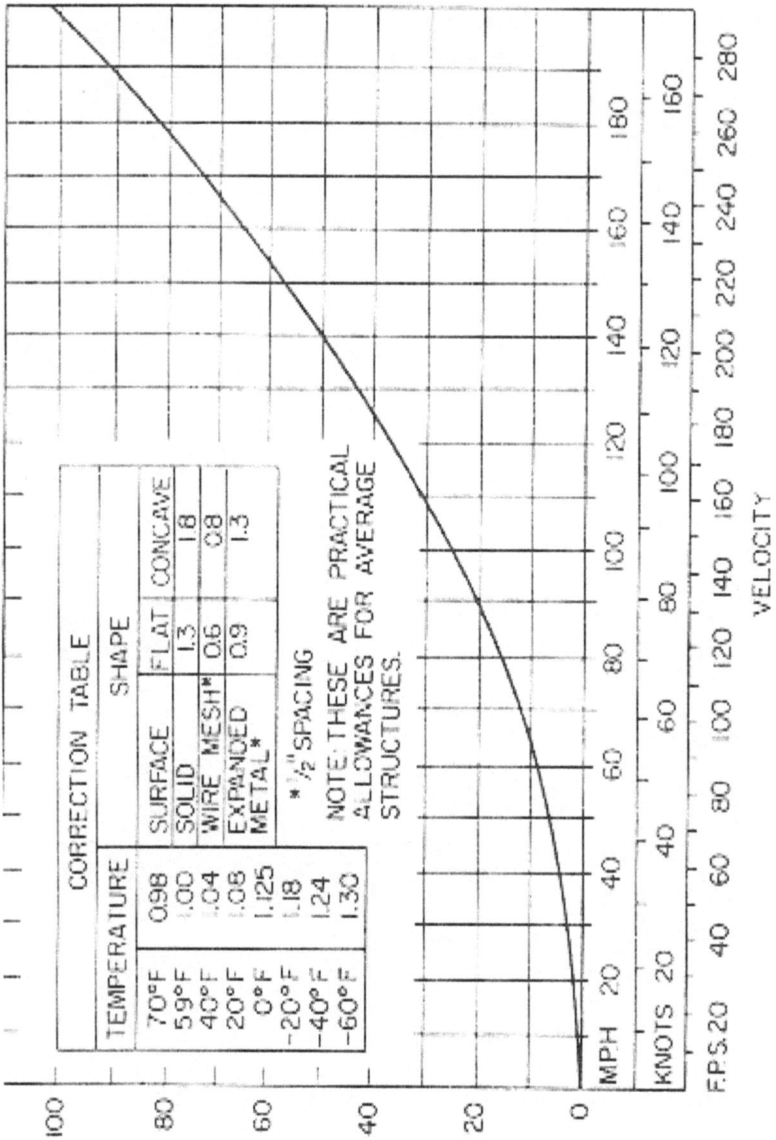
CORRECTION TABLE
TEMPERATURE
70°F 0.98
59°F 1.00
40°F 1.04
20°F 1.08
0°F 1.125
-20°F 1.18
-40°F 1.24
-60°F 1.30
SHAPE
SURFACE FLAT CONCAVE
SOLID 1.3 1.8
WIRE MESH* 0.6 0.8
EXPANDED METAL* 0.9 1.3
* ½" SPACING
NOTE: THESE ARE PRACTICAL ALLOWANCES FOR AVERAGE STRUCTURES.
100
80
60
40
20
0
MPH 20 40 60 80 100 120 140 160 180
KNOTS 20 40 60 80 100 120 140 160
F.P.S. 20 40 60 80 100 120 140 160 180 200 220 240 260 280
VELOCITY

## Power vs. Altitude

### ALTITUDE DE-RATING CURVE

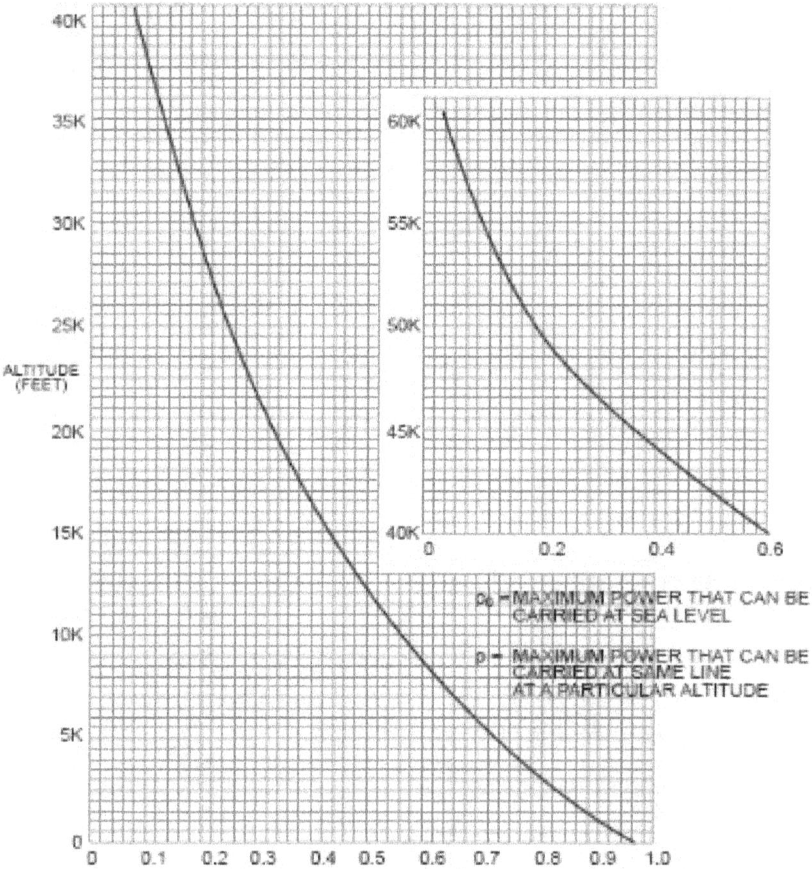

# Waveguide Ratings and Pressures

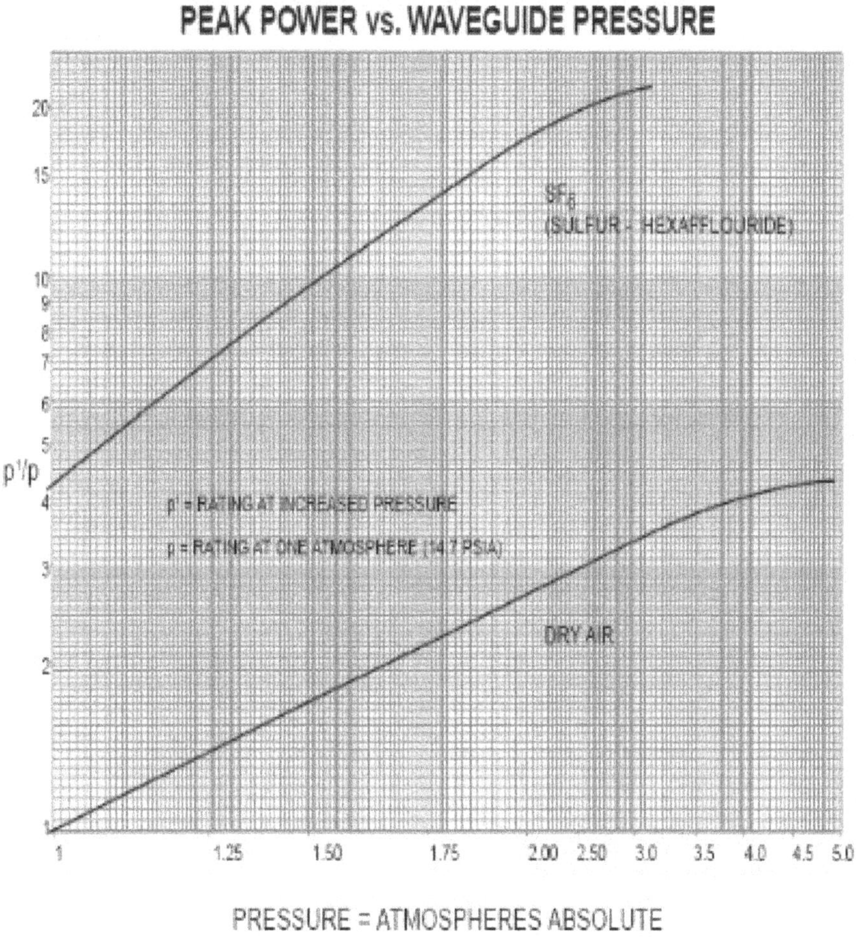

# Pressure and Temperature Tables

| ABSOLUTE PRESSURE | | | | | | | | | GAGE PRESSURE (2) | |
|---|---|---|---|---|---|---|---|---|---|---|
| cm of Hg | Torr or mm of Hg | Micron | Atmosphere | lb/in² | ton/ft² | gram/cm² | ft of H2O | in of Hg | lb/in² | in of Hg |
| 76 | 760 | 760000 | 1.0 | 14.70 | 1.06 | 1033 | 33.9 | 29.9 | 0 | 0 |
| 70 | 700 | 700000 | .921 | 13.53 | .975 | 952 | 31.2 | 27.6 | 1.16 | 2.36 |
| 60 | 600 | 600000 | .790 | 11.60 | .835 | 816 | 26.8 | 23.6 | 3.10 | 6.30 |
| 50 | 500 | 500000 | .659 | 9.67 | .696 | 680 | 22.3 | 19.7 | 5.03 | 10.2 |
| 40 | 400 | 400000 | .526 | 7.74 | .557 | 545 | 17.8 | 15.7 | 6.97 | 14.2 |
| 30 | 300 | 300000 | .395 | 5.80 | .417 | 408 | 13.4 | 11.8 | 8.90 | 18.1 |
| 20 | 200 | 200000 | .263 | 3.87 | .278 | 272 | 8.92 | 7.87 | 10.8 | 22.0 |
| 10 | 100 | 100000 | .132 | 1.94 | .139 | 136 | 4.46 | 3.94 | 12.8 | 26.0 |
| 5 | 50 | 50000 | .066 | .967 | .070 | 68.0 | 2.23 | 1.97 | 13.7 | 27.9 |
| 1 | 10 | 10000 | .013 | .194 | .014 | 13.6 | .446 | .394 | 14.5 | 29.5 |
| 0.1 | 1 | 1000 | .001 | .019 | .001 | 1.36 | .045 | .039 | 14.68 | 29.88 |
| 0 | 0 | 0 | 0 | 0 | 0 | 0 | 0 | 0 | 14.70 | 29.92 |

(1) Positive pressure measured from absolute zero.
(2) Negative pressure (or vacuum) measured fm atmospheric pressure.

## PRESSURE EQUIVALENTS

| Millitorr, or Micron | Torr or mm of Hg | |
|---|---|---|
| 1000 | = 1.0 | = $10^{0}$ |
| 100 | = 0.1 | = $10^{-1}$ |
| 10 | = 0.01 | = $10^{-2}$ |
| 1.0 | = 0.001 | = $10^{-3}$ |
| 0.5 | = 0.0005 | = $5 \times 10^{-4}$ |
| 0.1 | = 0.0001 | = $1 \times 10^{-4}$ x or $10^{-4}$ |
| 0.01 | = 0.00001 | = $10^{-5}$ |
| 0.001 | = 0.000001 | = $10^{-6}$ |

# Temperature and Pressure

| [illegible] U.S. Standard Atmosphere | | | | | | | | | | From N.A.C.A. Report No. 538 | |
|---|---|---|---|---|---|---|---|---|---|---|---|
| ALTITUDE | PRESSURE | | | TEMPERATURE | | ALTITUDE | PRESSURE | | | TEMPERATURE | |
| (Feet) | In Hg | mm Hg | P.S.I. | °C | °F | (Feet) | In Hg | mm Hg | P.S.I. | °C | °F |
| -1,000 | [illegible] | 737.9 | 15.25 | 17.0 | 52.2 | 32,000 | 7.91 | 201.0 | 3.89 | 42.4 | 55.9 |
| -500 | [illegible] | [illegible] | 14.97 | 16.0 | 60.7 | 33,000 | 7.72 | 196.4 | 3.80 | -52.4 | -62.7 |
| 0 | [illegible] | [illegible] | 14.70 | 15.0 | [illegible] | [illegible] | [illegible] | 191.9 | [illegible] | [illegible] | [illegible] |
| 500 | 29.38 | 746.4 | 14.43 | 14.0 | 57.2 | 34,000 | 7.36 | 187.4 | 3.65 | 52.4 | 52.8 |
| 1,000 | [illegible] | 732.9 | 14.10 | 13.0 | 55.4 | 34,000 | 7.10 | 180.0 | 3.48 | -53.4 | -64.1 |
| [illegible] | [illegible] | [illegible] | 13.88 | 12.0 | [illegible] | [illegible] | 7.04 | 170.7 | 3.46 | [illegible] | -66.0 |
| 2,000 | 27.52 | 706.6 | 13.67 | 11.0 | 51.5 | [illegible] | [illegible] | 1.060 | 3.40 | [illegible] | 55.0 |
| 2,500 | 27.31 | 693.6 | 13.41 | 10.0 | 50.0 | 35,000 | 5.87 | 174.5 | 3.372 | -55.0 | -65.0 |
| 3,000 | [illegible] | [illegible] | 13.19 | 9.1 | 48.1 | 36,000 | 5.71 | 170.4 | 3.282 | -55.0 | -66.0 |
| 3,500 | 26.32 | 668.6 | 12.95 | 8.1 | [illegible] | [illegible] | [illegible] | [illegible] | [illegible] | [illegible] | [illegible] |
| 4,000 | 25.84 | 656.5 | 12.70 | 7.1 | 44.5 | 37,000 | 5.50 | 165.4 | 3.14 | 55.0 | 55.0 |
| 4,500 | [illegible] | 644.2 | 12.45 | 6.1 | 43.2 | 37,000 | 5.24 | 150.5 | 3.227 | -55.0 | -65.0 |
| 5,000 | [illegible] | [illegible] | 12.23 | 5.1 | [illegible] | 38,000 | [illegible] | [illegible] | 2.994 | -55.0 | -66.0 |
| 5,500 | 24.43 | 620.6 | 12.00 | 4.1 | 39.4 | [illegible] | [illegible] | [illegible] | [illegible] | [illegible] | [illegible] |
| 6,000 | 23.98 | 609.0 | 11.77 | 3.1 | 37.5 | 39,000 | 5.51 | 140.8 | 2.862 | 55.0 | 65.0 |
| 6,500 | [illegible] | 597.6 | 11.56 | 2.1 | 35.2 | 39,000 | 5.40 | [illegible] | 2.783 | -55.0 | -65.0 |
| [illegible] | [illegible] | [illegible] | 11.34 | 1.1 | [illegible] | 40,000 | 5.54 | 140.7 | 2.72 | -55.0 | -66.0 |
| 7,500 | 22.55 | 579.5 | 11.12 | 0.1 | 32.2 | [illegible] | 5.41 | 138.4 | 2.66 | [illegible] | 55.0 |
| 8,000 | 22.22 | 564.4 | 10.90 | -0.6 | 30.67 | 41,000 | 5.25 | 134.2 | 2.595 | 55.0 | 65.0 |
| 8,500 | [illegible] | 553.7 | 10.70 | [illegible] | 29.7 | 41,000 | 5.16 | 131.0 | 2.535 | -55.0 | -65.0 |
| 9,000 | [illegible] | 543.2 | 10.50 | [illegible] | 27.2 | 42,000 | 7.04 | 127.9 | 2.47 | -55.0 | -67.0 |
| 9,500 | [illegible] | [illegible] | 10.30 | [illegible] | [illegible] | [illegible] | 3.19 | 125.1 | 2.425 | [illegible] | 18.0 |
| 10,000 | 20.58 | 522.6 | 10.10 | -4.8 | 23.4 | 42,000 | 4.80 | 122.0 | 2.36 | -55.0 | -33.0 |
| 10,500 | 20.17 | 512.5 | 9.91 | -5.6 | [illegible] | 43,000 | 4.69 | 119.1 | 2.304 | -55.0 | -66.0 |
| 11,000 | [illegible] | [illegible] | [illegible] | -6.6 | [illegible] | [illegible] | [illegible] | 116.3 | [illegible] | [illegible] | [illegible] |
| 11,500 | 19.40 | 492.5 | 9.52 | -7.5 | 18.5 | 44,000 | 4.44 | 113.5 | 2.195 | 55.0 | 55.0 |
| 12,000 | 9.03 | 483.0 | 9.33 | -3.6 | 6.2 | 45,000 | 4.28 | 110.6 | 2.14 | -55.0 | -33.0 |
| 12,500 | 18.65 | 473.6 | 9.17 | -9.6 | [illegible] | 45,000 | 4.26 | 108.9 | 2.094 | -55.0 | -66.0 |
| 13,000 | [illegible] | 464.6 | [illegible] | 10.8 | [illegible] | [illegible] | 3.10 | 106.7 | [illegible] | [illegible] | 12.0 |
| 13,500 | 17.92 | 455.4 | 8.81 | 11.7 | [illegible] | 46,000 | 4.06 | 103.2 | 2.02 | 55.0 | 55.0 |
| 14,000 | 7.57 | 446.4 | 8.63 | -12.7 | 9.1 | 47,000 | 3.97 | 100.7 | 243 | -55.0 | -65.0 |
| 14,500 | 17.22 | 437.5 | 8.46 | -13.7 | 7.8 | 47,000 | 0.070 | 90.11 | 1.90 | -55.0 | -66.0 |
| 15,000 | [illegible] | 428.6 | 8.29 | 14.7 | [illegible] | [illegible] | 3.701 | [illegible] | [illegible] | [illegible] | 12.0 |
| 15,500 | 16.54 | 420.2 | 8.12 | 15.7 | 3.7 | 48,000 | 3.608 | 93.73 | 1.813 | 55.0 | 55.0 |
| 16,000 | 6.21 | 411.3 | 7.95 | -16.7 | 1.9 | 49,000 | 3.605 | 91.57 | 772 | -55.0 | -65.0 |
| 16,500 | 15.88 | 403.5 | 7.80 | -17.7 | 0.1 | 49,000 | 3.59 | 89.1 | 735 | -55.0 | -66.0 |
| 17,000 | 15.56 | 395.5 | 7.64 | 18.7 | 1.7 | [illegible] | 3.380 | [illegible] | 1.661 | [illegible] | 12.0 |
| 17,500 | 3.22 | 387.0 | 7.48 | -19.7 | -3.5 | 51,000 | 3.276 | 83.22 | 1.610 | 55.0 | 55.0 |
| 18,000 | 14.94 | 379.4 | 7.34 | -20.7 | -5.3 | 52,000 | 3.124 | 79.34 | 533 | -55.0 | -65.0 |
| 18,500 | [illegible] | [illegible] | 7.19 | -21.7 | -7.1 | [illegible] | 2.978 | 75.51 | 1.465 | -55.0 | -66.0 |
| 19,000 | 14.33 | 364.0 | 7.04 | -22.6 | 8.7 | 54,000 | 2.839 | 72.12 | 1.395 | 55.0 | 55.0 |
| 19,500 | 14.04 | 356.5 | 6.90 | -23.6 | -10.5 | 55,000 | 2.707 | 68.76 | 1.33 | -55.0 | -33.0 |
| 20,000 | [illegible] | 349.1 | 6.76 | -24.6 | -12.3 | 56,000 | 2.581 | 65.57 | 1.268 | -55.0 | -66.0 |
| 20,500 | 13.46 | 341.9 | 6.61 | -25.6 | [illegible] | 57,000 | 2.460 | 62.49 | 1.209 | -55.0 | -66.0 |
| 21,000 | 13.18 | 334.7 | 6.48 | -26.6 | -15.9 | 58,000 | 2.346 | 59.58 | 1.152 | 55.0 | 55.0 |
| [illegible] | [illegible] | 327.7 | 6.34 | -27.6 | -17.7 | [illegible] | 2.236 | 56.81 | 1.10 | -55.0 | -66.0 |
| 22,000 | 12.62 | 320.6 | 6.21 | -28.6 | -19.5 | 60,000 | 2.132 | 54.16 | 1.048 | 55.0 | 55.0 |
| 22,500 | 12.32 | 314.1 | 6.08 | -29.6 | -21.3 | 61,000 | 2.033 | 51.62 | 1.000 | -55.0 | -33.0 |
| 23,000 | 9.12 | 307.1 | 5.94 | -30.6 | -23.1 | 62,000 | 1.938 | 49.22 | 0.952 | -55.0 | -66.0 |
| [illegible] | 11.82 | 301.0 | [illegible] | -31.6 | -24.9 | [illegible] | 1.847 | 46.92 | 0.908 | [illegible] | [illegible] |
| 24,000 | 11.59 | 294.4 | 5.70 | -32.5 | -26.5 | 64,000 | 1.761 | 44.73 | 0.865 | 55.0 | 55.0 |
| 24,500 | 11.34 | 288.1 | 5.58 | -33.5 | -28.3 | 65,000 | 1.679 | 42.55 | 0.825 | -55.0 | -33.0 |
| 25,000 | 11.10 | 281.9 | 5.45 | -34.5 | -30.1 | 66,000 | 1.601 | 40.56 | 0.786 | -55.0 | -66.0 |
| [illegible] | 10.86 | 275.8 | [illegible] | -35.5 | -31.9 | 67,000 | 1.526 | 38.75 | 0.749 | -55.0 | -66.0 |
| 26,000 | 10.62 | 269.8 | 5.22 | -36.5 | -33.7 | 68,000 | 1.455 | 36.96 | 0.715 | 55.0 | 55.0 |
| 26,500 | 10.32 | 263.2 | 5.11 | -37.5 | -35.5 | 69,000 | 1.387 | 35.23 | 0.68 | -55.0 | -33.0 |
| 27,000 | 10.12 | 258.1 | 4.99 | -38.5 | -37.1 | 70,000 | 1.299 | 33.78 | 0.648 | -55.0 | -66.0 |
| [illegible] | 9.95 | 252.6 | 4.88 | -39.5 | -39.1 | 71,000 | 1.261 | 32.03 | 0.619 | [illegible] | [illegible] |
| 28,000 | 9.72 | 246.9 | 4.78 | -40.5 | -40.9 | 72,000 | 1.202 | 30.52 | 0.590 | 55.0 | 55.0 |
| 28,500 | 9.50 | 241.4 | 4.27 | -41.5 | -42.7 | 73,000 | 1.146 | 29.10 | 0.562 | -55.0 | -33.0 |
| 29,000 | 9.28 | 235.9 | 4.57 | -42.5 | -44.5 | 74,000 | 1.093 | 27.77 | 0.537 | -55.0 | -66.0 |
| [illegible] | 9.08 | 230.6 | 4.46 | -43.5 | -46.3 | [illegible] | 1.041 | 26.45 | 0.511 | [illegible] | [illegible] |
| 30,000 | 8.88 | 225.6 | 4.36 | -44.4 | -47.9 | 76,000 | 0.993 | 25.22 | 0.488 | 55.0 | 55.0 |
| 30,500 | 8.68 | 220.5 | 4.27 | -45.4 | -49.7 | 77,000 | 0.946 | 24.04 | 0.465 | -55.0 | -33.0 |
| 31,000 | 8.48 | 215.5 | 4.17 | -46.4 | -51.5 | 78,000 | 0.902 | 22.92 | 0.443 | -55.0 | -66.0 |
| 31,500 | 8.29 | 210.6 | 4.07 | -47.4 | -53.3 | 79,000 | 0.860 | 21.84 | 0.422 | [illegible] | [illegible] |
| 32,000 | 8.10 | 205.5 | 3.98 | -48.4 | -55.1 | 80,000 | 0.820 | 20.83 | 0.403 | 55.0 | 55.0 |

## Far Field Calculations for Various Frequencies (Meters)

| Diameter - Meters | 100 MHz | 1 GHz | 10 GHz | 100 GHz |
|---|---|---|---|---|
| 1 | 0.67 | 6.67 | 66.67 | 666.67 |
| 2 | 2.67 | 26.67 | 266.67 | 2666.56 |
| 3 | 6 | 60 | 600 | 6000 |
| 4 | 10.67 | 106.67 | 1066.67 | 10666.67 |
| 5 | 16.67 | 166.67 | 1666,67 | 16666.67 |
| 6 | 24 | 240 | 2400 | 24000 |
| 7 | 32.67 | 326.67 | 3266.67 | 32666.67 |
| 8 | 42.67 | 426.67 | 4266.67 | 42666.67 |
| 9 | 54 | 540 | 5400 | 54000 |
| 10 | 66.67 | 666.67 | 6666.67 | 66666.67 |
| 11 | 80.67 | 806.67 | 8066.67 | 80666.67 |
| 12 | 96 | 960 | 9600 | 96000 |
| 13 | 112.67 | 1126.67 | 11266.67 | 12666.67 |
| 14 | 130.67 | 1306.67 | 13066.67 | 130666.67 |
| 15 | 150 | 1500 | 15000 | 150000 |
| 16 | 170.67 | 1706.67 | 17066.67 | 170666.67 |
| 17 | 192.67 | 1926.67 | 19266.67 | 192666.67 |
| 18 | 216 | 2160 | 21600 | 216000 |
| 19 | 240.67 | 2406.67 | 24066.67 | 240666.67 |
| 20 | 266.67 | 2666.67 | 26666.67 | 266666.67 |

## Far Field Calculations for Various Frequencies (Feet)

| Diameter - Feet | 100 MHz | 1 GHz | 10 GHz | 100 GHz |
|---|---|---|---|---|
| 1 | 0.2 | 2.04 | 20.35 | 203.46 |
| 2 | 0.81 | 8.16 | 81.38 | 813.84 |
| 3 | 1.83 | 18.37 | 183.11 | 1831.13 |
| 4 | 3.25 | 32.65 | 325.53 | 3255.34 |
| 5 | 5.08 | 51.02 | 508.65 | 5086.47 |
| 6 | 7.32 | 73.47 | 732.45 | 7324.52 |
| 7 | 9.96 | 100 | 996.95 | 9969.48 |
| 8 | 13.01 | 130.61 | 1302.14 | 13021.36 |
| 9 | 16.46 | 165.31 | 1648.02 | 16480.16 |
| 10 | 20.33 | 204.08 | 2034.59 | 20345.88 |
| 11 | 24.59 | 246.94 | 2461.85 | 24618.51 |
| 12 | 29.27 | 293.88 | 2929.81 | 29298.07 |
| 13 | 34.35 | 344.9 | 3438.45 | 34384.54 |
| 14 | 39.84 | 400 | 3987.79 | 39877.92 |
| 15 | 45.73 | 459.18 | 4577.82 | 45778.23 |
| 16 | 52.03 | 522.45 | 5208.55 | 52085.45 |
| 17 | 58.74 | 589.8 | 5879.96 | 58799.59 |
| 18 | 65.85 | 661.22 | 6592.07 | 65920.65 |
| 19 | 73.37 | 736.73 | 7344.86 | 73448.63 |
| 20 | 81.3 | 816.33 | 8138.35 | 81383.52 |

## Directivity and Beamwidth vs. Diameter at 1 GHz

| Diameter - Meters | Directivity (dBi) | Beamwidth, Degrees |
|---|---|---|
| 1 | 20.4 | 17.19 |
| 2 | 26.42 | 8.6 |
| 3 | 29.94 | 5.73 |
| 4 | 32.44 | 4.3 |
| 5 | 34.38 | 3.44 |
| 6 | 35.96 | 2.87 |
| 7 | 37.3 | 2.46 |
| 8 | 38.46 | 2.15 |
| 9 | 39.48 | 1.91 |
| 10 | 40.4 | 1.72 |
| 11 | 41.23 | 1.56 |
| 12 | 41.98 | 1.43 |
| 13 | 42.68 | 1.32 |
| 14 | 43.32 | 1.23 |
| 15 | 43.92 | 1.15 |
| 16 | 44.48 | 1.07 |
| 17 | 45 | 1.01 |
| 18 | 45.5 | 0.96 |
| 19 | 45.97 | 0.9 |
| 20 | 46.42 | 0.86 |

## Directivity and Beamwidth vs. Diameter at 10 GHz

| Diameter - Meters | Directivity (dBi) | Beamwidth, Degrees |
|---|---|---|
| 1 | 40.4 | 1.72 |
| 2 | 46.42 | 0.86 |
| 3 | 49.94 | 0.57 |
| 4 | 52.44 | 0.43 |
| 5 | 54.38 | 0.34 |
| 6 | 55.96 | 0.29 |
| 7 | 57.3 | 0.25 |
| 8 | 58.46 | 0.21 |
| 9 | 59.48 | 0.19 |
| 10 | 60.4 | 0.17 |
| 11 | 61.23 | 0.16 |
| 12 | 61.94 | 0.14 |
| 13 | 62.68 | 0.13 |
| 14 | 63.32 | 0.12 |
| 15 | 63.92 | 0.11 |
| 16 | 64.48 | 0.11 |
| 17 | 65 | 0.1 |
| 18 | 65.5 | 0.1 |
| 19 | 65.97 | 0.09 |
| 20 | 66.42 | 0.09 |

## Losses in Coax over Frequency (dB per 100')

| Type | Size In. | 150 MHz | 220 MHz | 450 MHz | 900 MHz | 1.5 GHz | 2 GHz | 5.8 GHz |
|---|---|---|---|---|---|---|---|---|
| LDF6-50 | 1.55 | 0.34 | - | 0.62 | 0.91 | 1.22 | 1.45 | 2.5 |
| LMR-1700 | 1.67 | 0.35 | 0.43 | 0.43 | 0.63 | 1.27 | 1.5 | - |
| Heliax LDF5 | 1.09 | 0.46 | - | 0.83 | 1.23 | 1.66 | 1.97 | - |
| LMR-1200 | 1.2 | 0.48 | 0.59 | 0.86 | 1.26 | 1.69 | 1.99 | - |
| LMR-900 | 0.87 | 0.62 | 0.76 | 1.1 | 1.6 | 2.12 | 2.49 | - |
| LMR-600 | 0.59 | 0.96 | 1.18 | 1.72 | 2.5 | 3.31 | 3.9 | 7.3 |
| Heliax FSJ4 | 0.63 | 0.85 | - | 1.51 | 2.2 | 2.93 | 3.45 | - |
| LMR-500 | 0.5 | 1.22 | 1.49 | 2.17 | 3.13 | 4.13 | 4.84 | - |
| Heliax FSJ4 | 0.52 | 1.29 | - | 2.32 | 3.38 | 4.5 | 5.31 | - |
| LMR-400 | 0.41 | 1.5 | 1.8 | 2.7 | 3.9 | 5.1 | 6 | 10.8 |
| Belden 9913 | 0.41 | 1.6 | 1.9 | 2.8 | 4.2 | 5.6 | 6.7 | 13.8 |
| Ultra-Link | 0.41 | 1.5 | - | 2.7 | 4.19 | - | 6.7 | - |
| RG213 RG214 | 0.41 | 2.8 | 3.5 | 5.2 | 8 | 10.1 | 15.2 | 28.6 |
| Heliax FSJ1 | 0.3 | 2.23 | - | 3.93 | 5.69 | 7.47 | 8.73 | - |
| LMR-240 | 0.24 | 3 | 3.7 | 5.3 | 7.6 | 9.9 | 11.5 | 20.4 |
| ProFlex 800 | 0.24 | - | - | 7.8 | - | - | - | - |
| Belden RG8X | 0.24 | 4.7 | 6 | 8.6 | 12.8 | 15.9 | 23.1 | 40.9 |
| LMR-200 | 0.2 | 4 | 4.8 | 6.9 | 9.9 | 12.9 | 15 | - |
| Ultra-Link | 0.2 | 5.1 | - | 9.5 | 14 | - | 36 | - |
| RG-58 | 0.2 | 6.2 | 7.4 | 10.6 | 16.5 | 21.1 | 32.2 | 51.6 |
| LMR-100 | 0.15 | 8.9 | 10.9 | 15.8 | 22.8 | 30 | 35 | - |

**Coax Equations:** $Z=\frac{138}{(\sqrt{(Er)})}*log10\left(\frac{D}{d}\right)$ **and** $V=\frac{1}{\sqrt{(\epsilon r)}}$

where Z= impedance,V = Velocity of Propagation,D=Outer Diameter, d = Inner Diameter and Er = Dielectric Constant

# Coax Data

### Attenuation - db/100 feet

| Belden # | Impedance | 100 MHz | 400 MHz | 1000 MHz | OD | V Factor |
|---|---|---|---|---|---|---|
| 9880 | 50 | 1.3 | 2.8 | 4.5 | 0.39 | 0.82 |

This is Thicknet Ethernet cable. Most is marked "Style 1478" and has a #12 solid center conductor and 4 shields (2 braid/2 foil).

### Attenuation - db/100 feet

| Belden # | Impedance | 100 MHz | 400 MHz | 1000 MHz | OD | V Factor |
|---|---|---|---|---|---|---|
| 8240 | 50 | 4.9 | 11.5 | 20 | 0.195 | 0.66 |
| 8267 | 50 | 2.2 | 4.7 | 8 | 0.405 | 0.66 |
| 8208 | 50 | | | 9 | 0.405 | 0.66 |
| 9258 | 50 | 3.7 | 8 | 12.8 | 0.242 | 0.78 |
| 9913 | 50 | 1.3 | 2.8 | 4.5 | 0.405 | 0.82 |
| 9914 | 50 | | | 9 | 0.403 | 0.66 |

### Attenuation - db/100 feet

| Hardline | Impedance | 100 MHz | 400 MHz | 1000 MHz | OD | V Factor |
|---|---|---|---|---|---|---|
| 1/2" | 50 | 0.8 | 1.8 | 3 | 0.500 | 0.66 |
| 1/2" | 75 | 0.93 | 2.2 | 3.6 | 0.500 | 0.66 |
| 3/4" | 50 | 0.66 | 1.49 | 2.4 | 0.750 | 0.66 |
| 3/4" | 75 | 0.7 | 1.55 | 2.6 | 0.750 | 0.66 |
| 7/8" | 50 | 0.55 | 1.3 | 2.3 | 0.875 | 0.66 |

| | | | | | | |
|---|---|---|---|---|---|---|
| 7/8" | 75 | 0.5 | 1.35 | 2.25 | 0.875 | 0.66 |

**Attenuation - db/100 feet**

RG # Impedance 100 MHz 400 MHz 1000 MHz OD V
Factor

| | | | | | | |
|---|---|---|---|---|---|---|
| 4 /U | 50 | | | | | |
| 5 B/U | 50 | 0 | | | 0.332 | 0.66 |
| 6 /U | 75 | 2.1 | 5 | 6.9 | 0.270 | 0.78 |
| 6 A/U | 75 | | | 11 | 0.332 | 0.78 |
| 7 /U | 95 | | | | | |
| 8 /U | 50 | 1.8 | 4.7 | 6.9 | 0.405 | 0.66 |
| 8 A/U | 50 | | | 9 | 0.405 | 0.66 |
| 8 /X | 50 | 3.7 | 8 | 12.8 | 0.242 | 0.78 |
| 9 /U | 51 | 2.2 | 4.7 | 8.9 | 0.420 | 0.66 |
| 10 A/U | 50 | | | | 0.475 | 0.66 |
| 11 /U | 75 | 2 | 4.2 | 6.8 | 0.405 | 0.66 |
| 11 A/U | 75 | 9 | 0.41 | 0.66 | | |
| 12 A/U | 75 | 2.15 | 4.7 | 8.2 | 0.475 | 0.66 |
| 13 A/U | 75 | 2.2 | 4.6 | 8 | 0.425 | 0.66 |
| 14 A/U | 52 | 1.4 | 3.1 | 5.8 | 0.545 | 0.66 |
| 15 /U | 76 | | | | | |
| 16 /U | 52 | | | | | |
| 17 A/U | 52 | 0.81 | 1.9 | 3.8 | 0.870 | 0.66 |
| 18 /U | 52 | 0.81 | 1.9 | 3.8 | 0.870 | 0.66 |
| 19 /U | 52 | 0.7 | 1.5 | 3.5 | 1.120 | 0.66 |
| 20 /U | 52 | 0.7 | 1.5 | 3.5 | 1.195 | 0.66 |

| | | | | | | |
|---|---|---|---|---|---|---|
| 21 /U | 53 | | | | 0.332 | 0.66 |
| 22 B/U | 95 | 3 | 9.5 | | 0.420 | 0.66 |
| 23 /U | 125 | | | | | |
| 24 /U | 125 | | | | | |

| RG # | Impedance | 100 MHz | 400 MHz | 1000 MHz | OD | V Factor |
|---|---|---|---|---|---|---|
| 25 /U | 48 | | | | | |
| 26 /U | 48 | | | | | |
| 27 /U | 48 | | | | | |
| 28 /U | 48 | | | | | |
| 28 A/U | 50 | | | | | |
| 29 /U | 53.5 | | | | 0.184 | 0.66 |
| 30 /U | 58 | | | | | |
| 31 /U | 51 | | | | | |
| 32 /U | 51 | | | | | |
| 33 /U | 51 | | | | | |
| 34 /U | 71 | | | | 0.630 | 0.66 |
| 35 /U | 71 | 1 | 2.5 | 4.5 | 0.940 | 0.66 |
| 36 /U | 69 | | | | | |
| 37 /U | 52.5 | | | | | |
| 38 /U | 52.5 | | | | | |
| 39 /U | 72.5 | | | | | |
| 40 /U | 72.5 | | | | | |
| 41 /U | 67.5 | | | | | |

| | | | | | | |
|---|---|---|---|---|---|---|
| 42 /U | 78 | | | | | |
| 43 /U | 95 | | | | | |
| 44 /U | 50 | | | | | |
| 45 /U | 50 | | | | | |
| 46 /U | 50 | | | | | |

**Attenuation - db/100 feet**

| RG # | Impedance | 100 MHz | 400 MHz | 1000 MHz | OD | V Factor |
|---|---|---|---|---|---|---|
| 47 /U | 50 | | | | | |
| 48 /U | 53 | | | | | |
| 54 A/U | 58 | 4 | 8 | 12 | 0.250 | 0.66 |
| 55 B/U | 53.5 | 4.3 | 8.8 | 16.5 | 0.206 | 0.66 |
| 56 /U | 53.5 | | | | | |
| 57 /U | 95 | | | | 0.625 | 0.66 |
| 58 A/U | 50 | 4.9 | 11.5 | 20 | 0.195 | 0.66 |
| 58 C/U | 50 | 4.9 | 11.5 | 20 | 0.195 | 0.66 |
| 59 B/U | 75 | 3.4 | 7 | 11.1 | 0.242 | 0.66 |
| 60 /U | 50 | | | | | |
| 62 A/U | 93 | 2.7 | 5.4 | 8.3 | 0.242 | 0.84 |
| 63 /U | 125 | | | | 0.405 | 0.84 |
| 64 /U | 48 | | | | | |
| 65 /U | 950 | 20 | 50 | | 0.405 | |
| 66 /U | 69 | | | | | |
| 71 B/U | 93 | 1.9 | 3.2 | 8.5 | 0.250 | 0.84 |
| 72 /U | 150 | | | | | |

| | | | | | | |
|---|---|---|---|---|---|---|
| 73 /U | 25 | | | | | |
| 74 /U | 50 | | | | | 0.62 |
| 76 /U | 50 | | | | | |
| 77 /U | 48 | | | | | |
| 78 /U | 48 | | | | | |
| 79 B/U | 125 | 0.48 | 0.84 | | 0.475 | 0.84 |
| 80 /U | 51 | | | | | |

| <u>RG #</u> | Impedance | 100 MHz | 400 MHz | 1000 MHz | OD | V Factor |
|---|---|---|---|---|---|---|
| 81 /U | 52 | | | | | |
| 82 /U | 52 | | | | | |
| 83 /U | 35 | | | | | |
| 84 /U | 71 | 1 | 2.5 | 4.5 | 1.000 | 0.66 |
| 85 /U | 71 | 1 | 2.5 | 4.5 | 1.565 | 0.66 |
| 86 /U | 205 | | | | | |
| 87 A/U | 50 | 0.43 | | | | |
| 88 /U | 50 | | | | | |
| 89 /U | 125 | | | | | |
| 90 /U | 50 | | | | | |
| 91 /U | 50 | | | | | |
| 92 /U | 50 | | | | | |
| 93 /U | 50 | | | | | |
| 94 /U | 50.5 | | | | | |

| | | | | | | |
|---|---|---|---|---|---|---|
| 95 /U | 50 | | | | | |
| 96 /U | 50 | | | | | |
| 97 /U | 50 | | | | | |
| 98 /U | 50 | | | | | |
| 99 /U | 50 | | | | | |
| 100 /U | 35 | 0.24 | 0.66 | | 0.242 | 0.66 |
| 101 /U | 70 | | | | | |
| 102 /U | 140 | | | | | |
| 108 A/U | 78 | 26.2 | | | 0.235 | |

| RG # | Impedance | 100 MHz | 400 MHz | 1000 MHz | OD | V Factor |
|---|---|---|---|---|---|---|
| 109 /U | 76 | | | | | |
| 111 /U | 95 | | | | | |
| 114 /U | 185 | | | | 0.405 | 0.66 |
| 115 /U | 50 | | | | 0.375 | |
| 116 /U | 50 | | | | 0.490 | |
| 117 /U | 185 | | | | | |
| 118 /U | 50 | | | | | 0.78 |
| 119 /U | 50 | | | | 0.465 | |
| 120 /U | 50 | | | | | |
| 121 /U | 50 | | | | | |
| 122 /U | 50 | 7 | 15.2 | 25 | 0.160 | 0.66 |
| 124 /U | 73 | | | | | |

| | | | | | | |
|---|---|---|---|---|---|---|
| 125 /U | 150 | | | | | |
| 126 /U | 50 | | | | 0.280 | |
| 128 /U | 50 | | | | | |
| 130 /U | 95 | | | | 0.625 | 0.66 |
| 131 /U | 95 | | | | | |
| 140 /U | 75 | | | 13 | 0.233 | |
| 141 /U | 50 | 3.2 | 6.9 | 13 | 0.190 | |
| 142 /U | 50 | 3.9 | 8.2 | 13.5 | 0.206 | |
| 143 /U | 50 | | | | 0.325 | |
| 144 /U | 75 | | | | 0.410 | |
| 156 /U | 50 | | | | 0.540 | |
| 157 /U | 50 | | | | 0.725 | |

<u>RG #</u> Impedance 100 MHz 400 MHz 1000 MHz OD V
Factor

| | | | | | | |
|---|---|---|---|---|---|---|
| 158 /U | 25 | | | | 0.730 | |
| 161 /U | 70 | | | | 0.090 | |
| 164 A/U | 75 | | | | 0.870 | 0.66 |
| 165 /U | 50 | | | | 0.410 | |
| 174 /U | 50 | 8.9 | 17.5 | 28.2 | 0.101 | 0.66 |
| 178 B/U | 75 | 10.5 | 28 | 46 | 0.075 | |
| 179 B/U | 75 | 10 | 16 | 24 | 0.105 | |
| 180 B/U | 95 | 5.7 | 10.7 | 17 | 0.145 | 0.66 |
| 188 A/U | 50 | 9.8 | 15.8 | 25 | 0.110 | 0.66 |
| 190 /U | 50 | | | | 0.700 | |

| | | | | | | |
|---|---|---|---|---|---|---|
| 191 /U | 25 | | | | 1.460 | |
| 195 /U | 95 | 9.8 | 15.8 | 25 | 0.155 | 0.66 |
| 196 A/U | 50 | 9.8 | 15.8 | 25 | 0.080 | 0.66 |
| 209 /U | 50 | | | | 0.750 | |
| 210 /U | 93 | 3.1 | | | 0.240 | |
| 211 /U | 50 | | | | 0.730 | |
| 212 /U | 50 | 1.6 | 3.6 | 8.8 | 0.336 | 0.66 |
| 213 /U | 50 | 2.2 | 4.7 | 8 | 0.405 | 0.66 |
| 214 /U | 50 | 2.2 | 4.7 | 8 | 0.425 | 0.66 |
| 215 /U | 50 | 2.2 | 4.6 | 9 | 0.475 | 0.66 |
| 216 /U | 75 | | | | 0.425 | 0.66 |
| 217 /U | 50 | 1.4 | 3.1 | 5.8 | 0.545 | 0.66 |
| 218 /U | 50 | 0.81 | 1.9 | 3.8 | 0.870 | 0.66 |
| 219 /U | 50 | 0.81 | 1.9 | 3.8 | 0.870 | 0.66 |

RG # Impedance 100 MHz 400 MHz 1000 MHz OD V
Factor

| | | | | | | |
|---|---|---|---|---|---|---|
| 220 /U | 50 | 0.7 | 1.5 | 3.5 | 1.120 | 0.66 |
| 221 /U | 50 | 0.7 | 1.5 | 3.5 | 1.195 | 0.66 |
| 223 /U | 50 | 4.5 | 9.2 | 14.3 | 0.212 | 0.66 |
| 224 /U | 50 | 1.5 | 3 | 6 | 0.615 | |
| 225 /U | 50 | 7.5 | | | 0.430 | |
| 226 /U | 50 | | | | 0.500 | |
| 227 /U | 50 | | | | 0.490 | |
| 228 /U | 50 | | | | 0.795 | |
| 279 /U | 75 | | | | 0.145 | |

| | | | | | | |
|---|---|---|---|---|---|---|
| 280 /U | 50 | | | | 0.480 | |
| 281 /U | 50 | | | | 0.750 | |
| 301 /U | 50 | | | | 0.245 | |
| 302 /U | 75 | 3.9 | 8 | 12.8 | 0.206 | |
| 303 /U | 50 | 9.8 | 15.8 | 25 | 0.170 | 0.66 |
| 304 /U | 50 | | | | 0.280 | |
| 307 /U | 75 | | | | 0.270 | |
| 316 /U | 50 | 10.4 | 16.5 | 31 | 0.102 | 0.66 |
| 393 /U | 50 | 2.1 | 4.4 | 7.5 | 0.360 | |
| 400 /U | 50 | 3.1 | 8.1 | 13 | 0.171 | |
| 403 /U | 50 | 13.6 | 26.5 | 45 | 0.116 | |
| 404 /U | 50 | 16.3 | 32.4 | 68 | 0.116 | |
| 405 /U | 50 | 22 | | | 0.090 | |

# Waveguide Characteristics

| Designation (a)=aluminum, (b)=brass, (c)=copper, (s)=silver | | | | $f_L - f_U$ (GHz) | $f_{CO}$ (GHz) | Inside Width (in) | Inside Height (in) |
|---|---|---|---|---|---|---|---|
| WR | U.S. Mil. ___/U | British Mil. | IEC | | | | |
| WR975 | RG204 (a) | | | 0.75-1.12 | | 9.750 | 4.875 |
| WR770 | RG205 (a) | | | 0.96-1.45 | | 7.700 | 3.850 |
| WR650 | RG69 (b)<br>RG103 (a) | WG6 | | 1.12-1.70 | | 6.500 | 3.250 |
| WR510 | | | | 1.45-2.20 | | 5.100 | 2.550 |
| WR430 | RG104 (b)<br>RG105 (a) | WG8 | | 1.70-2.60 | | 4.300 | 2.150 |
| WR340 | RG112 (b)<br>RG113 (a) | WG9A | | 2.20-3.30 | | 3.400 | 1.700 |
| WR284 | RG48 (b)<br>RG75 (a) | WG10 | | 2.60-3.95 | 2.08 | 2.840 | 1.340 |
| WR229 | RG340 (c)<br>RG341 (a) | WG11A | R40 | 3.30-4.90 | 2.577 | 2.290 | 1.145 |
| WR187 | RG49 (b)<br>RG95 (a) | WG12 | R48 | 3.95-5.85- | 3.156 | 1.872 | 0.872 |
| WR159 | RG343 (c)<br>RG344 (a) | WG13 | R58 | 4.90-7.05 | 3.705 | 1.590 | 0.795 |
| WR137 | RG50 (b)<br>RG106 (a) | WG14 | R70 | 5.850-8.200 | 4.285 | 1.372 | 0.622 |
| WR112 | RG51 (b)<br>RG68 (a) | WG15 | R84 | 7.050-10.000 | 5.26 | 1.122 | 0.497 |
| WR90 | RG52 (b)<br>RG67 (a) | WG16 | R100 | 8.20-12.40 | 6.56 | 0.900 | 0.400 |
| WR75 | RG346 (c)<br>RG347 (a) | WG17 | | 10.0-15.0 | 7.847 | 0.750 | 0.375 |
| WR62 | RG91 (b)<br>RG349 (a) | WG18 | | 12.40-18.00 | 9.49 | 0.622 | 0.311 |
| WR51 | RG352 (c)<br>RG351 (a) | WG19 | | 15.00-22.00 | 11.54 | 0.510 | 0.255 |
| WR42 | RG53 (b)<br>RG121 (a) | WG20 | | 18.00-26.5 | 14.08 | 0.420 | 0.170 |
| WR34 | RG354 (c) | | | 20.0-33.0 | 17.28 | 0.340 | 0.170 |
| WR28 | RG96 (s)<br>RG271 (c) | WG22 | | 26.50-40.00 | 21.1 | 0.280 | 0.140 |
| WR22 | RG97 (s) | WG23 | | 33.00-50.00 | | 0.224 | 0.112 |
| WR19 | | WG24 | | 40.00-60.00 | | 0.188 | 0.094 |
| WR15 | RG98 (s) | WG25 | | 50.00-75.00 | | 0.148 | 0.074 |
| WR12 | RG99 (s) | WG26 | | 60.00-90.00 | | 0.122 | .061 |
| WR10 | | WG27 | | 75.00-110.0 | | 0.100 | 0.050 |
| WR8 | RG138 (s) | WG28 | | 90.00-140.0 | | 0.080 | 0.040 |
| WR7 | RG136 (s) | | | 110.0-170.0 | | 0.065 | 0.0325 |
| WR4 | RG137 | | | 170.0-260.0 | | 0.043 | 0.0215 |
| WR3 | RG139 (s) | | | 220.0-325.0 | | 0.0340 | 0.0170 |

**Table 1**

**Characteristics of Commonly Used Transmission Lines**

| RG or Type | Part Number | $Z_0$ Ω | VF % | Cap. pF/ft | Cent. Cond. AWG | Diel. | Shield | Jacket | OD in. | Max V (RMS) | Matched Loss (dB/100) 1 MHz | 10 | 100 | 1000 |
|---|---|---|---|---|---|---|---|---|---|---|---|---|---|---|
| RG-6 | Belden 8215 | 75 | 66 | 20.5 | #21 Solid | PE | FC | PE | 0.275 | 2700 | 0.4 | 0.8 | 2.7 | 9.8 |
| RG-8 | TMS LMR400 | 50 | 85 | 23.9 | #10 Solid | FPE | FC | PE | 0.405 | 600 | 0.1 | 0.4 | 1.3 | 4.1 |
| RG-8 | Belden 9913 | 50 | 84 | 24.6 | #10 Solid | ASPE | FC | P1 | 0.405 | 600 | 0.1 | 0.4 | 1.3 | 4.5 |
| RG-8 | WM CQ102 | 50 | 84 | 24.0 | #9.5 Solid | ASPE | S | P2 | 0.405 | 600 | 0.1 | 0.4 | 1.3 | 4.5 |
| RG-8 | DRF-BF | 50 | 84 | 24.5 | #9.5 Solid | FPE | FC | PEBF | 0.405 | 600 | 0.1 | 0.5 | 1.6 | 5.2 |
| RG-8 | WM CQ106 | 50 | 82 | 24.5 | #9.5 Solid | FPE | FC | P2 | 0.405 | 600 | 0.2 | 0.6 | 1.8 | 5.3 |
| RG-8 | Belden 9914 | 50 | 82 | 24.8 | #10 Solid | TFE | FC | P1 | 0.405 | 3700 | 0.1 | 0.5 | 1.6 | 6.0 |
| RG-8 | Belden 8237 | 52 | 66 | 29.5 | #13 Flex | PE | S | P1 | 0.405 | 3700 | 0.2 | 0.6 | 1.9 | 7.4 |
| RG-8X | TMS LMR240 | 50 | 84 | 24.2 | #15 Solid | FPE | FC | PE | 0.242 | 300 | 0.2 | 0.8 | 2.5 | 8.0 |
| RG-8X | WM CQ118 | 50 | 82 | 25.0 | #16 Flex | FPE | S | P2 | 0.242 | 300 | 0.3 | 0.9 | 2.8 | 8.4 |
| RG-8X | Belden 9258 | 50 | 80 | 25.3 | #16 Flex | TFE | S | P1 | 0.242 | 300 | 0.3 | 1.0 | 3.3 | 14.3 |
| RG-9 | Belden 8242 | 51 | 66 | 30.0 | #13 Flex | PE | D | P2N | 0.420 | 3700 | 0.2 | 0.6 | 2.1 | 8.2 |
| RG-11 | Belden 8213 | 75 | 78 | 17.3 | #14 Solid | FPE | S | PE | 0.405 | 600 | 0.2 | 0.4 | 1.5 | 5.4 |
| RG-11 | Belden 8238 | 75 | 66 | 20.5 | #18 Flex | PE | S | P1 | 0.405 | 600 | 0.2 | 0.7 | 2.0 | 7.1 |
| RG-58C | TMS LMR200 | 50 | 83 | 24.5 | #17 Solid | FPE | FC | PE | 0.195 | 300 | 0.3 | 1.0 | 3.2 | 10.5 |
| RG-58 | WM CQ124 | 53.5 | 66 | 28.5 | #20 Solid | PE | S | P2N | 0.195 | 1400 | 0.4 | 1.3 | 4.3 | 14.3 |
| RG-58 | Belden 8240 | 53.5 | 66 | 28.5 | #20 Solid | PE | S | P1 | 0.193 | 1400 | 0.3 | 1.1 | 3.8 | 14.5 |
| RG-58A | Belden 8219 | 50 | 78 | 26.5 | #20 Flex | FPE | S | P1 | 0.198 | 300 | 0.4 | 1.3 | 4.5 | 18.1 |
| RG-58C | Belden 8262 | 50 | 66 | 30.8 | #20 Flex | PE | S | P2N | 0.195 | 1400 | 0.4 | 1.4 | 4.9 | 21.5 |
| RG-58A | Belden 8259 | 50 | 66 | 30.8 | #20 Flex | PE | S | P1 | 0.193 | 1400 | 0.4 | 1.5 | 5.4 | 22.8 |
| RG-59 | Belden 8212 | 75 | 78 | 17.3 | #20 Solid | TFE | S | PE | 0.242 | 300 | 0.6 | 1.0 | 3.0 | 10.9 |
| RG-59B | Belden 8263 | 75 | 66 | 20.5 | #23 Solid | PE | S | P2N | 0.242 | 1700 | 0.6 | 1.1 | 3.4 | 12.0 |
| RG-62A | Belden 9269 | 93 | 84 | 13.5 | #22 Solid | ASPE | S | P1 | 0.260 | 750 | 0.3 | 0.9 | 2.7 | 8.7 |
| RG-62B | Belden 8255 | 93 | 84 | 13.5 | #24 Solid | ASPE | S | P2N | 0.260 | 750 | 0.3 | 0.9 | 2.9 | 11.0 |
| RG-63B | Belden 9857 | 125 | 84 | 9.7 | #22 Solid | ASPE | S | P2N | 0.405 | 750 | 0.2 | 0.5 | 1.5 | 5.8 |
| RG-142B | Belden 83242 | 50 | 69.5 | 29.2 | #18 Solid | TFE | D | TFE | 0.195 | 1400 | 0.3 | 1.1 | 3.9 | 13.5 |
| RG-174 | Belden 8216 | 50 | 66 | 30.8 | #26 Solid | PE | S | P1 | 0.101 | 1100 | 1.9 | 3.3 | 8.4 | 34.0 |
| RG-213 | Belden 8267 | 50 | 66 | 30.8 | #13 Flex | PE | S | P2N | 0.405 | 3700 | 0.2 | 0.6 | 2.1 | 8.2 |
| RG-214 | Belden 8268 | 50 | 66 | 30.8 | #13 Flex | PE | D | P2N | 0.425 | 3700 | 0.2 | 0.6 | 1.9 | 8.0 |
| RG-216 | Belden 9850 | 75 | 66 | 20.5 | #18 Flex | PE | D | P2N | 0.425 | 3700 | 0.2 | 0.7 | 2.0 | 7.1 |
| RG-217 | M17/79-RG217 | 50 | 66 | 30.8 | #9.5 Solid | PE | D | P2N | 0.545 | 7000 | 0.1 | 0.4 | 1.4 | 5.2 |
| RG-218 | M17/78-RG218 | 50 | 66 | 29.5 | #4.5 Solid | PE | S | P2N | 0.870 | 11000 | 0.1 | 0.2 | 0.8 | 3.4 |
| RG-223 | Belden 9273 | 50 | 66 | 30.8 | #19 Solid | PE | D | P2N | 0.212 | 1700 | 0.4 | 1.2 | 4.1 | 14.5 |
| RG-303 | Belden 84303 | 50 | 69.5 | 29.2 | #18 Solid | TFE | S | TFE | 0.170 | 1400 | 0.3 | 1.1 | 3.9 | 13.5 |
| RG-316 | Belden 84316 | 50 | 69.5 | 29.0 | #26 Solid | TFE | S | TFE | 0.098 | 900 | 1.2 | 2.7 | 8.3 | 29.0 |
| RG-393 | M17/127-RG393 | 50 | 69.5 | 29.4 | #12 Solid | TFE | D | TFE | 0.390 | 5000 | 0.2 | 0.5 | 1.7 | 6.1 |
| RG-400 | M17/128-RG400 | 50 | 69.5 | 29.4 | #20 Solid | TFE | D | TFE | 0.195 | 1900 | 0.4 | 1.1 | 3.9 | 13.2 |
| LMR500 | TMS LMR500 | 50 | 85 | 23.9 | #7 Solid | FPE | FC | PE | 0.500 | 2500 | 0.1 | 0.3 | 0.9 | 3.3 |
| LMR600 | TMS LMR600 | 50 | 86 | 23.4 | #5.5 Solid | FPE | FC | PE | 0.590 | 4000 | 0.1 | 0.2 | 0.8 | 2.7 |
| LMR1200 | TMS LMR1200 | 50 | 88 | 23.1 | #0 Tube | FPE | FC | PE | 1.200 | 4500 | 0.04 | 0.1 | 0.4 | 1.3 |
| **Hardline** | | | | | | | | | | | | | | |
| ½" | CATV Hardline | 50 | 81 | 25.0 | #5.5 | FPE | SM | none | 0.500 | 2500 | 0.05 | 0.2 | 0.8 | 3.2 |
| ½" | CATV Hardline | 75 | 81 | 16.7 | #11.5 | FPE | SM | none | 0.500 | 2500 | 0.1 | 0.2 | 0.8 | 3.2 |
| ⅞" | CATV Hardline | 50 | 81 | 25.0 | #1 | FPE | SM | none | 0.875 | 4000 | 0.03 | 0.1 | 0.6 | 2.9 |
| ⅞" | CATV Hardline | 75 | 81 | 16.7 | #5.5 | FPE | SM | none | 0.875 | 4000 | 0.03 | 0.1 | 0.6 | 2.9 |
| LDF4-50A | Heliax – ½" | 50 | 88 | 25.9 | #5 Solid | FPE | CC | PE | 0.630 | 1400 | 0.05 | 0.2 | 0.6 | 2.4 |
| LDF5-50A | Heliax – ⅞" | 50 | 88 | 25.9 | 0.355" | FPE | CC | PE | 1.090 | 2100 | 0.03 | 0.10 | 0.4 | 1.3 |
| LDF6-50A | Heliax – 1¼" | 50 | 88 | 25.9 | 0.516" | FPE | CC | PE | 1.550 | 3200 | 0.02 | 0.08 | 0.3 | 1.1 |
| **Parallel Lines** | | | | | | | | | | | | | | |
| TV Twinlead | | 300 | 80 | 5.8 | #20 | PE | none | P1 | 0.500 | | | | | |
| Transmitting Tubular | | 300 | 80 | 5.8 | #20 | PE | none | P1 | 0.500 | 8000 | 0.09 | 0.3 | 1.1 | 3.9 |
| Window Line | | 450 | 91 | 4.0 | #18 | PE | none | P1 | 1.000 | 10000 | 0.02 | 0.08 | 0.3 | 1.1 |
| Open Wire Line | | 600 | 92 | 1.1 | #12 | none | none | none | varies | 12000 | 0.02 | 0.06 | 0.2 | 0.7 |

**Approximate Power Handling Capability** (1:1 SWR, 40°C Ambient):

| | 1.8 MHz | 7 | 14 | 30 | 50 | 150 | 220 | 450 | 1 GHz |
|---|---|---|---|---|---|---|---|---|---|
| RG-58 Style | 1350 | 700 | 500 | 350 | 250 | 150 | 120 | 100 | 50 |
| RG-59 Style | 2300 | 1100 | 800 | 550 | 400 | 250 | 200 | 130 | 90 |
| RG-8X Style | 1830 | 840 | 560 | 360 | 270 | 145 | 115 | 80 | 50 |
| RG-8/213 Style | 5900 | 3000 | 2000 | 1500 | 1000 | 600 | 500 | 350 | 250 |
| RG-217 Style | 20000 | 9200 | 6100 | 3900 | 2900 | 1500 | 1200 | 800 | 500 |
| LDF4-50A | 38000 | 18000 | 13000 | 8200 | 6200 | 3400 | 2800 | 1900 | 1200 |
| LDF5-50A | 67000 | 32000 | 22000 | 14000 | 11000 | 5900 | 4800 | 3200 | 2100 |
| LMR500 | 12000 | 6000 | 4200 | 2800 | 2200 | 1200 | 1000 | 700 | 450 |
| LMR1200 | 39000 | 19000 | 13000 | 8800 | 6700 | 3800 | 3100 | 2100 | 1400 |

**Legend:**

| | | | |
|---|---|---|---|
| ASPE | Air Spaced Polyethylene | P1 | PVC, Class 1 |
| BF | Flooded direct bury | P2 | PVC, Class 2 |
| CC | Corrugated Copper | PE | Polyethylene |
| D | Double Copper Shields | S | Single Shield |
| DRF | Davis RF | SM | Smooth Aluminum |
| FC | Foil/Copper Shields | TFE | Teflon |
| FPE | Foamed Polyethylene | TMS | Times Microwave Systems |
| Heliax | Andrew Corp Heliax | WM | Wireman |
| N | Non-Contaminating | ** | Not Available or varies |

# Atmospheric Attenuation

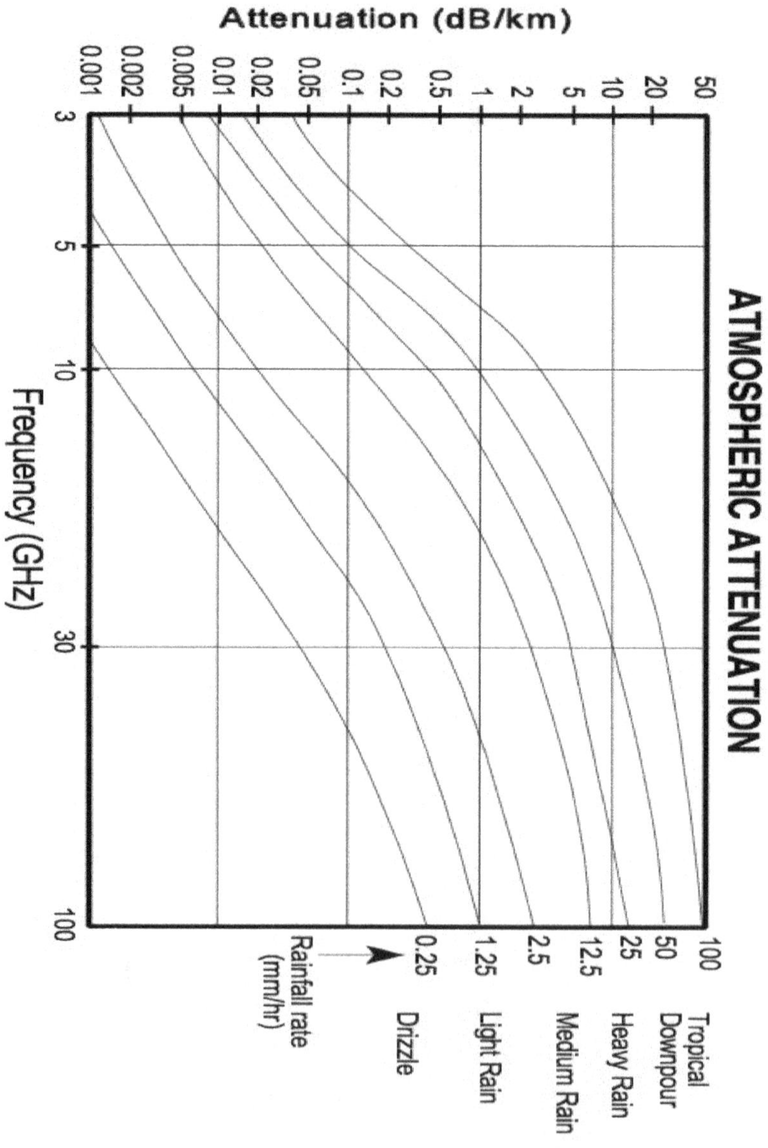

# Atmospheric Absorption vs. Frequency

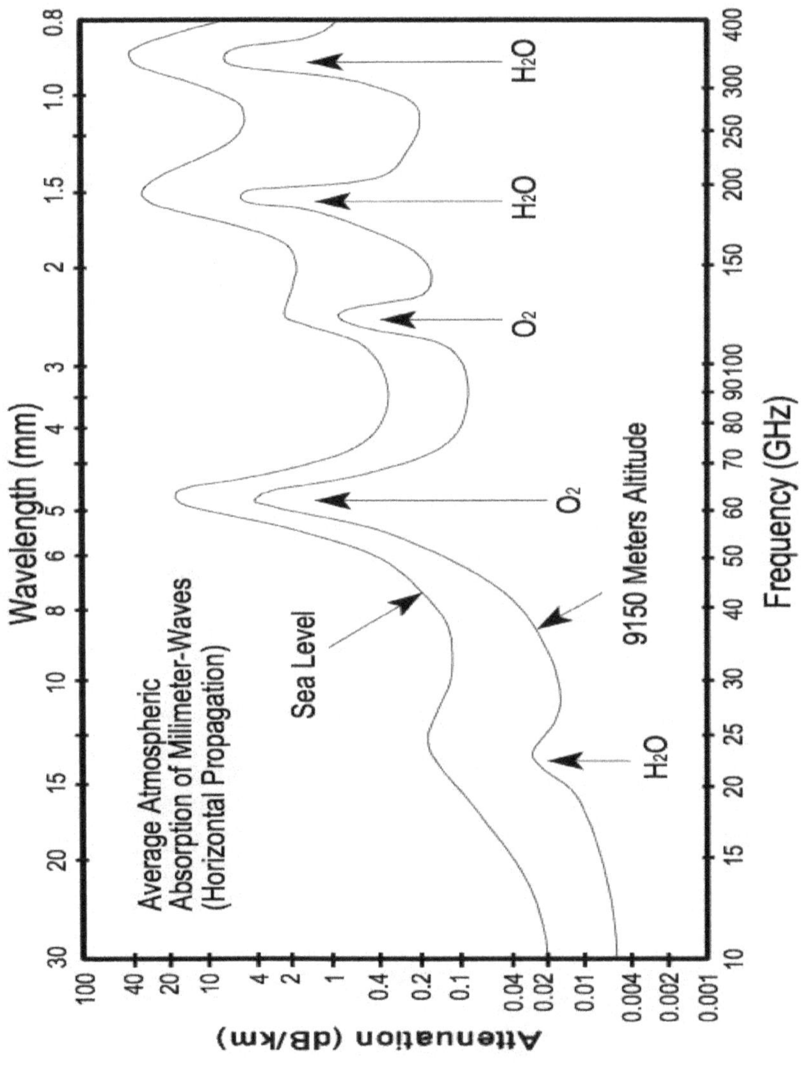

# Sky Noise at Various Frequencies

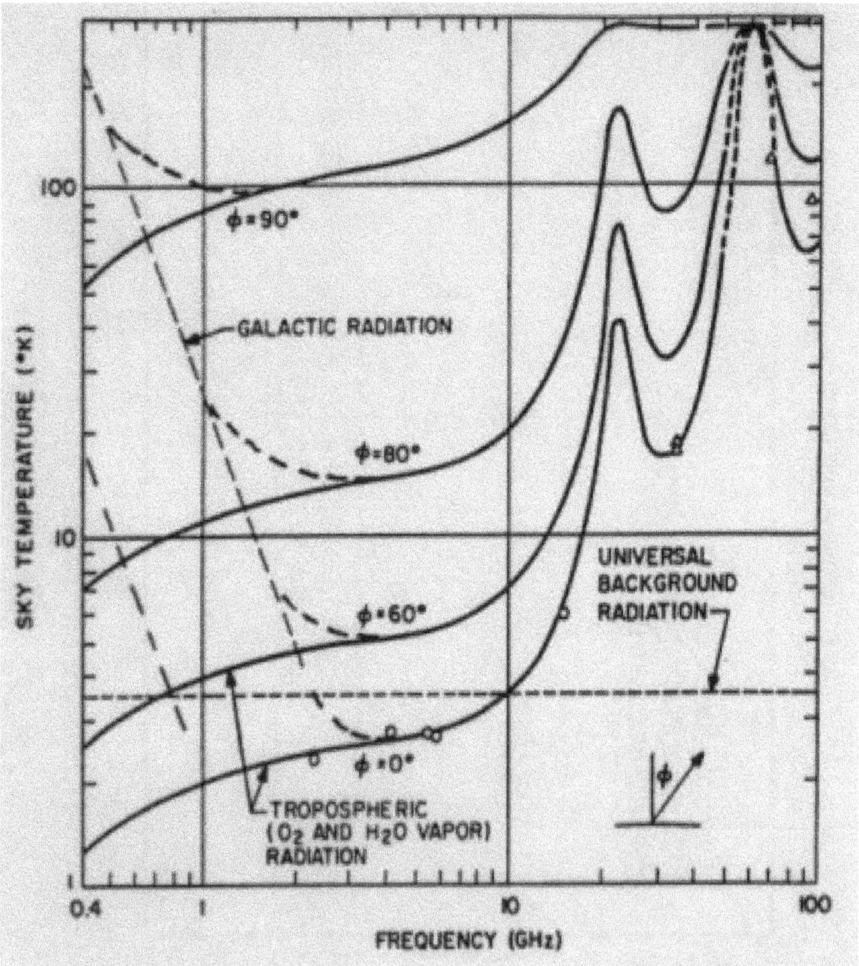


Sky Temperature ($O_2$ and $H_2O$)

# Rain Attenuation at Various Frequencies

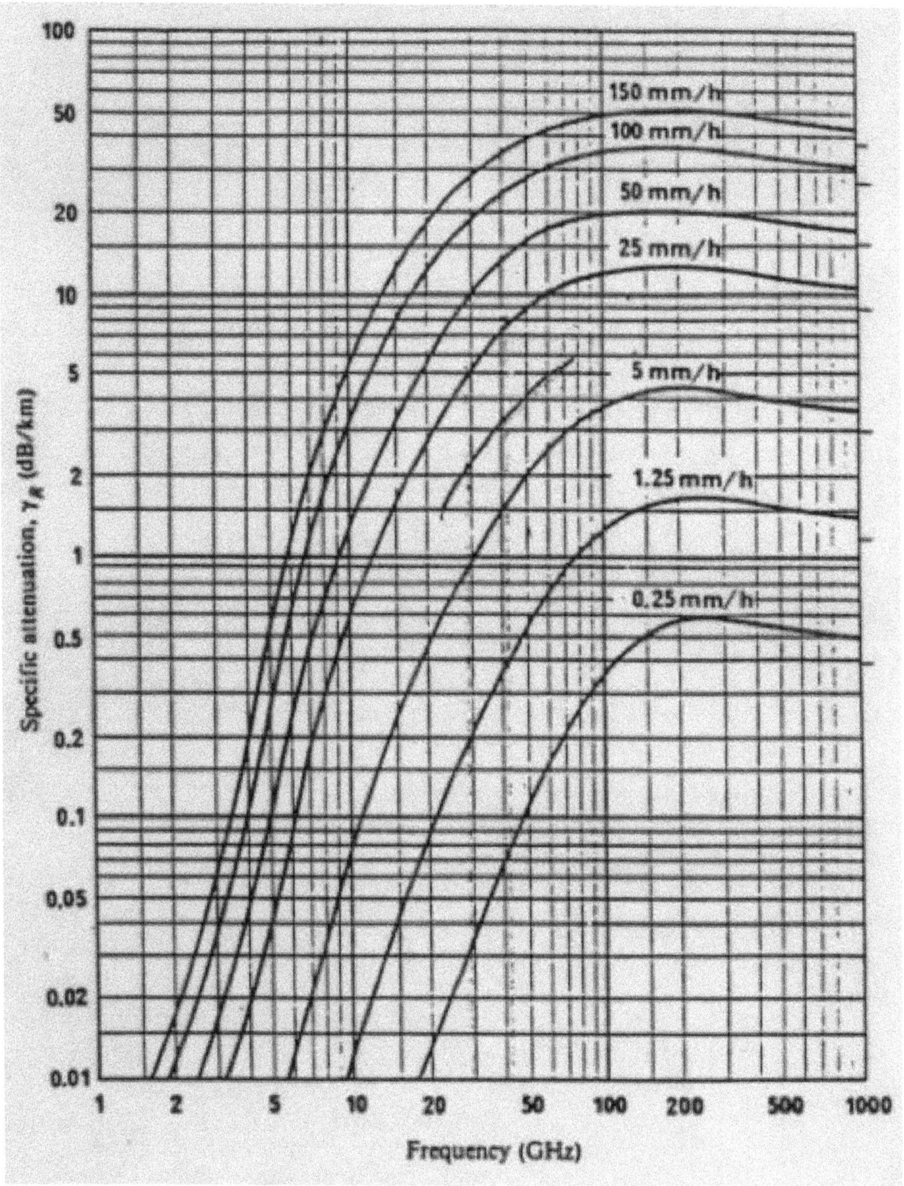

## Sky Noise at various Elevation Angles vs

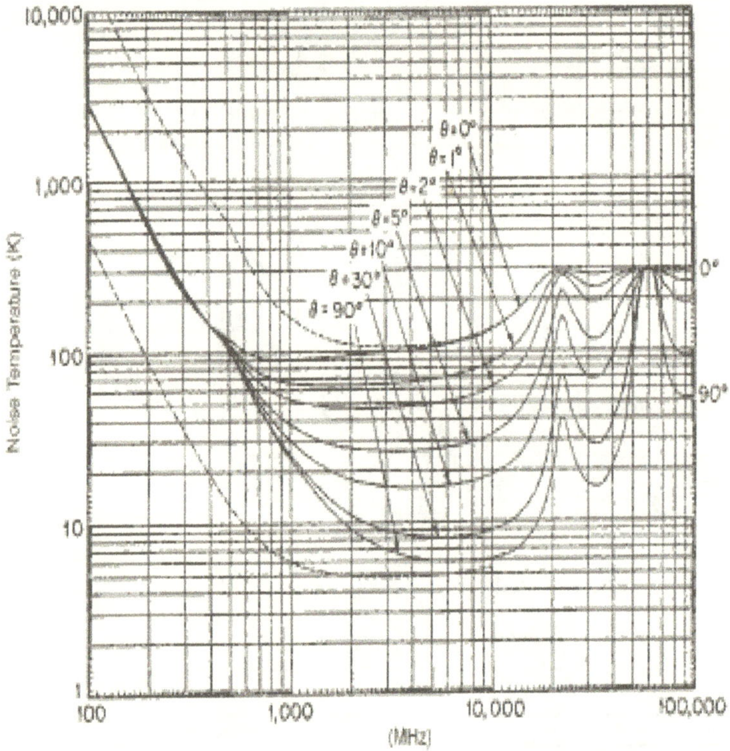


An approximation for solar noise can be found by evaluating the following equation:

$$Tsun = 120000 * F^{(-0.75)} \quad [144]$$

Where:

Tsun = Degrees K

F = Frequency in GHz

# Conductor and Resistor Properties

| Material | ρ (Ω·m) at 20°C | σ (S/m) at 20°C | Temperature coefficient ($K^{-1}$) |
|---|---|---|---|
| Carbon (graphene) | $1.00 \times 10^{-8}$ | $1.00 \times 10^{8}$ | −0.0002 |
| Silver | $1.59 \times 10^{-8}$ | $6.30 \times 10^{7}$ | 0.0038 |
| Copper | $1.68 \times 10^{-8}$ | $5.96 \times 10^{7}$ | 0.003862 |
| Annealed copper | $1.72 \times 10^{-8}$ | $5.80 \times 10^{7}$ | 0.00393 |
| Gold | $2.44 \times 10^{-8}$ | $4.10 \times 10^{7}$ | 0.0034 |
| Aluminum | $2.82 \times 10^{-8}$ | $3.50 \times 10^{7}$ | 0.0039 |
| Calcium | $3.36 \times 10^{-8}$ | $2.98 \times 10^{7}$ | 0.0041 |
| Tungsten | $5.60 \times 10^{-8}$ | $1.79 \times 10^{7}$ | 0.0045 |
| Zinc | $5.90 \times 10^{-8}$ | $1.69 \times 10^{7}$ | 0.0037 |
| Nickel | $6.99 \times 10^{-8}$ | $1.43 \times 10^{7}$ | 0.006 |
| Lithium | $9.28 \times 10^{-8}$ | $1.08 \times 10^{7}$ | 0.006 |
| Iron | $1.00 \times 10^{-7}$ | $1.00 \times 10^{7}$ | 0.005 |
| Platinum | $1.06 \times 10^{-7}$ | $9.43 \times 10^{6}$ | 0.00392 |
| Tin | $1.09 \times 10^{-7}$ | $9.17 \times 10^{6}$ | 0.0045 |
| Carbon steel (1010) | $1.43 \times 10^{-7}$ | $6.99 \times 10^{6}$ | |
| Lead | $2.20 \times 10^{-7}$ | $4.55 \times 10^{6}$ | 0.0039 |
| Titanium | $4.20 \times 10^{-7}$ | $2.38 \times 10^{6}$ | 0.0038 |
| Grain oriented electrical steel | $4.60 \times 10^{-7}$ | $2.17 \times 10^{6}$ | |
| Manganin | $4.82 \times 10^{-7}$ | $2.07 \times 10^{6}$ | 0.000002 |
| Constantan | $4.90 \times 10^{-7}$ | $2.04 \times 10^{6}$ | 0.000008 |
| Stainless steel | $6.90 \times 10^{-7}$ | $1.45 \times 10^{6}$ | |
| Mercury | $9.80 \times 10^{-7}$ | $1.02 \times 10^{6}$ | 0.0009 |
| Nichrome | $1.10 \times 10^{-6}$ | $6.7 \times 10^{5}$ | 0.0004 |
| GaAs | $1.00 \times 10^{-3}$ to | $1.00 \times 10^{-8}$ to | |

| | | | |
|---|---|---|---|
| | $1.00\times10^{8}$ | $10^{3}$ | |
| Carbon (amorphous) | $5.00\times10^{-4}$ to $8.00\times10^{-4}$ | $1.25\times10^{3}$ to $2\times10^{3}$ | −0.0005 |
| Carbon (graphite) | $2.50\times10^{-6}$ to $5.00\times10{-6}$ ∥basal plane $3.00\times10{-3}$ ⊥basal plane | $2.00\times10^{5}$ to $3.00\times10^{5}$ ∥basal plane $3.30\times10^{2}$ ⊥basal plane | |
| PEDOT:PSS | $1.00\times10^{-3}$ to $1.00\times10^{-1}$ | $1\times10^{1}$ to $4.6\times10^{3}$ | |
| Germanium | $4.60\times10^{-1}$ | 2.17 | −0.048 |
| Sea water | $2.00\times10^{-1}$ | 4.80 | |
| Swimming pool water | $3.33\times10^{-1}$ to $4.00\times10^{-1}$ | 0.25 to 0.30 | |
| Drinking water | $2.00\times10^{1}$ to $2.00\times10^{3}$ | $5.00\times10^{-4}$ to $5.00\times10^{-2}$ | |
| Silicon | $6.40\times10^{2}$ | $1.56\times10^{-3}$ | −0.075 |
| Wood (damp) | $1.00\times10^{3}$ to $1.00\times10^{4}$ | $10^{-4}$ to $10^{-3}$ | |
| Deionized water | $1.80\times10^{5}$ | $5.50\times10^{-6}$ | |
| Glass | $1.00\times10^{11}$ to $1.00\times10^{15}$ | $10^{-15}$ to $10^{-11}$ | |
| Hard rubber | $1.00\times10^{13}$ | $10^{-14}$ | |
| Wood (oven dry) | $1.00\times10^{14}$ to $1.00\times10^{16}$ | $10^{-16}$ to $10^{-14}$ | |
| Sulfur | $1.00\times10^{15}$ | $10^{-16}$ | |
| Air | $1.30\times10^{16}$ to $3.30\times10^{16}$ | $3\times10^{-15}$ to $8\times10^{-15}$ | |
| Carbon (diamond) | $1.00\times10^{12}$ | $\sim10^{-13}$ | |
| Fused quartz | $7.50\times10^{17}$ | $1.30\times10^{-18}$ | |
| PET | $1.00\times10^{21}$ | $10^{-21}$ | |
| Teflon | $1.00\times10^{23}$ to $1.00\times10^{25}$ | $10^{-25}$ to $10^{-23}$ | |

# Galvanic Corrosion Potential Tendency

| Potential Tendency for Galvanic Corrosion | | | | | | | | | |
|---|---|---|---|---|---|---|---|---|---|
| The higher the number, the greater the potential.<br>0-3: ☐ Minimal<br>4-5: ☐ Marginal<br>6-12: ■ High - Avoid if possible, or use sealants and paint barriers to prevent dissimilar material contact. | Mg Alloys | Zn, Galvanized Steel | Pure Al, 5000 & 6000 series Al | Cd, and Cd plating | 2000 & 7000 series Al | Low Alloy Steels | Cu, Brass, Bronze | Monel, Ni, Inconel, PH SS | Ti, 300 SS, Graphite |
| Mg Alloys | 0 | 1 | 2 | 4 | 5 | 6 | 10 | 11 | 12 |
| Zn, Galvanized Steel | | 0 | 1 | 3 | 4 | 5 | 9 | 10 | 11 |
| Pure Al, 5000 & 6000 series Al | | | 0 | 2 | 3 | 4 | 8 | 9 | 10 |
| Cd, and Cd plating | | | | 0 | 1 | 2 | 5 | 6 | 7 |
| 2000 & 7000 series Al | | | | | 0 | 1 | 5 | 6 | 7 |
| Low Alloy Steels | | | | | | 0 | 4 | 5 | 6 |
| Cu, Brass, Bronze | | | | | | | 0 | 1 | 2 |
| Monel, Ni, Inconel, PH SS | | | | | | | | 0 | 1 |
| Ti, 300 SS, Graphite | | | | | | | | | 0 |

# Galvanic Corrosion Potential

Galvanic corrosion potential is a measure of how dissimilar metals will corrode when placed against each other in an assembly. Metals close to one another on the chart generally do not have a strong effect on one another, but the farther apart any two metals are separated, the stronger the corroding effect on the one higher in the table. The Anodic end is where the corrosion occur.

| Element Action | Standard Electrode Potential(Volts) | |
|---|---|---|
| Lithium | -3.045 | Anodic or Active |
| Potassium | -2.920 | " |
| Sodium | -2.712 | " |
| Magnesium | -2.340 | " |
| Beryllium | -1.700 | " |
| Aluminum | -1.670 | " |
| Manganese | -1.050 | " |
| Zinc | -0.762 | " |
| Chromium | -0.744 | " |
| Iron; Mild Steel | -0.440 | " |
| Cadmium | -0.402 | " |
| Yellow Brass | -0.350 | " |
| 50 – 50 Tin-Lead Solder | -0.325 | " |
| Cobalt | -0.277 | " |
| Nickel | -0.250 | " |
| Tin | -0.136 | " |
| Lead | -0.126 | " |
| Hydrogen Reference Electrode | 0.000 | |
| Titanium | 0.055 | Cathodic or Passive |
| Copper | 0.340 | " |
| Mercury | 0.789 | " |
| Silver | 0.799 | " |
| Carbon | 0.810 | " |
| Platinum | 1.200 | " |
| Gold | 1.420 | " |
| Graphite | 2.250 | " |

Cathodic end where no corrosion occurs

## Dielectric Properties

| Material | Dielectric Constant | Loss Tangent |
|---|---|---|
| ABS | 3.25 | .02 |
| Air | 1.00 | Weather Dependent |
| Alsimag 243 | 6.28 | .0037 |
| Alsimag 393 | 3.7 | .001 |
| Alumina | 8.8 | .0033 |
| Bakelite | 5.40 | .06 |
| Balsa Wood | 1.30 | .050 |
| Beryllium Oxide | 6.70 | .003 |
| Borosilicate Glass | 4.40 | .0047 |
| Cloth Office Partition | 1.20 | |
| Concrete | 4.50 | .0111 |
| Corn Oil | 2.60 | .0077 |
| Cottonseed Oil | 2.64 | .06682 |
| Duroid 5650 | 2.65 | .003 |
| Eccofoam | 1.18 | .004 |
| Epoxy Resin RN-48 | 3.52 | .0142 |
| Fiberglass, lam. BK-174 | 5.3 | .046 |
| Hexel Honeycomb | 1.09 | .003 |
| Lexan | 2.86 | .006 |
| Sandy Soil | 2.55 | .0062 |
| Gallium Arsenide | 12.88 | .0004 |
| Glass, Ceramic | 6.0 | .0050 |
| Glass, Soda Lime | 6.0 | .02 |
| Glass, Window | 6.5 | |
| Liquid Crystal Polymer | 2.9 | .002 |
| Magnesium Oxide | 9.65 | .003 |
| Mica | 5.4 | .0003 |
| Nylon | 2.4 | .0083 |
| Olive Oil | 2.46 | .0610 |
| Paper | 3-4 | .0125-.0333 |

| | | |
|---|---|---|
| Plexiglas | 2.76 | .014 |
| Polymide (Kapton) | 3.7 | .004 |
| Polystyrene | 2.56 | .00007 |
| Polyethylene | 2.25 | .0002 |
| Polyvinyl Chloride, W-176 | 3.53 | .07 |
| Porcelain | 5.08 | .0075 |
| PTFE | 2.1 | . |
| 00015-.0003 | | |
| Quartz, Fused | 3.8 | .0001 |
| Rexolite | 2.52 | .0005 |
| Silica, Fused | 3.78 | .0002 |
| Silicon | 11.7 | |
| Silicon Carbide | 10.8 | .003 |
| Silicon Dioxide | 4.1 | .001 |
| Silicon Nitride | 5.5 | .003 |
| Snow | 1.5 | |
| Steatite 410 | 5.77 | .0007 |
| Styrofoam | 1.03 | .001 |
| Teflon (PTFE) | 2.1 | .002 |
| Titania | 100 | .0003 |
| Polyester | 4.0 | .005 |
| Water | 77 | .157 |
| Wood | 1.2-5 | .004-.4167 |

# Appendix 9 – The Electromagnetic Spectrum

## Frequency Multipliers

| | |
|---|---|
| Hz (Hertz) | 1 Hz |
| Khz (Kilohertz) | 1,000 Hz |
| MHz (Megahertz) | 1,000,000 Hz |
| GHz (Gigahertz) | 1,000,000,000 Hz |
| THz (Terahertz) | 1,000,000,000,000 Hz |

## Radio Band Designations

| Frequency | Wavelength | Radio Band Designation |
|---|---|---|
| 30 – 300 Hz | 10 – 1 Mm | ELF (Extremely Low Frequency) |
| 300 – 3000 Hz | 1000 – 100 Km | ULF (Ultra Low Frequency) |
| 3 – 30 KHz | 100 – 10 Km | VLF (Very Low Frequency) |
| 30 – 300 KHz | 10 – 1 Km | LF (Low Frequency) |
| 300 – 3000 KHz | 1000 – 100 m | MF (Medium Frequency) |
| 3 – 30 MHz | 100 – 10 m | HF (High Frequency) |
| 30 – 300 MHz | 10 – 1 m | VHF (Very High Frequency) |
| 300 – 3000 MHz | 100 – 10 cm | UHF (Ultra High Frequency) |
| 3 – 30 GHz | 10 – 1 cm | SHF (Super High Frequency) |
| 30 – 300 GHz | 10 – 1 mm | EHF (Extremely High Frequency) |

# IEEE Radar Band Designations

| Frequency | Wavelength | Band Designation |
|---|---|---|
| 1 – 2 GHz | 30 – 15 cm | L Band |
| 2 – 4 GHz | 15 -7.5 cm | S Band |
| 4 – 8 GHz | 7.5 – 3.75 cm | C Band |
| 8 – 12 GHz | 3.75 – 2.50 cm | X Band |
| 12 – 18 GHz | 2.5 – 1.67 cm | Ku Band |
| 18 – 27 GHz | 1.67 – 1.11 cm | K Band |
| 27 – 40 GHz | 11.1 – 7.5 mm | Ka Band |
| 40 – 75 GHz | 7.5 - 4 mm | V Band |
| 75 – 110 GHz | 4 – 2.73 mm | W Band |
| 110 – 300 GHz | 2.73 - 1 mm | Mm Band |
| 300 - 3000 GHz | 1 – 0.1 mm | U mm Band |

# Satellite TVRO Band Designations

| Frequency | TVRO Band |
|---|---|
| 1700 – 3000 MHz | S Band |
| 3700 – 4200 MHz | C Band |
| 10.9 – 11.75 GHz | Ku1 Band |
| 11.75 – 12.5 GHz | Ku2 Band (DBS) |
| 12.5 – 12.75 GHz | Ku3 Band |
| 18.0 – 20.0 GHz | Ka Band |

## Military Electronic Countermeasures Band Designations

| Frequency | IEEE Band |
|---|---|
| 30 – 250 MHz | A Band |
| 250 – 500 MHz | B Band |
| 500 – 1,000 MHz | C Band |
| 1 – 2 GHz | D Band |
| 2 – 3 GHz | E Band |
| 3 – 4 GHz | F Band |
| 4 – 6 GHz | G Band |
| 6 – 8 GHz | H Band |
| 8 – 10 GHz | I Band |
| 10 – 20 GHz | J Band |
| 20 – 40 GHz | K Band |
| 40 – 60 GHz | L Band |
| 60 – 100 GHz | M Band |

## Traffic Radar Frequencies

| Band | Frequency |
|---|---|
| S | 2.455 GHz |
| X | 10.525 Ghz (+/- 25 Mhz) |
| Ku | 13.45 GHz |
| K | 24.13 GHz (+/- 100 MHz) |
| K | 24.5 GHz (+/- 100 MHz) |
| Ka | 33.4 – 36.0 GHz (13 Channels, 200 MHz each) |

| | |
|---|---|
| IR - Infrared | 332 THz (Laser Radar, 904 mm) |

## Military Radar Bands

| Radar Band | Frequency |
|---|---|
| HF | 3- 30 MHz |
| VHF | 30 – 300 MHz |
| UHF | 300 – 1000 MHz |
| L | 1 – 2 GHz |
| S | 2 – 4 GHz |
| C | 4 – 8 GHz |
| X | 8 – 12 GHz |
| Ku | 12 – 18 GHz |
| K | 18 – 27 GHz |
| Ka | 27 – 40 GHz |
| mm | 40 – 300 GHz |

## International Telecommunications Union Radar Bands

| ITU Radar Band | Frequency |
|---|---|
| VHF | 138 – 144 MHz<br>216 – 225 MHz |
| UHF | 420 – 450 MHz<br>890 – 942 MHz |
| L | 1.215 – 1.400 GHz |
| S | 2.3 – 2.5 GHz, 2.7 – 3.7 GHz |
| C | 5.250 – 5.925 GHz |
| X | 8.500 – 10.680 GHz |

| | |
|---|---|
| Ku | 13.4 – 14.0 GHz, 15.7 – 17.7 GHz |
| K | 24.05 – 24.25 GHz |
| Ka | 33.4 – 36.0 GHz |

## The Electromagnetic Spectrum

| Nomenclature | Frequency | Wavelength |
|---|---|---|
| Radio | 3 Hz – 300 GHz | 10 Mm – 1 mm |
| Infrared - Far | 300 GHz – 1 THz | 1000 – 30 um |
| Infrared - Middle | 1 – 100 THz | 30 – 3 um |
| Infrared - Near | 100 – 400 THz | 3 - .75 um |
| Visible - Red | 389 – 482 THz | 770 – 622 nm |
| Visible - Orange | 482 – 502 THz | 622 – 597 nm |
| Visible - Yellow | 502 – 520 THz | 597 – 577 nm |
| Visible - Green | 520 – 610 THz | 577 – 492 nm |
| Visible - Blue | 610 – 659 THz | 492 – 455 nm |
| Visible - Violet | 659 – 769 THz | 455 – 390 nm |
| Ultraviolet – A (Least Harmful) | 750 – 1000 THz | 400 – 300 nm |
| Ultraviolet – B (More Harmful, absorbed by ozone) | 952 – 1071 THz | 315 – 280 nm |
| Ultraviolet – C (Most Harmful, absorbed by air) | 1071 – 3000 THz | 280 – 100 nm |
| Near UV ("Black Light") | 750 – 1000 THz | 400 – 300 nm |
| Far UV | 1000 – 1500 THz | 300 – 200 nm |
| Vacuum UV | 1500 – 3000 THz | 200 – 100 nm |
| X-Ray | 10^17 – 10^19 Hz | 10^-9 – 10^-11 m |
| Gamma Ray | 10^19 – 10^21 Hz | 10^-11 – 10^-13 m |

# Frequencies of Interest

| Frequency | Usage |
|---|---|
| 20 KHz | Submarine Communications |
| 60 KHz | Time Signal to set clocks (WWVB) |
| 100 KHz | Loran (Radio Location) |
| 500 KHz | International Marine Distress |
| .550 to 1.7 MHz | Commercial AM Band |
| 1.8 – 2.0 MHz | 180 Meter Ham Band |
| 2.182 MHz | International Marine Distress |
| 2.5 MHz | WWV Time Signals |
| 2.850 – 3.155 MHz | Aeronautical Mobile HF, Commercial Aircraft |
| 3.4 – 3.5 MHz | Aeronautical Mobile HF, Commercial Aircraft |
| 3.5 – 4.0 MHz | 80 Meter Amateur Radio Band |
| 4.65 – 4.75 MHz | Aeronautical Mobile HF, Commercial Aircraft |
| 5.0 MHz | WWV Time Signals |
| 5.45 – 5.73 MHz | Aeronautical Mobile HF, Commercial Aircraft |
| 5.95 – 6.20 MHz | 49 Meter Shortwave Band |
| 6.525 – 6.765 MHz | Aeronautical Mobile HF, Commercial Aircraft |
| 7.0 – 7.2 MHz | 40 Meter Amateur Radio Band |
| 7.1 – 7.3 MHz | 41 Meter Shortwave Band |
| 8.815 – 9.04 MHz | Aeronautical Mobile HF, Commercial Aircraft |
| 8.992 MHz | Military HF Voice |
| 9.50 – 9.95 MHz | 31 Meter Shortwave Band |
| 10.0 MHz | WWV Time Signals |
| 10.005 – 10.1 MHz | Aeronautical Mobile HF, Commercial Aircraft |
| 10.780 MHz | Eastern Rocket Test Range, Cape Canaveral |
| 11.175 MHz | Military HF Voice |

| | |
|---|---|
| 11.6 – 12.1 MHz | 25 Meter Shortwave Band |
| 13.2 – 13.36 MHz | Aeronautical Mobile HF, Commercial Aircraft |
| 13.6 – 13.8 MHz | 22 Meter Shortwave Band |
| 13.927 MHz | Military Affiliate Radio System (MARS) |
| 14.0 – 14.5 MHz | 20 Meter Amateur Radio Band |
| 15.0 MHz | WWV Time Signals |
| 15.01 – 15.1 MHz | Aeronautical Mobile HF, Commercial Aircraft |
| 15.1 – 15.3 MHz | 19 Meter Shortwave Band |
| 17.5 – 19.9 MHz | 16 Meter Shortwave Band |
| 20.39 MHz | Eastern Rocket Test Range, Cape Canaveral |
| 21 MHz | Peak of Jovian Decametric Radiation |
| 21.0 – 21.7 MHz | 15 Meter Amateur Radio Band |
| 27 – 27.2 MHz | Citizen's Band |
| 28 – 29 MHz | 10 Meter Amateur Radio Band |
| 30-80 MHz | Access Broadband over Power Line (BPL) |
| 35 – 36 MHz | Paging |
| 37.5 – 38.25 MHz | Radio Astronomy |
| 40-42 MHz | Industrial Scientific and Medical (ISM) |
| 43-44 MHz | Paging |
| 49 MHz | Meteor Scatter Optimal Frequency |
| 30 – 50 MHz | Military HF Band |
| 50 – 54 MHz | 6 Meter Amateur Radio Band |
| 54 – 72 MHz | Wireless Microphone |
| 72 – 76 MHz | Radio Control Radio Service |
| 73 – 74.6 MHz | Radio Astronomy |
| 75 MHz | Aviation Marker Beacon |
| 76 – 88 MHz | Wireless Microphone |
| 88 – 108 MHz | Commercial FM Band |

| | |
|---|---|
| 108 – 118 MHz | Aviation VOR and Localizer Navigation Band |
| 118 – 136 MHz | Aviation Communication Band |
| 121.5 MHz | Aviation Distress Frequency |
| 129.9 MHz | Extended Range Aviation Frequency |
| 130.7 MHz | Extended Range Aviation Frequency |
| 137 – 138 MHz | Mobile Satellite, Weather Instruments/Radar/Satellites |
| 144 – 148 MHz | 2 Meter Amateur Radio Band |
| 148 – 150 MHz | Mobile Satellite |
| 154 – 173 | Maritime |
| 174 - 216 | Biomedical Telemetry Devices |
| 200 – 400 | Military |
| 406 – 406.1 | Aviation and Nautical Distress Frequency |
| 410 – 440 MHz | Pulsar Emissions Peak |
| 440 – 450 MHz | .7 Meter Amateur Radio Band |
| 462 – 467 MHz | General Mobile Radio Service (GMRS) |
| 828 – 890 MHz | Cell Phone |
| 842 – 860 MHz | Cell Phone Trunking |
| 902 – 928 MHz | ISM Band, Spread Spectrum, Radar |
| Below 960 | Ultra Wide Band (UWB) |
| 980 MHz – 1 GHz | Aviation DME Frequencies |
| 1.030 GHz | Aircraft Transponder Downlink |
| 1.090 GHz | Aircraft Transponder Uplink |
| 1.227 GHz | L2 GPS |
| 1.300 GHz | ATC Radar |
| 1.420 GHz | Hydrogen Line, Radio Astronomy |
| 1.530-1.559 GHz | Iridium - Receive |
| 1.575 GHz | L1 GPS (Commercial Use) |

| | |
|---|---|
| 1.610-1.626 GHz | Iridium - Transmit |
| 1.645 GHz | Inmarsat, Mobilsat |
| 1.660 GHz | Carbon Dioxide Line, Radio Astronomy |
| 1.80 – 2.10 GHz | PCS Cell Phone |
| 1.99 – 10.6 GHz | Ultra Wide Band |
| 2.2 – 2.4 GHz | Space Telemetry |
| 2.4 – 2.5 GHz | ISM Band, Microwave Oven |
| 2.9 – 3.0 GHz | Automatic Vehicle Identification Service |
| 3.1 – 10.6 GHz | Ultra Wide Band |
| 3.8 – 4.2 GHz | C Band Backyard Satellite Downlink |
| 5.15 – 5.35 GHz | Vehicle Radar Systems |
| 5.47 – 5.825 GHz | Vehicle Radar Systems |
| 5.65 – 5.925 GHz | ISM Band, Cordless Phone |
| 6 – 6.5 GHz | C Band Satellite Uplink |
| 8 – 9 GHz | Space Command and Control |
| 9.3 GHz | Aviation Weather Radar |
| 11.7 – 12.2 GHz | Ku Band Backyard Satellite, DBS |
| 13.5 GHz | Marine Radar |
| 20 GHz | Ku Band Backyard Dish |
| 24.0 – 24.25 GHz | ISM Band |
| 30 GHz | Oxygen Line |
| 59.3 – 64 GHz | ISM Band |
| 60 GHz | Water Line |
| 93 GHz | Millimeter Wave Radar |
| 80 – 120 GHz | Multiple Radio Astronomy Spectral Emissions |
| 116 – 123 GHz | ISM Band |
| 180 – 230 GHz | Multiple Radio Astronomy Spectral Emissions |
| 241 – 248 GHz | ISM Band |

| | |
|---|---|
| 241 – 275 GHz | Radio Astronomy |
| 300 GHz - 1 THz | Sub Millimeter Radio Astronomy Band |

## Television Channel Frequencies

| Channel Number | Frequency (MHz) | Channel Number | Frequency (MHz) |
|---|---|---|---|
| 2 | 54 - 60 | 36 | 602 - 608 |
| 3 | 60 - 66 | 37 | 608 - 614 |
| 4 | 66 - 72 | 38 | 614 - 620 |
| 5 | 76 - 82 | 39 | 620 - 626 |
| 6 | 82 - 88 | 40 | 626 - 632 |
| 7 | 174 - 180 | 41 | 632 - 638 |
| 8 | 180 - 186 | 42 | 638 - 644 |
| 9 | 186 - 192 | 43 | 644 - 650 |
| 10 | 192 - 198 | 44 | 650 - 656 |
| 11 | 198 - 204 | 45 | 656 - 662 |
| 12 | 204 - 210 | 46 | 662 - 668 |
| 13 | 210 - 216 | 47 | 668 - 674 |
| 14 | 470 - 476 | 48 | 674 - 680 |
| 15 | 476 - 482 | 49 | 680 - 686 |
| 16 | 482 - 488 | 50 | 686 - 692 |
| 17 | 488 - 494 | 51 | 692 - 698 |
| 18 | 494 - 500 | 52 | 698 - 704 |
| 19 | 500 - 506 | 53 | 704 - 710 |
| 20 | 506 - 512 | 54 | 710 - 716 |

| | | | |
|---|---|---|---|
| 21 | 512 - 518 | 55 | 716 - 722 |
| 22 | 518 - 524 | 56 | 722 - 728 |
| 23 | 524 - 530 | 57 | 728 - 734 |
| 24 | 530 - 536 | 58 | 734 - 740 |
| 25 | 536 - 542 | 59 | 740 - 746 |
| 26 | 542 - 548 | 60 | 746 - 752 |
| 27 | 548 - 554 | 61 | 752 - 758 |
| 28 | 554 - 560 | 62 | 758 - 764 |
| 29 | 560 - 566 | 63 | 764 - 770 |
| 30 | 566 - 572 | 64 | 770 - 776 |
| 31 | 572 - 578 | 65 | 776 - 782 |
| 32 | 578 - 584 | 66 | 782 - 788 |
| 33 | 584 - 590 | 67 | 788 - 794 |
| 34 | 590 - 596 | 68 | 794 - 800 |
| 35 | 596 - 602 | 69 | 800 – 806 |

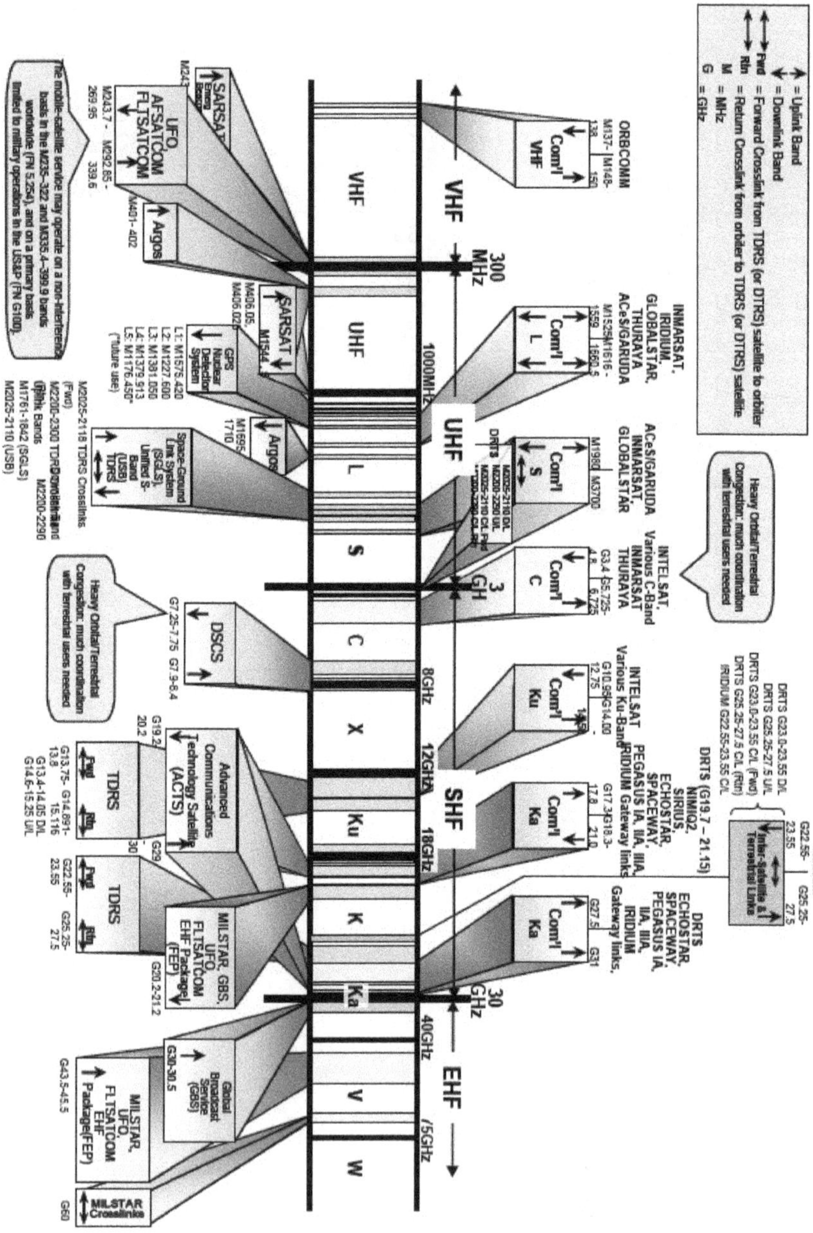
VHF
UHF
SHF
EHF
300 MHz
3 GHz
30 GHz
VHF
UHF
L
S
C
X
Ku
K
Ka
V
W

| Spectrum of Electromagnetic Radiation | | | | |
|---|---|---|---|---|
| Region | Wavelength (Angstroms) | Wavelength (centimeters) | Frequency (Hz) | Energy (eV) |
| Radio | $> 10^9$ | $> 10$ | $< 3 \times 10^9$ | $< 10^{-5}$ |
| Microwave | $10^9 - 10^6$ | $10 - 0.01$ | $3 \times 10^9 - 3 \times 10^{12}$ | $10^{-5} - 0.01$ |
| Infrared | $10^6 - 7000$ | $0.01 - 7 \times 10^{-5}$ | $3 \times 10^{12} - 4.3 \times 10^{14}$ | $0.01 - 2$ |
| Visible | $7000 - 4000$ | $7 \times 10^{-5} - 4 \times 10^{-5}$ | $4.3 \times 10^{14} - 7.5 \times 10^{14}$ | $2 - 3$ |
| Ultraviolet | $4000 - 10$ | $4 \times 10^{-5} - 10^{-7}$ | $7.5 \times 10^{14} - 3 \times 10^{17}$ | $3 - 10^3$ |
| X-Rays | $10 - 0.1$ | $10^{-7} - 10^{-9}$ | $3 \times 10^{17} - 3 \times 10^{19}$ | $10^3 - 10^5$ |
| Gamma Rays | $< 0.1$ | $< 10^{-9}$ | $> 3 \times 10^{19}$ | $> 10^5$ |

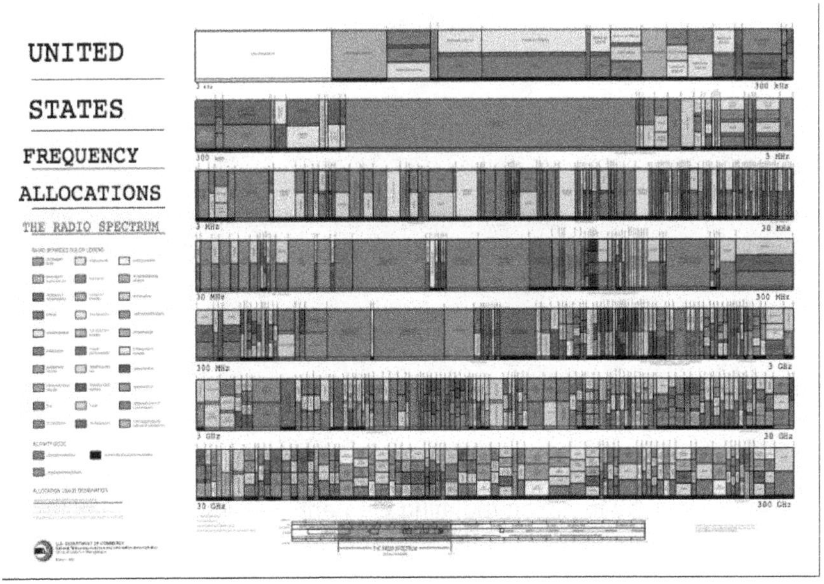

# Appendix 10 – ADP Computer Program

## Introduction

At the Shoemakerlabs.com web site is a series of Excel and Open Office spreadsheets designed to enable an engineer to quickly design an antenna or evaluate a particular antenna's performance. There is also attached a series of routines to allow the evaluation of many antenna design related activities, including far field, gain, bandwidth calculations along with many others.

These routines are all first order approximations from which a design engineer can start a particular antenna construction. These routines do not take into effect such factors as reactive loading from surrounding materials or mutual coupling etc.

Notes are included.

## Design:

### *Dipole*

Calculates simple dipole dimensions. Typically this dimension will resonate at a slightly lower frequency due to surrounding reactive influences. Trim to optimize performance using network analyzer or other suitable device.

### *Monopole*

½ Wavelength in length. Use ground plane at least 1 wavelength in extent. Tune by trimming radiating element. Design initially for slightly lower frequency of operation, then trim.

## Patch

Reasonably simple approach to designing microstrip patch. There are many influences in this type of design including amount of ground plane from edge of patch to edge of ground plane, should be at least 2 board thicknesses. Design for slightly lower frequency of operation and trim as necessary to optimize performance. Width varies impedance, length varies resonant frequency. Use Smith Chart option on network analyzer to best design this type of antenna.

## Microstrip Line

Standard formulas used here for design. Best approach is to make a series of lines on representative dielectric material with connectors on both end. Vary line width a few percent above and below and etch onto material. Check each line for impedance and insertion loss. Start with 50 ohms and find optimal width. Use this as standard and ratio as necessary for other impedance requirements.

## Microstrip Transformer

Used for combining lines of different impedances. Options on transformer include rectangular, tapered and curved (see Chapter 6). For rectangular transformers, use ¼ wavelength for long side, for tapered and curved, use ½ wavelength or more for proper operation. Asymmetric values can be used as well by using one half the value of the combined impedances of two lines to couple into one.

## Loop

This is a design routine for small loops, typically used for direction finding purposes.

### *Helix*

Standard design routine for end fire helix using ground plane.

### *Yagi*

Standard design routine for yagi based on Beasely's formulas.

### *Rod*

Requires knowledge of the ferrite rod material, includes calculations for capacitance to optimize resonant properties.

### *Trailing Wire*

Originally used for aircraft antennas operating at very low frequencies. This procedure can be used for designing random long wire antennas for shortwave frequencies.

### *Loaded Monopole*

Designs monopoles with loading coils. This shortens the monopole which is very useful for lower frequencies of operation.

### *Inverted "V"*

One of the more useful antennas for shortwave operations. This design creates omni-spherical patterns and can handle significant power levels.

### *Discone*

Very useful for wide band omni surveillance operations. This programs calculates optimum dimensions but relies on the designer to obtain the final matching.

### *Log Periodic*

Used for wide band directional uses. Can be designed for

operations in the HF band to several Gigahertz. Check chapter in book for more details.

## *Cubical Quad*

Used mostly by the Amateur Radio community. Simple, good performance and low cost.

## *Quarter Wave Monopole*

This antenna requires a ground plane but is simple and has a very uniform pattern and gain profile.

## *Rhombic*

Classic design of efficient antenna, used for broadcast in the past. Calculates values in both meters and feet.

## *Spiral*

Can be used for 2 arm or 4 arm design.

## *Array*

Calculates size requirements for array, elemental spacing and final size.

## *"L"*

Designs wire antenna in "L" configuration, useful for shortwave and Ham radio work.

## *Loaded Dipole*

For limited space applications, normally used for shortwave or Ham radio work.

## *G5RV*

A popular, multiband antenna with vertical and horizontal elements. Useful for shortwave and Ham radio work.

### *Coaxial Collinear*

Antenna with applications for cellular phone base stations, wind profiling radar systems. Elements are ½ wavelength separated by ½ wavelength in phase creating linear array with good gain and simple construction. Beam steers with change in frequency, useful for downtilt operations or simple scanning radar systems.

### *Bowtie*

Wideband antenna used in the past for UHF TV reception. Now sometimes used in phased arrays.

### *"F"*

Simple, compact antenna used in laptops for Wi-Fi operations. Other uses include cellular and Ham radio frequencies.

### *Horn*

Use for standard gain applications in antenna testing, radio astronomy and phased arrays.

### *Slot*

Uses Babinet's principle where slot in metal ground plane can be used as a compliment to a dipole.

### *Corner Reflector*

Used in UHF applications typically. Have been used in the past for wideband HF radio astronomy.

### *Backfire*

Very efficient antenna with good bandwidth qualities. This antenna is used extensively in military applications and in phased array.

### *Parabolic Dish*

Ubiquitous applications from satellite to radio astronomy to point to point communications.

## Analysis:

### *Wavelength*

Calculates full and half wavelength sizes in metric and English units.

### *Beamwidth*

Calculates beamwidth in degrees from aperture size. Note that asymmetric apertures with have resultant asymmetric beamwidths.

### *% Bandwidth*

Based on difference between high and low frequencies. Also calculates number of octaves.

### *Gain*

Based on total area of radiating aperture. Does not compensate for various aperture efficiencies. Calculations does not work with apertures such as yagis and log periodics.

### *Gain from Beam*

Based on measured beam sizes in orthogonal planes. Does not compensate for variations in aperture efficiencies.

### *Required Gain*

Inverted process of determining gain. Given required gain, calculates aperture size and area. Based on frequency and gain inputs.

## *Patterns*

Plots simple rectangular and polar patterns based on number of elements and inter-element phase shifts.

## *Efficiency*

Based on measured gain and size of aperture.

## *System Temperature*

Derived from low noise amplifier noise figure and other parameters. Useful for determining quality of radar, radio astronomical or satellite system.

## *RF Safety*

Determines watts per square area, which is used in determining total absorbed power by humans. Various countries and regulatory agencies have determined maximum safe RF exposure. This number is typically between 1 and 10 microwatts per square meter over a specified period of time (nominally 1 minute).

## *Field Strength*

Calculates watts per meter and volts per meter based on distance, transmitter power, transmitter antenna gain and frequency.

## *Far Field*

Determines distance from antenna to source where all beam details are non changing. Near field is below this value and is sometimes known as the Fresnel region.

## *Link Analysis*

Based on distance, receiver and transmitter sensitivities and power levels, this calculation determines how much power will be delivered to receiver terminals. Very useful for any type of communications work.

### *Spacing*

Calculates maximum spacing for intended beam offset from normal to a flat phased array. Spacing more than that recommended by this routine could result in grating lobe and high amplitude sidelobes.

### *Radar*

Uses the standard radar formula to predict expected signal return to the receiver. Includes antenna size, noise temperature, bandwidth and other parameters.

### *Radio Astronomy*

Uses standard radio astronomy formula to calculate expected signal level to receiver based on noise temperature and antenna qualities.

### *IMD*

Calculated Intermodulation Distortion Products for RF systems. Should always be considered during design phase to minimize in band signals that could compromise operational quality.

### *Radar 1*

Uses radar equation with emphasis on atmospheric returns, used for meteorological applications including wind profiler radar systems.

### *Return Loss*

Converts VSWR to Return Loss and vice versa. Also calculates miss match losses.

### *dB Conversions*

Converts dBi to dBd to Nepers. Also converts watts to dB

### *Path Loss*

Determines loss based on nautical miles, statute miles, kilometers, feet and meters.

### *Resonance*

Determines resonant frequency based on inductance and capacitance.

### *Coax Impedances*

Determines coaxial impedance based on diameters of inner conductor and outer shield.

### *Two Wire Impedances*

Determines wire line impedances based on wire diameter and separation distance.

### *Microstrip Impedances*

Calculates microstrip line impedance based on width, dielectric constant of material and loss tangent.

### *Steering Losses*

Based on element separation, this calculation determines maximum steering angle. Use this in conjunction with the pattern characteristics of the element.

### *Stripline Impedances*

Calculates impedance based on strip width, separation and characteristics of the material.

### *Wind Loading*

Calculates pressure on mounting mechanisms like pedestals based on wind speed, reflector characteristics and temperature.

### *I squared R losses*

Conductor losses can be determined and their effect on signal attenuation with this calculation.

### *G over T Calculations*

Vitally important for determining the quality of a satellite link. Includes gain and temperature inputs.

### *Microstrip Transformer*

When two or more microstrip lines are connected, impedances must be matched. This calculator determines the impedance and thus the width of the ¼ wave transformer that will best connect the lines.

### *Inter-element Spacing*

Based on maximum required scan angle, this calculator determines the maximum inter-element spacing for best performance.

### *CSR*

This calculates the size of an aperture based on the required gain and efficiency.

### *Noise Figure*

Converts noise temperature (degrees K) to dB and vice versa.

### *General Conversions*

Converts temperature, weights, lengths etc.

# Bibliography and References

[1] Baars, J.W.M., *The Paraboloidal Reflector Antenna in Radio Astronomy and Communications,* Springer, 2007

[2] Baker, B.N. And E.T. Copson, *The Mathematical Theory of Huygens' Principle,* Oxford University Press, 1939

[3] Bahl, I.J. and P. Bhartia, *Microstrip Antennas,* Artech House, 1980

[4] Balanis, C., *Antenna Theory, Analysis and Design*, John Wiley and Sons, 1982

[5] Balanis, *Antenna Theory*, Harper and Row

[6] Blake, *Antennas*, Artech House

[7] Born, M. and E. Wolf, *Principles of Optics,* $6^{th}$ edition, Oxford, Pergamon, 1980

[8] Brookner, *Practical Phased Array Antenna Systems*, Artech House

[9] Clarricoats and Olver, *Corrugated Horns for Microwave Antennas*, Peregrinus

[10] Collin, R.E. and F.J. Zucker, *Antenna Theory*, Part 1 and 2, Artech House

[11] Elliott, *Antenna Theory and Design,* Prentice-Hall

[12] Fradin, A.Z., *Microwave Antennas,* Oxford, Pergamon, 1961

[13] Hansen, R.C., *Microwave Scanning Antennas, Vol. I, II and III,* Academic Press

[14] Hansen, *Electrically Small, Superdirective, and Superconducting Antennas,* Wiley

[15] Harrington, R.F., *Field Computation by Moment Methods,* Macmillan, 1968

[16] Jansky, K.G., Radio waves from outside the solar system. *Nature* **132**, 66, 1933

[17] Jennison, R.C., *Introduction to Radio Astronomy,* London, Newnes, 1966

[18] King and Harrison, *Antennas and Waves,* MIT Press

[19] King and Smith, *Antennas in Matter,* MIT Press

[20] Kraus, *Antennas, 2nd Edition,* McGraw-Hill

[21] Kraus, *The Big Ear*

[22] Kraus and Marhefka, *Antennas, 3rd Edition,* McGraw-Hill

[23] Kumar, *Fixed and Mobile Terminal Antennas,* Artech House

[24] Law, *Shipboard Antennas,* Artech House

[25] Lo and Lee (eds.), *Antenna Handbook*, Van Nostrand Reinhold

[26] Lodge, O., *Signaling across Space without Wires,* 3rd Edtion, 1900.

[27] Love, A.W., (Editor), *Electromagnetic Horn Antennas,* New York, IEEE Press, 1976

[28] Luneburg, R.K., *Mathematical Theory of Optics,* Brown University Press, 1944

[29] Ma, *Theory and Application of Antenna Arrays,* Wiley

[30] Maxwell, J.C., *A Treatise on Electricity and Magnetism,* Oxford, 1873

[31] Maxwell, J.C., A Dynamical Theory of the Electromagnetic Field, *Royal Society Transactions,* **65,** 1865

[32] Milligan, T., *Modern Antenna Design.* New York; John Wiley and Sons, 2005

[33] Mailloux, *Phased Array Antenna Handbook,* Artech House

[34] Fujimoto, Henderson, Hirasawa and James, *Small Antennas,* Research Studies

[35] Fujimoto and James, *Mobile Antenna Systems Handbook,* Artech House

[36] Hertz, H.R., *Electric Waves,* Macmillan 1893, Dover 1962

[37] Lawson, *Yagi Antenna Design,* ARRL

[38] Love (ed.), *Electromagnetic Horn Antennas,* IEEE Press, 1976

[39] Love (ed.), *Reflector Antennas,* IEEE

[40] Reber, G., "Cosmic Static", *Proc. IRE,* **28**, 68-70, 1940.

[41] Rodge, Milne, Olver, Knight (eds.), *The Handbook of Antenna Design, Vol. 1 & 2*, Peter Peregrinus

[42] Rusch and Potter, *Analysis of Reflector Antennas,* Academic Press

[43] Rumsey, *Frequency Independent Antennas,* Academic Press

[44] Roubin and Bolomey, *Antennas,* Hemishpere

[45] Schelkunoff, S.A., Electromagnetic *Waves,* Van Nostrand, New York, 1943

[46] Schelkunoff, S.A., "A Mathematical Theory of Arrays," *Bell System Tech. J.,* **22**, 80-107, January 1943

[47] Skolnik, M.I., *Introduction to Radar Systems,* McGraw-Hill, 1980

[48] Silver (ed.), *Log Periodic Antenna Design Handbook,* Smith Electronics

[49] Silver, S., *Microwave Antenna Theory and Design,* MIT Radiation Lab

Series **12**, New York, McGraw-Hill, 1949

[50] Stratton, J.A., *Electromagnetic Theory,* New York, McGraw-Hill, 1941

[51] Straw (ed.), *The ARRL Antenna Book, 1-18th Editions,* ARRL

[52] Stutzman, W., G. Thiele, *Antenna Theory and Design*, 2nd Edition",John Wiley and Sons, 1988

[53] Thompson, A.R., J.M. Moran and G.W. Swenson, *Interferometry and Synthesis in Radio Astronomy,* 2nd Edition, New York, John Wiley, 2001

[54] Volakis (ed.), *Antenna Engineering Handbook*, 4th Edition, McGraw-Hill

[55] Wolff, *Antenna Analysis,* Artech House

[56] Weeks, *Antenna Engineering,* McGraw - Hill

[57] John Kraus and Ronald J Marhefka, "Antennas For all applications, 3rd edition", 2002, McGraw-Hill Higher Education

[58] Schlovski, Radio Astronomy

[59] Collins, *Antenna Design,*

Other Resources:

[60] QEX and QST Magazines

[61] APS Magazine, IEEE

[62] Fink, Antenna Handbook

[63] A. I. Nosich

[64] Wheeler, Small Antennas

[65] Kin-Lu Wong, Compact and Broadband Microstrip Antennas, Wiley, 2002

[66] CFR

[67] Pozar, Textbook and Software

[68] Johnk,  Textbook on Electro-magnetics

[69] Microwave Journal

[70] HP Application Notes

[71] Astrophysical Journal

[72] IEEE Microwave and RF Techniques  (?)

[73] The Handbook of Antenna Design, A.W. Rudge

[74] Warren Stutzman and Gary Thiele, "Antenna Theory and Design, 2nd Edition", 1988, John Wiley and Sons

[75] Basics of Measuring the Dielectric Properties of Materials, Agilent Application Note

[76] Ticra software for reflector design

www.qsl.net/n9zia/rlb.html for RL bridges

[77]  Wikepedia, the free encyclopedia

[78]  "The World's Smallest Radio", Ed Regis, Scientific American, March 2009

# Alphabetical Index

## About the Author

Kevin O. Shoemaker was born in New York City, raised in Texas, New Mexico, Pennsylvania and Florida, a son of a professor and actress. He attended St. Petersburg College in Florida and the University of Colorado, studying philosophy and physics.

He has held engineering jobs at NIST, NOAA, NCAR, Ball Aerospace, Lockheed / Martin, Raytheon, Harris and Orbital Sciences. These activities have all been associated with antenna design and test.

In addition, he has been a chief scientist at two companies, one private and one public, Phasar and ARC Wireless, in both cases the focus was on antenna design and manufacturing. Duties included both management and engineering. Later he worked on the design and testing of several wind profiler radars as well as air deployable meteorological sensor suites. After that he provided astronomical instrumentation for Southwest Research Institute and reviewed proposals for NASA.

Lately, he has consulted for the Discovery Channel's 'Storm Chasers' crew, working on antennas and radar components and is the president of an antenna company. He holds 9 patents.

Questions or Comments, Shoemakerlabs@gmail.com

www.ingramcontent.com/pod-product-compliance
Lightning Source LLC
Chambersburg PA
CBHW021426180326
41458CB00001B/156
* 9 7 8 0 9 8 1 5 0 9 2 8 0 *